REGRESSION DISCONTINUITY DESIGNS: THEORY AND APPLICATIONS

ADVANCES IN ECONOMETRICS

Series Editors: Thomas B. Fomby,
R. Carter Hill, Ivan Jeliazkov,
Juan Carlos Escanciano,
Eric Hillebrand, Daniel L. Millimet and
Rodney Strachan

Recent Volumes:

Volume 29: Essays in Honor of Jerry Hausman – Edited by Badi H. Baltagi, Whitney Newey, Hal White and R. Carter Hill

Volume 30: 30th Anniversary Edition – Edited by Dek Terrell and Daniel Millmet

Volume 31: Structural Econometric Models – Edited by Eugene Choo and Matthew Shum

Volume 32: VAR Models in Macroeconomics — New Developments and Applications: Essays in Honor of Christopher A. Sims – Edited by Thomas B. Fomby, Lutz Kilian and Anthony Murphy

Volume 33: Essays in Honor of Peter C. B. Phillips – Edited by Thomas B. Fomby, Yoosoon Chang and Joon Y. Park

Volume 34: Bayesian Model Comparison – Edited by Ivan Jeliazkov and Dale J. Poirier

Volume 35: Dynamic Factor Models – Edited by Eric Hillebrand and Siem Jan Koopman

Volume 36: Essays in Honor of Aman Ullah – Edited by Gloria González-Rivera, R. Carter Hill and Tae-Hwy Lee

Volume 37: Spatial Econometrics: Qualitative and Limited Dependent Variables – Edited by Badi H. Baltagi, James P. LeSage and R. Kelley Pace

ADVANCES IN ECONOMETRICS VOLUME 38

REGRESSION DISCONTINUITY DESIGNS: THEORY AND APPLICATIONS

Edited by

MATIAS D. CATTANEO

Department of Economics and Department of Statistics, University of Michigan, Ann Arbor, MI, USA

JUAN CARLOS ESCANCIANO

Department of Economics, Indiana University, Bloomington, IN, USA

emerald PUBLISHING

United Kingdom – North America – Japan
India – Malaysia – China

Emerald Publishing Limited
Howard House, Wagon Lane, Bingley BD16 1WA, UK

First edition 2017

British Library Cataloguing in Publication Data
A catalogue record for this book is available from the British Library

ISBN: 978-1-78714-390-6 (Print)
ISBN: 978-1-78714-389-0 (Online)
ISBN: 978-1-78714-729-4 (Epub)

ISSN: 0731-9053 (Series)

Printed and bound by CPI Group (UK) Ltd, Croydon, CR0 4YY

ISOQAR REGISTERED
Certificate Number 1985
ISO 14001

ISOQAR certified Management System, awarded to Emerald for adherence to Environmental standard ISO 14001:2004.

INVESTOR IN PEOPLE

CONTENTS

LIST OF CONTRIBUTORS

Otávio Bartalotti	Iowa State University, Ames, IA, USA
Quentin Brummet	United States Census Bureau, Washington, DC, USA
Gray Calhoun	Iowa State University, Ames, IA, USA
David Card	UC Berkeley, CA, USA
Giovanni Cerulli	IRCrES-CNR, Rome, Italy
Hanley Chiang	Mathematica Policy Research, Cambridge, MA, USA
Thomas D. Cook	Northwestern University, Evanston, IL, USA
Yingying Dong	UC Irvine, CA, USA
Brigham R. Frandsen	Brigham Young University, Provo, UT, USA
Sebastian Galiani	University of Maryland, College Park, MD, USA
Yang He	Iowa State University, Ames, IA, USA
Heinrich Hock	Mathematica Policy Research, Washington, DC, USA
Hugo Jales	Syracuse University, Syracuse, NY, USA
Luke Keele	Georgetown University, Washington, DC, USA
Yasemin Kisbu-Sakarya	Koc University, Istanbul, Turkey
Arthur Lewbel	Boston College, Chestnut Hill, MA, USA

David S. Lee	Princeton University, Princeton, NJ, USA
Scott Lorch	University of Pennsylvania, Philadelphia, PA, USA
Justin McCrary	UC Berkeley, CA, USA
Patrick J. McEwan	Wellesley College, Wellesley, MA, USA
Molly Passarella	University of Pennsylvania, Philadelphia, PA, USA
Zhuan Pei	Cornell University, Ithaca, NY, USA
Alexander Poulsen	Boston College, Boston, MA, USA
Jasjeet S. Sekhon	UC Berkeley, CA, USA
Yi Shen	University of Waterloo, Waterloo, ON, Canada
Dylan Small	University of Pennsylvania, Philadelphia, PA, USA
Yang Tang	Northwestern University, Evanston, IL, USA
Rocío Titiunik	University of Michigan, Ann Arbor, MI, USA
Brian Quistorff	Microsoft AI and Research, Redmond, WA, USA
Andrea Weber	Vienna University of Economics and Business, Austria
Zhengfei Yu	University of Tsukuba, Tsukuba, Japan

INTRODUCTION

The regression discontinuity (RD) design was introduced by Thistlethwaite and Campbell (1960) more than 50 years ago, but has gained immense popularity in the last decade. Nowadays, the design is well known and widely used in a variety of disciplines, including (but not limited to) most fields of study in the social, biomedical, behavioral, and statistical sciences. Many economists and other social scientists have devoted great efforts to advance the methodological knowledge and empirical practice concerning RD designs. Early reviews and historical perspectives are given by Cook (2008), Imbens and Lemieux (2008), and Lee and Lemieux (2010), but much progress has taken place since then. This volume of *Advances in Econometrics* seeks to contribute to this rapidly expanding RD literature by bringing together theoretical and applied econometricians, statisticians, and social, behavioral, and biomedical scientists, in the hope that these interactions will further spark innovative practical developments in this important and active research area.

This volume collects 12 innovative and thought-provoking contributions to the RD literature, covering a wide range of methodological and practical topics. Many of these chapters touch on foundational methodological issues such as identification and interpretation, implementation, falsification testing, or estimation and inference, while others focus on more recent and related topics such as identification and interpretation in a discontinuity-in-density framework, empirical structural estimation, comparative RD methods, and extrapolation. Considered together, these chapters will help shape methodological and empirical research currently employing RD designs, in addition to providing new bases and frameworks for future work in this area.

The following sections provide a more detailed discussion of the 12 contributions forming this volume of *Advances in Econometrics*. To this end, we first give a very brief overview of the state-of-the-art in the analysis and interpretation of RD designs by offering a succinct account of the RD literature. Although our overview covers a large number of classical and recent papers, it is surely incomplete, as this literature continues to grow and expand rapidly. Our goal here is not to provide a comprehensive review of the literature, but rather to set the ground for describing how each of the contributions in this volume fits in the broader RD literature.

Overview of the Literature

The RD design is arguably one of the most credible and internally valid non-experimental research designs in observational studies and program evaluation. The key distinctive features underlying all RD designs are that, for each unit under study, (i) treatment is assigned based on an observed variable X_i, usually called running variable, score or index, and (ii) the conditional probability of treatment status, which equals the probability of treatment assignment under perfect compliance, changes abruptly or discontinuously at a known cutoff value c on the support of the running variable. Therefore, in RD designs, treatment assignment occurs via hard-thresholding: each unit is assigned to the control group if $X_i < c$, and to treatment group if $X_i \geq c$. The most standard RD setting also assumes that the running variable is continuously distributed in a neighborhood of the cutoff value, with a positive density. In this canonical RD framework, the two basic parameters of interest are the average treatment effect at the cutoff (interpreted as an intention-to-treat parameter under non-compliance), and the probability limit of a two-stage treatment effect estimator at the cutoff when compliance is imperfect (interpreted as a local average treatment effect at the cutoff under additional assumptions). Most popular estimation and inference methods in applied work rely on local polynomial regression techniques based on large sample approximations.

Many departures from the canonical RD design have been proposed in the literature, spanning a wide range of possibilities. For example, researchers have considered different RD designs (e.g., multi-cutoff RD or geographic RD), different population parameters (e.g., kink RD or distributional RD), different estimators and inference procedures (e.g., randomization inference or empirical likelihood), and even different departures from the underlying canonical assumptions (e.g., measurement error or discretely valued running variable). Furthermore, many new methodologies have been developed in recent years covering related problems such as graphical presentation techniques, falsification/validation methods, and treatment effect extrapolation approaches.

Our succinct overview of the classical and recent literature on RD designs is organized in four main categories: (i) Identification, Interpretation, and Extrapolation; (ii) Presentation, Falsification, and Robustness Checks; (iii) Estimation and Inference; and (iv) Software. We then summarize and discuss the new contributions in this volume of *Advances in Econometrics* by placing them in context relative to these four categories and the associated references.

Finally, a large list of references to empirical applications employing RD designs may be found in Lee and Lemieux (2010), Cattaneo, Keele, Titiunik, and Vazquez-Bare (2016, supplemental appendix), and references therein.

Identification, Interpretation, and Extrapolation

Hahn, Todd, and van der Klaauw (2001) were the first to formally discuss identification of average treatment effects at the cutoff in the so-called Sharp and Fuzzy RD designs, that is, in RD settings with perfect and imperfect treatment compliance, respectively. They employed the potential outcomes framework to analyze the RD design, and gave conditions based on continuity of conditional expectation functions at the cutoff, guaranteeing large sample identification of the treatment effect parameters of interest. Lee (2008) also studied identification in sharp RD designs, focusing on the interpretation of the estimand in a context where imperfect manipulation of the running variable prevents units from precisely sorting around the cutoff determining treatment assignment. In his imperfect manipulation setting, Lee established continuity of conditional expectations and distribution functions, and offered an heuristic interpretation of RD designs as local randomized experiments. Together, these two cornerstone contributions provided general frameworks for analyzing and interpreting RD designs, which led to widespread methodological innovation in the RD literature.

Building on the above potential outcomes frameworks, and therefore focusing on large sample identification of average treatment effects at the cutoff via continuity assumptions on conditional expectations, more recent work has studied identification and interpretation of treatment effects in other RD designs. For example, Papay, Willett, and Murnane (2011) focus on RD designs with two or more running variables, Keele and Titiunik (2015) analyze geographic RD designs, Card, Lee, Pei, and Weber (2015) study regression kink designs (giving a causal interpretation to kink RD designs), Chiang and Sasaki (2016) focus on quantile kink RD designs, Cattaneo et al. (2016) investigate RD designs with multiple cutoffs, Choi and Lee (2016) consider interactions and partial effects in RD settings with two running variables, and Caetano and Escanciano (2015) exploit the presence of additional covariates to identify RD marginal effects. See also

Calonico, Cattaneo, Farrell, and Titiunik (2016) for a discussion of the potential benefits and pitfalls of employing additional pre-intervention covariates in the RD design. Many other empirical problems are at present being placed in the context of, or formally connected to, different variants of the RD design.

Cattaneo, Frandsen, and Titiunik (2015) present an alternative causal framework to analyze RD designs, introducing and formalizing the notation of local randomization. This framework is conceptually and methodological distinct from the more standard continuity-based framework employed by the papers discussed previously. In their local randomization framework, the goal is to formalize the idea of a local randomized experiment near the cutoff by embedding the RD design in a classical, Fisherian causal model, thereby giving interpretation and justification to randomization inference and related classical experimental methods. This alternative approach was later extended by Cattaneo, Titiunik, and Vazquez-Bare (2017a), where methodological and empirical comparisons between the two causal inference frameworks (continuity and local randomization) are also given.

Finally, a very recent strand of the RD literature has focused on the important question of extrapolation. It is by now well recognized that an important limitation of modern identification approaches at or near the cutoff is that the resulting estimates and inference results are not easily transferable to other populations beyond those having running variables near the cutoff. There are now a few recent papers trying to address this issue: Angrist and Rokkanen (2015) employ a local conditional independence assumption to discuss extrapolation via variation in observable characteristics, Dong and Lewbel (2015) look at local extrapolation via marginal treatment effects and an exclusion restriction in a continuity-based RD framework, Cattaneo et al. (2016) exploit variation in multiple cutoffs to extrapolate RD treatment effects also using an exclusion restriction in a continuity-based RD framework, Bertanha and Imbens (2016) exploit variation induced by imperfect compliance in fuzzy RD designs, Cattaneo, Keele, Titiunik, and Vazquez-Bare (2017) exploit variation in multiple cutoffs but allowing for possible selection into cutoffs, and Rokkanen (2016) employs a factor model for extrapolation of RD treatment effects.

Presentation, Falsification, and Robustness Checks

One of the main virtues of the RD design is that it can be easily and intuitively presented and falsified in empirical work. Automatic, optimal graphical presentation via RD Plots is discussed and formally studied in Calonico, Cattaneo, and Titiunik (2015a). These recent methods offer graphical tools for summarizing the RD design as well as for informally testing its plausibility, which can also be done formally using some of the estimation and inference methods discussed further below.

McCrary (2008) proposed a very interesting and creative falsification method specifically tailored to RD designs. This falsification test looks at whether there is a discontinuity in the density of the running variable near the cutoff, the presence of which is interpreted as evidence of "manipulation" or "sorting" of units around the cutoff. This test is implemented empirically by comparing the estimated densities of the running variable for control and treatment units separately. McCrary's originally implementation used smoothed-out histogram estimators via local polynomial techniques. More recently, Otsu, Xu, and Matsushita (2014) proposed a density test based on empirical likelihood methods, and Cattaneo, Jansson, and Ma (2016a) developed a density test based on a novel local polynomial density estimator that avoids preliminary tuning parameter choices.

Another more standard, but also quite common, falsification approach in RD designs looks at whether there is a null RD treatment effect on pre-intervention covariates or placebo outcomes. The presence of a non-zero RD treatment effect on such variables would provide evidence against the design. This idea follows standard practices in the analysis of experiments, and was first formalized in a continuity-based framework by Lee (2008). Any estimation and inference method for RD designs can be used to implement this falsification approach, employing the pre-intervention covariate or placebo outcome as the outcome variable. For example, the robust bias-corrected local polynomial methods of Calonico, Cattaneo, and Titiunik (2014b) and local randomization methods of Cattaneo et al. (2015) are readily applicable, as well as other methods, all briefly discussed below. As a complement to these estimation and inference methods, Canay and Kamat (2016) recently developed a permutation testing approach for equality of control and treatment distributions, and Ganong and Jäger (2016) also recently developed a different permutation-based approach for kink RD designs.

Estimation and Inference

Local polynomial methods are by now widely accepted as the default technique for the analysis of RD designs. Global polynomial regressions are useful for presentation and graphical analysis (Calonico, Cattaneo, & Titiunik, 2015a), but not recommended for actual estimation and inference of RD treatment effects (Gelman & Imbens, 2014). See also Wing and Cook (2013) for a related discussion of parametric methods in RD designs.

For point estimation purposes, conventional local polynomial methods were originally suggested by Hahn et al. (2001), and later Porter (2003) provided an in-depth large sample analysis in the specific RD context. Building on this work, Imbens and Kalyanaraman (2012) developed mean-squared-error (MSE) optimal bandwidth selectors for local-linear RD estimators in sharp and fuzzy designs. Employing this MSE-optimal bandwidth selector when implementing the corresponding local polynomial estimator gives an MSE-optimal RD treatment effect estimator, which is commonly used in modern empirical work.

For inference purposes, Calonico, Cattaneo, and Titiunik (2014b, CCT hereafter) pointed out that the MSE-optimal local polynomial point estimator cannot be used for constructing confidence intervals in RD designs – or for conducting statistical inference more generally – because of the presence of a first-order misspecification bias. CCT developed new robust bias-corrected inference methods, based on both removing the first-order misspecification bias present in the MSE-optimal RD estimator and adjusting the standard errors accordingly to account for the additional variability introduced by the bias correction. This new method of nonparametric inference for RD designs works very well in simulations, and was also shown to deliver uniformly valid inference (Kamat, 2017) as well as higher-order refinements (Calonico, Cattaneo, & Farrell, 2017a, 2017b). In addition, Calonico et al. (2017a) develop new bandwidth selection procedures specifically tailored to constructing confidence intervals with small coverage errors in RD designs. See Cattaneo and Vazquez-Bare (2016) for an accessible discussion on bandwidth selection and related neighborhood selection methods.

More recently, Calonico et al. (2016) studied identification, estimation and inference of average RD treatment effects when additional pre-intervention covariates are also included in the local polynomial estimation. This paper develops new optimal bandwidth selectors and valid robust bias-corrected inference methods valid under both heteroskedasticity and clustering in the data.

As an alternative to local polynomial methods, researchers also employ flexible methods near the cutoff. This approach is usually justified by assuming some form of local randomization or similar assumption for some neighborhood near the cutoff. Building on this intuitive and commonly employed approach, Cattaneo et al. (2015) and Cattaneo et al. (2017a) present a formal local randomization framework for RD designs employing ideas and methods from the classical analysis of experiments literature. For estimation and inference, Neyman's and Fisher's methods are introduced and developed for RD designs, though Fisherian inference (also known as randomization inference) is recommended due to the likely small sample sizes encountered in the neighborhoods near the cutoff where the local randomization assumption is most plausible. Keele, Titiunik, and Zubizarreta (2015) apply these ideas to geographic RD designs, combining them with a "matching" algorithm to incorporate pre-intervention covariates.

The methods above focus on estimation and inference of average treatment effects at or near the cutoff, under either a continuity-based or randomization-based framework. There are, of course, other methods (and parameters) of potential interest in the RD literature. For example, Otsu, Xu, and Matsushita (2015) discuss empirical likelihood methods for average treatment effects at the cutoff, Shen and Zhang (2016) discuss local polynomial methods for distributional treatment effects at the cutoff, Xu (2016) considers local polynomial methods for limited dependent outcome variable models near the cutoff, Bertanha (2017) considers estimation and inference of different average treatment effects in a multi-cutoff RD design, and Armstrong and Kolesar (2016a, 2016b) discuss nonparametric confidence interval estimation for the sharp average treatment effect at the cutoff. All these contributions employ a continuity-based framework at the cutoff, and therefore employ large sample approximations. In addition to the local randomization framework discussed above, another finite sample framework for the analysis of RD designs was recently introduced by Chib and Jacobi (2016), who employ Bayesian methods in the context of fuzzy RD designs.

Last but not least, some recent research has focused on different departures from the canonical assumptions employed for methodological and practical research. For example, Lee and Card (2008) study RD designs where the running variable is discrete and the researcher employs linear regression extrapolation to the cutoff, Dong (2015) focuses on RD settings where the underlying running variable is continuous but the researcher only observes a discretized version, Lee (2017) studies the issue of classical measurement error in the running variable, Feir, Lemieux, and Marmer (2016) explore the consequences of having weak instruments in the context of fuzzy RD designs, and Dong (2017) studies the implications of non-random sample selection near the cutoff.

Software

Many of the methodological and practical contributions mentioned above are readily available in general purpose software in R and Stata, while other contributions previously discussed and many of the contributions included in this volume can also be implemented using already available software. Calonico, Cattaneo, and Titiunik (2014a, 2015b) and Calonico, Cattaneo, Farrell, and Titiunik (2017) give a comprehensive introduction to software implementing RD methods based on partitioning and local polynomial techniques, covering RD Plots, bandwidth selection, estimation and inference, and many other possibilities. Cattaneo, Jansson, and Ma (2016b) discuss software implementing discontinuity-in-density tests. Cattaneo, Titiunik, and Vazquez-Bare (2016) give a comprehensive introduction to software implementing RD methods based on a local randomization assumption, building on the classical analysis of experiments literature as well as more recent related developments. Finally, Cattaneo, Titiunik, and Vazquez-Bare (2017b) discuss power calculation and survey sample selection for RD designs based on local polynomial estimation and inference methods.

This R and Stata software is available at https://sites.google.com/site/rdpackages.

Contributions in this Volume

This volume of *Advances in Econometrics* includes 12 outstanding chapters on methodology and applications using RD designs. We now offer a brief overview of each of these contributions, and discuss how they fit into the RD literature presented previously.

Identification, Interpretation, and Extrapolation

The first six contributions in this volume are related to fundamental issues of identification, interpretation and extrapolation in RD designs. The first chapter, by Sekhon and Titiunik, discusses the connections and discrepancies between the continuity-based and randomization-based RD frameworks – the two main paradigms for the analysis and interpretation of RD designs. The authors introduce the different concepts in a familiar setting where potential outcomes are random (as opposed to being fixed as in the classical analysis of experiments literature), and then discuss at length the issues and features of each of the two most popular conceptual frameworks in RD designs. This chapter not only clarifies the underlying conditions many times implicitly imposed in each of these frameworks, but also gives the reader a unique opportunity to appreciate some of the underlying key differences between them.

The second chapter, by Jales and Yu, is truly thought-provoking. The authors introduce and discuss ideas of identification and interpretation in settings where a continuous (running) variable exhibits a discontinuity in its probability density function. They not only review several recent empirical papers where such a situation arises naturally, but also discuss in great detail how this reduced form feature can be used to identify useful parameters in several seemingly unrelated economic models. This chapter introduces the reader to these ideas and, perhaps more importantly, offers a general framework for analysis of economic situations where discontinuities in density functions are present. This contribution will surely spike further methodological research, both on identification as well as on estimation and inference.

The third chapter, by Lee and McCrary, provides another intellectually stimulating instance where identification and interpretation in RD designs can be naturally enhanced by employing economic theory. This outstanding chapter not only was (when originally written) one of the first to report a credible zero causal treatment effect of incarceration on recidivism, but also provides two remarkable and highly innovative methodological contributions. First, it illustrates how modern methodology in RD designs can be successfully adapted to incorporate the specific features of the empirical problem at hand (i.e., sample selection and non-random censoring). Second, it shows how an economic model can be used, together with reduced form estimates from the RD design, to estimate interesting and useful structural parameters, thereby offering a tight connection between reduced form and structural methods in RD contexts.

The fourth and fifth chapters in this volume are closely related to each other, both focusing on different aspects of geographic RD designs. The chapter by Keele, Lorch, Passarella, Small, and Titiunik offers an overview of research designs based on a geographically discontinuous treatment assignment leading to adjacent treated and untreated areas. The authors discuss how the availability of geo-referenced data affects the ability of researchers to employ this type of design in a pure (two-dimensional) RD framework. When researchers have access to the exact geographic location of each individual observation, the geographically discontinuous treatment assignment can be analyzed in a standard RD setup. In contrast, when information about geographic location is only available for aggregate units, these designs are better analyzed as RD designs with discrete running variables, if the aggregate units are sufficiently small, or otherwise as geographic "quasi-experiments," possibly after controlling for observable characteristics. The discussion and underlying issues are illustrated with an empirical application, which shows some of the acute internal validity challenges that are typical in research designs based on geographically discontinuous treatments (e.g., treated and control units continue to have systematic differences even after adjusting for observables or considering only geographically close units).

The chapter by Galiani, McEwan, and Quistorff illustrates similar internal validity challenges in geographic-quasi experiments, and also discusses challenges related to their external validity. To study both types of threats, the authors use data from an experimental study in development economics as benchmark. Their empirical study focuses on various geographic designs that compare treated units close to municipal borders to both experimental and non-experimental untreated groups. This analysis shows that the geographic quasi-experiment is unable to recover the experimental benchmark. This is related to both internal and external validity threats. First, there is empirical evidence of location-based sorting on observed (and possibly unobserved) variables, as treated and control units appear systematically different in at least one important covariate – this raises concerns about internal validity. Second, the exclusion of units far from the border in the geographic-quasi experiment is shown to lead to a covariate distribution that differs from the covariate distribution in the experimental sample. Because some of these covariates are potential moderators of the treatment effect, this raises concerns about the external validity of the geographic quasi-experiment effect. In sum, the discussion and results in Keele et al. and Galiani et al. suggest that research designs based on geographically discontinuous treatments offer exciting opportunities to evaluate policies and programs, but they are also vulnerable to considerable internal and external validity challenges, setting the ground for much needed future research in this area.

The sixth contribution, by Tang, Cook, Kisbu-Sakarya, Hock, and Chiang, focuses on the Comparative RD design, a recently introduced methodology that incorporates a placebo outcome variable to improve extrapolation of average treatment effects in sharp RD designs, in addition to aiding parametric estimation and inference. The authors present an insightful review of this novel methodology, and also illustrate its main practical features by employing an empirical application with an underlying randomized controlled trial component. This new methodology employs global parametric methods coupled with an outcome variable unaffected by treatment but observed over the full support of the running variable, to improve efficiency in parametric estimation and extrapolation in RD designs. In their empirical application, the Comparative RD design methodology performs well when compared to the results from the randomized controlled trial component.

Presentation, Falsification, and Robustness Checks

Two chapters in this volume are related to falsification and robustness checks in RD designs. The chapter by Frandsen investigates how the idea underlying the widely used McCrary's density discontinuity test for manipulation can be adapted and employed in settings where the running variable is discrete. This is a very important question, as many RD designs employ discrete running variables. The author develops a new manipulation test that employs finite sample distributional methods and is justified via large sample approximations and bounds on the underlying smoothness of unknown functions. This novel manipulation test complements existing tests, most of which are only valid when the running variable is continuously distributed, as well as the simple binomial tests also widely used in practice.

A second contribution in this volume to robustness checks in RD designs (and, by implication, extrapolation) is the chapter by Cerulli, Dong, Lewbel, and Poulsen. The authors introduce and discuss a new test for local stability of RD treatment effects. In particular, this chapter proposes to test for zero slope change in the average treatment effect at the cutoff, which is effectively equivalent to testing for a null kink RD treatment effect. The authors then argue that, whenever there is no change in the treatment effect of interest relative to the running variable near the cutoff, the usual sharp treatment effect is more stable and hence provides a more global result for units near the cutoff. A key feature of this idea is that it can be implemented quite easily using available modern methods for RD estimation and inference, which will surely contribute to the popularity of this test in empirical applications.

Estimation and Inference

The last four chapters in this volume focus attention on different aspects of estimation and inference in RD designs. In all cases, these chapters take a continuity-based approach, employ local polynomial methods, and either assess the empirical properties of recently proposed methods in the literature or develop new methods in practically relevant settings.

First, the chapter by Card, Lee, Pei, and Weber offers an insightful and thorough empirical study of the finite sample properties of the robust bias-corrected inference methods proposed by Calonico, Cattaneo, and Titiunik (2014b, CCT) in the context of regression kink designs (and, more generically, kink RD designs). Their paper offers several valuable lessons for practitioners hoping to employ the most recent methodological innovations in the RD design literature. In particular, the authors bring attention to issues related to (i) choice of polynomial order, (ii) bandwidth selection methods, and (iii) potential lack of precision of robust methods. These findings are not only important for empirical work, but also set the ground for future research and further methodological improvements.

Second, the chapter by Bartalotti and Brummet studies bandwidth selection for point estimation and inference when robust bias-correction methods are used, in a setting where generic clustering among units is possibly present. Building on CCT's recent work under random sampling, the authors develop a new MSE expansion for sharp RD designs under general clustering, and employ this approximation to obtain a new MSE-optimal bandwidth under clustered data. This bandwidth choice is different from the standard MSE-optimal choice obtained under random sampling, and can be used to construct an MSE-optimal RD local polynomial point estimator under general clustering. The authors also discuss the special case of clustering at the running variable level, which is common in empirical work and leads to important simplifications in the methodology. These new methods are highly relevant and very useful for empirical work employing RD designs.

Third, the chapter by Bartalotti, Calhoun, and He introduces a bootstrap inference method based on robust bias-correction techniques. Building on CCT's robust bias-correction approach, the authors develop a double wild bootstrap method where the first layer of bootstrap is used to approximate the misspecification bias and the second-layer is used to compute valid variance and distributional approximations taking into account the bias-correction first step. The authors also show that the first bootstrap layer gives a bias estimate that is equivalent to the analytic bias-correction proposed by CCT, up to simulation error. These results are not only useful for empirical work (i.e., they provide an alternative way of implementing CCT robust bias-correction methods), but also open the door for future research connecting bootstrapping methods and bias-correction in other RD designs settings (e.g., with clustered data or when including additional covariates).

Last but not least, the chapter by Pei and Shen studies RD settings where the running variable is measured with error, and provides alternative sufficient conditions guaranteeing identifiability of RD treatment effects when estimated using the mismeasured assignment variable, the treatment status, and the outcome variable. The authors study RD settings where the running variable is either discrete or continuous, thereby offering quite a complete analysis with wide applicability for empirical work. These results contribute to a recent literature on this topic, and more generally to the literature on departures from canonical assumptions in RD designs, briefly summarized above. The issue of mismeasured running variables is quite important in practice, and this chapter not only offers a clear introduction to this important problem, but also sets a framework for the analysis and interpretation of RD designs with measurement error. This chapter will surely motivate future work in this important research area.

ACKNOWLEDGMENTS

We thank Sebastian Calonico, Max Farrell, Michael Jansson, Xinwei Ma, Kenichi Nagasawa, Rocío Titiunik, and Gonzalo Vazquez-Bare for many enlightening and insightful discussions over the years. Cattaneo gratefully acknowledges financial support from the National Science Foundation through grants SES-1357561 and SES-1459931.

As part of this volume preparation, a research conference was held at the University of Michigan on May 19–20, 2016. Details on this conference can be found online (https://sites.google.com/site/aie38rdd/conference). Cattaneo thanks MITRE at the University of Michigan's Department of Economics and the National Science Foundation (through grants SES-1357561 and SES-1459931) for providing financial and logistical support, without which the *Advances in Econometrics* conference would have not been possible. In addition, we thank Alberto Abadie, Josh Angrist, Martha Bailey, Michael Jansson, Pat Kline, and Jeff Wooldridge for agreeing to serve in the conference scientific committee. We also thank Alberto Abadie, Tim Armstrong, Martha Bailey, Quentin Brummet, Federico Bugni, Gray Calhoun, Sebastian Calonico, Otavio Bartalotti, Max Farrell, Thomas Fomby, Margherita Fort, Brigham Frandsen, Sebastian Galiani, Andreas Hagemann, Nicolas Idrobo, Hugo Jales, Michael Jansson, Luke Keele, Lutz Kilian, Michal Kolesar, Arthur Lewbel, Xinwei Ma, Justin McCrary, Patrick McEwan, Walter Mebane, Jose Luis Montiel Olea, Kenichi Nagasawa, Zhuan Pei, Andres Santos, Rocío Titiunik, Gonzalo Vazquez-Bare, and Tim Vogelsang for attending and providing excellent feedback during the conference.

Last but not least, we are also very thankful to many of the scholars listed above, who generously agreed to serve as anonymous reviewers of the chapters published in this volume. Their invaluable and constructive comments and suggestions certainly improved the overall quality of this volume.

Matias D. Cattaneo
Juan Carlos Escanciano
Editors

REFERENCES

Angrist, J., & Rokkanen, M. (2015). Wanna get away? Regression discontinuity estimation of exam school effects away from the cutoff. *Journal of the American Statistical Association, 110*(512), 1331–1344.

Armstrong, T. B., & Kolesar, M. (2016a). Optimal inference in a class of regression models. arXiv:1511.06028.

Armstrong, T. B., & Kolesar, M. (2016b). A simple adjustment for bandwidth snooping. arXiv:1412.0267.

Bertanha, M. (2017). *Regression discontinuity design with many thresholds*. Working Paper, University of Notre Dame.

Bertanha, M., & Imbens, G. W. (2016). *External validity in fuzzy regression discontinuity designs*. Working Paper No. 20773, National Bureau of Economic Research.

Caetano, C., & Escanciano, J. C. (2015). *Identifying multiple marginal effects with a single binary instrument or by regression discontinuity*. Working Paper, Indiana University.

Calonico, S., Cattaneo, M. D., & Farrell, M. H. (2017a). *Coverage error optimal confidence intervals for regression discontinuity designs*. Working paper, University of Michigan.

Calonico, S., Cattaneo, M. D., & Farrell, M. H. (2017b). On the effect of bias estimation on coverage accuracy in nonparametric inference. *Journal of the American Statistical Association*, forthcoming.

Calonico, S., Cattaneo, M. D., Farrell, M. H., & Titiunik, R. (2016). *Regression discontinuity designs using covariates*. Working Paper, University of Michigan.

Calonico, S., Cattaneo, M. D., Farrell, M. H., & Titiunik, R. (2017). rdrobust: Software for regression discontinuity designs. *Stata Journal*, forthcoming.

Calonico, S., Cattaneo, M. D., & Titiunik, R. (2014a). Robust data-driven inference in the regression-discontinuity design. *Stata Journal, 14*(4), 909–946.

Calonico, S., Cattaneo, M. D., & Titiunik, R. (2014b). Robust nonparametric confidence intervals for regression-discontinuity designs. *Econometrica, 82*(6), 2295–2326.

Calonico, S., Cattaneo, M. D., & Titiunik, R. (2015a). Optimal data-driven regression discontinuity plots. *Journal of the American Statistical Association, 110*(512), 1753–1769.

Calonico, S., Cattaneo, M. D., & Titiunik, R. (2015b). rdrobust: An R package for robust nonparametric inference in regression-discontinuity designs. *R Journal, 7*(1), 38–51.

Canay, I. A., & Kamat, V. (2016). *Approximate permutation tests and induced order statistics in the regression discontinuity design.* Working Paper, Northwestern University.

Card, D., Lee, D. S., Pei, Z., & Weber, A. (2015). Inference on causal effects in a generalized regression kink design. *Econometrica, 83*(6), 2453–2483.

Cattaneo, M. D., Frandsen, B., & Titiunik, R. (2015). Randomization inference in the regression discontinuity design: An application to party advantages in the U.S. Senate. *Journal of Causal Inference, 3*(1), 1–24.

Cattaneo, M. D., Jansson, M., & Ma, X. (2016). *rddensity: Manipulation testing in stata.* Working Paper, University of Michigan.

Cattaneo, M. D., Jansson, M., & Ma, X. (2016a). *Simple local regression distribution estimators with an application to manipulation testing.* Working Paper, University of Michigan.

Cattaneo, M. D., Keele, L., Titiunik, R., & Vazquez-Bare, G. (2016). Interpreting regression discontinuity designs with multiple cutoffs. *Journal of Politics, 78*(4), 1229–1248.

Cattaneo, M. D., Keele, L., Titiunik, R., & Vazquez-Bare, G. (2017). *Extrapolating treatment effects in multi-cutoff regression discontinuity designs.* Working Paper, University of Michigan.

Cattaneo, M. D., Titiunik, R., & Vazquez-Bare, G. (2016). Inference in regression discontinuity designs under local randomization. *Stata Journal, 16*(2), 331–367.

Cattaneo, M. D., Titiunik, R., & Vazquez-Bare, G. (2017a). Comparing inference approaches for RD designs: A reexamination of the effect of head start on child mortality. *Journal of Policy Analysis and Management*, forthcoming.

Cattaneo, M. D., Titiunik, R., & Vazquez-Bare, G. (2017b). *Power calculations for regression discontinuity designs.* Working Paper, University of Michigan.

Cattaneo, M. D., & Vazquez-Bare, G. (2016). The choice of neighborhood in regression discontinuity designs. *Observational Studies, 2*, 134–146.

Chiang, H. D., & Sasaki, Y. (2016). *Causal inference by quantile regression kink designs.* arXiv:1605.09773.

Chib, S., & Jacobi, L. (2016). Bayesian fuzzy regression discontinuity analysis and returns to compulsory schooling. *Journal of Applied Econometrics, 31*(6), 1026–1047.

Choi, J., & Lee, M. (2016). *Regression discontinuity with multiple running variables allowing partial effects.* Working Paper.

Cook, T. D. (2008). "Waiting for Life to Arrive": A history of the regression-discontinuity design in psychology, statistics and economics. *Journal of Econometrics, 142*(2), 636–654.

Dong, Y. (2015). Regression discontinuity applications with rounding errors in the running variable. *Journal of Applied Econometrics, 30*(3), 422–446.

Dong, Y. (2017). Regression discontinuity designs with sample selection. *Journal of Business and Economic Statistics*, forthcoming.

Dong, Y., & Lewbel, A. (2015). Identifying the effect of changing the policy threshold in regression discontinuity models. *Review of Economics and Statistics*, *97*(5), 1081–1092.

Feir, D., Lemieux, T., & Marmer, V. (2016). Weak identification in fuzzy regression discontinuity designs. *Journal of Business & Economic Statistics*, *34*(2), 185–196.

Ganong, P., & Jäger, S. (2016). *A permutation test for the regression kink design*. Working Paper, Harvard University.

Gelman, A., & Imbens, G. W. (2014). *Why high-order polynomials should not be used in regression discontinuity designs*. NBER Working Paper No. 20405.

Hahn, J., Todd, P., & van der Klaauw, W. (2001). Identification and estimation of treatment effects with a regression-discontinuity design. *Econometrica*, *69*(1), 201–209.

Imbens, G., & Lemieux, T. (2008). Regression discontinuity designs: A guide to practice. *Journal of Econometrics*, *142*(2), 615–635.

Imbens, G. W., & Kalyanaraman, K. (2012). Optimal bandwidth choice for the regression discontinuity estimator. *Review of Economic Studies*, *79*(3), 933–959.

Kamat, V. (2017). On nonparametric inference in the regression discontinuity design. *Econometric Theory*, forthcoming.

Keele, L. J., & Titiunik, R. (2015). Geographic boundaries as regression discontinuities. *Political Analysis*, *23*(1), 127–155.

Keele, L. J., Titiunik, R., & Zubizarreta, J. (2015). Enhancing a geographic regression discontinuity design through matching to estimate the effect of ballot initiatives on voter turnout. *Journal of the Royal Statistical Society: Series A*, *178*(1), 223–239.

Lee, D. S. (2008). Randomized experiments from non-random selection in U.S. house elections. *Journal of Econometrics*, *142*(2), 675–697.

Lee, D. S., & Card, D. (2008). Regression discontinuity inference with specification error. *Journal of Econometrics*, *142*(2), 655–674.

Lee, D. S., & Lemieux, T. (2010). Regression discontinuity designs in economics. *Journal of Economic Literature*, *48*(2), 281–355.

Lee, M.-J. (2017). Regression discontinuity with errors in the running variable: Effect on truthful margin. *Journal of Econometric Methods*, *6*(1–8), 281–355.

McCrary, J. (2008). Manipulation of the running variable in the regression discontinuity design: A density test. *Journal of Econometrics*, *142*(2), 698–714.

Otsu, T., Xu, K.-L., & Matsushita, Y. (2014). Estimation and inference of discontinuity in density. *Journal of Business and Economic Statistics*, *31*(4), 507–524.

Otsu, T., Xu, K.-L., & Matsushita, Y. (2015). Empirical likelihood for regression discontinuity design. *Journal of Econometrics*, *186*(1), 94–112.

Papay, J. P., Willett, J. B., & Murnane, R. J. (2011). Extending the regression-discontinuity approach to multiple assignment variables. *Journal of Econometrics*, *161*(2), 203–207.

Porter, J. (2003). *Estimation in the regression discontinuity model*. Working Paper, University of Wisconsin.

Rokkanen, M. (2016). *Exam schools, ability, and the effects of affirmative action: Latent factor extrapolation in the regression discontinuity design*. Working Paper, Columbia University.

Shen, S., & Zhang, X. (2016). Distributional regression discontinuity: Theory and applications. *Review of Economics and Statistics*, *98*(4), 685–700.

Thistlethwaite, D. L., & Campbell, D. T. (1960). Regression-discontinuity analysis: An alternative to the ex-post facto experiment. *Journal of Educational Psychology*, *51*(6), 309–317.

Wing, C., & Cook, T. D. (2013). Strengthening the regression discontinuity design using additional design elements: A within-study comparison. *Journal of Policy Analysis and Management*, *32*(4), 853–877.

Xu, K.-L. (2016). *Regression discontinuity with categorical outcomes.* Working Paper, Indiana University.

ON INTERPRETING THE REGRESSION DISCONTINUITY DESIGN AS A LOCAL EXPERIMENT

Jasjeet S. Sekhon[a] and Rocío Titiunik[b]

[a]*UC Berkeley, Travers Department of Political Science and Department of Statistics, Berkeley, CA, USA*

[b]*Department of Political Science, University of Michigan, Ann Arbor, MI, USA*

ABSTRACT

We discuss the two most popular frameworks for identification, estimation and inference in regression discontinuity (RD) designs: the continuity-based framework, where the conditional expectations of the potential outcomes are assumed to be continuous functions of the score at the cutoff, and the local randomization framework, where the treatment assignment is assumed to be as good as randomized in a neighborhood around the cutoff. Using various examples, we show that (i) assuming random assignment of the RD running variable in a neighborhood of the cutoff implies neither that the potential outcomes and the treatment are statistically independent, nor that the potential outcomes are unrelated to the running variable in this neighborhood; and (ii) assuming local independence between the potential outcomes and the treatment does not imply the exclusion restriction that the score affects the outcomes only through the treatment

Regression Discontinuity Designs: Theory and Applications
Advances in Econometrics, Volume 38, 1–28

ISSN: 0731-9053/doi:10.1108/S0731-905320170000038001

indicator. Our discussion highlights key distinctions between "locally randomized" RD designs and real experiments, including that statistical independence and random assignment are conceptually different in RD contexts, and that the RD treatment assignment rule places no restrictions on how the score and potential outcomes are related. Our findings imply that the methods for RD estimation, inference, and falsification used in practice will necessarily be different (both in formal properties and in interpretation) according to which of the two frameworks is invoked.

Keywords: Regression discontinuity; local experiment; as-if random assignment; local randomization

The regression discontinuity (RD) design is a research strategy based on three main components – a score or "running variable," a cutoff, and a treatment. Its basic characteristic is that the treatment is assigned based on a known rule: all units receive a score value, and the treatment is offered to those units whose score is above a cutoff and not offered to those units whose score is below it (or viceversa).

The RD design has been available since the 1960s (Thistlethwaite & Campbell, 1960), but its popularity has grown particularly fast in recent years. In the last decade, an increasing number of empirical researchers across the social and biomedical sciences has turned to the RD design to estimate causal effects of treatments that are not, and often cannot be, randomly assigned. This growth in RD empirical applications has occurred in parallel with a rapid development of methodological tools for estimation, inference, and interpretation of RD effects. See Cook (2008), Imbens and Lemieux (2008), and Lee and Lemieux (2010) for early reviews, and the introduction to this volume by Cattaneo and Escanciano for a comprehensive list of recent references.

The recent popularity of the RD design was in part sparked by the work of Hahn, Todd, and van der Klaauw (2001), who translated the design into the Neyman–Rubin potential outcomes framework (Holland, 1986) and offered minimal conditions for nonparametric identification of average effects. These authors showed that, when all units comply with their assigned treatment, the average effect of the treatment at the cutoff can be identified under the assumption that the conditional expectations of the potential outcomes given the score are continuous (and other mild regularity conditions). This emphasis on continuity conditions for identification was a departure from other "quasi-experimental" research designs, which are typically based on independence or mean independence assumptions.

In the Neyman–Rubin framework, randomized experiments are seen as the gold standard, and quasi-experimental designs are broadly characterized in terms of the assumptions under which the (nonrandom) treatment assignment mechanism is as good as randomized. For example, in observational studies based on the unconfoundedness or "selection-on-observables" assumption, the treatment is as good as randomly assigned, albeit with unknown distribution, after conditioning on observable covariates (Imbens & Rubin, 2015). Similarly, instrumental variables (IV) designs can be seen as randomized experiments with imperfect compliance, and difference-in-difference designs compare treatment and control groups that, on average – and except for time-invariant characteristics that can be removed by differencing – differ only on treatment status. In all these cases, a treatment and a control group are well defined and, under certain assumptions, can be compared to identify the average treatment effect for the population of interest (or a subpopulation thereof).

In this context, the results derived by Hahn et al. (2001) set the RD design apart. In contrast to most other quasi-experimental designs, RD identification was established only for the average treatment effect at a single point: the cutoff. This implied that, unlike differences-in-differences or selection-on-observable designs, the RD design could not be accurately described by appealing to a comparison between a treatment and a control group. Given the RD assignment rule, under perfect compliance it is impossible for treated and control units to have the same score value. Moreover, when the running variable is continuous, the probability of seeing an observation with a score value exactly equal to the cutoff is zero. Thus, in the continuity-based RD setup, identification of the average treatment effect at the cutoff necessarily relies on extrapolation, as there are neither treated nor control observations with score values exactly equal to the cutoff.

The influential contribution by Lee (2008) changed the way RD was perceived, and aligned the interpretation of RD designs with experimental (and the other quasi-experimental) research designs. Lee (2008) argued that in an RD setting where the score can be influenced by the subjects' choices and unobservable characteristics, treatment status can be interpreted to be as good as randomized in a local neighborhood of the cutoff as long as subjects lack the ability to precisely determine the value of the score they receive – that is, as long as their score contains a random chance component. Lee's framework captured the original ideas in the seminal article by Thistlethwaite and Campbell (1960), who called a hypothetical experiment where the treatment is randomly assigned near the cutoff an "experiment for which the regression-discontinuity analysis may be regarded as a substitute" (Thistlethwaite & Campbell, 1960, p. 310).

The interpretation of RD designs as local experiments developed by Lee (2008) has been very influential, both conceptually and practically. Among other things, it established the need to provide falsification tests based on predetermined covariates just as one would do in the analysis of experiments, a practice that has now been widely adopted and has increased the credibility of countless RD applications (Caughey & Sekhon, 2011; Eggers et al., 2015; Hyytinen, Meriläinen, Saarimaa, Toivanen, & Tukiainen, 2015). Moreover, it provided an intuitive interpretation of the RD parameter that allowed researchers to think about treatment and control groups instead of an effect at a single point where there are effectively no observations.

The claim that the RD treatment assignment rule is "as good as randomized" in a neighborhood of the cutoff can be interpreted in at least two ways. In one interpretation, it means that there should be no treatment effect on predetermined covariates at the cutoff, and that the validity of the underlying RD assumptions can be evaluated by testing the null hypothesis that the RD treatment effect is zero on predetermined covariates. In another interpretation, it means that the treatment is (as good as) randomly assigned near the cutoff, and estimation and inference for treatment effects (and covariate balance tests) can be carried out using the same tools used in experimental analysis. The first interpretation does not imply the second because one can test for covariate balance (at the cutoff) under the usual continuity assumptions. While the first interpretation has resulted in an increased and much needed focus on credibility and falsification, the second has been the source of considerable confusion. Our goal is to discuss the source of such confusion in detail. In doing so, we clarify crucial conceptual distinctions within the local randomization RD framework, which in turn elucidate the differences and similarities between this framework and the more standard continuity-based approach to RD analysis.

We explore both the relationship between continuity and local randomization assumptions in RD designs, and the implications of adopting an RD framework based on an explicit local randomization assumption. Our chapter builds on prior studies that have considered this issue. Hahn et al. (2001)

first invoked a local randomization assumption for identification of RD effects under noncompliance. More recently, Cattaneo, Frandsen, and Titiunik (2015) formalized an analogous assumption using a Fisherian, randomization-based RD framework, which was extended by Cattaneo, Titiunik, and Vazquez-Bare (2016, 2017). Various super-population versions of the local randomization RD assumption were also proposed by Keele, Titiunik, and Zubizarreta (2015) and Angrist and Rokkanen (2015), and more recently de la Cuesta and Imai (2016) discussed the relationship between local randomization and continuity RD assumptions.

We make two main arguments. First, we show that the usual RD continuity assumptions are not sufficient for a literal local randomization interpretation of RD designs – that is, not sufficient to ensure that, near the cutoff, the potential outcomes are independent of the treatment assignment and unrelated to the running variable. This has been argued previously (Cattaneo et al., 2015; de la Cuesta & Imai, 2016); we simply provide a stylized example to further illustrate the main issues. Second, contrary to common practice, we show that the assumption that the treatment is randomly assigned among units in a neighborhood of the cutoff *cannot* be used to justify analyzing and interpreting RD designs as actual experiments. The restrictions imposed by the treatment assignment rule in a sharp RD design rule out a local experiment where one randomly assigns treatment status without changing the units' score values. Instead, one could assume that score values are randomly assigned in a neighborhood of the cutoff, a randomization model that is consistent with the RD assignment mechanism. However, even under this model, interpreting and analyzing the RD design as an experiment will generally result in invalid inferences because the score can affect the potential outcomes directly in addition to through treatment status.

In particular, we show that (i) assuming random assignment of the RD running variable in a neighborhood of the cutoff does not imply that the potential outcomes and the treatment assignment are statistically independent or that the potential outcomes are unrelated to the running variable in this neighborhood, and (ii) assuming local independence between the potential outcomes and the treatment assignment does not imply the exclusion restriction that the score affects the outcomes only via the treatment assignment indicator. Our discussion makes clear that the RD treatment assignment rule need not place any restrictions on the ways in which the score influences the potential outcomes and shows that, in local randomization RD settings, statistical independence and random assignment are conceptually different.

The discussion that follows focuses on the sharp RD design, in which units' compliance with treatment assignment is perfect, and the probability of receiving treatment changes from zero to one at the cutoff. We do not explicitly discuss fuzzy RD settings, where compliance is imperfect and the decision to take treatment is endogenous, because most of our arguments and conclusions apply equally to both sharp and fuzzy RD designs. When our discussion must be modified for the fuzzy case, we note it explicitly. Throughout, we refer to the standard RD design based on continuity identification assumptions as the *continuity-based* RD design; and to the RD design based on a local randomization assumption as the *local randomization* or *randomization-based* RD design.

1. THE CONTINUITY-BASED RD FRAMEWORK

In this and the subsequent sections, we assume that we have a random sample $\{Y_i(1), Y_i(0), X_i\}_{i=1}^{n}$, where X_i is the score on the basis of which a binary treatment T_i is assigned according to the rule $T_i = \mathbb{1}(X_i \geq c)$ – or $T_i = \mathbb{1}(X_i \leq c)$ depending on the example – for a known constant c, $Y_i(1)$ is the potential outcome under the treatment condition, and $Y_i(0)$ the potential outcome under the untreated or control condition. For every unit i, we observe either $Y_i(0)$ or $Y_i(1)$, so the observed sample is $\{Y_i, X_i\}_{i=1}^{n}$ where $Y_i := T_i Y_i(1) + (1 - T_i) Y_i(0)$. We assume throughout that all moments we employ exist, and that the density of X_i is positive and continuous at the cutoff or in the intervals we consider.

The continuity-based framework is based on the identification conditions and estimation methods first proposed by Hahn et al. (2001). The authors proposed the following assumption:

Assumption 1 (Continuity). The regression functions $\mathbb{E}[Y_i(1)|X_i = x]$ and $\mathbb{E}[Y_i(0)|X_i = x]$ are continuous in x at the cutoff c.

Under this assumption, they showed that

$$\tau_{\text{CB}}^{\text{RD}} \equiv \mathbb{E}[Y_i(1) - Y_i(0)|X_i = c] = \lim_{x \downarrow c} \mathbb{E}[Y_i|X_i = x] - \lim_{x \uparrow c} \mathbb{E}[Y_i|X_i = x]. \quad (1)$$

Thus, when the focus is on mean effects, the target parameter in the continuity-based RD framework is $\tau_{\text{CB}}^{\text{RD}}$ – the average treatment effect at cutoff. The identification result in Eq. (1) states that, under continuity, this parameter – which depends on potential outcomes that are fundamentally unobservable, can be expressed as the difference between the right and left limits of the average observed outcomes at the cutoff. Estimation is therefore concerned with constructing appropriate estimators for $\lim_{x\uparrow c} \mathbb{E}[Y_i|X_i = x]$ and $\lim_{x\downarrow c} \mathbb{E}[Y_i|X_i = x]$.

The most commonly used approach to estimate these limits is to rely on local polynomial methods, fitting two polynomials of the observed outcome on the score – one for observations above the cutoff, the other for observations below it – using only observations in a neighborhood of the cutoff, with kernel weights assigning higher weights to observations closer to the cutoff. Because these fitted polynomials are approximations to the unknown regression functions $\mathbb{E}[Y_i(1)|X_i = x]$ and $\mathbb{E}[Y_i(0)|X_i = x]$ near the cutoff, the choice of neighborhood, commonly known as *bandwidth*, is crucial. Given the order of the polynomial – typically one – the bandwidth controls the quality of the approximation, with smaller bandwidths reducing the bias of the approximation and larger bandwidths reducing its variance – see Cattaneo and Vazquez-Bare (2016) for an overview of RD neighborhood selection.

Although the technical details of local polynomial estimation and inference are outside the scope of our discussion, we highlight several issues that are central to distinguishing the continuity-based approach from the local randomization approach we discuss next. In the continuity-based RD design:[1]

- The target parameter, $\tau_{\text{CB}}^{\text{RD}}$, is an average effect at a single point, rather than in an interval.
- The functional form of the regression functions $\mathbb{E}[Y_i(0)|X_i = x]$ and $\mathbb{E}[Y_i(1)|X_i = x]$ is unknown, and is locally approximated by a polynomial for estimation and inference.
- In general, the polynomial approximation will be imperfect. The approximation error is controlled by the bandwidth sequence, h_n: the smaller h_n, the smaller the error.
- Local polynomial estimation and inference are based on large-sample results that require the bandwidth to shrink to zero as the sample size increases. For example, consistent estimation of $\tau_{\text{CB}}^{\text{RD}}$ requires $h_n \to 0$ and $nh_n \to \infty$.
- Local polynomial methods require smoothness conditions on the underlying regression functions $\mathbb{E}[Y_i(0)|X_i = x]$ and $\mathbb{E}[Y_i(1)|X_i = x]$ in order to control the leading biases of the local polynomial RD estimators. These smoothness conditions are stronger than the continuity assumption required for identification of $\tau_{\text{CB}}^{\text{RD}}$.

- The bandwidth h_n plays no role in the identification of $\tau_{\text{CB}}^{\text{RD}}$.
- Neither the continuity assumptions required for identification nor the smoothness assumptions required for estimation and inference are implied by the RD treatment assignment rule. See Sekhon and Titiunik (2016) for further discussion.

1.1. Continuity Does Not Imply Local Randomization

We now consider a stylized example that shows that continuity of the potential outcomes regression functions (as in Assumption 1) does not imply that the treatment can be seen as locally randomly assigned in a literal or precise sense. We assume that we have a sample of n students who take a mathematics exam in the first quarter of the academic year. Each student receives a test score X_i in the exam, which ranges from 0 to 100, and students whose grade is equal to or below 50 receive a double dose of algebra instruction ($T_i = 1$) during the second quarter – while students with $X_i > 50$ receive a single dose ($T_i = 0$). The outcome of interest is the test score Y_i obtained in another mathematics exam taken at the end of the second quarter, also ranging between 0 and 100. We assume that test scores are fine enough so that they have no mass points and can be treated as continuous random variables.

To illustrate our argument, we assume that a student's expected grade in the second quarter under the control condition given her score in the first quarter, $\mathbb{E}[Y_i(0)|X_i]$, is simply equal to her score in the first quarter, X_i. We also assume that the double-dose algebra treatment effect, τ, is constant for all students. The potential outcomes regression functions are therefore:

$$\begin{aligned} \mathbb{E}[Y_i(0)|X_i] &= X_i, \\ \mathbb{E}[Y_i(1)|X_i] &= X_i + \tau. \end{aligned} \tag{2}$$

This model is of course extremely simplistic, but we adopt it because it allows us to illustrate the difference between the treatment assignment mechanism and the outcome model in a straightforward way. We now assume that the initial grade X_i is entirely determined by each student's fixed inherent (and unobservable) ability a_i – for example, $X_i = g(a_i)$ for some strictly increasing function $g(\cdot)$. Given this setup, the assignment of treatment according to the rule $\mathbb{1}(X_i \leq 50)$ is as far from random as can be conceived, since it assigns all students of lower ability to the treatment group and all students of higher ability to the control group, inducing a complete lack of common support in the distribution of ability between the groups. Despite this severe selection into treatment based on ability, the effect of the treatment at the cutoff is readily identifiable, because the regression functions in Eq. (2) are continuous at the cutoff (and everywhere else).

In this context, what does it mean to say that the RD design induces as-if randomness near the cutoff? No matter how small the neighborhood around the cutoff, the average ability in the control group is always higher than in the treatment group: the effect of X on $\mathbb{E}[Y_i(j)|X_i]$ near the cutoff is always nonzero – $d\mathbb{E}[Y_i(j)|X_i = x]/dx = 1$ for $j = 0, 1$ and for all x. Thus, the continuity of the regression functions is entirely compatible with a very strong relationship between outcome and score. This means that continuity is not sufficient to guarantee the comparability of units on either side of the cutoff, even in a small neighborhood around it. In other words, for any $w > 0$, one can always conceive a data generating process such that the distortion induced by ignoring the relationship between X and Y in the window $[c - w,\ c + w]$ is arbitrarily large (a uniformity argument).

Thus, from an identification point of view, it is immediate to see that the continuity condition in Assumption 1 (which ensures identification in the super population) does not imply that the usual finite-sample "random assignment" identification assumptions hold. We will discuss the latter type of assumptions in detail in the following section, but we now consider one possibility in the context of the example in Eq. (2). Imagine that in this example we wish to invoke a mean independence assumption in a small neighborhood $W = [c - w,\ c + w]$ around the cutoff, with $w > 0$, such as $\mathbb{E}[Y_i(j)|T_i, X_i \in W] = \mathbb{E}[Y_i(j)|X_i \in W]$ for $j = 0, 1$; recall that in this example $T_i = \mathbb{1}(X_i \leq c)$.

Focusing, for example, on the potential outcome under control, and given Eq. (2), we have

$$\begin{aligned}\mathbb{E}[Y_i(0)|T_i = 0, X_i \in W] &= \mathbb{E}[Y_i(0)|X_i > c, X_i \in [c - w, c + w]] \\ &= \mathbb{E}[Y_i(0)|X_i \in (c, c + w]] \\ &= \mathbb{E}[X_i|X_i \in (c, c + w]]\end{aligned}$$

and

$$\begin{aligned}\mathbb{E}[Y_i(0)|X_i \in W] &= \mathbb{E}[Y_i(0)|X_i \in [c - w, c + w]] \\ &= \mathbb{E}[X_i|X_i \in [c - w, c + w]].\end{aligned}$$

In general, $\mathbb{E}[X_i|X_i \in (c, c+w]] \neq \mathbb{E}[X_i|X_i \in [c-w, c+w]]$, so the mean independence assumption that is typically invoked in experiments cannot be invoked in this case. In other words, this example shows that the continuity assumption is not enough to guarantee independence between the potential outcomes and the treatment near the cutoff, and consequently cannot be used to justify analyzing an RD design as one would analyze an experiment.

From an estimation point of view, imagine that we mistakenly decided to analyze this valid continuity-based RD design as an experiment in the fixed neighborhood $[c-w, c+w]$ around the cutoff, defining the parameter of interest as the average treatment effect in this window. To calculate this effect, we would simply compare the average treated–control difference in the observed outcomes in $[c-w, c+w]$. Unsurprisingly, this approach would lead to an incorrect answer, since

$$\begin{aligned}
&\mathbb{E}[Y_i|c-w \leq X_i \leq c] - \mathbb{E}[Y_i|c < X_i \leq c+w] \\
&\quad = \tau + \mathbb{E}[X_i|c-w \leq X_i \leq c] - \mathbb{E}[X_i|c < X_i \leq c+w] \\
&\quad < \tau,
\end{aligned}$$

where the last line follows from $\mathbb{E}[X_i|c-w \leq X_i \leq c] - \mathbb{E}[X_i|c < X_i \leq c+w] \in [-2w, 0)$ and the assumption that the density of the score is positive and continuous in $[c-w, c+w]$.[2] Thus, for any fixed window $W = [c-w, c+w]$, analyzing this RD design as one would analyze an experiment will lead to an incorrect treatment effect estimate.

In this example, the smaller w, the closer the naive difference-in-means estimate will be to the true effect τ. However, as we discuss below, in order for the local randomization RD framework to be conceptually distinct from the continuity-based framework, the neighborhood W must necessarily be conceived as fixed. And, given a fixed W, the lack of comparability between treated and control groups cannot be eliminated by increasing the sample size – a direct consequence of the fact that such lack of comparability is an identification problem rather than an estimation one.

In contrast, in the continuity-based framework, focusing on the average treated–control outcome difference in a neighborhood of the cutoff can be understood as approximating the unknown potential outcomes regression functions with a local constant fit. Such strategy will result in a possibly large approximation error; however, the error is entirely due to the estimation strategy and will vanish asymptotically. In this fundamental sense, an RD design based on a continuity assumption cannot be interpreted literally as a local experiment. In other words, assuming the conditional regression functions $\mathbb{E}[Y_i(1)|X_i]$ and $\mathbb{E}[Y_i(0)|X_i]$ are continuous in X_i at c is not enough to treat the RD design as a pure experiment near the cutoff. If we are to treat the RD design as a local experiment, we must effectively fix a window width, and thus change the parameter of interest.

2. THE LOCAL RANDOMIZATION RD FRAMEWORK

The simple example above shows that continuity of the conditional regression functions is not enough to justify analyzing or interpreting the RD design as an experiment in a neighborhood of the cutoff. We now consider a local randomization RD setup where we explicitly assume that the treatment is randomly assigned for all subjects with $X_i \in [c - w, c + w]$. As we discuss, formalizing precisely the local randomization RD framework is difficult, because the notion of random assignment near the cutoff can be interpreted in different ways.

2.1. Local Randomization of Score or Treatment?

Intuition suggests that simply assuming that the treatment is randomly assigned in a small neighborhood around the cutoff should be enough to analyze the RD design as a local experiment. However, this is not the case for two reasons: (i) the RD treatment assignment rule places strict restrictions on the type of random assignments that are conceivable, and (ii) the assignments that are conceivable do not imply an exclusion restriction that is always true in actual experiments.

We consider two possible scenarios, according to two different interpretations of what it means for the treatment to be locally randomly assigned. In the first scenario, the values of X_i stay constant but the treatment received is randomly changed for subjects near the cutoff. In the second scenario, the value of X_i is randomly assigned for all subjects near the cutoff. As we show, this is an important distinction that must be considered when formalizing the local randomization RD assumption.

2.1.1. Scenario 1: Treatment Status Randomly Assigned for all Subjects Near the Cutoff

The first way to understand locally random treatment assignment in the RD context is to imagine a situation where all subjects with score in a neighborhood of the cutoff – that is, with $X_i \in [c - w, c + w]$ – are randomly assigned to receive treatment or control. In the context of our education example introduced in the previous section, we could accomplish this if, for example, we randomly assigned all students who scored between 45 and 55 in the first exam to receive either a single or double dose of algebra, with every student receiving double dose with the same positive probability. Given our assumption that the test score in the first quarter is entirely determined by the students' ability, this assignment mechanism would break the relationship between ability and treatment status induced by the RD rule $\mathbb{1}(X_i \leq 50)$, because it would make receiving the treatment entirely independent of the grade obtained in the first exam.

However, this way of conceptualizing the randomization implies that the score X_i is unrelated to treatment status in the local neighborhood, which is incompatible with the treatment assignment rule. In particular, this assignment would imply a positive number of both treated and control subjects on each side of the cutoff in the neighborhood $[c - w, c + w]$, contradicting the RD treatment assignment rule $\mathbb{1}(X_i \leq c)$. This illustrates the general point that, in a sharp RD design, treatment status is deterministic given the score, and thus it is not possible to randomly assign the treatment without altering the values of the running variable.[3]

2.1.2. Scenario 2: Score Value Randomly Assigned for All Subjects Near the Cutoff

The alternative is to assume that the running variable, not treatment status, is randomly assigned near the cutoff – a manipulation that would be consistent with the sharp RD treatment rule. There are multiple ways in which one could conceive of such an experiment. For example, following our double-dose algebra example, we might believe that, even though exam performance is broadly influenced by ability, the precise grade received by each student involves some degree of randomness and arbitrariness, such that students whose grades are ten points or less apart should be of comparable ability. The school may therefore implement a two-stage process, where first all exams are graded, and for those students whose grades fall between 45 and 55, the original score is replaced with a uniform random number between 45 and 55. The justification for this two-stage procedure might be that it still assigns the treatment to those students who most need it (all students who get very low grades are guaranteed to receive the treatment) but it implements a transparent and fair process to assign the treatment to those students very near the cutoff whose observable characteristics may be indistinguishable.

Under this setup, the test score in the first exam, X_i, is still broadly determined by ability – so that a student who obtains a grade of 25 is on average of lower ability than a student who obtains a grade of 90 – but for two students whose grades are between 45 and 55, the second stage ensures that they have the same ex-ante probability of receiving treatment. This hypothetical random assignment of score values implies that the average level of ability (and any other predetermined confounder) of treated students with $X_i \in [45, 50]$ is the same as for control students with $X_i \in [50, 55]$. Note that this setup also assumes that the score X_i collected by the researcher is the second stage score for all students, so that the treatment rule $\mathbb{1}(X_i \leq 50)$ correctly distinguishes treated and control students. Such a two-stage rule is of course artificial, although a second-stage randomization has been used in some actual RD settings (Hyytinen et al., 2015).[4]

The crucial question is whether the assumption that the value of the score is (as-if) randomly assigned in a neighborhood of the cutoff – that is, the assumption that all units with score in $[c - w, c + w]$ have the same probability of receiving any score value x within this neighborhood and therefore the same probability of receiving treatment or control – is enough to justify analyzing and interpreting the RD design in this neighborhood as an actual experiment. At first glance, it might appear as if it is. We are assuming that, among units in a local neighborhood of the cutoff, the value of the score is randomly assigned; since the treatment is deterministically assigned based on this score, this guarantees that, on average, there are no differences in the predetermined characteristics of units above and below the cutoff within this window.

However, this reasoning ignores the possibility that the score itself may affect the potential outcomes, a phenomenon we have explicitly allowed in our example in Eq. (2). When we introduced this equation, we motivated it by imagining that students' unobservable ability affects both the grade obtained in the first exam, X_i, and the grade obtained in the second exam, Y_i. Our new assumption that the score is unrelated to students' predetermined characteristics in the neighborhood [45, 55] might seem at first to imply that the running variable must be unrelated to the regression functions in this neighborhood, as illustrated in Fig. 1(a). However, if the running variable X_i affects the potential outcomes through factors other than predetermined characteristics, the regression functions can follow Eq. (2) even when X_i is randomly assigned in a window around the cutoff – this situation is illustrated in Fig. 1(b).

A more precise way to denote the potential outcomes would be $Y_i(T_i, X_i)$, where the first argument captures the effect of the treatment

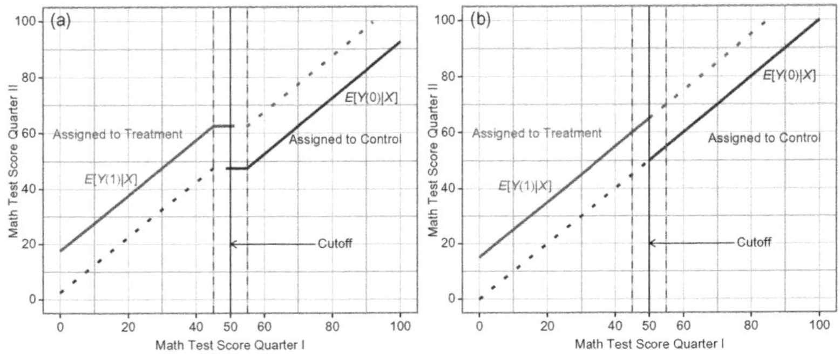

Fig. 1. Two Scenarios with Randomly Assigned Score. (a) Test Scores Locally Unrelated to Potential Outcomes. (b) Test Scores Locally Related to Potential Outcomes.

assignment – that is, of being on one side of the cutoff or the other – and the second argument captures the effect of the running variable on the outcome that occurs independently of the treatment status. This "direct" effect of X_i on Y_i would occur in our example if, for instance, a student's test score in the first exam had a reinforcing effect. If near the cutoff students who receive lower test scores were discouraged and put less effort in the second exam than students who receive higher scores (because, e.g., they feel stigmatized by being assigned to the treatment group and are frustrated by having been so close to the cutoff), X_i might be positively associated with the potential outcomes even in the absence of a treatment effect. Under this assumption, the model in Eq. (2) and Fig. 1(b) could still be true even if the score were randomly assigned in [45, 55].

Another way to see the distinction is to note that any simple experiment can be seen as an RD design in which the score is a random number and the cutoff is chosen to ensure the desired probability of treatment assignment. In our example, if instead of having instructors grade the exams we assigned each student a uniform random number between 0 and 100, the treatment assignment rule $\mathbb{1}(X_i \leq 50)$ would effectively become a rule that assigns double-dose algebra entirely randomly among students, with each student having a 50% probability of receiving treatment. In this case, however, there would be no need to add an extra argument in the regression functions for the random number that determines treatment, as the score used to randomly assign treatment would be a computer-generated pseudo-random number that is unrelated to the potential outcomes by construction.[5]

In contrast, in an RD design where the score is assumed to be randomly assigned in the local neighborhood $[c - w, c + w]$ but is not an arbitrarily generated number unrelated to the phenomena under study, nothing prevents the value of the running variable from affecting the potential outcomes directly. In other words, local random assignment of X_i does not guarantee the exclusion restriction that we typically take for granted in actual randomized experiments. This illustrates the distinction between assumptions about the assignment of the score X_i, and assumptions about the shape of the regression functions: assumptions about the law of the random variable X_i place no restrictions on the functions $\mathbb{E}[Y_i(1)|X_i]$, and $\mathbb{E}[Y_i(0)|X_i]$.

Note also that this phenomenon is analogous to the IV design, where random assignment of the instrument does not imply the exclusion restriction that is required to interpret the usual estimand as the average treatment effect for the compliers (Angrist, Imbens, & Rubin, 1996). The parallelism arises because in both IV and RD designs, researchers are interested in the effect of the treatment on the outcome, but there is a third variable (the score in RD, the instrument in IV) that induces a change in treatment status and can also affect the potential outcomes directly. In both cases, the way in which this third variable is assigned imposes no general restrictions on its relationship with the potential outcomes.

The analogy between IV and fuzzy RD designs has been long recognized in the continuity-based framework, where the similarities arise because of imperfect compliance with the treatment assignment rule in fuzzy RD settings. However, in the context of the local randomization RD framework, similarities occur even in the sharp RD case where compliance with treatment assignment is perfect. The analogy we point out is not related to treatment compliance but rather to the possible role of the score (instrument) as a determinant of the outcome irrespective of treatment assignment and/or status.

2.2. Formalizing the Local Randomization RD Assumption

We now attempt to formalize a local randomization RD assumption. The first such formalization was proposed by Cattaneo et al. (2015), who used a Fisherian randomization-based approach in which the potential outcomes are seen as fixed as opposed to random variables. Their proposed assumption has two parts. The first part states that the conditional distribution function of the score in the finite sample is the same for all units in the neighborhood of the cutoff. The second explicitly adopts an exclusion restriction that rules out any "direct" effects of the score on the potential outcomes and states that, within the window, the fixed potential outcomes depend on the running variable solely through the treatment indicator.[6] As our discussion above illustrates, the exclusion restriction plays a crucial role in the analogy between RD designs and experiments, an issue that has not been generally recognized by scholars outside of the Fisherian framework.

Since the continuity-based RD framework has been almost exclusively developed within a random sampling framework in which the potential outcomes are random variables, and the analogy between RD and experiments is often understood in these terms, we formalize the local randomization assumption using the random sampling setting introduced in Section 1. We thus retain the random sampling framework throughout the chapter. In particular, our formalization is based on statistical independence between the potential outcomes $Y_i(1)$ and $Y_i(0)$ and the binary treatment assignment indicator in a neighborhood of the cutoff, an assumption routinely invoked in experimental analysis and guaranteed by the random assignment of treatment in actual randomized experiments. A condition of this kind has been invoked in local randomization RD settings by, for example, Angrist and Rokkanen (2015), de la Cuesta and Imai (2016), and Keele et al. (2015).

Formally, we state this assumption as:

> **Assumption 2: (Super-population) Local Independence.** There exists a neighborhood around the cutoff c, $W = [c - w, c + w]$, $w > 0$, such that $Y_i(1), Y_i(0) \perp\!\!\!\perp T_i | X_i \in W$, where $\perp\!\!\!\perp$ denotes statistical independence and $T_i = \mathbb{1}(X_i \geq c)$ as defined above.

Under this assumption, the average treatment effect in the window W is identified by

$$\tau_{\text{LR}}^{\text{RD}} \equiv \mathbb{E}[Y_i(1) - Y_i(0) | X_i \in W] = \mathbb{E}[Y_i | T_i = 1, X_i \in W] - \mathbb{E}[Y_i | T_i = 0, X_i \in W]$$

If the window W is known, estimation can proceed, for example, simply by computing a difference-in-means between treated and control groups for observations in this window.

Under Assumption 2, the local randomization RD framework has the following features:

- The target parameter, $\tau_{\text{LR}}^{\text{RD}}$, is the average treatment effect in an interval of the support of the running variable, rather than at a point.

- Because the focus is on an average over an interval rather than at a point, approximation of the functional form of the regression functions $\mathbb{E}[Y_i(0)|X_i = x]$ and $\mathbb{E}[Y_i(1)|X_i = x]$ is not necessary for estimation.
- Knowledge of the window $W = [c - w, c + w]$ around the cutoff is necessary for identification of $\tau_{\text{LR}}^{\text{RD}}$.
- Estimation and inference methods assume the window is fixed as the sample size grows.

Thus, as we have defined it, the local randomization RD setup stands in contrast to the continuity-based framework described above. We highlight two distinctions in particular.

Remark 1 (Randomization in window vs. at the cutoff). Our characterization of the local randomization RD framework is explicit in stating a "random assignment" assumption that holds in an interval around the cutoff rather than at the cutoff point. This is in contrast to some formalizations of the local randomization RD framework that make assumptions at the cutoff point. The reason we focus on an interval rather than a point is because the requirement of statistical independence at a point is trivial: any random variable is independent of a constant, so the potential outcomes and the score will always be independent at the cutoff. In other words, an assumption of randomization or independence at the cutoff has no empirical or theoretical content, and thus cannot be used as a justification to analyze RD designs as experiments.

Naturally, one can still use the local randomization RD interpretation simply as an approximation device, where the window is not seen as the interval where a randomization condition holds but rather as the interval where the unknown regression functions are approximated – see Cattaneo et al. (2015, §6.5). But used in this heuristic way, the local randomization RD framework becomes in essence identical to the continuity-based framework: the parameter of interest becomes the treatment effect at the cutoff, and identification and estimation results are ultimately based on some kind of continuity condition (Canay & Kamat, 2016; Lee, 2008). ■

Remark 2 (The Role of Neighborhood). If the local randomization RD framework is understood in terms of a fixed window as in Assumption 2 and not as the heuristic approximation discussed in Remark 1, the role of the neighborhood in this approach is conceptually different from the role of the bandwidth in the continuity-based framework. In the latter framework, the bandwidth is used to control the bias and variance of the local polynomial approximation to the unknown regression functions; this implies, among other things, that optimal and valid estimation and inference will require choosing a different bandwidth for every outcome variable or covariate analyzed. In contrast, in the local randomization RD framework based on an assumption such as Assumption 2, the window is a fundamental piece of the research design, since it is the interval where the required identification assumption holds. Consequentially, in the local randomization framework, this single window is used to perform estimation and inference for all outcomes and covariates. See Cattaneo and Vazquez-Bare (2016) for further details on the role of neighborhood selection in RD estimation and inference. ■

3. THE DIFFERENCE BETWEEN RANDOM ASSIGNMENT AND INDEPENDENCE IN RD CONTEXTS

We now consider whether Assumption 2 is implied by the random assignment of the score near the cutoff and whether it implies that the exclusion restriction is satisfied in the RD context. An in depth consideration of these issues reveals subtle relationships between the concepts of random treatment assignment, statistical independence, and exclusion restriction in RD settings. Our discussion makes three main points: (i) the random assignment of the score does not imply local independence between treatment assignment and potential outcomes; (ii) the local statistical independence between treatment assignment and potential outcomes does not imply the exclusion restriction that holds by construction in experiments; and (iii) an experimental analysis is possible under local independence if the exclusion restriction fails, provided that the interpretation of the parameter is modified accordingly.

3.1. Random Assignment of Score Does Not Imply Local Independence

The local independence assumption as stated in Assumption 2 is not guaranteed to hold when we randomly assign the score near the cutoff – that is, when all subjects with $c - w \leq X_i \leq c + w$ face the same ex-ante probability of receiving every value of the score between $c-w$ and $c+w$. The reason is that, even if all subjects in the neighborhood of the cutoff have the same marginal distribution of the running variable, the potential outcomes and the running variable may be related in ways that violate the local independence assumption. If, as discussed above, higher test scores lead students to expend systematically more or less effort in future exams, the randomly assigned value of X_i will affect the potential outcomes directly even inside the local neighborhood $[c-w, c+w]$, inducing a relationship between the potential outcomes $Y_i(1)$, $Y_i(0)$ and the treatment indicator $\mathbb{1}(X_i \leq c)$ or $\mathbb{1}(X_i \geq c)$ that may violate independence.

Thus, unlike in the case of actual randomized experiments, in a local randomization RD design in the sense of Assumption 2, statistical independence and random assignment are conceptually different. The reason is that, as discussed above, the implicit randomization rule in a local randomization RD design does not and cannot manipulate treatment status directly; instead, the rule must randomly assign the score values. Therefore, if the randomly assigned score has a direct impact on the potential outcomes – so that, for example, high values of X_i lead to high values of the potential outcomes – the indicators $\mathbb{1}(X_i \leq c)$ and $\mathbb{1}(X_i \geq c)$ may fail to be statistically independent of the potential outcomes.

To see this more formally, simply consider the example in Eq. (2) introduced above. We already showed in Section 1 that this example violates Assumption 2; now simply assume that X_i is randomly assigned in a neighborhood of the cutoff, which is of course allowed by Eq. (2).

3.2. Local Independence Does Not Imply Exclusion Restriction

Moreover, the local independence assumption as stated in Assumption 2 is not a sufficient condition for the exclusion restriction that the score does not affect the potential outcomes except via the treatment indicator. That is, the local independence assumption does not imply that the score X_i and the regression functions $\mathbb{E}[Y_i(0) \mid X_i]$, $\mathbb{E}[Y_i(1) \mid X_i]$ are unrelated in a local neighborhood of the cutoff. Graphically, this means that regression functions need not be flat in the neighborhood $[c-w, c+w]$ even when Assumption 2 holds; in other words, local independence does not necessarily imply a scenario like the one illustrated in Fig. 1(a).

We briefly present an example to illustrate this point. It suffices to focus on one potential outcome, so we focus on $Y_i(1)$.

$$Y_i(1) = \mathbb{1}(X_i \in W) \cdot (1 - |X_i - c|) + \mathbb{1}(X_i \notin W) \cdot (1 - w) + \varepsilon_i \qquad (3)$$

for all i, with ε_i a random error independent of X_i, c the cutoff, $\mathbb{1}(X_i \leq c)$ the treatment rule, and $W = [c - w, c + w]$ as before. The regression function $\mathbb{E}[Y_i(1)|X]$ implied by this model for $Y_i(1)$ is shown in Fig. 2. To simplify notation, we redefine $\tilde{x} := x - c$ so that the cutoff for $\tilde{x}$ is normalized to zero. We also assume that the density of $\tilde{X}$, $f(\tilde{x})$, is symmetric around zero in $[-w, w]$.

We wish to show that in this setup, which clearly violates the exclusion restriction, the local independence assumption holds. Let $\tilde{W} = [-w, w]$. We want to show

$$\mathbb{P}\left[Y(1) \leq y, \tilde{X} \leq 0 \mid \tilde{X} \in \tilde{W}\right] = \mathbb{P}\left[Y(1) \leq y \mid \tilde{X} \in \tilde{W}\right] \cdot \mathbb{P}\left[\tilde{X} \leq 0 \mid \tilde{X} \in \tilde{W}\right].$$

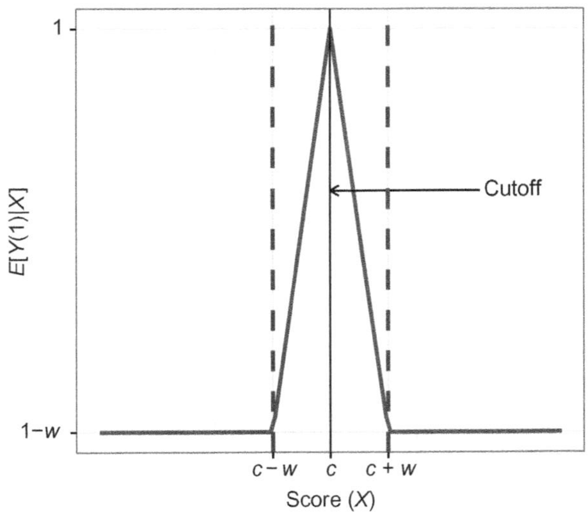

Fig. 2. Example Where Exclusion Restriction Is Violated But Local Independence Holds.

Under the assumptions we made, we can show that

$$
\begin{aligned}
&\mathbb{P}[Y(1) \leq y, \tilde{X} \leq 0 \mid \tilde{X} \in \tilde{W}] \\
&\quad = \mathbb{P}[Y(1) \leq y \mid -w \leq \tilde{X} \leq 0] \cdot \mathbb{P}[\tilde{X} \leq 0 \mid -w \leq \tilde{X} \leq w] \\
&\quad = \mathbb{E}[\,\mathbb{P}[Y(1) \leq y \mid \tilde{X}] \mid -w \leq \tilde{X} \leq 0] \cdot \mathbb{P}[\tilde{X} \leq 0 \mid -w \leq \tilde{X} \leq w] \\
&\quad = \frac{\int_{-w}^{0} F_\varepsilon(y - 1 + |\tilde{x}|) \cdot f(\tilde{x}) \cdot \mathrm{d}\tilde{x}}{\int_{-w}^{0} f(\tilde{x})\, \mathrm{d}\tilde{x}} \cdot \frac{\int_{-w}^{0} f(\tilde{x})\, \mathrm{d}\tilde{x}}{\int_{-w}^{w} f(\tilde{x})\, \mathrm{d}\tilde{x}} \\
&\quad = \frac{\frac{1}{2}\left(\int_{-w}^{0} F_\varepsilon(y - 1 + |\tilde{x}|) \cdot f(\tilde{x}) \cdot \mathrm{d}\tilde{x} + \int_{0}^{w} F_\varepsilon(y - 1 + |\tilde{x}|) \cdot f(\tilde{x}) \cdot \mathrm{d}\tilde{x}\right)}{\int_{-w}^{w} f(\tilde{x})\, \mathrm{d}\tilde{x}} \\
&\quad = \frac{\frac{1}{2}\int_{-w}^{w} F_\varepsilon(y - 1 + |\tilde{x}|) \cdot f(\tilde{x}) \cdot \mathrm{d}\tilde{x}}{\int_{-w}^{w} f(\tilde{x})\, \mathrm{d}\tilde{x}} \\
&\quad = \frac{1}{2}\mathbb{P}[Y(1) \leq y \mid \tilde{X} \in \tilde{W}]
\end{aligned}
$$

Since we assumed that the density of $\tilde{x}$ is symmetric around zero in the window, $\mathbb{P}\left[\tilde{X} \leq 0 \mid \tilde{X} \in \tilde{W}\right] = \frac{1}{2}$. Thus, we have shown that in this example

$$
\begin{aligned}
\mathbb{P}\left[Y(1) \leq y, \tilde{X} \leq 0 \mid \tilde{X} \in \tilde{W}\right] &= \frac{1}{2}\mathbb{P}\left[Y(1) \leq y \mid \tilde{X} \in \tilde{W}\right] \\
&= \mathbb{P}\left[\tilde{X} \leq 0 \mid \tilde{X} \in \tilde{W}\right] \cdot \mathbb{P}\left[Y(1) \leq y \mid \tilde{X} \in \tilde{W}\right]
\end{aligned}
$$

and therefore $Y(1)$ and $\mathbb{1}\left(\tilde{X} \leq 0\right)$ are independent in $\tilde{W}$.

The most important assumptions in this example are the symmetry of the density of the running variable X_i around the cutoff in the window $[c - w,\ c + w]$, the independence between the error term ε_i and X_i in this window, and the symmetry around the cutoff of the functional form that relates $Y(1)$ and X. The intuition behind the result is as follows: if X is symmetrically distributed in $[c - w,\ c + w]$, the value of the random variable $Y = 1 - |X - c| + \varepsilon$ will be the same regardless of whether $X > c$ or $X < c$; being above or below the cutoff does not provide any information regarding the particular value of $Y(1)$ that will be observed. Thus, the local independence condition between the potential outcome $Y_i(1)$ and the treatment indicator $\mathbb{1}(X \leq c)$ holds even though $Y(1)$ and X are related.

3.3. Experimental Analysis under Local Independence Will Capture 'Overall' Effect If Exclusion Restriction Fails

We have shown that local statistical independence as stated in Assumption 2 does not guarantee that the exclusion restriction holds. Although referring to the potential outcomes functions as $Y_i(1)$ and $Y_i(0)$ may give the impression that these functions are only affected by the treatment indicator but not the score itself, this notation is commonly used to broadly refer to the potential outcome under the treatment and control conditions, including everything that these conditions entail. In the context of actual randomized experiments, it is generally unnecessary to make the notation more explicit to let the potential outcomes depend on the particular randomization device used. But in contexts where the variable that induces variation in treatment assignment is an important determinant of the potential outcomes rather than an arbitrary device, generalizing the potential outcomes notation is necessary. For example, a more general potential outcomes notation has been used to allow for the direct effect of the instrument in IV setups (Angrist et al., 1996) and of the cutoff in multi-cutoff RD setups (Cattaneo, Keele, Titiunik, & Vazquez-Bare, 2016).

Following the notation we introduced briefly in Scenario 2 in Section 2, we use let $Y_i(T_i, X_i)$ denote the potential outcomes, now explicitly acknowledging that the potential outcomes may depend on the value of the score X_i directly in addition to through the treatment assignment indicator. Using this notation, Assumption 2 implies the mean independence condition $\mathbb{E}[Y_i(j, X_i)|T_i = j, X_i \in W] = \mathbb{E}[Y_i(j, X_i)|X_i \in W]$ for $j \in \{0, 1\}$. Given this generalization, we now consider whether the local randomization RD framework can be used in a context where Assumption 2 holds but the exclusion restriction fails.

Even if the exclusion restriction is violated in the sense that $Y_i(j, X_i) \neq Y_i(j, X_i')$ for $X_i \neq X_i'$, $j \in \{0, 1\}$, under Assumption 2 we have

$$\mathbb{E}[Y_i|T_i = 1, X_i \in W] = \mathbb{E}[Y_i(1, X_i)|T_i = 1, X_i \in W] = \mathbb{E}[Y_i(1, X_i)|X_i \in W],$$

$$\mathbb{E}[Y_i|T_i = 0, X_i \in W] = \mathbb{E}[Y_i(0, X_i)|T_i = 0, X_i \in W] = \mathbb{E}[Y_i(0, X_i)|X_i \in W],$$

leading to

$$\mathbb{E}[Y_i|T_i = 1, X_i \in W] - \mathbb{E}[Y_i|T_i = 0, X_i \in W] = \mathbb{E}[Y_i(1, X_i) - Y_i(0, X_i)|X_i \in W] = \tau_{\text{LR}}^{\text{RD}}.$$

Thus, in a broad sense, Assumption 2 justifies analyzing the RD design in the neighborhood of the cutoff as one would analyze an experiment because it allows identification of the average treatment effect $\tau_{\mathrm{LR}}^{\mathrm{RD}}$. However, if the exclusion restriction does not hold, the treated and control average outcomes within the local window will combine the effect of the treatment (e.g., a student receiving double-dose algebra or a party winning an election) on the outcome, *plus* the additional effect of the score on the outcome that would occur regardless of treatment status (e.g., students who receive higher grades may feel motivated to study more, political donors may wish to donate more money to localities where parties obtain higher vote shares, etc.). In this case, the parameter $\tau_{\mathrm{LR}}^{\mathrm{RD}}$ will not capture the (local) average effect of the treatment alone, but rather the effect of obtaining a value of the score in $[c, c + w]$ versus $[c - w, c)$, which will include, among other things, the effect of the treatment.

Using the more general notation, we can see that in a local randomization RD design where Assumption 2 holds but the exclusion restriction does not, the parameter $\tau_{\mathrm{LR}}^{\mathrm{RD}}$, unlike $\tau_{\mathrm{CB}}^{\mathrm{RD}}$, is not the average effect at any single point x:

$$\begin{aligned}\tau_{\mathrm{LR}}^{\mathrm{RD}} &= \mathbb{E}[Y_i(1) - Y_i(0)|X_i \in W] = \mathbb{E}[Y_i(1, X_i) - Y_i(0, X_i)|X_i \in W] \\ &\neq \mathbb{E}[Y_i(1, x) - Y_i(0, x)|X_i \in W].\end{aligned}$$

Imagine, for example, that $\mathbb{E}[Y_i(0, X_i)|X_i \in W] = \mu_0$, constant for every i in the window, and $Y_i(1, X_i)$ follows Eq. (3). In this case,

$$\tau_{\mathrm{LR}}^{\mathrm{RD}} = \mathbb{E}[1 - |X_i - c||X_i \in W] - \mu_0 = \frac{\int_{c-w}^{c+w}(1 - |x - c|)f(x)\,\mathrm{d}x}{\int_{c-w}^{c+w} f(x)\,\mathrm{d}x} - \mu_0,$$

which will take any value between $1 - w - \mu_0$ and $1 - \mu_0$, depending on the density of the running variable X inside the window. This example shows that, in a scenario where the exclusion restriction fails, the average effect $\tau_{\mathrm{LR}}^{\mathrm{RD}}$ will take different values according to the likelihood of observations concentrating in particular ranges of the running variable inside the window. For example, the same setup just described would lead to different values of $\tau_{\mathrm{LR}}^{\mathrm{RD}}$ if the observations within the window were uniformly distributed than if they were disproportionally concentrated near the cutoff. Thus, when the exclusion restriction fails and the regression functions are not constant in the window around the cutoff, the interpretation of the average effect $\tau_{\text{LR}}^{\text{RD}}$ differs from the interpretation it would have in a standard experiment.

4. CONCLUDING REMARKS

Our discussion highlights that the local randomization interpretation of the RD design, if taken literally, introduces conceptual distinctions that are absent in pure experimental designs. As we have shown, in the context of RD designs, the distinction between random assignment and statistical independence is consequential. This distinction is often meaningful in natural experiments (Sekhon & Titiunik, 2012). In an actual experiment, random assignment of treatment leads to statistical independence between treatment status and potential outcomes, because the "score" used to randomly assign subjects is a device (likely a computer-generated pseudo-random number) that is by construction arbitrary and unrelated to the potential outcomes or any systematic characteristic of the experimental subjects. As we observed, a randomized experiment can be recast as an RD design where the score is a randomly generated number and the cutoff is chosen to ensure the desired probability of receiving treatment. Thus, seen as functions of this random number, the regression functions are guaranteed to be constant (graphically flat), because the score is a randomly generated number that is by construction unrelated to the potential outcomes.

In contrast, in an RD design, the score used to assign treatment, even if its values are randomly allocated near the cutoff, are usually important determinants of the outcome of interest. Indeed, the importance of the RD score is often what motivates using it as the basis of treatment assignment in the first place. In a context where the score is meaningfully related to the outcome (past and future test scores, past and future vote shares, etc.), random assignment of the score value contains no information about the particular form of the regression functions $\mathbb{E}[Y_i(1)|X_i]$ and $\mathbb{E}[Y_i(0)|X_i]$. This is straightforward to see once one realizes that restrictions on the randomization distribution of the score X_i (and implicitly of the treatment assignment) are fundamentally different from restrictions on the shape of the regression functions – and more generally, on the conditional distribution of the potential outcomes given the score. No matter how much information we have about the way in which X_i is assigned, this information will generally be insufficient to determine how X_i and the potential outcomes are related.

Given the additional assumptions needed for the local randomization interpretation to hold, in most applications one should proceed using the continuity assumption alone. This is typically a plausible assumption if there are neither formal nor informal mechanisms for sorting – that is, for units to appeal and change the score value they were originally assigned in order to receive their preferred treatment condition. However, in practice, the methods used to estimate the continuous functions at the cutoff are consequential. Estimation is delicate because the functional forms are unknown, and one is estimating the trend at an endpoint (the cutoff) of measure zero. The example in Hyytinen et al. (2015) illustrates how inferences in RD designs are sensitive to the method used to estimate the continuous functions on both sides of the cutoff even when sample sizes are not small, and the design is valid by construction. More generally, the properties of RD estimation and inference based on local polynomial methods depend crucially on the bandwidth choice; for example, the commonly used mean-squared-error optimal bandwidth, though valid for point estimation, is too large for inference and requires either undersmoothing or robust methods to yield valid confidence intervals (Calonico et al., 2014). One appeal of the local randomization interpretation of RD is that it avoids these estimation and inference issues, but unfortunately, this interpretation requires additional assumptions that may not be plausible.

NOTES

1. For a general treatment of local polynomial methods, see Fan and Gijbels (1996) and Calonico, Cattaneo, and Farrell (2017) for recent higher-order results. For local polynomial methods applied specifically to the RD case, see Hahn et al. (2001), Porter (2003), Imbens and Kalyanaraman (2012), Calonico, Cattaneo, and Titiunik (2014), and Calonico et al. (2016).

2. If the treatment assignment rule were $\mathbb{1}(X_i \geq c)$, which is the more common definition of the treatment in the RD literature, we would have $\mathbb{E}[Y_i|c \leq X_i \leq c + w] - \mathbb{E}[Y_i|c - w \leq X_i < c] = \tau + \mathbb{E}[X_i|c \leq X_i \leq c + w] - \mathbb{E}[X_i|c - w \leq X_i < c]$, and $\mathbb{E}[X_i|c \leq X_i \leq c + w] - \mathbb{E}[X_i|c - w \leq X_i < c] \in (0, 2w]$.

3. Note that in a fuzzy RD design, it is possible to have both treated and control observations on either side of the cutoff. However, the simple random assignment we have described would still be incompatible with a fuzzy RD assignment because it would assign treatment with the same probability on either side of the cutoff, which would violate the assumption of discontinuity of the probability of treatment assignment at the cutoff.

4. The reason we use two stages to set up a scenario where the score is randomly assigned is because we want to create a scenario where ability – a predetermined

characteristic – is not systematically different between treated and control groups. Because of potential individual-level heterogeneity in ability within the window $[c - w, c + w]$, this cannot be guaranteed unless we assume a two-stage procedure or make an explicit assumption about the distribution of ability for subjects that fall in the neighborhood of the cutoff. Note also that the two-stage scenario ensures comparability of all predetermined characteristics near the cutoff, whereas a scenario based on a single stage must consider assumptions about all possible predetermined characteristics that affect the score.

To see this, consider the following alternative setup, where we assume that the score is a step function of ability, such that higher ability leads to higher grades but the relationship between both variables is constant in intervals. For example, we might imagine $X_i = f(a_i) + \varepsilon_i$, with $f(a_i) = f(\bar{a})$ for all i such that $c - w \leq X_i \leq c + w$. Then,

$$\begin{aligned}\mathbb{P}[T_i = 1 \leq c \mid c - w \leq X_i \leq c + w] &= \mathbb{P}[X_i \leq c \mid c - w \leq X_i \leq c + w] \\ &= \mathbb{P}[f(\bar{a}) + \varepsilon_i \leq c \mid c - w \leq X_i \leq c + w] \\ &= F_\varepsilon(c - f(\bar{a}) \mid c - w \leq X_i \leq c + w),\end{aligned}$$

where $F_\varepsilon(\cdot)$ is the CDF of ε, which is the same function for all individuals with $c - w \leq X_i \leq c + w$. Under this setup, all individuals in the local neighborhood of the cutoff have the same probability of receiving treatment and the same ability.

5. Note that this statement assumes that students are never informed about the particular random "grade" that they receive, as occurs in actual experiments.

6. Cattaneo et al. (2017) relax this restriction to allow the potential outcomes functions to depend on the score, and use that information to adjust the potential outcomes and build the randomization distribution of the desired test-statistic. Their approach clarifies that one could use a finite-sample randomization-based framework to analyze an RD design when the exclusion restriction is violated, provided one specifies a particular functional form for the potential outcome function $y_i(x, t)$.

ACKNOWLEDGMENTS

We are indebted to Matias Cattaneo, Anthony Fowler, Kellie Ottoboni, Nicolás Idrobo Rincón, Kosuke Imai, Joshua Kalla, Fredrik Sävje, Yotam Shem-Tov, and three anonymous reviewers for helpful comments and discussion. Sekhon gratefully acknowledges support from the Office of Naval Research (N00014-15-1-2367) and Titiunik gratefully acknowledges financial support from the National Science Foundation (SES 1357561).

REFERENCES

Angrist, J. D., Imbens, G. W., & Rubin, D. B. (1996). Identification of causal effects using instrumental variables. *Journal of the American Statistical Association, 91*(434), 444–455.

Angrist, J. D., & Rokkanen, M. (2015). Wanna get away? Regression discontinuity estimation of exam school effects away from the cutoff. *Journal of the American Statistical Association, 110*(512), 1331–1344.

Calonico, S., Cattaneo, M. D., & Farrell, M. H. (2017). On the effect of bias estimation on coverage accuracy in nonparametric inference. *Journal of the American Statistical Association, forthcoming.*

Calonico, S., Cattaneo, M. D., Farrell, M. H., & Titiunik, R. (2016). *Regression discontinuity designs using covariates.*

Calonico, S., Cattaneo, M. D., & Titiunik, R. (2014). Robust nonparametric confidence intervals for regression-discontinuity designs. *Econometrica, 82*(6), 2295–2326.

Canay, I. A., & Kamat, V. (2016). *Approximate permutation tests and induced order statistics in the regression discontinuity design.* Working Paper, Northwestern University.

Cattaneo, M. D., Frandsen, B., & Titiunik, R. (2015). Randomization inference in the regression discontinuity design: An application to party advantages in the U.S. Senate. *Journal of Causal Inference, 3*(1), 1–24.

Cattaneo, M. D., Keele, L., Titiunik, R., & Vazquez-Bare, G. (2016). Interpreting regression discontinuity designs with multiple cutoffs. *Journal of Politics, 78*(3), 1229–1248.

Cattaneo, M. D., Titiunik, R., & Vazquez-Bare, G. (2016a). Inference in regression discontinuity designs under local randomization. *The Stata Journal, 16*(2), 331–367.

Cattaneo, M. D., Titiunik, R., & Vazquez-Bare, G. (2017). Comparing inference approaches for RD designs: A reexamination of the effect of head start on child mortality. *Journal of Policy Analysis and Management, forthcoming.*

Cattaneo, M. D., & Vazquez-Bare, G. (2016). The choice of neighborhood in regression discontinuity designs. *Observational Studies, 2*, 134–146.

Caughey, D., & Sekhon, J. S. (2011). Elections and the regression discontinuity design: Lessons from close U.S. House Races, 1942–2008. *Political Analysis, 19*(4), 385–408.

Cook, T. D. (2008). "Waiting for Life to Arrive": A history of the regression-discontinuity design in Psychology, Statistics and Economics. *Journal of Econometrics, 142*(2), 636–654.

de la Cuesta, B., & Imai, K. (2016). Misunderstandings about the regression discontinuity design in the study of close elections. *Annual Review of Political Science, 19*, 375–396.

Eggers, A. C., Folke, O., Fowler, A., Hainmueller, J., Hall, A. B., & Snyder, J. M. (2015). On the validity of the regression discontinuity design for estimating electoral effects: New evidence from over 40,000 close races. *American Journal of Political Science, 59*(1), 259–274.

Fan, J., & Gijbels, I. (1996). *Local polynomial modelling and its applications.* New York, NY: Chapman & Hall/CRC.

Hahn, J., Todd, P., & van der Klaauw, W. (2001). Identification and estimation of treatment effects with a regression-discontinuity design. *Econometrica, 69*(1), 201–209.

Holland, P. W. (1986). Statistics and causal inference. *Journal of the American Statistical Association, 81*(396), 945–960.

Hyytinen, A., Meriläinen, J., Saarimaa, T., Toivanen, O., & Tukiainen, J. (2015). *Does regression discontinuity design work? Evidence from random election outcomes*. Working Paper No. 59. VATT Institute for Economic Research.

Imbens, G., & Lemieux, T. (2008). Regression discontinuity designs: A guide to practice. *Journal of Econometrics*, *142*(2), 615–635.

Imbens, G. W., & Kalyanaraman, K. (2012). Optimal bandwidth choice for the regression discontinuity estimator. *Review of Economic Studies*, *79*(3), 933–959.

Imbens, G. W., & Rubin, D. B. (2015). *Causal inference in statistics, social, and biomedical sciences*. Cambridge: Cambridge University Press.

Keele, L., Titiunik, R., & Zubizarreta, J. (2015). Enhancing a geographic regression discontinuity design through matching to estimate the effect of ballot initiatives on voter turnout. *Journal of the Royal Statistical Society: Series A (Statistics in Society)*, *178*(1), 223–239.

Lee, D. S. (2008). Randomized experiments from non-random selection in U.S. house elections. *Journal of Econometrics*, *142*(2), 675–697.

Lee, D. S., & Lemieux, T. (2010). Regression discontinuity designs in economics. *Journal of Economic Literature*, *48*(2), 281–355.

Porter, J. (2003). *Estimation in the regression discontinuity model*. Working Paper. University of Wisconsin at Madison.

Sekhon, J. S., & Titiunik, R. (2012). When natural experiments are neither natural nor experiments. *American Political Science Review*, *106*(01), 35–57.

Sekhon, J. S., & Titiunik, R. (2016). Understanding regression discontinuity designs as observational studies. *Observational Studies*, *2*, 174–182.

Thistlethwaite, D. L., & Campbell, D. T. (1960). Regression-discontinuity analysis: An alternative to the ex-post facto experiment. *Journal of Educational Psychology*, *51*(6), 309–317.

IDENTIFICATION AND ESTIMATION USING A DENSITY DISCONTINUITY APPROACH

Hugo Jales[a] and Zhengfei Yu[b]

[a]*Syracuse University, Syracuse, NY, USA*
[b]*University of Tsukuba, Tsukuba, Ibaraki, Japan*

ABSTRACT

This chapter reviews recent developments in the density discontinuity approach. It is well known that agents having perfect control of the forcing variable will invalidate the popular regression discontinuity designs (RDDs). To detect the manipulation of the forcing variable, McCrary (2008) developed a test based on the discontinuity in the density around the threshold. Recent papers have noted that the sorting patterns around the threshold are often either the researcher's object of interest or may relate to structural parameters such as tax elasticities through known functions. This, in turn, implies that the behavior of the distribution around the threshold is not only informative of the validity of a standard RDD; it can also be used to recover policy-relevant parameters and perform counterfactual exercises.

Keywords: Density function; discontinuity; density jump; sorting; effect structure

JEL classifications: C14; C21

Regression Discontinuity Designs: Theory and Applications
Advances in Econometrics, Volume 38, 29–72

ISSN: 0731-9053/doi:10.1108/S0731-905320170000038003

1. INTRODUCTION

Regression discontinuity designs (RDDs) are currently among the most popular research designs in applied economics research. This method exploits the discontinuous variation in the probability of treatment around a known cutoff to identify the local effect of the policy. A feature of this method that perhaps explains its popularity is that researchers need only to invoke mild continuity assumptions in conditional mean functions to identify the effects of interest. Moreover, similar to randomized trials, tests of the identifying assumptions can be used to verify the validity of the approach in various settings.

Often, a policy is set as a discontinuous function of an *endogenous* variable. As a result, agents may alter their choices to ensure that they remain on the desired side of the cutoff. Empirically, the distribution of these variables tends to present a discontinuity at the cutoff. For example, piecewise linear income tax schedules result in discontinuities (and bunching) in the density of earnings (Kleven, Landais, Saez, & Schultz, 2014; Kleven & Waseem, 2013; Saez, 2010). Firm size distributions exhibit similar behavior when companies are subject to discontinuous changes in tax liability at certain employment thresholds (Garicano, Lelarge, & Van Reenen, 2013). Imperfectly enforced price controls, such as the minimum wage, generate a similar pattern in the wage distribution (Doyle, 2007; Jales 2015; Meyer & Wise, 1983). Hospitals' levels of spending present discontinuities when reimbursement schedules are nonlinear functions of spending (Bajari, Hong, Park, & Town, 2011; Einav, Finkelstein, & Schrimpf, 2016). The presence of a "pass/fail" threshold may generate discontinuities in grade distributions when teachers have some degree of discretion in grading (Diamond & Persson, 2016).

The presence of discontinuities in the distribution around points at which individuals are subject to large changes in outcomes as a result of small changes in behavior suggests that these patterns may reveal important features of the agents' behavior. That is, discontinuities in distributions may be exploited for econometric identification. In a similar spirit to the RDDs, abnormal behavior in the distribution function around an otherwise continuous distribution is attributed to the discontinuous nature of a policy around a threshold. Importantly, contrary to the RDDs, outcome and forcing variables are usually the same object in these settings. This is a distinct feature of these problems; the researcher's object of interest is the change that the policy induces in the distribution of the forcing variable itself.

In this chapter, we review the recent developments in methods that exploit the behavior of the density of the forcing variable at and around a threshold to uncover agents' response to the policy or to estimate structural parameters. We refer to these methods as the density discontinuity approach.

Although both RDDs and the density discontinuity approach make use of policies or rules involving discontinuous changes at a known threshold, they differ in identification conditions and strategies. The key identification condition in RDDs is the agents' inability to perfectly control the forcing variable (Lee, 2008; Lee & Lemieux, 2010). RDDs become invalid if the discontinuous rule induces the agents to (precisely) manipulate the forcing variable around the threshold.[1] In contrast, the density discontinuity approach is well-suited to self-selection or sorting in the forcing variable. The forcing variable upon which the discontinuous rule or policy is based is regarded as an endogenous outcome variable in the density discontinuity approach. In RDDs, the requirement of imperfect control of the forcing variable implies that the density function of the forcing variable is continuous at the threshold (McCrary, 2008). Based on this, McCrary (2008) proposed the well-known manipulation test.[2] The rejection of the null hypothesis of continuity suggests the failure of RDDs. However, detecting a discontinuity in the density function at the threshold is precisely the start of the density discontinuity approach.[3] The density discontinuity approach exploits the relationship between sorting patterns around the threshold and changes in the density of the forcing variable to uncover the effects of the policy on the *forcing variable*. As we will show, it is often possible to relate the changes in the distribution of the forcing variable to structural parameters such as tax elasticities, employment effects, and other behavioral responses.

The density discontinuity approach also differs from a standard RDD in the following two respects. First, RDD relies on the assumption that the expectation of potential outcomes conditional on the forcing variable is continuous at the threshold (Hahn, Todd, & Van der Klaauw, 2001; Lee & Lemieux, 2010). In the density discontinuity approach, however, the outcome and forcing variables coincide and identification relies on a different type of smoothness condition: The variable of interest would have a continuous density function *had the discontinuous policy not been implemented.* Kleven and Waseem (2013) assumed that the density of earnings would be continuous under a baseline linear tax system. Doyle (2007) and Jales (2015) assumed that the wage density would be continuous in the absence of a minimum wage. Bajari et al. (2011) assumed that the density of health care expenditures would be continuous in the absence of discrete changes in the reimbursement rate. Second, in RDDs, the identified causal effect is expressed as the ratio of the jump in the conditional expectation of the outcome to the jump in the conditional probability of treatment at the threshold (Imbens & Lemieux, 2008). However, in the density discontinuity approach, the relationship between the structural parameters and the density jump is usually more context-specific and model-dependent. Researchers typically establish models that describe the agent's behavior around the threshold and then recover the structural parameters from their relationship with the observed density jump. For example, Doyle (2007) identified the probability of noncompliance with minimum wage policy as the ratio of the wage density just above the minimum wage to that just below the minimum wage. Other parameters such as the probability of workers receiving the minimum wage and the probability of disemployment are related to the density discontinuity in more complicated ways.

Although the density discontinuity approach is distinct from RDDs in conception and practice, interestingly, in some scenarios, the two approaches are connected via the sampling scheme.[4] In Section 3, we investigate the connection between density-based identification and RDDs and discuss the conditions under which the objects identified from a discontinuity in the density can be related to the objects typically identified in RDDs.[5] We study a problem in which the density discontinuity approach can be used to identify the sign and test the significance of the effect of the policy in a context where RDD is infeasible given the limitations of the data.

Inference in the density discontinuity approach typically involves estimation of the densities to the left and right of the threshold. This resembles the data binning approach of McCrary (2008), which tests for the discontinuity in the density of the forcing variable at the cutoff. Although in the density discontinuity approach, the parameter of interest may take a more complicated form than a simple difference (or a logarithmic difference) in terms of densities near the cutoff, an important building block is also the bias-corrected estimation of the density at the boundary (the cutoff). A vast amount of statistical literature has studied density estimation with bias reduction at the boundary (Cline & Hart, 1991; Cheng, 1994; Cheng, Fan, & Marron, 1997; Müller, 1991; Wand, Marron, & Ruppert, 1991; Zhang & Karunamuni, 1998; Zhang, Karunamuni, & Jones, 1999). Such work could be applied to inference in the density discontinuity approach. In this chapter, we review three well-developed approaches to test for or estimate discontinuity in density functions: the pre-binning approach proposed by McCrary (2008), the empirical likelihood approach developed by Otsu, Xu, and Matsushita (2013), and the recently introduced method of Cattaneo, Jansson, and Ma (2015), which employs a local polynomial smoothing of an empirical cumulative distribution function (CDF).

The organization of this chapter is as follows. Section 2 investigates identification in the density discontinuity approach. While identification strategies in the density discontinuity approach are context-specific, we attempt to uncover the similarities among them and to provide a general framework. Section 3 discusses the relationship between the density discontinuity approach and RDDs. Section 4 discusses inference in density discontinuity by reviewing the main approaches to test for or estimate the jump in the density function at a known point. The final section concludes.

2. IDENTIFICATION: A REVIEW OF SELECTED STUDIES

This section describes the assumptions typically used to achieve identification in the density discontinuity approach. We will review papers covering the fields of health, public, and labor economics. We discuss the analysis of the elasticity of earnings with respect to taxes conducted by Kleven and Waseem (2013), the response of hospitals to reimbursement structure as analyzed by Bajari et al. (2011), and the effects of the minimum wage on wages and employment studied by Doyle (2007) and Jales (2015). Despite the apparent dissimilarities across these topics, the identification arguments used in these papers are inherently connected. In all of these contexts, identification is achieved by a combination of (i) a smoothness condition, as is usually the case in RDD settings; (ii) limited spillovers, which essentially allows the researcher to use one of the sides of the threshold to uncover the underlying latent structure that would prevail in the absence of the policy; and (iii) an effect structure assumption, that is, a condition specifying the behavior of the individuals on the "affected" side of the threshold. In all of these applications, the density of the variable of interest will present anomalies around the threshold such as bunching at the threshold level or "missing mass" on one side of the cutoff.[6] The main differences among the papers we analyze concern, as we will see, how they model the behavior of the individuals on the affected side of the threshold. Variations of the effect structure assumption can be used to allow for intensive-margin responses, as in Bajari et al. (2011); extensive-margin responses, as in Doyle (2007); and response heterogeneity, as in Kleven and Waseem (2013).

2.1. *Kleven and Waseem*

A fundamental question in public economics is the extent to which individuals adjust their earnings in response to the tax structure. Empirical investigation of this question is subject to standard concerns regarding endogeneity: Individuals who are subject to different tax structures may differ in their behavior due to a *ceteris paribus* behavioral response to the tax system, or because they differ in other ways that are unobserved to the econometrician. The tax system, however, commonly contains the so-called notches: discrete changes in tax liability that occur after marginal changes in earnings (Slemrod, 2013). The discrete nature of the change in tax liability creates a dominated area, where the individual could increase both his consumption and his leisure by *reducing* his earnings until he locates himself on the "correct" side of the threshold. Moreover, the abrupt change in tax liability around the threshold suggests that analyzing the behavior around the threshold can shed light on the elasticity of earnings with respect to the tax structure.

2.1.1. *An Economic Model*

Individual preferences are given by $U(z) = z - T(z) - \left(n/\left(1 + 1/e\right)\right) \left(z/n\right)^{1+1/e}$, where z is the before-tax earnings, $T(z)$ is the tax liability, and n is an ability parameter. If faced with a linear tax system $T(z) = t \cdot z$, the solution to the individual's problem is given by $z = n(1 - t)^e$. Thus, in this setup, e represents the elasticity of earnings with respect to the marginal net-of-tax rate. If n is drawn from a continuous distribution, then the distribution of earnings will also be continuous.

Suppose that the tax system presents a notch at z^*.[7] That is, assume that $T(z) = t \cdot z + [\Delta T + \Delta t \cdot z]\mathbb{1}\{z > z^*\}$. If $\Delta T > 0$, then some individuals who choose earnings just above z^* under the linear tax system $T(z) = t \cdot z$ would be better off by bunching at z^*. This is the case because the potential marginal increase in utility given by the additional pre-tax earnings is more than compensated by the discrete decrease in post-tax earnings that occurs once one chooses earnings above z^*. In the absence of optimization frictions or heterogeneity in tax elasticities in the population, a tax notch will lead to a gap, that is, an area of zero mass, in the density of earnings.

Individuals who would choose $z < z^*$ under a strictly linear tax structure will still choose the same value of z under a system that presents a notch. That is, the model structure implies the absence of spillovers: The shape of the density of earnings for values below z^* is preserved after the introduction of the notch. The introduction of the notch also generates intensive-margin responses. These changes, however, will be negligible if the changes in the *marginal* tax rates are small, that is, if Δt is close to zero. Finally, there is a point z_H sufficiently far from z^* such that every individual who would choose earnings above z_H would still choose the same level of earnings in the presence of the notch in the tax structure.

It is worthwhile to consider how these results change in the presence of optimization frictions or unobserved heterogeneity in tax elasticities. Conditional on e, all the model results are unchanged. However, the unconditional distribution will be different: For values of z just above z^* all individuals will still react by bunching at z^*, since the notch still introduces an area of strictly dominated choices for *all* values of e. As we move further away from z^*, some individuals will bunch at z^* (large e), whereas others will not (small e). The presence of optimization frictions such as switching and attention costs, however, may prevent some individuals that could benefit from adjusting their behavior from bunching at z^*. These types of frictions explain why the density of earnings may not be strictly equal to zero in the dominated area.

2.1.2. Statistical Model

Assume that the econometrician observes a random sample from the distribution of Z, that is, the distribution of earnings that prevails in the presence of the notched tax system. Our goal is to infer the distribution of Z(0), that is, the distribution that would prevail in the counterfactual state defined as the absence of the tax notch at z^*. Based on the implications of the structural model, it seems reasonable to assume the following:

> **Assumption KW1.** *If* $Z(0) < z^*$, *then* $Z = Z(0)$.

> **Assumption KW2.** *If* $Z(0) > z^*$ *then* $Z = z^*$ *with probability* $\pi_m(z)$. *With the complementary probability* $\pi_d(z) = 1 - \pi_m(z)$, *we have that* $Z = Z(0)$, *that is, the individual does not react to the tax notch.*[8] *For all values of* $z > z_H$, $\pi_d(z) = 1$, *which means that* $Z = Z(0)$.

Under Assumptions KW1 and KW2, there is a relationship between the observed density of earnings $f_1(z)$ and the latent density of earnings that would prevail in the absence of notches $f_0(z)$. The relationship is given by

$$\begin{aligned} f(z) &= \mathbb{1}\{z < z^*\} f_0(z) + \delta(z - z^*) \int \pi_m(u) f_0(u)\, du \\ &\quad + \mathbb{1}\{z \in (z^*, z_H]\} \pi_d(z) f_0(z) + \mathbb{1}\{z > z_H\} f_0(z), \end{aligned} \tag{1}$$

where δ is the Dirac delta function. To achieve identification of the latent distribution, we need to specify a functional form condition on the latent distribution of earnings in the relevant interval. Define the relevant interval $\overline{A}$ as any set such that $[z^*, z_H] \subset \overline{A}$ and that $\Pr\{Z(0) \in \overline{A} \backslash [z^*, z_H]\} > 0$.[9]

> **Assumption KW3.** *In the relevant interval,* $\overline{A}, f_0(z)$, *the latent density of earnings, is equal to* $m(z; \beta_0)$ *for some* $\beta_0 \in \mathbb{R}^k$.

Fig. 1 displays graphically the relationship between the latent density and the observed density, approximated by a rescaled histogram. The observed density will present a point mass at z^*. Compared to the latent density, the observed density of earnings will present "missing mass" for values of z just above z^*. For values of z below z^* or sufficiently far above z^*, the observed and the latent densities are equal. This follows from the assumption of no spillovers, the fact that the gains from bunching at z^* are decreasing in the distance to z^*, and the implicit assumption that there are no extensive-margin responses. That is, the notch does not induce any individual to exit (or enter) the labor force.

2.1.3. Identification

Under Assumptions KW1, KW2, and KW3, the latent distribution of earnings and the effects of the notch tax structure on the earnings distribution are identified.

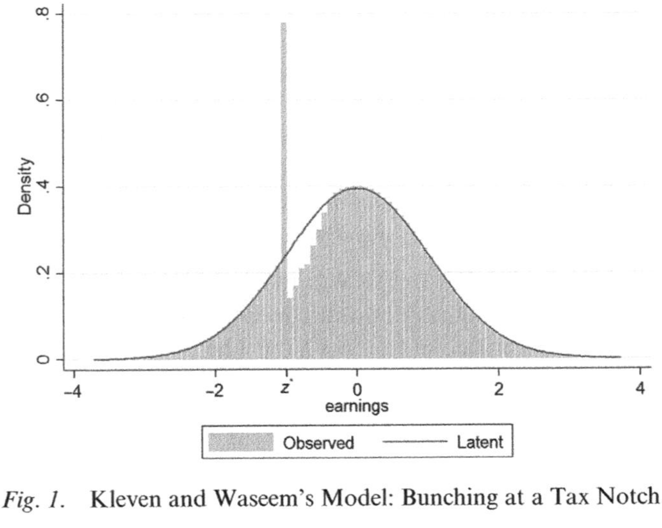

Fig. 1. Kleven and Waseem's Model: Bunching at a Tax Notch.

Identification of the latent density is trivial far from the notch point. For $z < z^*$ or $z > z_H$, $f_0(z) = f(z)$. For the values of z in the interval $[z^*, z_H]$, the densities differ. In this range, identification rests on the equality $f_0(z) = m(z; \beta_0)$. For example, assume that $f_0(z)$ is a polynomial of order k of z. Given that the function m is known, $f_0(z)$ is identified as long as β_0 is. Identification of β_0 follows from the fact that

$$\mathbf{E}_{Z \in \bar{A}}[(f(Z) - m(Z; \beta_0))^2 | Z \notin [z^*, z_H]] = 0.$$

Intuitively, $f_0(z)$ is identified in the affected area $[z^*, z_H]$ by extrapolating the observed pattern of $f(z)$ from the range where $f(z)$ and $f_0(z)$ coincide. The identification argument uses a moment condition that can be exploited for estimation.[10] However, the left-hand side in this moment condition contains a density. Thus, the dependent variable has to be estimated before it can be regressed on z to estimate β_0. This suggests a two-step procedure: Estimate the density of earnings, then choose β to minimize the distance between the estimated density and $m(z; \beta)$.[11] Kleven and Waseem used a histogram in the first step, then estimated β_0 by a flexible polynomial of z excluding the affected range. The counterfactual latent density in the affected range was then estimated by the predicted values obtained using the parameters from the second step.

This completes the identification of $f_0(z)$. The effects of the notch structure on the earnings distribution can directly be computed by comparing the observed density $f(z)$ and the counterfactual density $f_0(z)$. The probability of nonresponse, the fraction of individuals at earnings z that are nonresponsive to the notch, can then be identified from the relationship $\pi_d(z) = f(z)/f_0(z)$.

As the latent density is essentially an unobserved object in the affected area, the local functional assumption may be violated. In this case, the object recovered by this identification and estimation method is not exactly the latent density; it is an approximation of the latent density. Using standard results from Taylor expansions, we can bound the approximation errors in the extrapolated range. For example, if the density is estimated using a linear function, then the two factors that determine the magnitude of the approximation error are the curvature of the latent density and the size of the affected interval.

The identification argument seems to imply that the researcher must know the level z_H in advance. This, however, is not the case. Kleven and Waseem demonstrated that if the researcher begins with a guess of z_H that is too small, the estimated missing mass of earnings will be smaller than the bunching observed at z_H. This suggests a simple procedure, the convergence method: Begin with a guess of z_H arbitrarily close to z^* and progressively update your guess upward until the distance between the missing mass and the bunching mass is equal to zero.

Kleven and Waseem (2013) used this method to estimate the latent density of earnings. Since features of the observed density such as the bunching mass and the level of z_H are related to the latent density implicitly through the tax elasticity parameter e, one can use these relationships to construct estimates of the elasticity e. Their results reveal that tax elasticities are small and optimization frictions are large. However, the response to the tax notch is still substantial. The coexistence of large responses and small elasticities is only possible because, around the tax notch, the change in tax liability is of first order. That is, the observed bunching behavior is large because the change that the tax notch induces in the budget sets is substantial.

2.2. Minimum Wage

Meyer and Wise (1983), in a parametric framework, explored the concept of "missing workers" in the wage distribution to estimate the disemployment effects of the minimum wage. Doyle (2007) showed that identification of the employment effects can be achieved without relying on a functional form assumption for the latent distribution of wages. Doyle replaced the parametric functional form with a mild smoothness condition on the distribution of latent wages and showed that the probability of noncompliance – that is, the probability that a worker who would earn a wage below the minimum wage in the absence of the minimum wage policy will continue to earn a wage smaller than the minimum wage after the policy is introduced – can be recovered by the discontinuity in the density of wages at the minimum wage level. A feature of this strategy is that the effects of the minimum wage on employment can be identified even when the minimum wage presents no variation in the data, that is, when all the workers in the population are subject to the same level of the policy.

2.3. Doyle's Model

Let worker i be characterized by an observed wage W_i and a corresponding latent wage $W_i(0)$, which is defined as the wage that the worker would receive in the absence of the minimum wage. We will denote the minimum wage level by w^*. Denote by $F_0(w)$ ($f_0(w)$) the CDF (pdf) of latent wages. Similarly, denote by $F(w)$ ($f(w)$) the CDF (pdf) of observed wages.

In the presence of the minimum wage policy, the worker will receive a draw W_i from the distribution $F(w)$, which we will refer to as the distribution of observed wages. Assume that the econometrician observes a random sample of observed wages $\{W_i\}$ of size N from a population of interest. That is, only wages in the presence of the minimum wage policy are observed. The goal of the exercise is to recover the distribution of $W(0)$ from the information contained in the distribution of W. Let the following hold:

Assumption D1. *There are no spillovers from the minimum wage. This means that $W = W(0)$ when $W(0) > w^*$, where w^* denotes the minimum wage.*

Assumption D2. *If $W(0) < w^*$, then with probability π_m, the worker receives the minimum wage ($W = w^*$). With probability π_d, ($W = W(0)$), the worker's wage is the same as the latent wage (noncompliance). With the complementary probability, $\pi_u = 1 - \pi_m - \pi_d$, the worker becomes nonemployed ($W = \cdot$).*

Assumption D3. *The density of latent wages is continuous at w^*. That is, $\lim_{w \downarrow w^*} f_0(w) = \lim_{w \uparrow w^*} f_0(w)$.*

Assumption D3 exploits the fact that the underlying distribution of productivity and bargaining power would likely be smooth, which in turn would imply that, *in the absence of the minimum wage policy*, the distribution of wages would be continuous. Assumption D1 excludes effects that are often thought to be important in minimum wage settings: general equilibrium effects and local spillovers.[12] However, this assumption can be relaxed by strengthening Assumption D3 to a local functional form assumption such as Assumption KW3 in Kleven and Waseem (2013). Assumption D2 parametrizes the effects of the minimum wage in the bottom part of the wage distribution. It states that, as the minimum wage is introduced, a fraction of workers earn the minimum wage; a fraction of workers will become nonemployed; and another fraction of workers will not be affected by the policy.

It is helpful to consider the constraints imposed in Assumption D2 since at first glance it may appear to simply be a characterization, that is, just a list of all possible cases. Assumption D2 requires that (i) wages do not cross the minimum wage level, that is, $\Pr\{W > w^* | W(0) < w^*\} = 0$; (ii) the probabilities of noncompliance, earning the minimum wage and nonemployment are not heterogeneous across workers with the same latent wage; (iii) the probabilities of noncompliance, earning the minimum wage, and nonemployment are not functions of the workers' latent wage; and (iv) workers who do not respond to the policy earn a wage exactly equal to their latent wage.

Constraint (i) holds in simple microeconomic models of the labor market, such as the textbook neoclassical model. However, this constraint is violated in some search models and in models where workers of different skills are imperfect substitutes in the production function.[13] Constraint (ii) can be easily relaxed. When the minimum wage effect probabilities are heterogeneous in the population, the model will identify the natural extension of these parameters when they present heterogeneity in the population. That is, the model will identify the expected value of these (random) parameters in the population of affected workers. Constraint (iii) is at the heart of Doyle's identification strategy. By assuming that the probabilities of noncompliance are constant across latent wages, Doyle was able to infer the probability of noncompliance for *all* workers as long as he was able to recover the probability of noncompliance for workers who are close to the minimum wage level, which was done by exploiting the discontinuity in the observed density of wages at the minimum wage level. This assumption is clearly not innocuous since there are reasons to believe that workers who have latent wages close to the minimum wage level would be more likely to comply with the policy than workers whose wages are further away from it. Jales (2015) demonstrated that if one assumes that the probability of noncompliance is a small-order polynomial, such as a linear function of latent wages, then one can recover the slope of the noncompliance probability function by examining the discontinuity at the *derivative* of the wage density. Thus, although constraint (iii) is strong and plays an important role in the identification argument, one can relax it by strengthening the continuity assumption to hold for the derivatives of the latent wage density and exploiting the jumps in the derivative of the observed wage density to infer the changes in the noncompliance probabilities with respect to wages.

Under Assumptions D1 and D2, there is a relationship between the latent density of wages $f_0(w)$ and the observed density of wages $f(w)$. This relationship is given by

$$f(w) = \mathbb{1}\{w < w^*\}\frac{\pi_d f_0(w)}{c} + \delta(w - w^*)\pi_m \frac{F_0(w^*)}{c} + \mathbb{1}\{w > w^*\}\frac{f_0(w)}{c}, \quad (2)$$

where δ is a Dirac delta function. The rescaling factor $c \equiv 1 - F_0(w^*)\pi_u$ ensures that both the latent and the observed densities integrate to one in the presence of nonemployment effects. Note that the presence of extensive-margin responses (nonemployment effects) causes the observed and latent distributions to differ not only below the minimum wage, but everywhere. Fig. 2 graphically displays this relationship.

2.3.1. *Identification*

Consider the ratio of lateral limits of the observed density of wages around the minimum wage level:

$$\frac{\lim_{w\uparrow w^*} f(w)}{\lim_{w\downarrow w^*} f(w)} = \frac{\lim_{w\uparrow w^*} \pi_d f_0(w)/c}{\lim_{w\downarrow w^*} f_0(w)/c} = \pi_d,$$

where the first equality follows from Eq. (2), and the second equality is obtained using Assumption D3. Thus, the likelihood of noncompliance with the policy is identified by the discontinuity in the density of observed wages around w^*. Consider now the behavior of the CDF of observed wages around the minimum wage level:

$$\frac{\lim_{w\downarrow w^*} F(w)}{\lim_{w\uparrow w^*} F(w)} = \frac{\Pr\{W \leq w^*\}}{\Pr\{W < w^*\}} = \frac{\pi_d + \pi_m}{\pi_d}.$$

The left-hand side of this equation is clearly identified from data on observed wages. The right-hand side is a function of only one unknown, π_m.

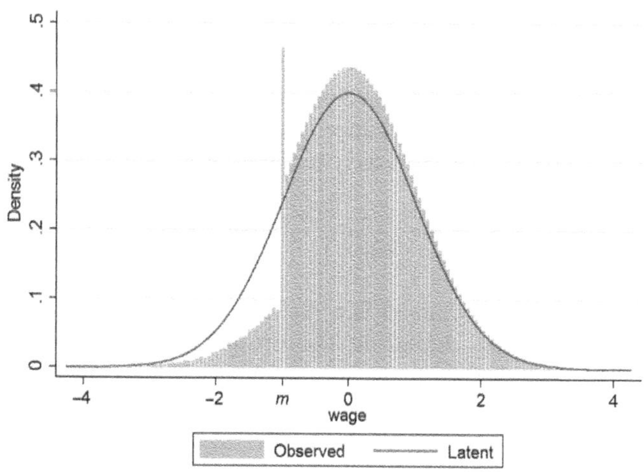

Fig. 2. Doyle's Model: Extensive–Margin Response and Imperfect Compliance.

This implies that π_m is identified, which also implies that $\pi_u = 1 - \pi_d - \pi_m$ is identified.

To recover the latent density of wages, one needs to recover $F_0(w^*)$. This is the case because the relationship between the observed and latent densities can be inverted once we know the rescaling factor c. To see this, note that $f_0(w) = f(w)c/\pi_d$ if $w < w^*$ and $f_0(w) = f(w)c$ if $w > w^*$. One way to identify $F_0(w^*)$ is to exploit the fact that

$$F_0(w^*) = \frac{\Pr\{W < w^*\}}{\Pr\{W < w^*\} + \pi_d \Pr\{W > w^*\}},$$

which follows from

$$\begin{aligned}\frac{\Pr\{W < w^*\}}{\Pr\{W < w^*\} + \pi_d \Pr\{W > w^*\}} &= \frac{\pi_d F_0(w^*)/c}{\pi_d F_0(w^*)/c + \pi_d(1 - F_0(w^*))/c} \\ &= \frac{F_0(w^*)}{F_0(w^*) + 1 - F_0(w^*)} = F_0(w^*).\end{aligned}$$

This implies that the latent distribution of wages can be recovered under Assumptions D1–D3. The discontinuity in the observed distribution around the minimum wage identifies the probability of noncompliance with the policy π_d. The discontinuity in the CDF of observed wages around the minimum wage identifies the ratio $(\pi_d + \pi_m)/\pi_d$. This, in turn, allows us to recover π_u, c, and, consequently, the entire latent wage distribution.

Employment effects identified by Doyle's method are related to the missing mass in the density of wages. That is, the employment effects are obtained by comparing the observed density of wages and the latent density that we can infer if noncompliance probabilities are constant and the latent density is continuous. It is easy to show that

$$\pi_u = \frac{F_0(w^*) - F(w^*)}{F_0(w^*)} \frac{1}{1 - F(w^*)}.$$

The first part of this expression is a measure of the missing mass of low-wage workers. It is the change in the proportion of workers who earn wages equal to or lower than the minimum wage level once the policy is introduced. The higher the value of this object is, the greater the number of workers who have "disappeared" from the wage distribution – over and above the mass accounted for by the bunching at w^* – and, consequently, the greater the unemployment probability π_u.

There are key differences between the approach taken by Doyle to analyze the effects of the minimum wage and the approach taken by Kleven and Waseem to analyze the effects of tax notches. Doyle assumed the continuity of the latent distribution, whereas Kleven and Waseem assumed a local functional form. The latter condition is slightly more restrictive than the former. However, the practical differences between these conditions are likely of limited relevance. On the other hand, the local functional form assumption made by Kleven and Waseem allowed them to specify a nonresponse probability as an (unspecified) fully flexible function of z.[14] The same object, π_d, is restricted to be a constant in Doyle's approach. Finally, the extensive-margin response is a central question in estimating minimum wage effects, whereas this type of response is less important in the effects of the tax structure on earnings. As a result, Kleven and Waseem could exploit the fact that $\int (1 - \pi_d(z)) f_0(z)\, dz = \Pr\{Z = z^*\}$. This essentially requires that all missing mass in the density to the right side of z^* can be accounted for by the bunching at z^*. This assumption is controversial in minimum wage settings since determining extensive-margin responses – the employment effect – is a central part of the investigation. The presence of employment effects generates an excess of missing mass in the density that cannot be explained by the bunching at w^*. Since densities are constrained to integrate to one, the employment effects induce a rescaling factor in the density. The rescaling factor makes the density of wages above the minimum wage greater than the latent density even under the assumption that wages above w^* are unaffected by the policy.

Doyle's model uses the concept of "missing mass" in the wage density to infer the employment effects of the minimum wage. In the next section, we discuss the extension to a dual economy in which the minimum wage is enforced only in the formal sector. Jales (2015) exploits the concept of an "excess of mass" to identify the probability of a worker moving from the formal to the informal sector: If many workers move from the formal to the informal sector as a result of the minimum wage policy, the cross-sectional distribution of wages will present missing mass for low wages in the formal sector and a corresponding *excess of mass* of low wages in the informal sector.

2.4. Jales' Model

Developing countries are often characterized by large informal sectors. That is, a substantial fraction of workers employed in the economy are not properly registered under the prevailing labor laws. Informal workers earn lower wages on average and do not have access to social security benefits such as severance payments. Naturally, they also do not pay labor taxes. Informality is thought to be associated with ineffective enforcement mechanisms and costly labor regulations such as labor taxes, job safety standards, overtime pay, and the minimum wage. Jales (2015) extended Doyle's model to a dual-economy setting in which workers can be employed in formal and informal sectors. This extension can be used to infer the role of the minimum wage in explaining the prevailing differences in the wage distribution and employment levels across sectors.

Let worker i be characterized by a pair of a wage (W_i) and sector (S_i), which is equal to one if the worker is employed in the formal sector and zero otherwise. Compliance with minimum wage legislation is enforced in the formal sector but not in the informal sector. This effectively means that the workers in the formal sector are not allowed to have wages below the minimum wage once the policy is introduced. If they remain employed in the presence of the policy, they must either move to an informal contract or comply with the policy by receiving a wage equal to w^*. In addition, for each worker, define a pair $(W_i(0), S_i(0))$ that denotes the counterfactual – or latent – wage and sector in the absence of the minimum wage. Finally, define $F_0(w)$ ($f_0(w)$) as the c.d.f (p.d.f) of $W(0)$ and $F(w)$ ($f(w)$) as the c.d.f (p.d.f) of observed wages (W_i or, in short notation, W_i). Let the following hold:

Assumption J1. *Workers whose latent wages would be above the minimum wage are not affected by the policy. That is,* $W = W(0)$ *and* $S = S(0)$ *when* $W(0) > w^*$.

Assumption J2. *For wages below the minimum wage* ($W(0) < w^*$), *we have the following: If* $S(0) = 0$, *then* $S = S(0)$. *Moreover, with probability* $\pi_d^{(0)}$, *the wage continues to be observed* ($W = W(0)$). *With the complementary probability* $\pi_m^{(0)} = 1 - \pi_d^{(0)}$, *the worker earns the minimum wage* ($W = w^*$). *If* $S(0) = 1$, *then with probability* $\pi_d^{(1)}$, *the wage continues to be observed* ($W = W(0)$), *meaning that the worker successfully transitions from the formal sector to the informal sector.*[15] *In this case, the observed sector will be* $S = 0$, *being different from the latent sector. With probability* $\pi_m^{(1)}$, *the worker earns the minimum wage* ($W = w^*, S = 1$). *With the complementary probability* ($\pi_u^{(1)} = 1 - \pi_d^{(1)} - \pi_m^{(1)}$), *the worker becomes unemployed* ($W = \cdot, S = \cdot$).

Assumption J3. *The density of latent wages and its first derivative are continuous at* w^*. *That is,* $\lim_{w \uparrow w^*} f_0(w) = \lim_{w \downarrow w^*} f_0(w)$, *and* $\lim_{w \uparrow w^*} f_0'(w) = \lim_{w \downarrow w^*} f_0'(w)$.

To complete the characterization of the latent economy, we need to define another object, $\Pr\{S(0) = 1|W(0) = w\}$:

> **Assumption J4.** *The conditional distribution of latent sector given the latent wage belongs to a parametric family* $\{\Lambda(w, \beta) : \beta \in B \subset \mathbb{R}^k\}$. *That is,* $\Pr\{S(0) = 1|W(0) = w\} = \Lambda(w, \beta_0)$ *for some* $\beta_0 \in B$. *Moreover,* $\Pr\{\Lambda(W(0), \beta_0) \neq \Lambda(W(0), \beta)|W(0) > w^*\} > 0$ *for all* $\beta \neq \beta_0$.

With the conditional probability of the latent sector (given the wage) and the marginal distribution of latent wages, we have completely specified the joint distribution of these variables.[16] The restrictive part of Assumption J4 is that the conditional probability of the latent sector given latent wages can be described by a parametric model. The first part of the above assumption states that there is a parameter β_0 for which the probability of the latent sector given the latent wage w is exactly equal to $\Lambda(w, \beta_0)$.

Assumption J2 is a sector-specific version of Assumption D2 imposed by Meyer and Wise (1983) and Doyle (2007). This assumption imposes several restrictions on how the minimum wage affects the bottom part of the joint distribution of sector and wages. These restrictions are similar to the restrictions imposed by Doyle's model discussed in the previous section. In particular, the assumption implies that (i) π is not indexed by i, that is, the probabilities that characterize the effects of the minimum wage are the same for all workers; (ii) $\nabla_w \pi = 0$, that is, the probabilities of noncompliance, unemployment, and earning the minimum wage are constants, as opposed to functions of workers' latent wage; and (iii) wages of sub-minimum-wage informal workers are not affected by the inflow of formal workers to that sector.

As discussed above, some of these restrictions simply ease the exposition, whereas others play a central role in the identification. Condition (i) is not central to the identification results. As in Doyle's model, if the probabilities are heterogeneous in the population, then we will identify the natural extension of these parameters in the presence of such heterogeneity. Condition (ii) plays a role in the identification. However, it can be relaxed if one is willing to impose smoothness conditions in the derivatives of the wage density. Condition (iii) is a standard "stable unit treatment value assumption" (Rubin, 1980) that rules out general equilibrium effects. Conditions (ii) and (iii) can be relaxed if one assumes that the latent wage distribution belongs to a known parametric family.

Under Assumptions J1, J2, and J4, there is a relationship between the observed and the latent wage distributions. The relationship is given by

$$f(w) = \mathbb{1}\{w < w^*\}\frac{\pi_d(w)f_0(w)}{c} + \delta(w - w^*)\frac{\int^{w^*} \pi_m(w)f_0(w)\,dw}{c} + \mathbb{1}\{w > w^*\}\frac{f_0(w)}{c}, \quad (3)$$

where $\pi_d(w) = \Lambda(w, \beta_0)\pi_d^{(1)} + (1 - \Lambda(w, \beta_0))\pi_d^{(0)}$, $\pi_m(w) = \Lambda(w, \beta_0)\pi_m^{(1)} + (1 - \Lambda(w, \beta_0))\pi_m^{(0)}$, $c \equiv 1 - \int^{w^*} \pi_u^{(1)}\Lambda(w, \beta_0)f_0(w)\,dw$, and δ is the Dirac delta function.

The unconditional probability of noncompliance will be a function of the latent wages even if the sector-specific parameters are not. This is the case because the unconditional probability of noncompliance is a weighted average of the sector-specific parameters weighted by the latent size of each sector for a given wage. As we move across different wages, the latent size of each sector will differ unless latent sector and wages are independent.

The sector-specific density of wages relates to their latent counterparts through the following relationships:

$$f(w|S = 1) = \delta(w - w^*)\frac{\pi_m^{(1)}F_0(w|S(0) = 1)}{c^{(1)}} + \mathbb{1}\{w > w^*\}\frac{f_0(w|S(0) = 1)}{c^{(1)}},$$

and

$$f(w|S = 0) = \mathbb{1}\{w < w^*\}\frac{\pi_d(w)f_0(w|S(0) = 0)}{(1 - \Lambda(w, \beta_0))c^{(0)}} + \delta(w - w^*)\frac{\pi_m^{(0)}F_0(w|S(0) = 0)}{c^{(0)}} + \mathbb{1}\{w > w^*\}\frac{f_0(w|S(0) = 0)}{c^{(0)}},$$

where $c^{(1)} \equiv 1 - F_0(w|S(0) = 1)(1 - \pi_m^{(1)})$ and $c^{(0)} \equiv 1 + \pi_d^{(1)}\int^{w^*} (\Lambda(w, \beta_0)/(1 - \Lambda(w, \beta_0)))f_0(w|S(0) = 0)\,dw$ ensure that both densities integrate to one. These relationships are displayed in Fig. 3. The observed conditional probability of formality given the wage is given by

$$\Pr\{S = 1|W = w\} = \mathbb{1}\{w = w^*\}\psi + \mathbb{1}\{w > w^*\}\Lambda(w, \beta_0).$$

That is, the model has a strong prediction for the conditional probability of formality with respect to the wage: This object is equal to zero for wages

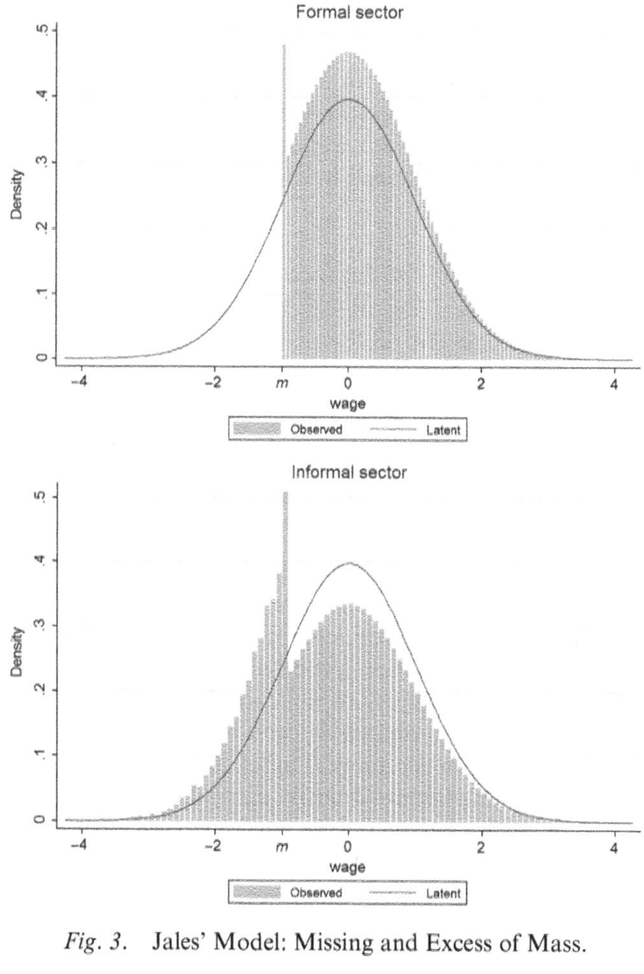

Fig. 3. Jales' Model: Missing and Excess of Mass.

below the minimum wage level. It is also equal to the latent probability for wages above the minimum wage level. At the minimum wage level, it is given by a constant $\psi = (\int^{w^*} \pi_m^{(1)} \Lambda(w, \beta_0) f_0(w)\, \mathrm{d}w)/(\int^{w^*} \pi_m(w) f_0(w)\, \mathrm{d}w)$.

The relationship between sector and wages in the presence of the minimum wage policy resembles the latent relationship between sector and wages for wages above w^*. This implies that the parameters that govern the conditional probability of *latent* sector given wage can be identified from the relationship between observed sector and wage above w^*. This result follows from the assumption of absence of spillovers and the implication of Assumption J2 that $\Pr\{W > w^*|W(0) < w^*\} = 0$. This result, together with the functional form assumption for the conditional probability of latent sector given wage, allows us to extrapolate the relationship between sector and wages recovered from data on wages above w^* toward the area where the latent relationship is unobserved in the presence of the policy.[17]

2.4.1. Identification

We will discuss identification for the special case in which Assumption J4 is strengthened to $\Lambda(w, \beta_0) = \lambda$ for all w.[18] Under this assumption, latent sector and wages are independent.

Consider the ratio of lateral limits of the observed density of wages around the minimum wage level:

$$\frac{\lim_{w\uparrow w^*} f(w)}{\lim_{w\downarrow w^*} f(w)} = \frac{\lim_{w\uparrow w^*} \pi_d(w) f_0(w)/c}{\lim_{w\downarrow w^*} f_0(w)/c} = \pi_d,$$

where the first equality follows from Eq. (3), and the second equality is obtained using Assumption J3. Note that this is not sufficient to recover the sector-specific probability of noncompliance, since in the unconditional distribution, the noncompliance probability is given by $\pi_d = \lambda\pi_d^{(1)} + (1-\lambda)\pi_d^{(0)}$. That is, the probability of noncompliance is a weighted average of the probability of noncompliance in each sector, weighted by the *latent* shares of each sector in the economy. Under independence between sector and wages, we can immediately identify π_m and π_u, and $F_0(w^*)$ since all assumptions of Doyle's model hold for the unconditional wage distribution. Moreover, it is possible to show that

$$\begin{aligned}\pi_m^{(1)} &= \frac{\Pr\{W = w^*|S = 1\}}{1 - \Pr\{W = w^*|S = 1\}} \frac{1 - F_0(w^*|S(0) = 1)}{F_0(w^*|S(0) = 1)} \\ &= \frac{\Pr\{W = w^*|S = 1\}}{1 - \Pr\{W = w^*|S = 1\}} \frac{1 - F_0(w^*)}{F_0(w^*)},\end{aligned}$$

where the last equality uses independence. Provided that $F_0(w^*)$ is identified, this equality demonstrates the identification of the sector-specific parameter $\pi_m^{(1)}$.

Identification of the latent share of the formal sector follows from $\Pr\{S(0) = 1\} = \Pr\{S(0) = 1|W(0) > w^*\} = \Pr\{S = 1|W > w^*\}$, where the first equality uses independence and the second equality uses the absence of spillovers.[19] Finally, the relationship between the identified aggregate parameters, the sector-specific parameters, and the latent shares of each sector can be used to solve for the remaining unknowns.[20] This completes the identification of the sector-specific parameters and the latent joint distribution of sector and wage.

Doyle (2007) and Jales (2015) found statistically significant disemployment effects of the minimum wage in the United States and Brazil, respectively. Jales (2015) also found that the relative size of the informal sector in Brazil grows substantially, from approximately 20% to 27% of the economy. This result is driven by disemployment effects in the formal sector and the movement of workers from the formal to the informal sector that occur as a response to the policy ($\pi_d^{(1)} \approx 10\%$). Overall, labor tax revenues decrease, as the employment effects and the movement of workers to the informal sector are large enough to dominate the increase in wages that the minimum wage induces in the formal sector.

2.5. Bajari, Hong, Park, and Town

The medical industry faces important informational constraints. Hospitals are often more informed than health insurers about the optimal spending levels for each patient. The asymmetry of information between the hospital and the insurer generates agency problems: Hospitals may under- or over-provide medical care to different patients as a response to the reimbursement schemes offered by the insurer. Reimbursement schemes often have nonlinearities such as "donut holes" – areas where the marginal reimbursement is zero. If hospitals react to reimbursement, then it seems plausible that they would choose their spending to locate themselves on the "right" side of the threshold, that is, the area where marginal reimbursement is larger. This suggests that the density of spending should be informative of the behavior of the hospitals. Bajari et al. (2011) developed a model and showed how to exploit the discontinuities in the density of spending to identify the parameters of interest. This section describes a simplified version of their model.[21]

2.5.1. Model

Consider a health insurer (principal) and a hospital (agent). The health insurer offers the principal a reimbursement scheme based on the agent's spending (q). The hospital observes the degree of severity in the patient's health shock ($\theta \geq 0$) and then chooses a level of spending. Assume that the reimbursement scheme is given by[22]

$$r(q) = \delta_0 + \mathbb{1}\{q \geq q^*\}\delta_1(q - q^*). \tag{4}$$

This is a fairly common reimbursement scheme. In this setup, there is a level of spending q^* at which the marginal reimbursement changes discontinuously.

Assume that hospitals maximize an indirect utility function given by $U(q, \theta) = v(q, \theta) - c(q) + \alpha r(q)$, where $c(q)$ is the hospital's cost function and $v(q, \theta)$ represents the patient's indirect utility. It seems reasonable to assume that the patient's utility is decreasing in the severity of the disease and increasing (and concave) in the hospital's care. To simplify the exposition, assume that $v(q, \theta) = -(1/2)q^2\theta^{-1} + kq - k_2\theta$. This specification implies that the hospital's decision takes into account the marginal benefits of increasing care on the patient's health and the marginal costs of providing care. Moreover, to simplify the discussion, assume that the hospital faces a constant cost of providing care, and thus, we have that $c(q) = c_0$.[23] Under a generic linear reimbursement scheme given by δq, the optimal level of care provided by the hospital is given by $\theta(k + \alpha\delta)$. Note that α is proportional to the responsiveness of hospitals' spending with respect to changes in the reimbursement scheme. Thus, when $\alpha = 0$ we have the case where the reimbursement does not enter the hospital utility function, which implies that the hospital spending does not depend on the reimbursement rate.

The optimal level of care provided by the hospital under the nonlinear reimbursement scheme is given by $q(\theta) = \mathbb{1}\{\theta < \theta^*\}\theta k + \mathbb{1}\{\theta > \theta^*\}\theta(k + \alpha\delta_1)$, where θ^* is a threshold of θ that determines the region of care where the hospital will choose to locate. The hospital is indifferent between locating itself in the lower reimbursement regions by spending $q(\theta^*) = \theta^* k$ or locating itself in the higher reimbursement region by spending $q(\theta^*) = \theta^*(k + \alpha\delta_1)$ when $\theta^* = 2q^*/(2k + \alpha\delta_1)$. If the distribution of health shocks is continuous, then the density of q will be continuous. The density of q will be given by

$$f(q) = \mathbb{1}\{q \leq q_L\} h\left(\frac{q}{k}\right)\frac{1}{k} + \mathbb{1}\{q \geq q_H\} h\left(\frac{q}{k + \alpha\delta_1}\right)\frac{1}{k + \alpha\delta_1}, \tag{5}$$

where $h(\theta)$ is the density of health shocks, $q_L \equiv \theta^* k$ and $q_H \equiv \theta^*(k + \alpha\delta_1)$.

Let $Q(0)$ be the level of spending chosen by the hospital under a flat reimbursement scheme, such as $r(q) = \delta_0$. Let Q be the level of spending chosen by the hospital under the observed, nonlinear reimbursement scheme. The nonlinear reimbursement scheme induces a gap in the density, an interval (q_L, q_H) of dominated choices where no optimizing hospital would decide to locate. Moreover, this model has another important feature: All outcomes are affected for $Q(0) > q^*$.[24] That is, $\Pr\{Q = Q(0)| \; Q(0) > q^*\} = 0$.[25] This contrasts with the models we discussed previously. The nonlinear reimbursement scheme (4) does not induce extensive-margin responses or bunching at the threshold. Instead, it generates an intensive-margin response.

There is a θ^*, a value of θ, at which the hospital is indifferent between locating on the two sides of q^*. The marginal benefit of spending is increasing in θ. Thus, the hospital chooses $q > q^*$ for all $\theta > \theta^*$, $q < q^*$ otherwise. Since the optimal level of care is a function of the marginal benefit of spending, the hospital is more responsive to changes in θ on the right-hand side of θ than on the left-hand side of θ^*. The relationship between the latent density of spending $Q(0)$ and the density that is observed under the nonlinear reimbursement scheme is shown in Fig. 4.

Contrary to the papers we have reviewed thus far, the density above q^* is not larger because individuals have moved to the other side of the threshold (bunching effect) or because they have disappeared from the distribution as a result of nonemployment effects (extensive-margin effect). The mechanism that generates the abnormal behavior of the density here is more subtle. Hospitals are more responsive to a patient's condition above q^* than they are below q^*. This causes marginal differences in θ, the patient's level of sickness, to generate larger differences in the level of hospital care if θ is such that the level of care selected by the hospital is above q^*. As a result, there will be a greater difference in the level of spending across individuals above q^* compared to the level we observe below q^*, that is, a larger sparsity in the distribution of spending. The intensive-margin response generated by the reimbursement scheme makes the hospital more responsive in terms of spending. If the density of θ is approximately flat in this area, this would lead to a decrease in the density of q above q^*.

The discontinuity in the marginal benefit of spending generates a gap in the density around q^*. Taking the ratio of q_H to q_L, we have that $q_H/q_L = (k + \alpha\delta_1)/k$. Thus, the magnitude of the gap in the density is proportional to the percentage change in the level, the intercept, of the marginal benefit curve (as a function of q).[26] If the marginal benefit of care

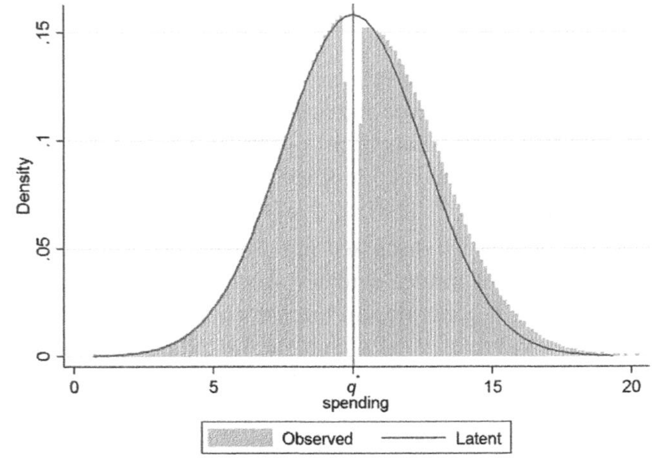

Fig. 4. Bajari, Hong, Park, and Town's Model: Intensive–Margin Responses.

in the patient's health is large (high k), the gap will be small. At the other extreme, if the marginal benefit of care on the patient's health is small (low k), then the change in marginal benefit of spending around q^* can be substantial, which will lead to a larger gap.[27]

2.5.2. Identification

This discussion suggests that the gap in the density around q^* can be informative of hospitals' behavior. This intuition is correct. Assume that the econometrician observes a random sample of Q, that is, the econometrician observes the hospitals' behavior under the nonlinear reimbursement scheme. The levels of q_L and q_H are identified directly from the observed distribution. Taking the ratio of q_H to q_L identifies the change in the level of a hospital's marginal utility function at q^*.

Under the model's structure, the latent density of f_0 can, in fact, be fully recovered from the data. For $q > q^*$, there are two differences in the density of Q relative to the density $Q(0)$. First, the rescaling factor is different. It is smaller for $q > q^*$ because the hospital is more responsive to changes in θ for $\theta > \theta^*$. Furthermore, the argument of $h(.)$ changes from $\theta/(k)$ to $\theta/(k + \alpha\delta_1)$. This is because, as the function that relates θ to $q(\theta)$ has a larger slope for $\theta > \theta^*$, the same value of q is a higher quantile of the distribution of $q(\theta)$ when $q(\theta)$ is given by θk than when $q(\theta)$ is given by

$\theta(k+\alpha\delta_1)$. To recover the density of $Q(0)$ for $q > q^*$, we must be able to infer which value of q would be chosen by the hospital in the counterfactual flat reimbursement scheme.

Identification of this object is facilitated by the fact that the gap in the density identifies the change in the intercept of the marginal utility of care. This factor is exactly the change in the argument of $h(.)$ that occurs when the hospital faces a different incentive. Denote by $F_0(q)$ and $f_0(q)$ the respective CDF and density of $Q(0)$ that would prevail under the flat reimbursement scheme. For $q < q_L, f_0(q) = f(q)$; thus, the latent density is trivially identified from the observed density. For $q > q_L$, $f_0(q) = f(q\frac{q_H}{q_L})\frac{q_H}{q_L}$. Thus, by rescaling the observed density by a factor equal to the magnitude of the gap around q^* *and* evaluating the observed density at a point different from q identifies the value of the latent density f_0 at q. This can be seen by writing the observed density as a function of the latent density:[28]

$$f(q) = \mathbb{1}\{q \leq q_L\} f_0(q) + \mathbb{1}\{q \geq q_H\} f_0\left(q\frac{k}{k+\alpha\delta_1}\right)\frac{k}{k+\alpha\delta_1}.$$

When $\alpha = 0$, we have that $\lim_{q\downarrow q_H} f(q) = \lim_{q\uparrow q_L} f(q)$ and $q_H = q_L = q^*$. These relationships suggest precisely the tests for the null hypothesis of no response of hospitals to the reimbursement scheme used in Bajari et al. (2011).

The presence of intensive-margin responses causes the argument by which the density is evaluated to change between the latent and observed distributions. Thus, it is more convenient in this case to work with the quantile function. Note that, in terms of the CDF

$$F(q) = \mathbb{1}\{q < q_L\} F_0(q) + \mathbb{1}\{q \in [q_L, q_H]\} F_0(q_L) + \mathbb{1}\{q > q_H\} F_0\left(q\frac{k}{k+\alpha\delta_1}\right).$$

Thus, we obtain

$$F^{-1}(\tau) = \mathbb{1}\{\tau \leq \tau^*\} F_0^{-1}(\tau) + \mathbb{1}\{\tau > \tau^*\} F_0^{-1}(\tau)\frac{k+\alpha\delta_1}{k},$$

where $\tau^* \equiv F_0^{-1}(q_L)$. The response of a hospital's spending to the reimbursement structure will reveal itself in the distribution of observed spending by a discontinuity in the level and derivatives of the quantile function. It is clear that rescaling the quantiles of the observed levels of spending Q by the factor $k/(k+\alpha\delta_1) = q_L/q_H$ for the values of $\tau > \tau^*$ recovers the original quantile function of the latent levels of spending $Q(0)$. From the quantile function, one can easily recover the corresponding CDF and the density.

Thus, the latent density of hospital spending that would prevail under a flat reimbursement scheme can be inferred from the density observed under the nonlinear reimbursement scheme. Bajari et al. (2011) estimated the magnitude of the density gap and the discontinuity of the density of spending around q^*.[29] Note that, despite the apparent dissimilarity, their model can be nested in our framework by modifying the effect structure assumption. That is, the statistical model that relates the latent density and the observed density is given by the following:

Assumption B1. *If* $Q(0) < q_L$, *then* $Q = Q(0)$.

Assumption B2. *If* $Q(0) > q_L$ *then* $Q = Q(0)(k + \alpha\delta_1)/k$ *with probability one.*

Assumption B3. *The density of latent hospital spending* $Q(0)$ *is continuous at* q_L. *That is,* $\lim_{q \downarrow q_L} f_0(q) = \lim_{q \uparrow q_L} f_0(q)$.

The unknown object in the effect structure assumption is not a probability as in the models we discussed thus far. Instead, it is the magnitude of the change in the hospital's spending, given by $(k + \alpha\delta_1)/k$, and the values of the boundaries $\{q_L, q_H\}$. Thus, the density discontinuity approach can be used to estimate the magnitude of intensive-margin responses. The structure imposed by the functional form assumptions for the patient and the hospital's indirect utility imply that the change in a hospital's behavior is constant, equal to the factor $(k + \alpha\delta_1)/k$ for all hospitals, and independent of the value of $Q(0)$, as long as $Q(0) > q_L$. This restriction can be relaxed, for example, by assuming a linear function that relates $Q(0)$ to Q for values of $Q(0) > q_L$. In this case, the discontinuity in the level of the quantile function identifies the gap, the change in the height of the function that maps θ to $q(\theta)$, and the discontinuity in the derivative of the quantile function identifies the change in the slope of the function that maps θ to $q(\theta)$.

Bajari et al. (2011) estimated the magnitude of the gap and the discontinuity in the quantile function around the cutoff for the case of organ transplants. Their results suggest that hospitals are indeed more responsive to a patient's health shocks when faced with a larger marginal reimbursement schedule.

3. RDD AND THE DENSITY DISCONTINUITY APPROACH

Thus far, we have argued that the behavior of the density of the forcing variable around the threshold can be useful in identifying the agents' response to certain policies. This approach has some similarities with the standard RDD. It is clear that in both approaches, identification is achieved from continuity assumptions: In RDDs, identification follows from an assumption of continuity in the conditional expectation of potential outcomes, whereas in the density discontinuity approach, the identification follows from an assumption of continuity of the latent density.

However, in contrast with the RDD, the parameters identified in the density discontinuity approach are typically model dependent. Furthermore, substantive knowledge regarding the sorting mechanisms must be embedded in the effect structure and spillover assumptions, whereas no such requirement is present in RDDs. In certain contexts, as noted by Einav et al. (2016), different models may be consistent with the bunching behavior of the data but imply considerably different structural parameters. Their findings illustrated the importance of model choices in translating the observed data pattern into economic objects for counterfactual analysis.[30] These findings suggest caution before attributing to density-based identification the same credibility given to regression discontinuity-type identification.

Thus, it is worthwhile to investigate whether there exists a scenario in which a discontinuity in the density would identify similar objects under the same assumptions as in an RDD. In other words, we ask whether there is a context in which a density-based identification would achieve the same degree of credibility as an RDD applied to the same problem. As we show below, the answer to this question is yes. In particular settings, the discontinuity in the density of the forcing variable *conditional on the outcome* identifies (up to a scaling factor) the same object, the local average treatment effect, as does the discontinuity in the conditional mean function in a standard RDD. We also argue that in certain contexts, only the former of these objects – that is, the discontinuity in the density – is estimable given the constraints in the sample.

For concreteness, we convey the main ideas with the following example. Suppose that the researcher is interested in the effects of a policy that restricts alcohol consumption. In particular, the goal is to recover the effects of the policy that prohibits individuals under the age of 21 from purchasing liquor. Let X be the age of the individual. Assume that each individual is constrained by the policy (denoted by $D = 1$) if and only if $X < x^* = 21$, that is, $D = \mathbb{1}\{X < x^*\}$. Following the standard potential outcome notations, let $Y(1)$ indicate whether an individual would consume alcohol when he is constrained by the policy, and $Y(0)$ indicate whether an individual would consume alcohol when he is not constrained by the policy. The observed behavior of the individual, denoted Y, is given by

$$Y = DY(1) + (1 - D)Y(0). \tag{6}$$

The researcher is interested in the local average treatment effect $\mathbf{E}[Y(1) - Y(0)|X = x^*]$, or some related parameters such as the treatment effect ratio $\mathbf{E}[Y(1)|X = x^*]/\mathbf{E}[Y(0)|X = x^*]$.[31] Assume that the following assumptions hold:

Assumption 1. $\Pr\{Y(0) = 1|X = x\}$ *and* $\Pr\{Y(1) = 1|X = x\}$ *are continuous in* x.

Assumption 2. *The density of* $X, f(x)$, *is positive and continuous at* x^*.

If the researcher observes a random sample of (Y, X), then this is a typical sharp RDD context. Therefore, the local average treatment effect is identified by the discontinuity of the observed conditional mean $\Pr\{Y = 1|X = x\}$ at $x = x^*$.

However, suppose that sampling scheme is different. Instead of observing a random sample of the pair (Y, X), the researcher only observes a random sample of the universe of individuals *who drink alcohol*. For each individual who drinks, the researcher observes his age. That is, the researcher has access to a random sample, representative of the population, of the distribution of $X_{|Y=1}$, whose density is given by

$$f(x|Y = 1) = \frac{\Pr\{Y = 1|X = x\}f(x)}{\Pr\{Y = 1\}}.$$

Observing the density of $X_{|Y=1}$ presents difficulties for identification. The standard approach in RDD would be to construct estimates of the conditional mean function on each side of the threshold and compare the differences at the limit. However, observing only a random sample of $X_{|Y=1}$, one cannot estimate the conditional mean function of $Y_{|X}$ since Y is trivially equal to one for all individuals in the data. In this setup, identification must come from the information contained in the density of X given $Y = 1$.

Fortunately, one can connect the differences in the limits of the conditional mean function to differences in the conditional density of X given the outcome. Under Assumptions 1 and 2, the density of $X_{|Y=1}$ will be discontinuous at x^* as long as the conditional mean function of $Y_{|X}$ is discontinuous:

$$\begin{aligned}\lim_{x\uparrow x^*} f(x|Y=1) - \lim_{x\downarrow x^*} f(x|Y=1) &= \frac{(Y^+ - Y^-)f(x^*)}{\mathbf{E}[Y]} \\ &= \frac{\mathbf{E}[Y(1) - Y(0)|X = x^*]f(x^*)}{\mathbf{E}[Y]},\end{aligned}$$

where $Y^+ = \lim_{x\downarrow x^*} \Pr\{Y = 1|X = x\}$, $Y^- = \lim_{x\uparrow x^*} \Pr\{Y = 1|X = x\}$. The first equation follows from Bayes' rule and Assumption 2, and the second equation follows from Assumption 1. This equation shows that in a sharp RDD with a binary outcome, the local average treatment effect is identified, up to a constant given by $f(x^*)/\mathbf{E}[Y]$, when the researcher observes the density of $X_{|Y=1}$, the density of the running variable *conditional on the outcome*. This is sufficient to test the significance of the policy and identify the sign of the effect. To obtain the magnitude of the local average treatment effect, the researcher must know or have access to external estimates of $f(x^*)/\mathbf{E}[Y]$. Furthermore, for the treatment effect ratio, we have that

$$\frac{\lim_{x\uparrow x^*} f(x|Y=1)}{\lim_{x\downarrow x^*} f(x|Y=1)} = \frac{\mathbf{E}[Y(1)|X = x^*]}{\mathbf{E}[Y(0)|X = x^*]}, \qquad (7)$$

and thus, the rescaling factors cancel out and the treatment effect ratio is identified.[32]

Thus, if the researcher observes a random sample of X with sampling weights given by $\Pr\{Y = 1|X\}$, a setup that will naturally emerge depending on how the data are collected, he can use the discontinuity in the density he observes to learn something about the effects of the policy. The discontinuity in the density will identify the limit of the ratio of the expected value of $Y(1)$ to the expected value of $Y(0)$ to the left of the threshold. This ratio equals the proportional change in the expected value of the outcome induced by the policy. This can be used to infer the presence and sign of the treatment effect. The local average treatment effect, however, can only be recovered if the researcher knows $\Pr\{Y = 1\}$ and the density of X at x^*.

RDD and the density discontinuity approach are also connected through their assumptions. These methods share the assumption of continuity of the density of the forcing variable. Moreover, the assumption of no spillovers used in the density discontinuity approach is implied by the observation equation (6) and the treatment assignment rule in the standard RDD. That is, in the RDD framework, Eq. (6) evaluated at $X \geq x^*$ implies $Y = Y(0)$, which is the equality labeled as "no-spillovers" in the density discontinuity approach.[33] Furthermore, embedding the RDD framework with the assumption that $\Pr\{Y(1) = 1|Y(0) = 0, X = x^*\} = 0$, the treatment effect ratio in Eq. (7) equals $\Pr\{Y(1) = 1|Y(0) = 1, X = x^*\}$.[34] Thus, the discontinuity in the density of X conditional on $Y = 1$ identifies the proportion of individuals at x^* who do not respond to the introduction of the policy. This parameter has a similar interpretation to the probabilities identified in Jales (2015), Doyle (2007), and Kleven and Waseem (2013). The assumption needed in the RDD framework to equalize the treatment effect ratio to the probability of nonresponse to the policy can be thought of as the RDD analogous of the effect structure assumption used in the density discontinuity approach.

In summary, in a sharp RDD with a binary outcome augmented with a continuity assumption on the density of the forcing variable, one can identify features of the treatment effect, such as the treatment effect ratio, from the discontinuity in the density of the forcing variable conditional on the outcome. This can be used to infer the size, test the null hypothesis of no treatment effect, and – with the aid of assumptions regarding the effect structure – recover other meaningful objects such as the probability of nonresponse to the policy.

4. INFERENCE ON THE DISCONTINUITY IN THE DENSITY

Section 2 shows that agents' responses and economic parameters can be identified as functions of the density to the left and right of a threshold, which is a potential discontinuity point. Denoting the threshold c, the main task of inference in the density discontinuity approach is to effectively estimate the densities

$$f^+(c) = \lim_{x \downarrow c} f(x) \quad \text{and} \quad f^-(c) = \lim_{x \uparrow c} f(x).$$

To focus on the essence and to be in line with the manipulation test in RDDs, we suppose that the researcher is interested in whether the density function $f(x)$ has a discontinuity at $x = c$ or is interested in measuring the jump size at the threshold c. In these scenarios, the parameter of interest is $\theta = f^+(c) - f^-(c)$. To be specific, in the density discontinuity testing problem, the null and alternative hypotheses are

$$H_0 : f^+(c) - f^-(c) = 0, \quad H_1 : f^+(c) - f^-(c) \neq 0.$$

The inference approaches discussed in this section are readily adaptable to scenarios in which the parameters of interest take other functional forms of density limits at the discontinuity point or densities at the boundary. For example, Doyle (2007) estimated the probability of noncompliance with the minimum wage policy from $f^+(c)/f^-(c)$, the ratio of density limits at the discontinuity. Bajari et al. (2011) recovered the responsiveness of health care supply to changes in reimbursements using $1/f^+(q_H) - 1/f^-(q_L)$, where q_H and q_L are boundary points.[35]

In this section, we discuss three main approaches available in the econometric literature for testing or estimating the discontinuity at a given point in the density function: the data binning approach of McCrary (2008), the empirical likelihood approach developed by Otsu et al. (2013), and the method recently introduced by Cattaneo et al. (2015), which implemented a local polynomial smoothing of an empirical CDF. Throughout this section, we assume that the underlying density function $f(x)$ has bounded left and right derivatives up to the order $p + 1$ at point c and has bounded derivatives up to the order $p + 1$ everywhere else, where p is the order of the local polynomial used by the researcher. Moreover, we assume that the researcher observes an i.i.d. sample $\{X_1, X_2, \ldots, X_n\}$.

While the inference procedure resembles the manipulation test in the RDD literature, a subtle deviation demands care in certain contexts. The expressions (2) and (3) in the preceding section imply that the distribution of the variable of interest exhibits a point mass at the threshold c, which makes $f(c)$ infinite. However, when we utilize the density discontinuity to identify parameters, the important objects are $f^+(c)$ and $f^-(c)$, irrespective of whether $f(c)$ is finite. In the estimation, by excluding the data points at c, the point mass will not contaminate the inference procedure. Estimating $\Pr\{X = c\}$ itself can be performed using the sample analog $(1/n) \sum_{i=1}^{n} \mathbb{1}\{X_i = c\}$.

The discontinuity test or estimator involves nonparametrically estimating the density just to the left and to the right of the potential discontinuity point, where the boundary problem arises in terms of considerably larger estimation bias than the interior points[36] (continuous points in the current setting). It is well known that local linear regression automatically reduces boundary bias.[37] Therefore, it is reasonable to adapt the local polynomial fitting to density estimation. However, in the density context, the regressand and the regressor are not obvious. Hence, the first step is to create the regressand and regressor. The local linear (or polynomial) regression is implemented in the second step.

McCrary (2008) created the regressand and regressor based on data binning. The centers of the bin were treated as the regressor, and the normalized bin counts were viewed as the regressand.[38] Let $\{Z_j\}_{j=1}^{J}$ be an equi-spaced grid of width b covering the support of X_i. That is,

$$\{Z_j\}_{j=1}^{J} = \left\{\ldots, c - \frac{5b}{2}, c - \frac{3b}{2}, c - \frac{b}{2}, c + \frac{b}{2}, c + \frac{3b}{2}, c + \frac{5b}{2}, \ldots\right\},$$

where the interval $[Z_1, Z_J]$ contains the support of X_i. For each j, define Y_j as the normalized counts of original data points falling into the bin $[Z_j - b/2, Z_j + b/2)$, that is,

$$Y_j = \frac{1}{nb}\sum_{i=1}^{n} \mathbb{1}\left\{X_i \in \left[Z_j - \frac{b}{2}, Z_j + \frac{b}{2}\right)\right\}\mathbb{1}\{X_i \neq c\}. \tag{8}$$

Note that the bin counts exclude the data points X_i's at the threshold c. Thereby, we avoid the possible mass point at c. In the second step of McCrary's approach, the estimators of $f^-(c)$ and $f^+(c)$, denoted $\widehat{f}^-(c)$ and $\widehat{f}^+(c)$, are obtained by applying local linear regressions for $(Y_j, Z_j)_{\{j:Z_j < c\}}$ and $(Y_j, Z_j)_{\{j:Z_j > c\}}$, respectively. Specifically, $\widehat{f}^-(c) = \widehat{\alpha}^-$ and $\widehat{f}^+(c) = \widehat{\alpha}^+$ are given by

$$\begin{aligned}\left(\widehat{\alpha}^-, \widehat{\beta}^-\right) &= \arg\min_{\alpha^-, \beta^-} \sum_{j:Z_j \leq c} K\left(\frac{Z_j - c}{h}\right)\left(Y_j - \alpha^- - \beta^-(Z_j - c)\right)^2, \\ \left(\widehat{\alpha}^+, \widehat{\beta}^+\right) &= \arg\min_{\alpha^+, \beta^+} \sum_{j:Z_j > c} K\left(\frac{Z_j - c}{h}\right)\left(Y_j - \alpha^+ - \beta^+(Z_j - c)\right)^2,\end{aligned} \tag{9}$$

where K is the kernel function with a bounded support and bounded second derivative and h is the bandwidth. This second step offers additional smoothing and boundary bias reduction.

The test of McCrary (2008) was based on the logarithmic difference $\ln\widehat{f}^{+}(c) - \ln\widehat{f}^{-}(c)$. Under the null hypothesis of no jump in the density at point c, the test statistic

$$\frac{\sqrt{nh}\left(\ln\widehat{f}^{+}(c) - \ln\widehat{f}^{-}(c)\right)}{\sqrt{\frac{24}{5}\left(\frac{1}{\widehat{f}^{-}(c)} + \frac{1}{\widehat{f}^{+}(c)}\right)}}$$

has approximately a standard normal distribution, provided that $h \to 0, nh \to \infty$, $nh^5 \to 0$ and $b/h \to 0$.[39] Note that this approach uses two tuning parameters, the bin size b in the first step (normalized bin counts) and the bandwidth h in the second step (local linear fitting). The performance of the estimators $\widehat{f}^{-}(c)$ and $\widehat{f}^{+}(c)$ is much more robust with respect to b than to h. Under the condition $b/h \to 0$, the leading bias and variance of the estimators no longer depend on b. The bias of $\widehat{f}^{-}(c)$ and $\widehat{f}^{+}(c)$ is of order h^2. In general, if the researcher uses the local polynomial of order p to fit the bin counts, the bias will be of order h^{p+1} Cheng et al. (1997, Theorem 1).[40] McCrary (2008) used the triangular kernel function $K(u) = (1 - |u|)\mathbb{1}\{|u| \leq 1\}$ in the local linear smoothing (9). Cheng et al. (1997) showed that the triangular kernel function minimizes the mean-square error of $\widehat{f}^{-}(c)$ and $\widehat{f}^{+}(c)$ among all nonnegative and Lipschitz continuous functions. McCrary (2008) also provided data-driven procedures for the choices of both tuning parameters b and h. Monte Carlo simulations demonstrated that the size property of the proposed discontinuity test is quite stable with respect to the choice of the bin size b.

The basic idea of Cheng (1994) and McCrary (2008) was to employ the local linear regression to a "rough" estimation of the density: the normalized bin counts. When the bin size b goes to zero, the normalized bin counts defined in Eq. (8) become the empirical density function

$$f_n(x) = \frac{1}{n}\sum_{i=1}^{n}\delta(x - X_i), \tag{10}$$

where δ is the Dirac delta function. Jones (1993) discussed an approach that applies local linear fitting to the empirical density function (10).

He further noted that this approach yields the same estimator of $f^+(c)$ and $f^-(c)$ as the kernel density estimator using some particular boundary kernels.[41] Doyle (2007) applied this equivalent boundary kernel method to Current Population Survey (CPS) data.

Recently, Cattaneo et al. (2015) employed a different strategy in the construction of a local polynomial estimator of the density at the boundary. The basic idea is to apply local polynomial fitting to the empirical CDF instead of the normalized bin counts (or the empirical density). In the first step, they constructed the (leave-one-out) empirical CDF evaluated at X_i as the regressand and treated the original data X_i as the regressor. In particular, to estimate $f^-(c)$, they generated the regressand as

$$Y_i^- = \frac{1}{n_0 - 1} \sum_{j=1, j \neq i}^{n} \mathbb{1}\{X_j < c\} \mathbb{1}\{X_j \leq X_i\},$$

where $n_0 = \sum_{i=1}^{n} \mathbb{1}\{X_i < c\}$. Analogously, to estimate $f^+(c)$, they generated the regressand as

$$Y_i^+ = \frac{1}{n_1 - 1} \sum_{j=1, j \neq i}^{n} \mathbb{1}\{X_j > c\} \mathbb{1}\{X_j \leq X_i\},$$

where $n_1 = \sum_{i=1}^{n} \mathbb{1}\{X_i > c\}$. An advantage of their approach is that they do not need the tuning parameter bin size b as McCrary (2008) did. In the second step, they implemented a local polynomial regression with order p for observations on each side of the potential discontinuity point c. This yields the estimators $\widehat{f^-}(c)$ and $\widehat{f^+}(c)$. Because the empirical CDF is generated as the regressand in the first step, it is the coefficient before the first-order term of the polynomial that corresponds to the densities $f^-(c)$ and $f^+(c)$. Specifically,

$$\widehat{f^-}(c) = \mathbf{e}_1' \widehat{\beta^-}, \quad \widehat{f^+}(c) = \mathbf{e}_1' \widehat{\beta^+},$$

where $\mathbf{e}_1 = (0, 1, 0, \ldots, 0)'$, and

$$\begin{aligned}
\widehat{\beta}^{-} &= \arg\min_{\mathbf{b}\in\mathbb{R}^{p+1}} \sum_{i:X_i\,<\,c} K\left(\frac{X_i-c}{h}\right)\left(Y_i^{-}-\mathbf{r}_p(X_i-c)'\mathbf{b}\right)^2, \\
\widehat{\beta}^{+} &= \arg\min_{\mathbf{b}\in\mathbb{R}^{p+1}} \sum_{i:X_i\,>\,c} K\left(\frac{X_i-c}{h}\right)\left(Y_i^{+}-\mathbf{r}_p(X_i-c)'\mathbf{b}\right)^2, \\
\mathbf{r}_p(u) &= \left(1,u,u^2,\ldots,u^p\right)'.
\end{aligned} \tag{11}$$

Cattaneo et al. (2015) showed that the bias of their estimators $\widehat{f}^{-}(c)$ and $\widehat{f}^{+}(c)$ are of order $O_P(h^p)$, where h is the bandwidth, and they applied $p=2$ and 3 in simulations. Based on the density estimates at the boundary, the test statistic of Cattaneo et al. (2015) is constructed as a normalized difference between $\widehat{f}^{-}(c)$ and $\widehat{f}^{+}(c)$. Asymptotic properties were studied in detail by Cattaneo et al. (2015). For empirical applications, they recommended a jackknife standard error estimator and an MSE-optimal bandwidth choice with robust bias correction. Cattaneo, Jansson, and Ma (2016) introduced the Stata commands implementing their discontinuity test.

Otsu et al. (2013) provided an empirical likelihood framework for inference in density discontinuity approaches. Both McCrary's binning-based local linear approach and the local likelihood approach[42] lead to a minimization or maximization problem. For the former, the minimization problem is described in Eq. (9), while for the latter, the estimators $\widehat{f}^{-}(c)=\widehat{a}_l$ and $\widehat{f}^{+}(c)=\widehat{a}_r$ are the solutions to the following maximization problems:

$$\begin{aligned}
\left(\widehat{a}_l,\widehat{b}_l\right) &= \arg\max_{a_l,b_l}\left\{\begin{array}{l}\frac{1}{n}\sum_{i:X_i\,<\,c} K\left(\frac{X_i-c}{h}\right)(a_l+b_l(X_i-c)) \\ -\int_{u\,<\,c} K\left(\frac{X_i-c}{h}\right)\exp(a_l+b_l(u-c))\end{array}\right\}, \\
\left(\widehat{a}_r,\widehat{b}_r\right) &= \arg\max_{a_r,b_r}\left\{\begin{array}{l}\frac{1}{n}\sum_{i:X_i\,>\,c} K\left(\frac{X_i-c}{h}\right)(a_r+b_r(X_i-c)) \\ -\int_{u\,>\,c} K\left(\frac{X_i-c}{h}\right)\exp(a_r+b_r+(u-c))\end{array}\right\}.
\end{aligned} \tag{12}$$

Viewing the first-order conditions of the above minimization or maximization problems (Eq. (9) or (12)) as the sample moment conditions for $\mathbf{E}[\widehat{f}^{-}(c)]$ and $\mathbf{E}[\widehat{f}^{+}(c)]$, one can construct the empirical likelihood function and perform inference for the parameter $\theta = f^{+}(c) - f^{-}(c)$. Otsu et al. (2013) also compared the finite sample performance of their empirical likelihood estimator based on the local linear fitting of the binned data and that based on the local linear likelihood approach. They recommended the latter, as it exhibited smaller bias as an estimator and larger power as a test. Naturally, the empirical likelihood approach circumvents estimation of the standard error. It also inherits some desirable properties, including the invariance to transformation of the parameter and automatic determination of the shape of confidence sets.

As we have seen, inference in density discontinuity is based primarily on density estimation at a known boundary. Apart from the local linear (polynomial) estimator, the statistical literature has proposed a variety of methods to reduce the boundary bias in density estimation. Some additional well-known approaches include the reflection method (Cline & Hart, 1991), boundary kernel method (Jones, 1993; Müller, 1991; Zhang & Karunamuni, 1998), transformation method (Ruppert & Wand, 1994; Wand et al., 1991), and their combinations.[43] The local linear (polynomial) smoothing approach has been popular because it is not only intuitive and easy to implement, but is also highly efficient.[44]

5. CONCLUSION

This chapter discusses recent developments concerning the density discontinuity approach. We argue that the economic environment often presents substantial changes in outcomes associated with small-order changes in individual choices. In this case, the distribution of the outcome variable will present abnormal behavior around the threshold, which can be used to identify elasticities and behavioral responses. The typical setting in the density discontinuity approach involves (i) a continuity assumption on the latent distribution, similar to the assumption of the smooth conditional expectation of potential outcomes in the standard RDD setting; (ii) the absence of spillovers, which restricts the responses of individuals to just one side of the threshold; and (iii) an effect structure assumption, which describes the behavior of the individuals on the affected side of the threshold. Different forms of this assumption have been invoked in the literature to capture extensive-margin responses (Doyle, 2007; Jales, 2015), intensive-margin responses (Bajari et al., 2011), and heterogeneity in bunching probabilities (Kleven & Waseem, 2013).

This survey is not exhaustive. In a parametric setting, Garicano et al. (2013) explored the fact that companies in France face a substantial increase in labor costs[45] if they employ more than 50 individuals. They showed that in the absence of frictions and measurement error,[46] the density of observed employment will have a gap above 50 employees and bunching below this threshold. They used this fact to recover the underlying density of employment that would prevail in the absence of this policy and evaluate the effect of such employment rationing on the productivity distribution of French companies. Saez (2010) exploited the bunching at kink points of the tax system to estimate the elasticity of earnings with respect to taxes. Kopczuk and Munroe (2015) studied the effects of discontinuous housing transaction taxes on the price distribution and the size of the market in the case of New York and New Jersey mansion taxes. These papers exemplify how to nest the reduced-form approach from a density discontinuity setting with a fully specified structural model. By doing so, one obtains parameters that can be linked directly to policy-invariant, economically interpretable behavioral responses.

We believe that the density discontinuity approach can be a valuable tool in the applied econometrician's toolbox, allowing for identification and estimation of economic parameters in problems plagued by endogeneity of choices and lack of variation in the treatment across individuals.

However, in contrast to the RDD, where continuity of the conditional mean functions of potential outcomes and lack of sorting are sufficient for identification, the density discontinuity approach forces the researcher to impose additional restrictions. These restrictions are codified in assumptions concerning the limits of spillovers and the effect structure. In this survey, we demonstrated how different problems will require different effect structure assumptions. We also showed that these distinct economic problems share the common feature of abnormal behavior in the density at the threshold and how that can be used for identification.

The economic structure imposed in the density discontinuity approach is usually application specific, and its plausibility has to be judged on a case-by-case basis. For example, the stable unit treatment value assumption implicitly assumed in Kleven and Waseem's model of earnings implies an absence of spillovers of the tax effects.[47] The same assumption is certainly more controversial in minimum wage settings. Interestingly, the need to impose additional economic structure to achieve identification narrows the gap between the reduced-form approach of RDDs, where the goal is to simply identify the local effects of the policy, and the traditional structural approach, where the goal is to identify parameters that characterize, *ceteris paribus*, policy-invariant behavioral responses.

To conclude, when discussing sorting and manipulation in the context of RDD, Lee and Lemieux, (2010) wrote: *so far we have discussed the sorting or manipulation issue as a potential problem or nuisance to the general program evaluation problem. But there is another way of viewing this sorting issue. (…) That is, economic behavior may be what is driving discontinuities in the frequency distribution (…) and those behavioral responses may be of interest. These cases (…) do not fit into the standard RD framework, but nevertheless can tell us something important about behavior, and further expand the kinds of questions that can be addressed by exploiting discontinuous rules to identify meaningful economic parameters of interest.*" As we have shown in this review, this branch of the literature has developed considerably in recent years, and the use of discontinuities in densities to identify structural and reduced-form parameters related to sorting have become increasingly common in applied work.

NOTES

1. Urquiola and Verhoogen (2009) provide an example of such manipulation in the context of enrollment, class-size, and student outcomes.
2. Frandsen (2013) designed a test for manipulation when the forcing variable is discrete.
3. Interestingly, the discontinuity at the density of the forcing variable can also be exploited to provide bounds for the parameters of interest in an RDD in the presence of manipulation of the forcing variable (Gerard, Rokkanen, & Rothe, 2015).
4. Lee and McCrary (2009) serves as such an example in studying how the juvenile criminal behavior responds to an increase in the punitiveness of sanctions at age of 18. They used panel data that recorded the birth date and the offense date for felony arrests in Florida from 1989 to 2002. In this setup, a discontinuity in the density of age *conditional on offending* at the age of 18 would suggest an effect of the policy. Instead, Lee and McCrary (2009) constructed the hazard rate of arrest at each age (in weeks). Then, they employed an RDD analysis to search for a discontinuous change in the hazard function at age of 18.
5. We are thankful to a referee for the suggestion to explore this relationship.
6. To the best of our knowledge, the first paper to explore the concept of "missing mass" to achieve identification was Meyer and Wise (1983).
7. Saez (2010) employed a similar model and method to recover tax elasticities from observed *kinks* in the tax system.

8. As discussed in the previous section, individuals may not react to the notch due to small tax elasticities or optimization frictions such as switching and attention costs.

9. When the relevant interval is the entire support, Assumption KW3 imposes a global functional form restriction in the distribution of $Z(0)$. Local functional form assumptions on the density of $Z(0)$ can be specified by choosing $\overline{A}$ to be a strict subset of the support of $Z(0)$.

10. Another way to identify β_0 is to exploit that the level and the derivatives of the density are identified at z^*. This can be used to identify β_0. Dong and Lewbel (2015) used a similar argument for the conditional expectation to identify the effect of changing the cutoff in RDDs.

11. In the data, however, individuals have the tendency to report round numbers such as multiples of 50 and 100. Kleven and Waseem address this particular form of measurement error by estimating the bunching probability by the excess of bunching that is observed above and beyond the bunching that one would expect from rounding alone. The expected bunching is estimated from the bunching observed at round numbers that are not at the tax notch.

12. This assumption effectively prevents wages higher up in the wage distribution from being affected by the introduction of the minimum wage. This assumption will be valid as long as the state variables present in the equation determining wages for workers of higher productivity are unchanged by the introduction of the minimum wage. In this case, the introduction of the minimum wage is equivalent to adding a nonbinding constraint to an otherwise identical optimization problem. One example of a setup where this assumption is violated is Flinn (2006).

13. An example of a model that is excluded by this assumption is that of Teulings (2000).

14. The functional form assumption also bypasses the need to assume away spillovers everywhere. One can instead specify an interval where spillovers are thought to be present and use functional form extrapolation arguments to identify the spillover effects of the policy.

15. The assumption that the wage remains exactly the same when the worker moves to the informal sector, that is, ($W = W(0)$), substantially simplifies the exposition. The same results hold when this assumption is replaced with one in which the worker draws a new wage from $f_0(w|S(0) = 1, W(0) < w^*)$. This modification does not change the main results of the model.

16. This joint distribution is not explicitly written as the joint distribution that follows after workers maximize utility given heterogeneous wage draws, labor tax costs, and probability of fines for informal contracts, although this can be done. One can regard this latent joint distribution as the resulting joint distribution that would prevail after workers have made their choices given their draws.

17. Functional form assumptions such as Assumption J4 are convenient to uncover effects on joint distributions. Diamond and Persson (2016) use a similar assumption when studying the effects of teacher's discretion on grades. First, they recover the effects of such discretion on the distribution of grades themselves. Then, they specify a functional form assumption for the relationship between grades and outcomes such as wages and extrapolate the estimates of this conditional mean

function into the area where manipulation occurs to learn about the effects of grade manipulation on these other outcomes.

18. See Jales (2015) for identification in the general case.

19. Note that limited spillovers can immediately be incorporated by changing the identification argument to $\Pr\{S(0)=1\} = \Pr\{S(0)=1|W(0)>km\} = \Pr\{S=1 |W>km\}$, where $k>1$ and km is the point above which spillovers are assumed to vanish.

20. The equations are given by: $\pi_m^{(0)} = (1/(1-\lambda))(\pi_m - \lambda\pi_m^{(1)}), \pi_d^{(0)} = 1-\pi_m^{(0)}$, $\pi_d^{(1)} = (1/\lambda)(\pi_d - (1-\lambda)\pi_d^{(0)})$, and $\pi_u^{(1)} = 1-\pi_d^{(1)} - \pi_m^{(1)}$.

21. To simplify the discussion, we use an explicit functional form for the utility function and assume that the reimbursement scheme presents only one discontinuity.

22. The reimbursement scheme here is a simplified version of that in Bajari et al. (2011), as it only includes the second kink considered in their paper. However, this retains the essence of the analysis, as the first kink, which is omitted here, was not used for identification in their paper.

23. It is straightforward to generalize the model to allow the cost of care to be a function of the level of spending. This generalization does not change the main results of the model.

24. In practice, some hospitals may face large costs of adjusting their spending levels. In this case, the density of spending may not present a gap around q^*. Thus, there will be "missing mass" in the dominated area, and the identification arguments need to be adapted accordingly. For the benefit of clarity in the exposition, we will retain the (perhaps oversimplified) model, which abstracts from measurement error or optimization frictions.

25. Note that in the models previously discussed, there was a fraction of individuals on both sides of the threshold that did not react to the policy. This fraction is equal to one on one side of the threshold if the model possesses no spillovers, and it is typically smaller than one on the affected side of the threshold. In Bajari et al. (2011), this fraction was zero on the affected side of the threshold, that is, no level of observed spending Q was equal to the counterfactual spending $Q(0)$ for $Q(0) > q^*$.

26. This result does not rely on assuming that the density of spending is flat around q^*. This result is also robust to incorporating a nontrivial marginal cost of providing care.

27. Here, we implicitly assume that the hospital places a weight equal to one on the patient's utility. Generalizing the model to allow for a different weight leads to the following result: When hospitals place a large (small) weight on the patient's utility, the gap will be small (large). See Bajari et al. (2011) for details.

28. This result is valid under the assumption the hospital's marginal utility of care under a linear reimbursement scheme is linear in q, given θ. Thus, the observed change in the hospital's marginal utility identified at q^* can be used to determine the change in the hospital's marginal benefit for values of $\theta > \theta^*$ (or $q > q^*$). Given the results on flexible functional forms for the response probabilities under Doyle's and Jales' models, it seems that this assumption can be relaxed if one is willing to rely on derivatives of the density.

29. Given the strong linearity of marginal utility assumption needed to recover the entire counterfactual density of spending and the fact that under measurement error or optimization frictions the density of observed spending may not present a gap at q^*, Bajari et al. (2011) focused on identifying the changes in spending that would occur under a marginal change in the threshold q^*. Their parameter of interest is similar to the object studied by Dong and Lewbel (2015) in the context of RDDs, the marginal threshold treatment effect.

30. Kleven (2016) noted that the assumptions needed to identify the latent distribution and, thereby, the effects of the policy on the observed distribution of the forcing variable are typically weaker than those needed to identify the behavioral or structural elasticities associated with an agent's behavior. Moreover, Kleven (2016) argues that this problem is not unique to the density discontinuity approach. It is, in fact, ubiquitous to the problem of relating reduced-form parameters to structural parameters.

31. Assume that $\mathbf{E}[Y(0)|X = x^*] \neq 0$ so the treatment effect ratio is well defined.

32. An alternative way to express this relationship is by writing Eq. (7) in terms of a log-difference: $\lim_{x\uparrow x^*} \ln f(x|Y = 1) - \lim_{x\downarrow x^*} \ln f(x|Y = 1) = \ln(\mathbf{E}[Y(1)|X = x^*])/(\mathbf{E}[Y(0)|X = x^*])$, so the left-hand side of this equation becomes the estimand of the McCrary test.

33. Note that the observation equation implicitly implies the "stable unit treatment value assumption" of Rubin (1980).

34. $\Pr\{Y(1) = 1|Y(0) = 0, X = x^*\} = 0$ is implied by this stronger assumption: $Y(1) \leq Y(0)$. This assumption rules out individuals that would drink if they were under a policy that prohibits them from doing so but would not drink if they were not under such a restriction.

35. The two points q_H and q_L satisfy $f^+(q_H) \equiv \lim_{x\downarrow q_H} f(x) > 0$, $\lim_{x\uparrow q_H} f(x) = 0$, $f^-(q_L) \equiv \lim_{x\downarrow q_L} f(x) = 0$ and $\lim_{x\uparrow q_L} f(x) > 0$. Let q^* be the point where the reimbursement rate changes discontinuously and the researcher observes a random sample of $\{q_i\}_{i=1}^n$. Bajari et al. (2011) estimated q_L and q_H by $\widehat{q}_L = \min\{q_i : q_i > q^*\}$ and $\widehat{q}_H = \max\{q_i : q_i \leq q^*\}$, and showed that $n(\widehat{q}_L)$ and $n(\widehat{q}_H)$ converge to two independent exponential random variables.

36. The standard kernel estimator is generally not consistent at the boundary. Although rescaling the kernel estimator by a multiplier will restore consistency, the bias is still of order h (h is the bandwidth), which is larger than the bias of the density estimator at the interior points (of order h^2) (Jones, 1993).

37. Let p be the order of the polynomial. If p is odd, the local polynomial estimator of the conditional mean has bias of the same order at the boundary as in the interior. (Both are $O_P(h^{p+1})$, where h is the bandwidth.) If p is even, the bias at the boundary is still of order h^{p+1} while the bias in the interior reduces to order h^{p+2}; see Fan and Gijbels (1996) for details.

38. Local linear density estimator based on data binning was introduced in Cheng (1994) to account for the boundary effect; see also Cheng (1997) and Cheng et al. (1997), among others.

39. If the researcher is particularly interested in estimating $\ln f^+(c)$ and $\ln f^-(c)$, McCrary (2008) noted that one can also first take log of the normalized bin counts Y_j and then apply the local linear smoothing.

40. When the polynomial with the pth order is used, the kernel function accordingly should have bounded derivatives up to order p.

41. See Eq. (3.5) in Jones (1993) for the formula of the boundary kernel.

42. The log density $\ln f(c)$ is approximated by a linear function around point c; see Loader (1996).

43. For example, Zhang, Karunamuni, and Jones (1999) proposed an estimator that involves reflecting a transformation of the data and estimating a density derivative using one-sided kernels.

44. Cheng et al. (1997) showed that the local polynomial estimator, weighted by the triangular kernel, is highly efficient in the sense that its linear minimax risk is very close to that yielded by the minimax optimal kernel.

45. For example, legislation requires a firm with 50 employees or more to establish a worker's council and a safety committee. See Garicano et al. (2013) for details.

46. Given the presence of measurement error in employment levels in the data, the observed density is hump-shaped around the threshold. Garicano et al. (2013) specified a normal distribution for the measurement error and then relied on maximum likelihood for the parameter estimation.

47. The absence of spillovers is implied by the fact that an individual's potential earnings are solely a function of his idiosyncratic skill level and thus invariant to policy changes.

REFERENCES

Bajari, P., Hong, H., Park, M., & Town, R. (2011). *Regression discontinuity designs with an endogenous forcing variable and an application to contracting in health care*. Discussion Paper, National Bureau of Economic Research.

Cattaneo, M. D., Jansson, M., & Ma, X. (2015). *Simple local regression distribution estimators with an application to manipulation testing*. Discussion Paper, University of Michigan.

Cattaneo, M. D., Jansson, M., & Ma, X. (2016). *Rddensity: Manipulation testing based on density discontinuity*. Discussion Paper, University of Michigan.

Cheng, M.-Y. (1994). On boundary effects of smooth curve estimators. Unpublished manuscript Series 2319, Institute for Statistics, University of North Carolina.

Cheng, M.-Y. (1997). Boundary aware estimators of integrated density derivative products. *Journal of the Royal Statistical Society: Series B (Statistical Methodology)*, *59*(1), 191–203.

Cheng, M.-Y., Fan, J., & Marron, J. S. (1997). On automatic boundary corrections. *The Annals of Statistics*, *25*(4), 1691–1708.

Cline, D. B. H., & Hart, J. D. (1991). Kernel estimation of densities with discontinuities or discontinuous derivatives. *Statistics: A Journal of Theoretical and Applied Statistics*, *22*(1), 69–84.

Diamond, R., & Persson, P. (2016). *The long-term consequences of teacher discretion in grading of high-stakes tests*. Discussion Paper, National Bureau of Economic Research.

Dong, Y., & Lewbel, A. (2015). Identifying the effect of changing the policy threshold in regression discontinuity models. *Review of Economics and Statistics*, *97*(5), 1081–1092.

Doyle, Jr., J. J. (2007). Employment effects of a minimum wage: A density discontinuity design revisited. Discussion Paper.

Einav, L., Finkelstein, A., & Schrimpf, P. (2016). *Bunching at the kink: Implications for spending responses to health insurance contracts.* Discussion Paper, National Bureau of Economic Research.

Fan, J., & Gijbels, I. (1996). *Local polynomial modelling and its applications: Monographs on statistics and applied probability* (Vol. 66). Chapman and Hall, London: CRC Press.

Flinn, C. J. (2006). Minimum wage effects on labor market outcomes under search, matching, and endogenous contact rates. *Econometrica, 74*(4), 1013–1062.

Frandsen, B. R. (2013). Party bias in union representation elections: Testing for manipulation in the regression discontinuity design when the running variable is discrete. Unpublished manuscript, Brigham Young University.

Garicano, L., Lelarge, C., & Van Reenen, J. (2013). *Firm size distortions and the productivity distribution: Evidence from France.* Discussion Paper, National Bureau of Economic Research.

Gerard, F., Rokkanen, M., & Rothe, C. (2015). *Partial identification in regression discontinuity designs with manipulated running variables.* Discussion Paper, Columbia University.

Hahn, J., Todd, P., & Van der Klaauw, W. (2001). Identification and estimation of treatment effects with a regression-discontinuity design. *Econometrica, 69*(1), 201–209.

Imbens, G. W., & Lemieux, T. (2008). Regression discontinuity designs: A guide to practice. *Journal of Econometrics, 142*(2), 615–635.

Jales, H. (2015). *Estimating effects of the minimum wage in a developing country: A density discontinuity design approach.* Center for Policy Research Working Papers 184, Center for Policy Research, Maxwell School, Syracuse University.

Jones, M. C. (1993). Simple boundary correction for kernel density estimation. *Statistics and Computing, 3*(3), 135–146.

Kleven, H. J. (2016). Bunching. *Annual Review of Economics, 8*(1), 435–464.

Kleven, H. J., Landais, C., Saez, E., & Schultz, E. (2014). Migration and wage effects of taxing top earners: Evidence from the Foreigners' Tax Scheme in Denmark. *Quarterly Journal of Economics, 129*(1), 333–378.

Kleven, H. J., & Waseem, M. (2013). Using notches to uncover optimization frictions and structural elasticities: Theory and evidence from Pakistan. *The Quarterly Journal of Economics, 128*(2), 669–723.

Kopczuk, W., & Munroe, D. (2015). Mansion tax: The effect of transfer taxes on the residential real estate market. *American Economic Journal: Economic Policy, 7*(2), 214–257.

Lee, D., & McCrary, J. (2009). *The deterrence effect of prison: Dynamic theory and evidence.* Discussion Paper, Department of Economics, Center for Economic Policy Studies, Princeton University.

Lee, D. S. (2008). Randomized experiments from nonrandom selection in US House elections. *Journal of Econometrics, 142*(2), 675–697.

Lee, D. S., & Lemieux, T. (2010). Regression discontinuity designs in economics. *Journal of Economic Literature, 48*, 281–355.

Loader, C. R. (1996). Local likelihood density estimation. *The Annals of Statistics, 24*(4), 1602–1618.

McCrary, J. (2008). Manipulation of the running variable in the regression discontinuity design: A density test. *Journal of Econometrics, 142*(2), 698–714.

Meyer, R. H., & Wise, D. A. (1983). Discontinuous distributions and missing persons: The minimum wage and unemployed youth. *Econometrica, 51*(6), 1677–1698.

Müller, H.-G. (1991). Smooth optimum kernel estimators near endpoints. *Biometrika*, *78*(3), 521–530.

Otsu, T., Xu, K.-L., & Matsushita, Y. (2013). Estimation and inference of discontinuity in density. *Journal of Business and Economic Statistics*, *31*(4), 507–524.

Rubin, D. B. (1980). Discussion of "Randomization analysis of experimental data: The Fisher randomization test" by Basu. *Journal of the American Statistical Association*, *75*(371), 591–593.

Ruppert, D., & Wand, M. (1994). Multivariate locally weighted least squares regression. *Annals of Statistics*, *22*(3), 1346–1370.

Saez, E. (2010). Do taxpayers bunch at kink points? *American Economic Journal: Economic Policy*, *2*(3), 180–212.

Slemrod, J. (2013). Buenas notches: Lines and notches in tax system designe. *Journal of Tax Research*, *11*(3), 259.

Teulings, C. N. (2000). Aggregation bias in elasticities of substitution and the minimum wage paradox. *International Economic Review*, *41*(2), 359–398.

Urquiola, M., & Verhoogen, E. (2009). Class-size caps, sorting, and the regression-discontinuity design. *The American Economic Review*, *99*(1), 179–215.

Wand, M. P., Marron, J. S., & Ruppert, D. (1991). Transformations in density estimation. *Journal of the American Statistical Association*, *86*(414), 343–353.

Zhang, S., Karunamuni, R., & Jones, M. (1999). An improved estimator of the density function at the boundary. *Journal of the American Statistical Association*, *94*(448), 1231–1240.

Zhang, S., & Karunamuni, R. J. (1998). On kernel density estimation near endpoints. *Journal of Statistical Planning and Inference*, *70*(2), 301–316.

THE DETERRENCE EFFECT OF PRISON: DYNAMIC THEORY AND EVIDENCE*

David S. Lee[a,b] and Justin McCrary[b,c]

[a]*Princeton University, Princeton, NJ, USA*
[b]*NBER, Cambridge, MA*
[c]*UC Berkeley, Berkeley, CA, USA*

ABSTRACT

Using administrative, longitudinal data on felony arrests in Florida, we exploit the discontinuous increase in the punitiveness of criminal sanctions at 18 to estimate the deterrence effect of incarceration. Our analysis suggests a 2% decline in the log-odds of offending at 18, with standard errors ruling out declines of 11% or more. We interpret these magnitudes using a stochastic dynamic extension of Becker's (1968) model of criminal behavior. Calibrating the model to match key

*Earlier drafts of this paper circulated as NBER Working Paper # 11491, "Crime, Punishment, and Myopia" and Princeton Industrial Relations Section Working Paper # 550, "Deterrence Effect of Prison: Dynamic Theory and Evidence."

Regression Discontinuity Designs: Theory and Applications
Advances in Econometrics, Volume 38, 73–146

ISSN: 0731-9053/doi:10.1108/S0731-905320170000038005

empirical moments, we conclude that deterrence elasticities with respect to sentence lengths are no more negative than −0.13 for young offenders.

Keywords: Prison; deterrence; self-control

JEL classifications: D9; K4

1. INTRODUCTION

Crime continues to be an important social and economic issue in the United States. While crime rates have fallen in the recent past, the cost of controlling crime has not. From 1970 to 2006, criminal justice system expenditures as a share of national income increased 112%, and the ratio of criminal justice employees to the population grew 107% (LEAA 1972; U.S. DOJ 2006). Over the same period, the incarcerated fraction of the population increased 373%, making the U.S. incarceration rate the highest in the world (Chaddock, 2003; Maguire & Pastore, 2001).

Perhaps because of these trends, the economics of crime has emerged as an active area of research. Recent studies have investigated a wide range of potential factors, including the effects of police and incarceration (Chalfin & McCrary, 2017; Di Tella & Schargrodsky, 2004; Drago, Galbiati, & Vertova, 2009; Helland & Tabarrok, 2007; Klick & Tabarrok, 2005; Levitt, 1996; 1997; 1998), conditions in prisons (Katz, Levitt, & Shustorovich, 2003), gun ownership (Ayres & Donohue, 2003; Cook & Ludwig, 2000; Duggan, 2001; Lott, 1998), parole and bail institutions (Abrams & Rohlfs, 2011; Kuziemko, 2007), education (Lochner & Moretti, 2004), social interactions and peer effects (Bayer, Hjalmarsson, & Pozen, 2009; Case & Katz, 1991; Gaviria & Raphael, 2001; Glaeser, Sacerdote, & Scheinkman, 1996; Kling, Ludwig, & Katz, 2005), and family circumstances and structure (Donohue & Levitt, 2001; Glaeser & Sacerdote, 1999).

Economists have also considered the returns to education among recent prison releasees (Western, Kling, & Weiman, 2001), the impact of criminal histories on labor market outcomes (Grogger, 1995; Kling, 2006; Mueller-Smith, 2015), the impact of wages and unemployment rates on crime (Grogger, 1998; Raphael & Winter-Ebmer, 2001), the strategic interplay between violent and property crime (Silverman, 2004), the effect of incarceration on the supply of crime in the economy (Freeman, 1996; 1999), and optimal law enforcement (Eeckhout, Persico, & Todd, 2010; Polinsky & Shavell, 2000).

One of the key questions for both the literature as well as policymakers is the extent to which more punitive criminal justice sanctions can deter criminal behavior. This notion is at the core of the seminal model of Becker (1968) and of many less formal treatments, such as the classic works by Beccaria (1764) and Bentham (1789), widely cited in the law and economics and criminology literatures.

However, despite a good deal of empirical research, magnitudes of the deterrence effect of criminal justice sanctions remain somewhat uncertain. Our own focus is on the deterrence effect of prison. The most prominent papers addressing this question yield a somewhat wide range of elasticities. On the one hand, assuming exogeneity of changes in the punitiveness of criminal sanctions across U.S. states, Levitt (1998) finds crime elasticities with respect to punishments as large as −0.40, and assuming exogeneity of sentence enhancements given to early prison releasees, Drago, Galbiati, and Vertova (2009) find elasticities as high as −0.74 for a population of Italian offenders. On the other hand, Helland and Tabarrok (2007) compare convicted defendants who become exposed to the threat of the California "three-strikes" statute to observationally equivalent acquitted defendants, and find elasticities of −0.06, an order of magnitude smaller.

In this chapter, we use a different identification strategy to isolate the deterrence effects of long prison sentences. Specifically, we take advantage of the following fact: when an individual is charged with a crime that occurs before his 18th birthday, his case is handled by the juvenile courts.[1] If the offense is committed on or after his 18th birthday, however, his case must be handled by the adult criminal court, which is known to administer more punitive criminal sanctions.[2] Thus, when a minor turns 18, there is an immediate increase in the expected cost of participating in crime. We argue that while other determinants of criminal offending may change rapidly with age, they do not change discontinuously at 18. This allows us to attribute any discontinuous drop in offense rates at 18 to a behavioral response to adult criminal sanctions, relative to juvenile criminal sanctions.

Our identification strategy is conceptually distinct from a standard regression discontinuity (RD) design, which relies on assumptions of imprecise control, and hence the smoothness of the density of the RD running variable to identify treatment effects (Lee, 2008; Lee & Lemieux, 2010). Here, it is precisely the measured discontinuity in the density of age at offending that we are attributing to a deterrence effect.

Our analysis takes advantage of a large, person-level, longitudinal data set covering the universe of felony arrests in Florida between 1995 and 2002. This is a major improvement over publicly available data sets, which only capture age in years and are not longitudinal. Publicly available data sets thus make it difficult to distinguish deterrence from incapacitation: a lower arrest rate for 18-year olds could be entirely driven by more offenders being imprisoned, and thus being unable to commit crimes.[3] In contrast, our data are longitudinal and furnish information on exact dates of birth and offense. We use the precise timing of arrests relative to 18 to isolate a pure deterrence effect, as opposed to an incapacitation effect.

Our discontinuity analysis yields small, precise point estimates: a 1.8% decline in the log-odds of offending at 18, with standard errors that can statistically rule out declines of 11% or more. Our estimates are consistently small across crime categories and across types of jurisdictions within Florida.

These findings can be interpreted directly as an evaluation of at least three different types of policy reforms: (1) reducing the age of criminal majority, (2) increasing the rate at which juveniles are transferred to the adult criminal court, thereby increasing the expected sanction that a juvenile faces, or (3) increasing adult sentences, leaving juvenile sentences fixed. The first two of these policy reforms have been enacted in recent years in multiple states and are currently part of ongoing criminal justice policy discussions (National Research Council, 2001; Snyder & Sickmund, 2006), and the third policy reform dominated the criminal justice landscape during the 1980s (Levitt, 1998). Our results suggest that these reforms have limited deterrence effects for youthful offenders. To extrapolate our results to evaluate a broader range of policy reforms of interest, and to connect our empirical results to structural parameters of offender behavior, we develop a stochastic dynamic model of crime. The model retains at its core the essence of the static Becker model of criminal behavior, but places the individual in a dynamic setting. The model makes precise the link between our discontinuity estimates of the intertemporal behavioral response to an anticipated increase in punishments and other policy-relevant deterrence elasticities. We calibrate the model to match easily attainable empirical quantities. This calibration exercise leads us to three additional conclusions.

First, our model and estimates are inconsistent with "patient" time preferences (e.g., annual discount factors of 0.95) and suggest much smaller discount factors. Essentially, the small change in behavior at 18 is consistent with offenders having short time horizons, leading them to perceive little difference between nominally long and short incarceration periods. Second, a moderately sized (e.g., −0.25) elasticity of crime with respect to the probability of apprehension is consistent with our estimates, but requires the discount factor to be extremely small, which in turn implies a small elasticity with respect to sentence lengths. This result follows because our model leads to implied bounds for deterrence elasticities for police and prison. Finally, the most negative elasticity with respect to sentence lengths that is consistent with our estimates and model is −0.13.

In addition to estimating deterrence parameters, we also present RD evidence on the incapacitation effect of adult sanctions. We find that being arrested just after 18 leads to an immediate (within 30 days) subsequent reduction in offending, relative to being arrested just before 18. We interpret this as evidence of the incapacitation effect of adult sanctions. Overall, from our main analysis as well as this supplementary evidence on incapacitation, we conclude that if lengthening prison sentences leads to significant crime reduction, it is likely operating through a direct, "mechanical" incapacitation effect, rather than through a behavioral response to the threat of punishment.

The remainder of the chapter is organized as follows. Section 2 discusses related studies on deterrence. Section 3 describes our identification strategy and approach to estimation, while Section 4 describes our data in detail. Section 5 presents our main results. In Section 6, we develop our stochastic dynamic model, calibrate it, and interpret our reduced-form magnitudes through the lens of this model. Section 7 concludes. Two appendices provide further detail on our data and theoretical model.

2. EXISTING LITERATURE

This section briefly describes a number of prominent studies that are most related to our analysis. These recent studies attempt to address an important problem that arises in isolating the causal impact of more punitive criminal sanctions on criminal behavior: crime control policies (e.g., policing, sentencing) are often suspected to be endogenous responses to current crime levels and trends. The problem of policy endogeneity has long been recognized in the literature (Donohue & Wolfers, 2008; Ehrlich, 1973; Ehrlich, 1987; Levitt, 2004; Nagin, 1998). To circumvent policy endogeneity, recent studies exploit arguably exogenous variation in the punitiveness of criminal sanctions (i.e., prison sentences).[4] The elasticities resulting from these analyses range from somewhat small (−0.06) to somewhat large (−0.74).

The largest magnitudes come from the analyses of Levitt (1998) and Drago, Galbiati, and Vertova (2009). The highly cited study of Levitt (1998) is the most closely related to our own. The analysis uses a state-level panel, regressing juvenile crime rates on a measure of punitiveness of the juvenile justice system (number of delinquents in custody per 1,000 juveniles), including state and year effects, as well as other control variables. It also disaggregates the data by age cohort, and implements a difference-in-difference strategy, whereby the changes in adult–juvenile relative crime rates are compared between states that had higher and lower increases in adult–juvenile relative punitiveness over time.[5] Levitt (1998) concludes that juveniles are significantly responsive to criminal sanctions, reporting effects that imply an elasticity of −0.40 for violent crime.[6]

Another study finding large elasticities of the deterrence effect of prison is that of Drago et al. (2009), which examines the crime impact of a 2006 Italian clemency act. This statute led to the release of prisoners whose crimes were committed prior to May of 2006, subject to some minor exceptions. Importantly, the statute contained a sentence enhancement provision: for releasees who were re-arrested within five years of release and subsequently sentenced to more than two years, their sentence would be augmented by the amount of time that was remaining on their first sentence at the time of clemency. The chapter documents the empirical relation between re-arrest within 7 months of release and the time remaining on the sentence length at time of clemency, controlling for other observables, such as the original sentence length. The results suggest an elasticity of crime with respect to sentence length of −0.74 at 7 months follow-up and approximately −0.45 at 12 months follow-up.[7] The identification strategy is based on sentence enhancements, and this quantity is interpreted as representing a deterrence effect. Given that upon release, the difference in age between those with little or much time remaining on their sentences are similar, the key identifying assumption here is that those who are arrested earlier in life have the same propensity to commit crime as those arrested later in life.

At the other end of the range of elasticities, Helland and Tabarrok (2007) and Iyengar (2008) find generally small behavioral responses of criminals to California's three-strikes law. This statute requires that those previously convicted of two "strikeable offenses" who are subsequently convicted of any felony must receive a prison sentence of 25 years to life, with a further requirement that parole can begin no earlier than completion of 80% of the sentence.[8] Helland and Tabarrok (2007) assess the impact of the three-strikes law using data on prisoners released in 1994, the first year the statute was in effect. The chapter compares the recidivism pattern between those previously convicted of two strikeable offenses and those previously tried for two strikeable offenses, but convicted for only one of those offenses. The estimates suggest that three-strikes lowered the incidence of crime by about 20%. Since the increase in expected sentence lengths associated with three-strikes is at least 300%, this implies an elasticity estimate on the order of −0.07.[9] Using a different data set and approach, Iyengar (2008) also studies the three-strikes statute and arrives at a set of estimates that imply similar elasticity estimates, on the order of −0.10.[10]

These papers differ in several different ways, making it hard to know precisely why the elasticity estimates are so different. The papers analyze data from different jurisdictions (the U.S. as a whole, Italy, and California) and for different time periods. They also employ different methods for measuring crime. Levitt (1998) measures crime using crime aggregates, while Drago et al. (2009); Helland and Tabarrok (2007), and Iyengar (2008) measure crime using recidivism. However, perhaps the most important difference in these papers is the assumptions employed. Arguably, the most plausible set of assumptions invoked are those in Iyengar (2008), who assumes that the *ordering* of criminal history is idiosyncratic – that is, the criminal propensity of individuals with a theft then a burglary on their record equals that of those with a burglary then a theft on their record. In our view, this assumption is still stronger than the smoothness assumptions we employ, however, is more plausible than the difference-in-difference approaches employed by the other three papers.

As noted, our approach differs from those in the previous literature by exploiting smoothness assumptions, which, combined with our high-frequency data, allows us to separate deterrence from incapacitation effects of sanctions. The use of annual data (cf., Levitt, 1998) has the potential to conflate deterrence and incapacitation effects. The scope for conflation of these concepts is large when differences between adult and juvenile incarceration rates appear within a year of the 18th birthday. In Section 5, we show empirically that differences in incapacitation rates emerge rapidly and are evident within even 30 days after the 18th birthday.

It is important to note that our estimates are most relevant for individuals arrested, but not necessarily convicted, on felony charges in Florida prior to age 17. These juveniles are likely to have high propensities for crime, and thus our sample is likely to be more comparable to the Italian ex-prisoners in Drago et al. (2009) or the Californian repeat offenders in

Helland and Tabarrok (2007), than to the aggregate data utilized in Levitt (1998), which also includes first-time and low-propensity offenders.

Note that our approach is conceptually distinct from the standard RD design, which relies on an assumption that individuals do not precisely manipulate their running variable around the threshold of interest. In the standard RD context, a discontinuity in the density of the running variable constitutes evidence against the validity of the RD design (McCrary, 2008). In our main analysis, the discontinuity in the density of age at offense measures deterrence and is itself the object of interest. We utilize a more standard RD analysis in our measurement of incapacitation, as discussed in Section 5.

The final distinctive feature of our analysis is that we interpret our reduced-form estimates through the lens of a stochastic dynamic model of criminal behavior. We thus expand on Becker's (1968) model and explicitly consider the dynamic nature of the problem, much in the spirit of the theoretical work of Lochner (2004).[11]

3. IDENTIFICATION AND ESTIMATION

Our identification strategy exploits the fact that in the United States, the severity of criminal sanctions depends discontinuously on the age of the offender at the time of the offense. In all 50 U.S. states, offenders younger than a certain age, typically 18, are subject to punishments determined by the juvenile courts. The day the offender turns 18, however, he is subject to the more punitive adult criminal courts. The criminal courts are known to be more punitive in a number of ways. A significant difference is that the expected length of incarceration is significantly longer when the offender is treated as an adult, rather than as a juvenile.[12] We quantify this difference in Section 5. To identify the deterrence effect of adult prison, we follow a cohort of youths from Florida longitudinally, and examine whether there is a discontinuous drop in their offense rates when they turn 18. Our data contain exact dates of birth and offense, allowing us to analyze the timing of offenses at high frequencies. A high-frequency approach is important, because it allows us differentiate between secular age effects (Grogger, 1998; Levitt & Lochner, 2001) and responses to sanctions.

Our approach does not require the determinants of criminal behavior to be constant throughout the individual's life. Instead, it relies on the arguably plausible assumption that determinants of criminal propensity other than the severity of punishments do not change discontinuously at 18. We are arguing that on the day of the 18th birthday, there is no discontinuous change in the ability of law enforcement to apprehend an offender; no "jump up" in wages; no discontinuous change in the distribution of criminal opportunities; and so on. There are some exceptions to this (e.g., the right to vote and the right to gamble change discontinuously at 18), but we deem these to be negligible factors in an individual's decision whether to commit crime.

We emphasize that we only believe that "all other factors" are roughly constant when examining offense rates in relatively short intervals (e.g., one day, or one week). Indeed, the determinants of criminal behavior change significantly at a year-to-year frequency, as is apparent from the age distribution of arrestees at annual frequencies (see Fig. A1).[13] In the age range of 17, 18, or 19, for example, youth are graduating from high school, starting new jobs, and developing physiologically and psychologically in ways that could affect underlying criminal propensities (Grogger, 1998; Levitt & Lochner, 2001). Since this research design relies heavily on the continuity of all factors aside from sanctions, in Sections 4 and 5, we assess a number of alternative ways in which data issues or police discretion in reporting could affect our interpretation of the discontinuity in arrest rates as a response to more punitive sanctions.

Implementing our research design is straightforward. Here, we describe the basic idea behind our approach, and later describe minor adjustments associated with implementation. Suppose we have a sample of N individuals, and we can track their subsequent offending behavior, starting at age 17. Then for each week, we can calculate the number of individuals arrested for the first time since 17, as a fraction of those who are still at risk of doing so. If n_1, n_2, and n_3 are the number of individuals who are arrested in the first, second, and third weeks after their 17th birthday, respectively, then nonparametric estimates of the hazard of arrest are given by $\widehat{h}(1) = n_1/N$, $\widehat{h}(2) = n_2/(N - n_1)$, $\widehat{h}(3) = n_3/(N - n_1 - n_2)$. We refer to these as local average estimates of the hazard of offense. The empirical question to be addressed is whether the hazard drops off precipitously at age 18, as seems intuitive and as is predicted by deterrence theory, as we discuss below.

It is convenient to summarize these averages and the corresponding discontinuity with a flexible parametric form. To do this, we estimate a panel data logit model, as suggested in Efron (1988). Specifically, we organize the data set into an unbalanced panel, with N observations for the first period,

$N - n_1$ for the second, $N - n_1 - n_2$ for the third, and so on. We then estimate the logit model

$$\mathbb{P}(Y_{\text{it}} = 1 | X_t, D_t) = F\left(X_t'\alpha + D_t\theta\right), \tag{1}$$

where Y_{it} is the indicator for arrest for person i in period t, $X_t \equiv \left(1, t - t_0, (t - t_0)^2, \ldots, (t - t_0)^q\right)'$, q is the order of the polynomial, t_0 is the week of the 18th birthday, D_t is 1 if $t \geq t_0$ and is 0 otherwise, and $F(z) = \exp(z)/(1 + \exp(z))$.[14]

Below, we compare the predicted values from the logit model to the local average hazard estimates $\widehat{h}(t)$, and report a likelihood ratio test for the restrictions imposed by the logit form.[15] As will be clear from the empirical results below, these models do a good job of providing a parsimonious but accurate fit to the functional form suggested by the local averages. The reduced-form parameter of interest is θ, the discontinuous change in the log-odds of committing an offense when the youth turns 18 and immediately becomes subject to the adult criminal courts.

In a later section, we estimate the incapacitation effect of adult sanctions. For this, we focus on the timing of the second arrest since the 17th birthday, viewed as a function of age at first arrest since 17.[16] Specifically, we compare how quickly a re-arrest occurs after being arrested just before age 18, to how quickly it occurs if arrested just after 18. If the marginal adult takes longer to re-offend, this provides some indirect evidence of an incapacitation effect at work. While this is not the main target of our analysis, it nevertheless provides some information on the incapacitation mechanism. This is particularly useful for quantifying the sentence length elasticity and for calibrating the economic model that we develop below. The implementation of this analysis follows a more standard RD design, where the running variable is the age at first arrest since 17, the "treatment" is whether that arrest occurred before or after age 18, and the dependent variable is whether the second arrest occurs within, for example, 30 days.

4. DATA AND SAMPLE

In this section, we describe our data, our main analysis sample, and how various features of our data affect our estimation. We also discuss why our sample is unlikely to be affected by differential reporting at age 18.

For ease of exposition, we defer to Section 5 a more detailed discussion of expungement of juvenile records.

4.1. Main Analysis Sample

Our analysis uses an administrative database maintained by the Florida Department of Law Enforcement (FDLE). Essentially, the data consist of all recorded felony arrests in the state of Florida from 1989 to 2002. The database includes exact date of birth, gender, and race for each person. For each arrest incident, there is information on the date of the offense, the date of arrest, the county of arrest, the type of offense, whether or not the individual was formally charged for the incident, and whether or not the incident led to a conviction and prison term. Importantly, the data are longitudinal: each arrest incident is linked to a person-level identifier.[17] The raw data, therefore, can be described as a database of individuals, each with an associated arrest history.

From this database, we focus on three key arrest events that define our sample and outcomes of interest: (1) the first arrest recorded in our administrative data, which we refer to as the "baseline arrest," (2) the first arrest since the 17th birthday (we call this the "first arrest"), and (3) the second arrest since the 17th birthday (we call this the "second arrest"). Our main analysis sample is defined by those whose baseline arrest occurs prior to the 17th birthday, for a total of $N = 64,073$ individuals.[18] To examine deterrence we examine the incidence of the "first arrest" at age 18, and to provide some evidence on incapacitation, we examine the time between the "first arrest" and "second arrest" as a function of the age at "first arrest."

More specifically, for our analysis on deterrrence, we implement the estimation approach described in Section 3, with the following modifications and considerations:

- Our last date of observation is December 31, 2002, leading to some individuals being censored. The standard way to adjust for this is to compute hazards as $\widehat{h}(1) = n_1/N$, $\widehat{h}(2) = n_2/(N - n_1 - m_1)$, $\widehat{h}(3) = n_3/(N - n_1 - n_2 - m_1 - m_2)$, and so on, where m_1 and m_2 are the numbers of individuals who are 1 and 2 weeks into their 17th year on December 31, 2002 (i.e., censored), respectively. Following Efron (1988), we construct the unbalanced panel for the logit in an analogous way.
- We note that $N = 64,703$ is larger than the true "at risk" population, since at any age $t > 17$, an individual could still be incarcerated for

the "baseline arrest" or could be incarcerated because of a non-Index crime arrest that occurred between age 17 and t.[19] This fact by itself will not generate a discontinuity in our estimated hazards, as long as the true "at risk" population is evolving smoothly in age, particularly at age 18. There is an institutional reason for a discontinuity in the number "at risk" at age 21; Florida law mandates that no individuals above age 21 can be held in a juvenile correctional facility. However, there does not appear to be such a reason for an effect at age 18.

- In terms of the estimated magnitude of our discontinuity, this means our logit parameter θ estimates

$$\ln\left(\frac{\rho h}{1-\rho h}\right) - \ln\left(\frac{\rho h_J}{1-\rho h_J}\right) = \ln(h) - \ln(h_J) + \ln(1-\rho h) - \ln(1-\rho h_J),$$

where h and h_J are the arrest hazards for an 18.02- and 17.98-year old, respectively, and $\rho \in (0, 1)$ reflects the over-estimation of the risk set. Thus, if $h_J > h$, as would be consistent with deterrence, then

$$\ln\left(\frac{h}{1-h}\right) - \ln\left(\frac{h_J}{1-h_J}\right) \leq \theta < \ln(h) - \ln(h_J) < 0.$$

In practice, the difference between the upper and lower bounds is scant. Consequently, there is little loss in thinking of θ as estimating the percentage change in the hazard as a youthful offender transitions to adulthood.

- Our main analysis sample will include individuals who may be more likely be affected by the increase in sanctions. In particular, it seems likely that for this group, there is a perceived positive net benefit to criminal activity. After all, an individual in our sample has already been arrested at least once by 17, indicating that at least one crime was deemed worthwhile. By contrast, those who have not been arrested as of 17 could potentially include many youth who have virtually no chance of committing a serious crime. For these near-certain law-abiders, it would be mechanically impossible for their criminal activity to decline after 18.
- It is plausible that those who have already been arrested by age 17 are more likely to understand that there is a difference between the juvenile and adult criminal justice systems; they may even have been warned about this fact upon their "baseline arrest." Glassner et al. (1983) provide anecdotal evidence to support this viewpoint.[20]
- Our main analysis sample is not likely to be affected by expungement or sealing of criminal records. In Florida, as in most states, it is possible to

have criminal records sealed or expunged. If juvenile records were systematically missing relative to adult records, then we would be biased against finding a deterrence effect. But our sample, which requires having at least one juvenile record, necessarily consists of those who have not expunged their entire juvenile arrest record. We discuss this in greater detail in Section 5.

Table 1 reports some summary statistics for our main analysis sample. We begin with 64,073 individuals whose baseline arrest occurred prior to 17. As is common in criminal justice data sets, 80–90% of these arrestees are male, and roughly 50% are non-white. The first two columns present information on the baseline arrest and the first arrest since 17. Age at the baseline arrest is about 15, and the most common category of offense is property crime, followed by violent crime. Individuals are distributed evenly among small, medium, and large counties.[21] Slightly less than half of our main analysis sample is observed reoffending after age 17. The sex and gender composition of arrestees is similar at baseline arrest and first

Table 1. Summary Statistics, Estimation Sample.

Variable	Main Analysts Sample		
	Baseline Arrest	First Arrest	Young Arrestees
	(1)	(2)	(3)
Non-white	0.42	0.52	0.48
Male	0.80	0.88	0.88
Age	14.93	18.39	18.14
	(1.39)	(1.24)	(0.56)
Arrested in small county	0.28	0.31	0.31
Arrested in medium county	0.36	0.35	0.35
Arrested in large county	0.36	0.34	0.34
Index crime, violent	0.26	0.25	0.24
Index crime, property	0.51	0.33	0.39
Non-index crime, drug	0.07	0.22	0.20
Non-index crime, non-drug	0.15	0.19	0.17
Number of persons	64,073	30,938	163,037

Notes: Standard deviations in parentheses. Young arrestees are those arrested between 17 and 19. The number of arrests observed for young arrestees is 247,037.

arrest since 17, as is the county-size distribution. Offenses are distributed somewhat more evenly among the four crime types described at first arrest since 17 than at baseline arrest. For comparison, the final column reports the same statistics for all arrests where the individual was 17 or 18 years old at the time of arrest. From the means, it is apparent that our main estimation sample is broadly representative of this larger arrest population.

4.2. Measurement and Reporting Discontinuities

Our approach requires that criminality be measured in a smooth fashion near 18. There are some potential threats to continuous measurement, which we now discuss. One issue is that crimes are not necessarily defined similarly for juveniles and adults (e.g., truancy, corruption of a minor, statutory rape, and so on). To avoid definitional problems, we consider only crimes that are well defined for both juveniles and adults, such as burglary.

Even for crimes that are defined the same way for juveniles and adults, it is still possible that police officers could exercise discretion in executing an arrest that would lead to discontinuous measurement. For example, a police officer might view possession of small amounts of marijuana as forgivable for youth and might overlook the incident upon learning that the individual was still a minor. This would result in a discontinuous increase at 18 in the probability of observing criminality in arrest records.[22] Alternatively, one could imagine that the officer would view possession of marijuana as forgivable, but would want to "teach a lesson" to the offender, as long as the cost of the lesson were not too great. This would result in a discontinuous decrease at 18 in the probability of observing criminality.

To avoid such problems with discontinuous measurement of criminality at 18, we focus on arrests for so-called Index crimes: murder, rape, robbery, assault, burglary, and theft, including motor vehicle theft. We are confident that these felonies are sufficiently serious that an arresting officer would not overlook an offense. An additional benefit of focusing on Index crimes is that these are the crimes most commonly studied in the literature.

Finally, even if police officers themselves do not exercise discretion regarding making an arrest, administrative records may be more complete for adult arrests than juvenile arrests. The data from Florida do not suffer from this problem for the post-1994 period, due to an important criminal justice reform, the 1994 Juvenile Justice Reform Act, which requires that police departments forward records of serious juvenile felony arrests to the FDLE (see Section 5).

5. EVIDENCE ON DETERRENCE AND INCAPACITATION

This section presents our main empirical results. We address two issues. First, we quantify the deterrence effect of adult criminal sanctions, relative to juvenile criminal sanctions, by studying the pattern of offending around the 18th birthday. Second, we quantify the incapacitation effect of adult criminal sanctions, relative to juvenile criminal sanctions, by studying the change around the 18th birthday in durations between arrests.

5.1. Evidence on Deterrence

Our main result is summarized in Fig. 1. The top panel of this figure is an empirical hazard function for being arrested, between ages 17 and 19, among those arrested at least once before their 17th birthday. Each open circle represents those arrested in a given week as a proportion of those not yet arrested. For example, the first circle shows that about 0.005 of the main sample is arrested within a week of their 17th birthday. Of those who still not had been arrested by their 18th birthday, almost 0.0025 are arrested in the week of their 18th birthday. The solid line gives predicted probabilities of arrest, based on maximum likelihood estimates of the logit in Eq. (1).[23] The figure shows little indication of a systematic drop in arrest rates at the age of 18. The arrest probability literally does fall between the week before and after the 18th birthday, but that drop does not appear to be unusual, as compared to typical week-to-week differences.

For comparison, the bottom panel of Fig. 1 plots the analogous weekly arrest probabilities for those who were arrested at least once before their 19th birthday. We track the arrest records for individuals in this "placebo" sample for two years, from 19 to 21. The arrest probabilities are smooth in age, as would be expected since this age range is well past 18, and there is no legal significance to being 20 years old. The top and bottom panels of the figure are quite similar.

Table 2 reports estimated discontinuities in arrest probabilities at 18, based on the logit model of Eq. (1). These estimates support the inference

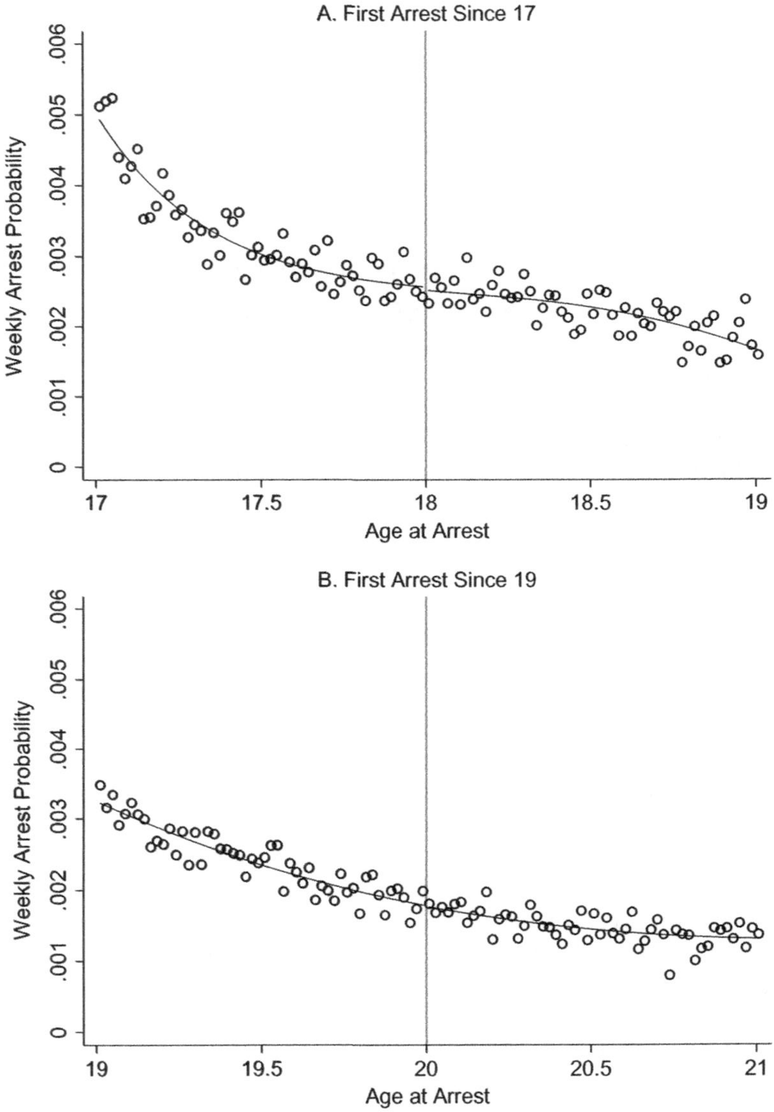

Fig. 1. Criminal Propensity Estimates by Age. (a) First Arrest Since 17 and (b) First Arrest Since 19. *Notes*: Panel (a) shows estimates of the hazard for index crime arrest for all those arrested at least once for any felony prior to 17 between 1995 and 1998. Open circles are weekly nonparametric estimates of the hazard, computed as the number offending in the given week, as a fraction of those who have neither reoffended nor been censored, as of the given week. Solid line presents a smoothed estimate based on a logit model, allowing for a jump at 18 (see text for details). Panel (b) presents a falsification analysis pertaining to the arrest hazard for all those arrested at least once prior to 19 between 1995 and 1998.

Table 2. Discontinuity Estimates of Deterrence.

Variable	(1)	(2)	(3)	(4)	(5)	(6)	(7)	(8)
Discontinuity estimate	−0.018	−0.017	−0.017	−0.018	−0.017	−0.017	−0.017	−0.017
	(0.047)	(0.047)	(0.047)	(0.047)	(0.047)	(0.047)	(0.047)	(0.047)
Non-white		0.545	0.503	0.576	0.533	0.542	0.540	0.540
		(0.017)	(0.017)	(0.017)	(0.018)	(0.018)	(0.018)	(0.018)
County size, baseline (relative to large county)								
Small			0.250		0.255	0.247	0.250	0.250
			(0.021)		(0.021)	(0.021)	(0.021)	(0.021)
Medium			0.081		0.087	0.087	0.086	0.086
			(0.021)		(0.021)	(0.021)	(0.021)	(0.021)
Type of crime, baseline								
Index crime, violent				0.020	0.009	0.016	0.016	0.016
				(0.028)	(0.028)	(0.028)	(0.028)	(0.028)
Index crime, property				0.191	0.190	0.202	0.205	0.205
				(0.026)	(0.026)	(0.026)	(0.026)	(0.026)
Non-index drug				−0.153	−0.154	−0.207	−0.213	−0.213
				(0.042)	(0.042)	(0.042)	(0.042)	(0.042)
Controls for age at baseline arrest?	N	N	N	N	N	Y	Y	Y
Order of polynomial in age at baseline arrest					1	3	5	
Log-likelihood	−96,006	−95,491	−95,420	−95,412	−95,339	−95,237	−95,178	−95,175

Notes: Standard errors in parentheses. Table presents coefficients from a logit model for being arrested for an index crime since 17. In addition to controls described, each model controls for a cubic polynomial in age at current arrest, relative to 18. Estimates are based on a panel of 4,928,226 observations pertaining to 64,703 persons.

suggested by Fig. 1: the drop in arrests at 18 is small in magnitude and statistically insignificant. The estimated discontinuity is roughly −0.018, with an estimated standard error of about 0.047.[24] These estimates imply that we can statistically rule out values of θ more negative than −0.11 at the 5% level of significance.[25]

In Section 6, we present evidence that the expected sentence length facing an adult arrestee is roughly 230% greater than that facing a juvenile arrestee. The small decline in arrest rates at 18 thus suggests an elasticity of crime with respect to expected sentence lengths no more negative than −0.048. In Section 6, we compare these magnitudes to the predictions from a dynamic economic model of criminal behavior.

The estimated discontinuity is robust to changes in specification, corroborating the smoothness assumptions required for our approach. Moving from left to right in Table 2, we control for an increasing number of covariates measured as of the time of the baseline arrest. Column (1) gives our most parsimonious model, controlling only for a juvenile/adult dummy, and a cubic polynomial in age at current arrest; column (8) gives our most complex model, adding controls for race, size of county in which the baseline arrest occurred, offense type of baseline arrest, and a quintic polynomial in age at baseline arrest. In each column, the added controls are good predictors of the probability of arrest, but in no case does including additional controls significantly affect the estimated discontinuity.

Table A3 explores the sensitivity of the estimates to functional form. It reports the estimated θ for different orders of the polynomial, ranging from a linear to a quintic polynomial in time and allowing for interactions of the polynomial with the juvenile/adult dummy. The models are also tested against an unrestricted specification, where the polynomial and the dummy are replaced with a full set of week-dummies. Overall, the linear and quadratic specifications are apparently too restrictive, and can be statistically rejected by a test against the unrestricted model. For richer specifications, including ones that include a linear term interacted with juvenile/adult status, the point estimates range from −0.065 to 0.029, with none of the estimates being statistically significant. A similar pattern is found when baseline covariates are included.

Our approach to estimation and inference follows Efron (1988) and thus is rooted in parametric models applied to local neighborhoods. Arguing in favor of this method is that our parametric testing approach allows us to verify that the local averages are not inconsistent with the parametric model fitted to them, as described above. However, an alternative approach to estimation and inference in general RD contexts is linear estimation applied to potentially smaller local neighborhoods (e.g., Calonico, Cattaneo and Titiunik 2014; Hahn, Todd and van der Klaauw 2001).

A challenge to comparing our results to the nonparametric inference results of Calonico, Cattaneo and Titiunik (2014) is that their results are limited to linear models – and hence do not apply directly to nonlinear models such as logit. Nonetheless, as is by now widely appreciated (c.f., Angrist and Pischke, 2009), regression and logit marginal effects rarely differ substantially. Here, the parameter θ can be accurately approximated as $(h - h_J)/h_J$. This suggests a natural way to connect our approach to that of Calonico et al. (2014): (1) estimate h_J using local linear regression and the local averages from Fig. 1 before the 18th birthday, (2) estimate the difference $h - h_J$ using local linear regression and all the local averages, and (3) scale the estimate of the difference $h - h_J$ by $1/h_J$ using the estimate from (1).[26] This approach can be thought of as an alternative approach to column (1) of Table 2, since those estimates rely only on the local averages plotted in Fig. 1.

In step (1) from above, we use a bandwidth of 135 days because that is what is implied by the bandwidth selector employed by `rdrobust` in step (2), below. Local linear estimation of the hazard with this bandwidth leads to a boundary estimate of $1/\widehat{h}_J = 1/0.00247 \approx 404$. In step (2), local linear estimation with quadratic models for bias adjustment leads to a point estimate (standard error) of −0.018 (0.065).[27] For comparison, our parametric point estimate (standard error) from column (1) is −0.018 (0.047).

In the approach of Calonico et al. (2014), we adjust the point estimate for the possibility of bias; doing so leads to an adjusted estimate of −0.038. Those authors present an estimated standard error that additionally incorporates the uncertainty due to the fact that the bias is estimated. This flavor of standard error is somewhat higher: 0.075, or about 17% larger than the local linear standard error ignoring bias estimation. Consequently, and in summary, while the Efron (1988) parametric approach suggests a point estimate (standard error) of −0.018 (0.047), a local linear framework suggests −0.018 (0.075).[28]

These comparisons clarify that changes to estimation and inference approaches generally corroborate the magnitudes from our more parametric analysis. Point estimates are quite similar, with second-decimal differences arising only in the context of bias adjustment. Even those differences are not large relative to the range of sampling uncertainty. Changes to standard errors are more meaningful and yet are also qualitatively similar to what was obtained from the more parametric approach.

Below, we set aside econometric considerations to consider some potential threats to the validity of our interpretation of these discontinuities as reflecting deterrence effects.

5.2. Transfers of Juveniles to the Adult Criminal Court

The first potential threat to the validity of our interpretation is the possibility of a lack of a discontinuity in the "treatment." That is, while all adults are handled by the criminal courts, and most minors are handled by the juvenile courts, all states allow a juvenile offender to be transferred to the criminal courts to be tried as an adult (Government Accounting Office, 1995).[29] In principle, prosecutors could be more likely to request that a juvenile case be transferred to the criminal justice system when the arrestee is almost 18. In the extreme case, all arrestees aged 17.8 or 17.9 could be transferred to the adult court, which would result in no discontinuous jump in the punitiveness of criminal sanctions and no "treatment."

Our data allow us to rule out this possibility empirically. The top panel of Fig. 2 plots the probability of being formally charged as an adult as a function of the age at the first post-17 arrest. Each open circle represents the individuals against whom a formal prosecution was filed, expressed as a fraction of those who were arrested in that particular week.[30] There is a striking upward discontinuity at the age of 18.[31] Those who are arrested just before their 18th birthday have about a 20% chance of being formally prosecuted as an adult, while those arrested just after their 18th birthday have a 60 to 70 percent chance. The latter probability is not 1, because not all arrestees will have formal charges filed against them.

The bottom panel of Fig. 2 provides further evidence of a discontinuity in the treatment, using a different measure of punishment. It plots the probability that the arrestee is eventually convicted and sentenced to either state prison or a county jail. Again, the figure shows a flat relationship between this measure of punitiveness and the age at arrest. There is a noticeable jump at age 18, from about 0.03 to 0.17.

We quantify these discontinuities in Table 3, which reports coefficient estimates from different OLS regressions. In columns (1) through (5) the dummy variable for whether the individual was prosecuted is regressed on the juvenile/adult status dummy, a cubic polynomial in age at arrest, as well as its interaction with the juvenile/adult dummy, and other covariates. Columns (6) through (8) further include the age at the baseline arrest as an additional control. Across specifications, the discontinuity estimate of

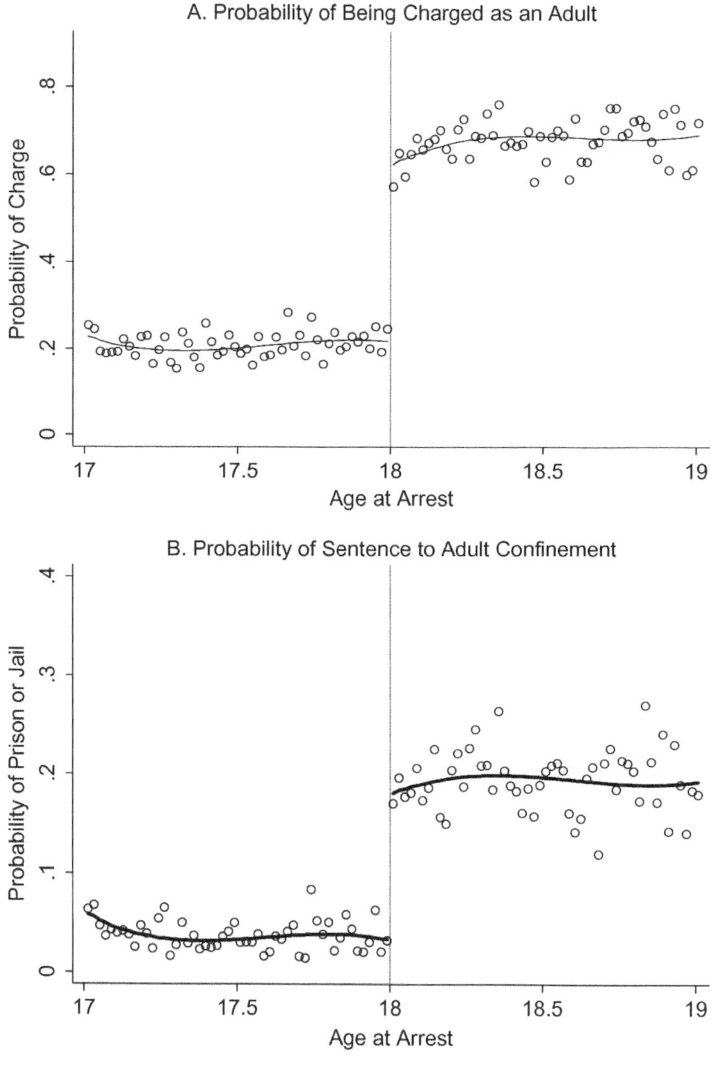

Fig. 2. Transfer and Punitiveness by Age. (a) Probability of Being Charged as an Adult and (b) Probability of Sentence to Adult Confinement. *Notes*: Panel (a) shows the probability of being charged as an adult, as a function of age at first arrest since 17. Sample is identical to that for Fig. 1. Open circles are nonparametric local averages, computed as the sample proportion charged as an adult for individual arrested in the given week. Solid line is based on a flexible polynomial model (see text for details). Panel (b) shows analogous figure for whether arrestee was sentenced to adult confinement.

Table 3. Discontinuity in Probability of Being Charged in Adult Court.

Variable	(1)	(2)	(3)	(4)	(5)	(6)	(7)	(8)
Estimated discontinuity	0.4042	0.4035	0.4041	0.4037	0.4043	0.4051	0.4052	0.4054
	(0.0235)	(0.0235)	(0.0235)	(0.0235)	(0.0235)	(0.0235)	(0.0235)	(0.0235)
Log discontinuity	*1.1056*	*1.1044*	*1.1054*	*1.1048*	*1.1058*	*1.1070*	*1.1073*	*1.1075*
Non-white		−0.0274	−0.0294	−0.0300	−0.0320	−0.0294	−0.0296	−0.0295
		(0.0057)	(0.0058)	(0.0058)	(0.0059)	(0.0059)	(0.0059)	(0.0059)
Size of county of baseline arrest (relative to large county)								
Small			0.0141		0.0140	0.0129	0.0128	0.0129
			(0.0072)		(0.0072)	(0.0071)	(0.0071)	(0.0072)
Medium			0.0199		0.0198	0.0195	0.0194	0.0195
			(0.0069)		(0.0069)	(0.0069)	(0.0069)	(0.0069)
Type of crime, baseline arrest (relative to non-index non-drug)								
Index crime, violent				0.0055	0.0052	0.0060	0.0060	0.0060
				(0.0095)	(0.0095)	(0.0095)	(0.0095)	(0.0095)
Index crime, property				−0.0031	−0.0028	−0.0015	−0.0017	−0.0018
				(0.0085)	(0.0085)	(0.0085)	(0.0085)	(0.0085)
Non-index drug				0.0228	0.0229	0.0131	0.0127	0.0127
				(0.0120)	(0.0120)	(0.0121)	(0.0121)	(0.0121)
Controls for age at baseline arrest?	N	N	N	N	N	Y	Y	Y
Order of polynomial in age at baseline arrest						1	3	5
R^2	0.2144	0.2152	0.2155	0.2154	0.2157	0.2173	0.2174	0.2174

Notes: Standard errors in parentheses. Log discontinuity is difference in log probabilities and is calculated from the presented difference estimate using a baseline rate of 0.2 for the marginal juvenile. Delta method standard errors for the log discontinuity are in each instance approximately 0.039.

about 0.40 is relatively stable. Table A4 is an analogous table for the probability of being convicted as an adult and sentenced to prison or jail.

Overall, the evidence strongly suggests that juvenile transfers to the criminal court are not prevalent enough to eliminate a sharp discontinuity in the punitiveness of criminal sanctions at 18.

5.3. Age-Based Law Enforcement Discretion

Another possibility is that offenses committed by juveniles and adults have different likelihoods of being recorded in our data. For example, it is possible that law enforcement may exercise discretion in formally arresting an individual, based on age. Suppose that the probability of arresting an individual, conditional on the same offense, is substantially higher for an 18.1-year old than a 17.9-year old. Then it is theoretically possible that the small effects we observe are a combination of a negative deterrence effect and a positive and offsetting jump in the arrest probability.

There are a number of reasons why we believe this is not occurring in our data. First, our analysis focuses on very serious crimes, where it seems unlikely that an officer would be willing to release a suspect without an arrest, purely on the basis of the individual's age. For example, all Index crimes involve a victim. We suspect that the pressure to capture a suspect is too great for officers to be willing to release an individual suspected of committing an Index crime. By contrast, for relatively less serious crimes such as misdemeanors or drug possession, it is more plausible that officers might exercise discretion in making the arrest.

Second, each individual in our main estimation sample already has a recorded formal arrest as of age 17, when we begin following their arrest experiences. Thus, it seems unlikely that the law enforcement agency will exercise leniency in recording an arrest: if a juvenile is apprehended just before his 18th birthday, it is too late to do anything to keep the youth's felony arrest record clean.

Third, our analysis focuses on arrests since 1994, the year of Florida's Juvenile Justice Reform Act (JJRA), which requires that felonies and some misdemeanors committed by juveniles be forwarded to the state for inclusion in the criminal history records maintained by the FDLE.[32] The impact of this law on the prevalence of juvenile records is shown in Fig. A2. This figure show the number of juvenile arrests as a proportion of all arrests in the FDLE arrest data, by month from 1989 to 2002. There is a marked discontinuity in the ratio at October 1994, the month the JJRA took effect.

This feature of Florida's criminal justice system was a major factor influencing us to use Florida as our study state.

Finally, if juvenile and adult arrests had different likelihoods of being recorded in our data, we would expect to observe significant heterogeneity in the estimated θ, by different groups of individuals, and different crime types, since it is likely that any off-setting measurement problems will vary by characteristics of the individual, as well as by crime type. Fig. 3 provides evidence contrary to this prediction. The top panel of the figure disaggregates the arrest probabilities from the top panel of Fig. 1 into two components: property and violent crime. The figure shows that the estimated discontinuity is essentially the same for the two categories of crime.

We also estimate θ separately by sub-groups defined by key correlates of arrest propensities, and find no evidence of significant negative effects for some groups being masked by positive effects of other sub-groups. Table 4 reports estimates from interacting the juvenile/adult dummy with race, size of county of the baseline arrest, and offense of the baseline arrest. The estimates for these different sub-groups range from −0.07 to 0.09. These estimates are generally of small magnitude; moreover, none of the 20 are statistically significant. Finally, we fail to reject the null hypothesis that the interaction effects are all zero in Table 2, and this holds for all specifications considered.

Although our analysis focuses on Index crimes, for completeness we show the results for all remaining offenses in the bottom panel of Fig. 3. We consider the potential for arrest discretion to be the most serious for these non-Index crimes, which include "victimless" offenses such as drug possession. Here, the cubic polynomial predictions do show a small perverse discontinuity, although the local averages do not reveal an obviously compelling jump at age 18. Still, if law enforcement discretion is of particular concern, a conservative approach would be to discount the results for non-Index crimes.

5.4. Expungement of Juvenile Records

A third possibility is that the ability of individuals to expunge and seal juvenile arrest records could generate downward biased estimates of arrest rates for juveniles, and hence mask any true deterrence effects. Florida law allows individuals who successfully complete a juvenile diversion program to apply to have all juvenile records expunged (Fla. Stat. 943.0582). Apart from this provision, Florida law also mandates that juvenile arrest histories be expunged when the individual turns 24.[33] Our choice of sample, however, circumvents these two expungement provisions in the following ways.

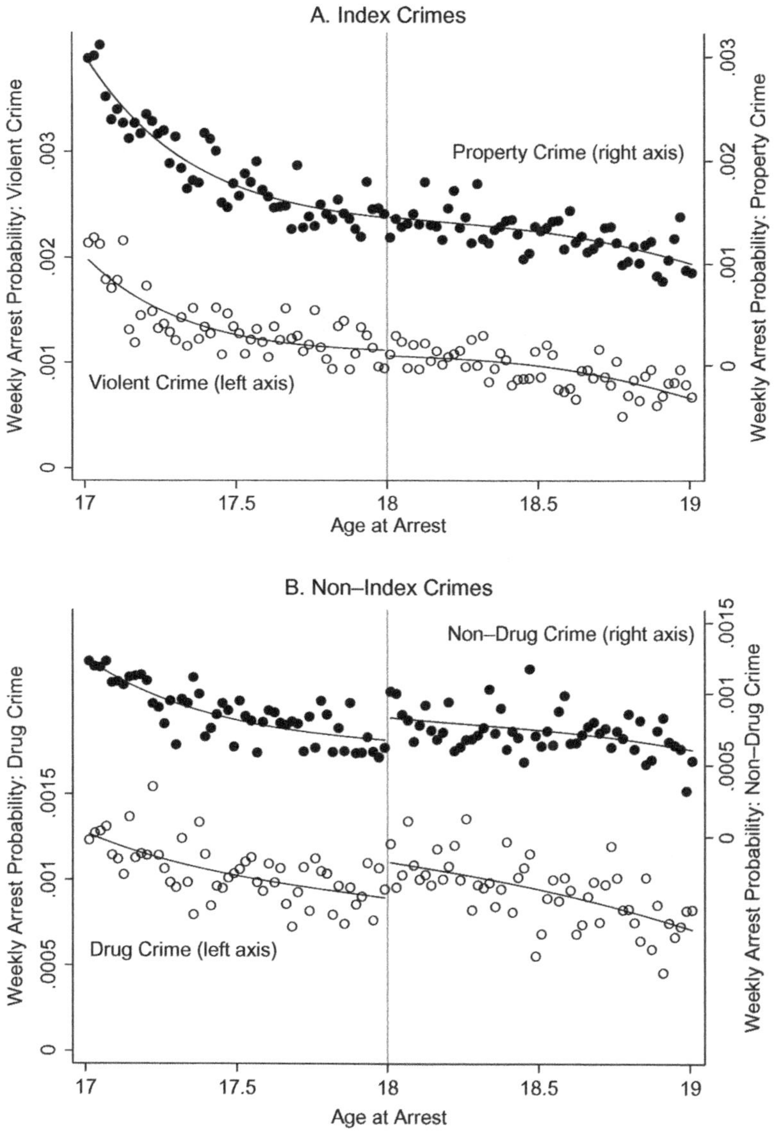

Fig. 3. Criminal Propensity by Type of Offense. (a) Index Crimes and (b) Non-Index Crimes. *Notes*: Figure is analogous to Fig. 1. Top panel shows re-arrest, disaggregated for murder, rape, robbery, and aggravated asault (violent crime) and burglary, larceny, and motor vehicle theft (property crime). Bottom panel shows re-arrest for a non-index felony offense.

Table 4. Heterogeneity in Discontinuity Estimates.

Subsample	(1)	(2)	(3)	(4)	(5)	(6)	(7)	(8)
All arrestees	−0.0180	−0.0171						
	(0.0474)	(0.0474)						
White arrestees			0.0240	0.0240				
			(0.0670)	(0.0670)				
Non-white arrestees			−0.0592	−0.0590				
			(0.0670)	(0.0670)				
Baseline crime was in:								
Large county					0.0292	0.0300		
					(0.0821)	(0.0821)		
Medium county					−0.0214	−0.0208		
					(0.0807)	(0.0807)		
Small county					−0.0622	−0.0619		
					(0.0834)	(0.0834)		
Baseline crime was:								
Index, violent							0.0553	0.0556
							(0.0930)	(0.0930)
Index, property							−0.0696	−0.0683
							(0.0646)	(0.0646)
Non-index, drug							0.0919	0.0952
							(0.1960)	(0.1959)
Non-index, non-drug							−0.0032	−0.0025
							(0.1250)	(0.1250)
Log-likelihood	−96,005.7	−95,174.8	−95,490.1	−95,173.8	−95,835.6	−95,170.6	−95,954.9	−95,165.0

Controls for race?	N	Y	Y	Y	N	Y	N	Y
Controls for county size?	N	Y	N	Y	Y	Y	N	Y
Controls for Baseline crime type?	N	Y	N	Y	N	Y	Y	Y
Controls for Age at baseline arrest?	N	Y	N	Y	N	Y	N	Y
Order of polynomial in age at baseline arrest		5		5		5		5
Test for equality of discontinuity estimates degrees of freedom			0.771 [1]	0.770 [1]	0.612 [2]	0.620 [2]	1.590 [3]	1.582 [3]
p-Value			0.380	0.380	0.736	0.733	0.662	0.663

Notes: Standard errors in parentheses. Table presents discontinuity estiamtes for different groups estimated from logit models. Odd-numbered columns include only those controls appropriate to testing the treatment interaction of interest. For example, column (3) includes controls for (i) white, (ii) a cubic polynomial in age, as in Table II, (iii) the interaction of the same cubic polynomial with the indicator for white, (iv) the interaction of an indicator for being above 18 with the indicator for white (estimate for whites shown), and (v) the interaction of the indicators for being above 18 and non-white (estimate for non-whites shown). Even-numbered columns additionally include the richest set of controls considered in Table II. The final row of the table tests for the equality of the interacted treatment effects.

First, our estimation sample is restricted to those committing baseline crimes before age 17 but subsequent to January 1, 1995. Therefore, the individuals are not yet 24 by the end of our sample frame, and thus will not be subject to the time-activated expungement. Second, a requirement for inclusion in our sample is an observed arrest record prior to age 17. Thus, by construction, the individuals in our main estimation sample did not have their complete juvenile arrest history expunged.

To illustrate the importance of this sample choice to avoiding problems related to expungement, we present the time profile of arrests for the individuals who are excluded from our main analysis: those who were not observed as arrested prior to turning 17. Some of these individuals' first real arrest will occur before 18, and for some it will occur after 18. If there is an opportunity for the former group to later expunge their juvenile records, then a positive discontinuity in the number of arrests should occur at age 18.

This is the pattern found in Fig. 4. This figure is a stacked histogram, where the combined total represents the total number of people who are

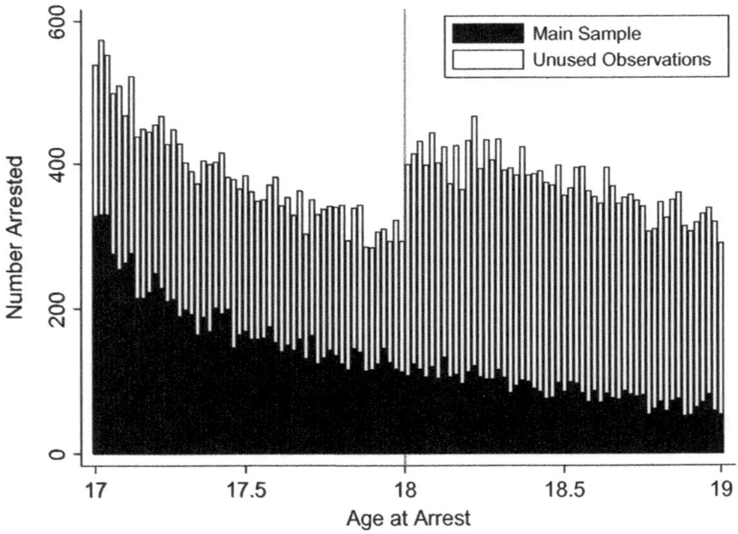

Fig. 4. Number of Arrests by Age: Main Sample versus Unused Observations. *Notes*: Figure is a stacked histogram (shaded and light rectangles sum to total). Heights refer to the number of individuals arrested. "Main sample" refers to the sample underlying Fig. 2 and pertains to those arrested at least once for any offense prior to 17 between 1995 and 1998. "Unused observations" pertains to all other felony arrests.

arrested for the first time since turning 17. The histogram is comprised of two populations, those arrests corresponding to our estimation sample (the dark bars) and the remaining, unused observations (the light bars). For the total, there is a striking positive discontinuity at age 18. But this discontinuity is entirely concentrated in the unused sample (the upper part of the stacked graph).[34]

5.5. Evidence on Incapacitation

Up to this point, we have examined the evidence for a deterrence effect of adult criminal sanctions, relative to juvenile criminal sanctions. We now ask: What is the incapacitation effect of treating an apprehended offender as an adult instead of a juvenile? To answer this, we use a standard RD design, where the "treatment" of adult status is a discontinuous function of the running variable of age at "first arrest." As described in Lee (2008), if there is imprecise sorting around the age of majority, then being treated as an adult has statistical properties analogous to a randomized experiment.[35] Indeed, in the analysis above we are unable to detect strong evidence of such sorting behavior. In essence, our results on deterrence imply that this design passes the test of manipulation of the running variable suggested in McCrary (2008), enabling credible assessment of incapacitation effects despite analyzing a selected sample of arrests as opposed to all individuals, arrested or not.

The RD design for detecting incapacitation effects is illustrated in Fig. 5. The top panel of the figure plots the probability that the "second arrest" occurs within a specific window of time since the "first arrest," as a function of the age at "first arrest." Specifically, the leftmost open circle indicates that among those whose "first arrest" occurs the week after their 17th birthday, the probability of re-arrest within the subsequent 30 days is about 15 percent. Among those whose "first arrest" occurs just before their 18th birthday, the probability of re-arrest within the subsequent 30 days is about 20 percent.

There is a sharp discontinuity in the probability of re-arrest at age 18, with 18.02-year olds having a probability of re-arrest within 30 days of about 10 percent. A natural explanation for this difference is that being handled by the adult criminal court leads to a longer period of custody than does being processed as a juvenile. The 17.98-year old is released earlier and hence has a greater opportunity to re-offend within any short time window, compared to the 18.02-year old.

The solid circles and open triangles plot the same kind of graph, except that we examine the probability of a re-arrest occurring within 120

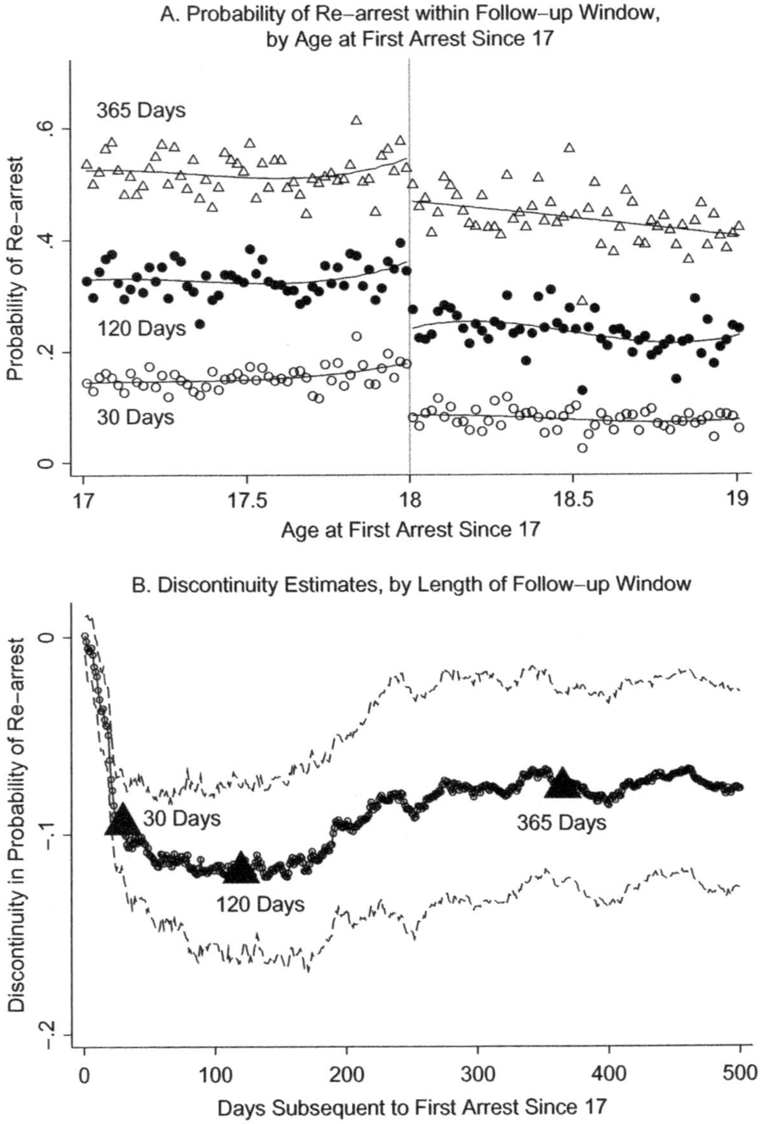

Fig. 5. Incapacitation Effects of Adult Sanctions. (a) Probability of Re-arrest within Follow-Up Window by Age at First Arrest Since 17 and (b) Discontinuity Estimates, by Length of Follow-Up Window. *Notes*: Panel (a) presents probability of re-arrest within 30, 120, and 365 days of first arrest since 17, by age in weeks. Local averages by week, computed as sample proportions, are accompanied by a smoothed estimate based on a flexible polynomial model (see text for details). Panel (b) plots estimated discontinuity in probability of re-arrest, for follow-up lengths ranging from 1 to 500 days, with twice pointwise standard error bands.

and 365 days after the "first arrest." The probabilities are higher for 120 and 365 days, since the probability of re-arrest increases with the window width.

The length of the follow-up period is arbitrary. The bottom panel of Fig. 5 plots the profile of discontinuity estimates using follow-up lengths ranging from 1 to 500 days. For example, the estimates at 30, 120, and 365 days in the bottom panel, emphasized with large solid triangles, correspond to the discontinuity estimates from the top panel of the figure. Overall, the bottom panel shows that already by 20 days, there is a large divergence in the cumulative number of arrests between those who are arrested as a 17.98-year old and those arrested at age 18.02, for example.[36] This divergence continues to grow, slowing down at around 100 days after the initial arrest.[37]

6. PREDICTED EFFECTS FROM A DYNAMIC MODEL OF CRIME

In this section, we interpret the magnitudes of our estimated deterrence effects through the lens of an economic model of criminal behavior. We develop a dynamic extension of Becker's (1968) model of crime. We first consider how large of a discontinuity we should expect given our model, which can be calibrated to readily available sample means, and standard assumptions. We then use our model to draw a precise link between our discontinuity estimates, which identify a particular set of policy responses as discussed above, and other policy elasticities of interest that measure deterrence effects in a broader way. In doing so, we employ both standard parametric assumptions as well as consider a nonparametric approach where we impose a monotonicity restriction on the distribution of criminal benefits. We show how that nonparametric restriction allows us to provide informative bounds on the broader policy elasticities that are not directly identified by our study. This nonparametric analysis delivers the most negative deterrence elasticities that are consistent with both our model and our reduced-form estimates.

6.1. The Model

The essence of Becker's model of crime is that an individual weighs the expected benefits (e.g., monetary or otherwise) against the expected costs (e.g., fines, disutility from being incarcerated). This basic notion can be captured in a discrete-time dynamic model, in which the individual faces a criminal opportunity every period and chooses between committing the offense and abstaining. If he commits the offense, there is a chance that he will be apprehended and incarcerated for a random number of periods before being released.

The model described below closely resembles the canonical job search model McCall (1970). The elements of the model are as follows:

- In each period, the individual is presented with a criminal opportunity. The value of this opportunity, or benefit, is given by the non-negative continuous random variable B with finite mean, cumulative distribution function $F(\cdot)$, and density $f(\cdot)$. [38]
- After observing the value of the current period's criminal opportunity, the individual chooses to offend or abstain. In a given period t, if he abstains, he receives flow utility $u_t = a$. If he offends and is not caught, he obtains the flow utility $u_t = a + B$. If he is caught, however, he will be incarcerated beginning immediately and perhaps for some time. Apprehension occurs with probability p. While incarcerated, he receives flow utility $a - c$, where c is a positive per-period utility cost of being incarcerated.[39]
- The duration of incarceration is given by the non-negative discrete random variable S, which has probability function $\{\pi_s\}_{s=1}^{\infty}$ and finite mean $\mathbb{E}[S]$.
- The individual knows the structure of the problem as outlined, the probabilities p and $\{\pi_s\}_{s=1}^{\infty}$, the distribution of criminal opportunities $F(\cdot)$, and the parameters a and c. He chooses to offend or abstain in each period he is free, seeking to maximize the expected present discounted value of utility, $E_t\left[\sum_{\tau=t}^{\infty} \delta^{\tau-t} u_\tau\right]$. Here, E_t is the expectation operator conditional on information available as of period t, δ is the discount factor, and u_t is either $a - c$, a, or $a + B$, depending on the agent's choices, whether he has been apprehended for any crimes committed, and whether he is currently detained.

The Bellman equation for this problem is

$$V(b) = \max \left\{ \begin{array}{l} a + \delta \mathbb{E}[V(B)], p \sum_{s=1}^{\infty} \pi_s \left[(a-c) \dfrac{1-\delta^s}{1-\delta} + \delta^s \mathbb{E}[V(B)] \right] \\ +(1-p)[a + b + \delta \mathbb{E}[V(B)]] \end{array} \right\}. \quad (2)$$

The first argument of the max function is the payoff from abstaining: receipt of the flow utility a and the expected continuation payoff. The second argument of the max function is the expected payoff from committing the offense. If caught and incarcerated for s periods, the payoff is $(a-c)(1+\delta+\delta^2+\cdots+\delta^{s-1})+\delta^s\mathbb{E}[V(B)]$. If not caught, the individual's payoff is the best possible: $a+b+\delta\mathbb{E}[V(B)]$.

The individual's optimal strategy is characterized by a "reservation benefit," b^*, such that when $B>b^*$, he commits the crime, and when $B<b^*$ he abstains.[40] Of course, when $B=b^*$, the individual is indifferent. Consequently, we may calculate the reservation benefit by equating the two arguments of the max function above. Doing so shows that

$$b^* = c\frac{p}{1-p}\left[1+\sum_{s=1}^{\infty}\pi_s\frac{\delta-\delta^s}{1-\delta}\left(1+\frac{(1-\delta)\mathbb{E}[V(B)]-a}{c}\right)\right]. \tag{3}$$

We make the following observations about this expression:

- When $\delta=0$ and the individual is entirely present-focused, the expression reduces to $b^*=c\frac{p}{1-p}$. In that special case, crime will be committed whenever the ratio of the net benefit to net cost, B/c, exceeds the odds of apprehension, $\frac{p}{1-p}$. This is precisely the notion put forth in a standard (and static) Becker model of crime. It is intuitive that when the individual completely discounts the future, the length of incarceration, S, is irrelevant.
- Similarly, $b^*=c\frac{p}{1-p}$ if the punishment is only 1 period long (i.e., $\pi_s=0$ for $s=2,3,\ldots$).
- The second term inside the square braces thus captures the increased expected cost to offending due to a period of incarceration longer than 1 period. During incarceration, the offender bears the per-period cost c, but also is not able to exercise the option of committing offenses during this period. This lost value is summarized by the term $((1-\delta)\mathbb{E}[V(B)]-a)/c$.
- The Theory Appendix shows that the annuitized value of the dynamic program can be expressed as

$$(1-\delta)\mathbb{E}[V(B)] = a+(1-F(b^*))(1-p)\mathbb{E}[B-b^*|B>b^*]. \tag{4}$$

It is intuitive that the annuitized option value of the dynamic program, the second term in Eq. (4), is the chances of encountering a worthwhile crime times the chances of avoiding apprehension times the expected benefit of those worthwhile crimes.

We obtain the following intuitive comparative statics (proof in Theory Appendix). The crime rate, which is given by $1 - F(b^*)$, decreases with:

- Higher discount factor, δ. As the individual places more weight on future utility, the penalty of incarceration poses a higher cost.
- Higher per-period direct utility cost to incarceration, c.
- Higher apprehension rates, p.
- Longer sentence lengths. Any rightward shift in the distribution of the incarceration length S (i.e., the new distribution first-order stochastically dominates the old) decreases offending.

6.2. Benchmark Model Calibration and Predicted Values for θ: A Parametric Analysis

In this section, we calibrate the above model to produce predictions on the size of the reduced-form estimates of θ from our logit estimation. This exercise answers the question: "In this economic model of crime, how large of a θ would we have expected to observe?" In the calibration of our model, we let $a = 0$ and $c = 1$ and we take each period to be a single day.[41] Because each period is a day, the model's implied arrest probabilities for the boundary adult and juvenile, $p(1 - F(b^*))$ and $p(1 - F(b_J^*))$, respectively, where b_J^* is the juvenile analogue of b^*, correspond to our empirical estimate of θ as

$$\begin{aligned}\theta &= \ln\left(\frac{\mathbb{P}_{18.02}(\text{Arrest})}{1-\mathbb{P}_{18.02}(\text{Arrest})}\right) - \ln\left(\frac{\mathbb{P}_{17.98}(\text{Arrest})}{1-\mathbb{P}_{17.98}(\text{Arrest})}\right) \\ &= \ln\left(\frac{1-(1-p(1-F(b^*)))^7}{(1-p(1-F(b^*)))^7}\right) - \ln\left(\frac{1-(1-p(1-F(b_J^*)))^7}{(1-p(1-F(b_J^*)))^7}\right),\end{aligned} \tag{5}$$

where $\mathbb{P}_{17.98}(\text{Arrest})$ and $\mathbb{P}_{18.02}(\text{Arrest})$ are the predictions from our reduced-form empirical models for the probabilities of arrest one week prior to and after the 18th birthday, respectively. That is, the left-hand side of Eq. (5) can be thought of as "measurement" and the right-hand side can be theory of as "theory."

To tighten this connection, we combine Eqs. (3) and (4) to obtain an implicit equation for b^*, and similarly obtain an implicit equation for b_J^*:

$$b^* = \frac{p}{1-p}\left[1 + \sum_{s=1}^{\infty} \pi_s \frac{\delta - \delta^s}{1-\delta}\left\{1 + \left(1 - F(b^*)\right)(1-p)\mathbb{E}[B - b^* | B >, b^*]\right\}\right], \quad (6)$$

$$b_J^* = \frac{p}{1-p}\left[1 + \sum_{s=1}^{\infty} \pi_s^J \frac{\delta - \delta^s}{1-\delta}\left\{1 + \left(1 - F(b^*)\right)(1-p)\mathbb{E}[B - b^* | B > b^*]\right\}\right]. \quad (7)$$

The sole difference in these two expressions is the distribution of sentence lengths. Intuitively, this is because the term $(1 - F(b^*))(1 - p)$ $\mathbb{E}[B - b^* | B > b^*]$ is related to the expected continuation payoff and so pertains to adulthood for both the boundary juvenile and the boundary adult.

In our benchmark calibration, we (1) let the adult and juvenile sentence lengths S and S_J be distributed as a discretized exponential (geometric), with $\mathbb{E}[S] = 207$ (days) and $\mathbb{E}[S_J] = 63$, (2) let the probability of apprehension be $p = 0.08$ and the daily discount factor be $\delta = 0.95^{1/365}$, and (3) let the criminal opportunity, B, be distributed exponentially with mean parameter $\lambda = 4.56$. With these magnitudes and parameterizations, Eqs. (6) and (7) are used to solve for b^* and b_J^*, respectively, which yields the daily arrest hazards $p(1 - F(b^*))$ and $p\left(1 - F\left(b_J^*\right)\right)$, respectively. These are then used to produce a predicted θ using Eq. (5).[42]

The magnitudes we use for these parameters have some empirical backing, as we now describe:

- Our model suggests that we can use the evidence from Fig. 5 to estimate the gap $\mathbb{E}[S] - \mathbb{E}[S_J]$. Specifically, we compute the time between "first arrest" and "second arrest" for 17.98- and 18.02-year olds. According to our model, that difference is an estimate of $\mathbb{E}[S] - \mathbb{E}[S_J]$.[43] Using the same data and similar approach as the RD analysis in Fig. 5, we estimate a Weibull duration model, where the duration is a function of a cubic in age and a dummy variable for adulthood.[44] This yields estimated durations of approximately 962 and 818 for adults and juveniles, respectively, and a gap of $\mathbb{E}[S] - \mathbb{E}[S_J] = 144$ days.
- We obtain $\mathbb{E}[S_J] = 9.05 \times 7 \approx 63$ from Table A1, which reports average incarceration lengths from a completely different data source (see Data Appendix for discussion). This approach suggests $\mathbb{E}[S] = 63 + 144 = 207$.

We note that this different data source yields $\mathbb{E}[S] = 46.48 \times 7 \approx 325$ days, which implies a difference $\mathbb{E}[S] - \mathbb{E}[S_J]$ of 262. We opt for a conservative calibration by choosing the smaller gap of 144 days. The predicted θ would be yet larger if we were instead to increase this gap above 144 days.

- Table A2 provides estimates of clearance rates and reporting rates for Index crimes in 2002. These numbers suggest $p = 0.08$.[45]
- Finally, we choose $\lambda = 4.56$ so that the implied b^* (cf., Eq. (6)) is consistent with our estimate of the adult arrest hazard of $p(1 - F(b^*)) = 0.0013$, which is computed as follows. Using the same Weibull estimates from above, we have $1/(p(1 - F(b^*))) = 962 - \mathbb{E}[S] = 962 - 207 = 755$. The implied daily arrest hazard is $p(1 - F(b^*)) = 1/755$, or about 0.0013.

The results of our calculations for the benchmark calibration are presented in the first row of Table 5, which shows that this set of parameters would predict a discontinuity estimate of −2.76. As noted above, our standard errors are small enough to statistically rule out values of θ more negative than −0.11. Thus, in the context of the model outlined, a benchmark calibration suggests that the data are very much at odds with long time horizons. The rest of Table 5 reports our exploration of the sensitivity of the predicted θ to different parameter values for the model. We vary each parameter one at a time, while maintaining the values for the other parameters in the benchmark calibration in the top row.

Inspection of Table 5 reveals that while the predicted θ increases in magnitude with increasing p, from −2 to −3.74 as p ranges from 0.025 to 0.400, none of the predicted values for θ are close to our point estimate of −0.018. Indeed, none are close to the outer edge of the two-sided confidence region for θ, −0.11. Note that the extreme values of p that we consider, 0.025 and 0.400, are at odds with the evidence summarized in Table A2. The conclusion that our estimates are much smaller than the benchmark prediction seems robust to assumptions regarding p.

Table 5 also presents estimates based on shifts to the distribution of sentence lengths. As noted, our benchmark calibration assumes sentences are distributed as a discretized exponential (geometric). The exponential distribution is a special case of the Weibull distribution, with the shape parameter of the Weibull equal to 1. To explore the sensitivity of the predicted θ to the assumed shape of the distribution, we vary the shape parameter for the juvenile and adult sentence length distribution over 0.25, 0.40, 2.0, and 4.0. For each such shape parameter, k_J and k_A, we adjust the Weibull scale

Table 5. Predicted Reduced Form Discontinuities: Parametric Calibration.

Panel A. Baseline Parameterization

Probability of Apprehension, p	Annual Discount Factor; $\delta^{1/365}$	Expected Sentences, in Weeks		Benefit Distribution Scale Parameter, λ	Predicted Reduced Form Discontinuity, θ
		Juveniles	Adults		
0.08	0.95	9.03	29.57	4.56	−2.76

Panel B. Effects of Changes in Individual Parameters, Relative to Baseline

Probability of Apprehension		Juvenile Sentences Parameter		Adult Sentences Parameter	
p	θ	k_J	θ	k_A	θ
0.025	−2.00	0.250	−2.95	0.250	−2.22
0.050	−2.46	0.400	−2.78	0.400	−2.70
0.200	−3.33	2.000	−2.76	2.000	−2.77
0.400	−3.74	4.000	−2.76	4.000	−2.78
Annual Discount Factor		Benefit Distribution Parameter			
$\delta^{1/365}$	θ	k	θ		
0.500	−2.50	0.250	−1.03		
0.250	−2.26	0.400	−1.79		
0.100	−2.01	2.000	−3.55		
0.010	−1.57	4.000	−3.84		

Notes: In Panel A, all Weibull shape parameters k, k_J, and k_A are equal to 1. In panel B, Weibull shape and other parameters vary, as specified. Throughout, when k_J varies, l_J is adjusted to hold $E[S_J]$ constant, and when k_A varies, l_A is adjusted to hold $E[S]$ constant. Each entry of the table matches the adult hazard by adjusting the benefit distribution scale parameter, λ. Details in text.

parameters to match the means $\mathbb{E}[S_J] = 63$ and $\mathbb{E}[S] = 207$. To gain intuition about these changes, note that when the shape parameter is 0.25, the density is more convex and skewed, while when the shape parameter is 4, the shape of the distribution is more akin to that of the chi-square distribution. The table shows that the shape of the sentence length distribution affects our predictions only negligibly. For example, with the shape parameter for the juvenile distribution, k_J , equal to 0.25, the predicted θ is -2.95 and with a shape parameter of 4.0, the predicted θ is -2.76. Results for the impact of the adult shape parameter, k_A, are similar.[46]

The final row of Table 5 shows the effect on the predicted θ of changes to the discount factor. When the discount factor declines from 0.95 to 0.01 on an annual basis, the magnitude of the predicted θ falls from -2.50 to -1.57. This is a notable pattern, because – unlike the case of p, $\mathbb{E}[S]$, and $\mathbb{E}[S_J]$ – we have no independent information on the discount factor. The relevant discount factor is not that of a typical individual, but that of a marginal offender – namely, those individuals for whom the economic and social environment is such that offending is contemplated but not assured – and thus could be much smaller than 0.95. We note that in some settings, such as the market for subprime loans, the literature has documented behaviors that are consistent with extremely small discount factors (e.g., Adams, Einav, & Levin, 2009).[47] Perhaps most importantly, discount factors could be much smaller when arrestees are drug-users, a common pattern. For example, urinanalysis evidence indicates that a large fraction of arrestees test positive for drug use. In 2000, 61.8% of Fort Lauderdale and 62.8 percent of Miami arrestees tested positive for one or more of the following: cocaine, marijuana, opiates, methamphetamine, or PCP (National Institute of Justice, 2003). These considerations suggest the prevalence of short time horizons and low discount factors among the relevant subpopulation. In our model, as is intuitive, θ is small when the discount factor is. Present-oriented individuals may respond little to threatened future punishments.

Finally, we consider the impact of changes in the shape of the criminal benefit distribution, the other element of the model for which we have no independent information. In the benchmark calibration, this distribution was assumed to be exponential. Generalizing to a Weibull distribution, we vary the shape parameter, k, over 0.25, 0.40, 2.0, and 4.0. When $k = 4.0$, many individuals are on the margin of the crime participation decision, but when $k = 0.25$, few individuals are on the margin. We see that as the shape parameter falls from 4.0 to 0.25, the predicted θ falls in magnitude from -3.84 to -1.03.

We conclude from these analyses that a reasonable benchmark parametric model with standard "patient" discount factors predicts much larger discontinuities in offending at age 18 than we observe in our data. However, given that the predictions seem somewhat sensitive to the discount factor and the shape of the benefit distribution, when we consider the policy implications of our estimates below, we relax the parametric assumptions on the benefit distribution, and present a bounding analysis.

6.3. Connecting θ to Broader Policy Elasticities

In the introduction, we outlined a set of specific policies that are addressed by our reduced-form estimates above (e.g., changing the age of criminal majority). However, much of the literature focuses on broader policy parameters, such as elasticities of crime with respect to policing and expected sentence lengths. Intuitively, our reduced-form estimates are not *un*related to these policy parameters, but the nature of the connection is vague, imprecise, and qualitative. The feature of a behavioral model is that it makes these connections specific, precise, and even quantitative. In this subsection, we connect our empirical work to these broader policy parameters, with a particular focus on minimizing the assumptions necessary to make that connection.

In elasticity terms, the responsiveness of the crime rate to sentence length and police increases under our model can be shown to be

$$\eta_{\mathbb{E}[S]} \equiv \frac{\partial(1-F(b^*))}{\partial\mathbb{E}[S]}\frac{\mathbb{E}[S]}{1-F(b^*)} = -\frac{pf(b^*)}{h}\mathbb{E}[S]\frac{\partial b^*}{\partial\mathbb{E}[S]} = -\left[\left(\frac{h_J-h}{h}\right)\Big/\left(\frac{\nu-\nu_J}{\nu}\right)\right]\eta_\nu\left(\frac{1}{1+\nu h}\right)\frac{f(b^*)}{f(\overline{b})}, \tag{8}$$

$$\eta_p \equiv \frac{\partial(1-F(b^*))}{\partial p}\frac{p}{1-F(b^*)} = -\frac{pf(b^*)}{h}p\frac{\partial b^*}{\partial p} = -\left[\left(\frac{h_J-h}{h}\right)\Big/\left(\frac{\nu-\nu_J}{\nu}\right)\right]\kappa\left(\frac{1}{1+\nu h}\right)\frac{f(b^*)}{f(\overline{b})}, \tag{9}$$

respectively, where $h = p(1-F(b^*))$, $h_J = p\left(1-F\left(b_J^*\right)\right)$, $\nu = \sum_{s=1}^{\infty}\pi_s\frac{\delta-\delta^s}{1-\delta}$, $\nu_J = \sum_{s=1}^{\infty}\pi_s^J\frac{\delta-\delta^s}{1-\delta}$, $\eta_\nu \equiv \frac{\partial\nu}{\partial\mathbb{E}[S]}\frac{\mathbb{E}[S]}{\nu}$, $\kappa \equiv \left(c\frac{p}{(1-p)^2}(1+\nu)+\nu hE[B-b^*|B>b^*]\right)\Big/\left(c\frac{p}{1-p}\nu+\nu hE[B-b^*|B>b^*]\right)$, and $\overline{b}\in\left(b_J^*,b^*\right)$ is such that $f\left(\overline{b}\right) = \frac{F(b^*)-F\left(b_J^*\right)}{b^*-b_J^*}$ (see Theory Appendix for detailed derivation).

Since $\theta \approx -(h_J-h)/h$, Eqs. (8) and (9) show that our reduced-form discontinuity estimate is proportional to the broader policy elasticities $\eta_{\mathbb{E}[S]}$ and η_p. This is then the result that confirms the intuition that our reduced-form estimates are relevant to broader policy parameters. The additional factors in Eqs. (8) and (9) sharpen this connection and raise the possibility of quantifying the distinction between our reduced-form estimates and these broader policy parameters. We now provide discussion regarding these additional factors.

The second factor of Eqs. (8) and (9), $(\nu - \nu_J)/\nu$, gives the percent increase in the expected discounted number of periods of incarceration when crossing the age 18 threshold. This term reflects the influence of δ as well as the change of punishment regime from S_J to S. For patient individuals, $(\nu - \nu_J)/\nu$ is approximately equal to $(\mathbb{E}[S] - \mathbb{E}[S_J])/\mathbb{E}[S]$, but for present-oriented individuals it may be much smaller.[48]

The third factors are unique to each elasticity. In the case of $\eta_{\mathbb{E}[S]}$, the third factor η_ν reflects the percentage change in ν brought on by a percentage change in the expected number of periods of incarceration. In the case of η_p, the third factor κ is a quantity that reflects the percentage change in p that has the equivalent impact as a percentage change in ν.

The fourth factor is an offsetting adjustment that accounts for the following effect: when the expected costs of offending rise, the option value of being free is reduced, which in turn dampens the per-period cost of incarceration.[49] Finally, the fifth term reflects the shape of the benefit distribution: $f(b^*)/f(\overline{b}) = 1$ if the density of B is flat between b_J^* and b^*, but more generally this ratio may differ from 1.

We can calibrate our benchmark model to ask what policy elasticities our estimates imply. For example, we can adjust the discount factor so that the predicted θ matches -0.11, the outer edge of our confidence interval. This yields a daily discount factor of 0.71, which predicts elasticities of $\eta_{\mathbb{E}[S]} = -0.049$ and $\eta_p = -4.44$. We could alternatively have kept δ fixed and attempted to match the predicted θ to the observed θ by adjusting the shape parameter k of the benefit distribution. Since it is arbitrary which parameter to adjust, in the next subsection, we consider a wide range of discount factors, while relaxing the parametric restriction on the benefit distribution.

6.4. Nonparametric Bounds on Policy Elasticities

We now explore the range of policy elasticities that are consistent with our discontinuity estimates, relaxing the parametric restriction on the shape of the criminal benefit distribution.

We first note that it is always possible to *construct* a distribution for B that is both consistent with any negative value of θ, yet yields policy elasticities of zero or $-\infty$. One needs to simply set the density $f(b^*)$ to zero or arbitrarily large. This fact is not unique to our dynamic model: this will be the case with any model that has a distribution of criminal benefits and a threshold above which crime occurs.

It is thus more informative and plausible to consider some mild restrictions on the benefit distribution. We focus on the assumption that the density of benefits is weakly declining, or that more valuable opportunities are more rare. This assumption has some economic and conceptual justification, going back at least to Stigler (1970) and Viscusi (1986). Intuitively, as these authors have argued, opportunities that are valuable to the offender are typically costly for the victim. This provides incentives for victims to take greater precaution regarding more valuable criminal opportunities (e.g., deadbolt locks, car alarms, and mace). It is thus plausible that more valuable opportunities are scarce, relative to less valuable opportunities.

In the Theory Appendix, we show that the nonparametric assumption of a declining density delivers an upper bound on the magnitude of the policy elasticities $\eta_{\mathbb{E}[S]}$ and η_p. In particular, we show that

$$\eta_{\mathbb{E}[S]} > \overline{\eta}_{\mathbb{E}[S]} \equiv -\left[\left(\frac{h_J - h}{h}\right) \Big/ \left(\frac{\nu - \nu_J}{\nu}\right)\right] \eta_\nu \left(\frac{1}{1 + \nu h}\right), \tag{10}$$

$$\eta_p > \overline{\eta}_p \equiv -\left[\left(\frac{h_J - h}{h}\right) \Big/ \left(\frac{\nu - \nu_J}{\nu}\right)\right] \overline{\kappa} \left(\frac{1}{1 + \nu h}\right), \tag{11}$$

where $\overline{\kappa} = \left(c\frac{p}{(1-p)^2}(1+\nu) + \nu h\overline{E}\right) \Big/ \left(c\frac{p}{1-p}\nu + \nu h\overline{E}\right)$ and $\overline{E}$ is given in the Theory Appendix.[50] Both $\overline{\eta}_{\mathbb{E}[S]}$ and $\overline{\eta}_p$ can be computed using information on discount factors, the distribution of sentence lengths, and our empirical findings on arrest rates for youthful offenders at age 17.98 and 18.02. Other than assuming a declining density, no further information regarding the distribution of criminal benefits is required to obtain these bounds; this is the sense in which this might be described as a nonparametric restriction.

Table 6 reports our estimates of the elasticity bounds $\overline{\eta}_{\mathbb{E}[S]}$ and $\overline{\eta}_p$ for our point estimate of $\theta = -0.018$ and for the edge of our (two-sided) 95 percent confidence interval, $\theta = -0.11$, using various values of the discount factor.

Table 6. Bounds on Policy Elasticities.

Annual	$\theta = -0.11$		$\theta = -0.018$	
Discount Factor	$\bar{\eta}_p$	$\bar{\eta}_{E[S]}$	$\bar{\eta}_p$	$\bar{\eta}_{E[S]}$
10^{-8}	−0.760	−0.060	−0.117	−0.009
0.0010	−0.379	−0.071	−0.057	−0.011
0.0025	−0.349	−0.073	−0.052	−0.011
0.0050	−0.326	−0.075	−0.048	−0.012
0.0100	−0.302	−0.077	−0.044	−0.012
0.0250	−0.271	−0.082	–	–
0.0500	−0.248	−0.086	–	–
0.1000	−0.223	−0.091	–	–
0.2500	−0.190	−0.101	–	–
0.5000	−0.163	−0.113	–	–
0.9500	−0.136	−0.130	–	–

Notes: Table presents lower bounds on the elasticity of crime with respect to police, $\bar{\eta}_p$, and with respect to expected sentence lengths, $\bar{\eta}_{E[S]}$, for various values of the annual discount factor, δ, under two assumptions regarding the reduced form parameter ($\theta = -0.11$ and $\theta = -0.018$). Not all values of δ are consistent with the assumption of a declining density function for the distribution of criminal benefits; for values of δ inconsistent with this assumption, elasticity columns contain a dash.

Using the most negative value of $\theta = -0.11$, the sentencing elasticity ranges from −0.130 to −0.060, and the policing elasticity ranges from very large negative values to −0.136. If we use our point estimate, the sentencing elasticity $\bar{\eta}_{\mathbb{E}[S]}$ varies only slightly, from −0.013 to −0.010, while the probability of apprehension elasticity η_p varies from −0.117 to −0.044.

Interestingly, for the case of $\theta = -0.018$, there are some values of δ that cannot be rationalized within the framework of the model. That is, as we show in the Theory Appendix, we find that if $\theta = -0.018$, then there exists no $\mathbb{E}[B - b^*|B > b^*]$ that is consistent with both the model and the restriction of a declining density.

Overall, the table points to two policy conclusions. First, if one believes that the probability of apprehension elasticity is large, then this implies very low discount factors, and consequently very low sentence length elasticities. Second, across the entire range of permissible discount factors, all of the sentence length elasticities are generally small in magnitude.

7. CONCLUSION

Over the past 40 years, the incarceration rate in the United States has soared, from 93 persons in prisons per 100,000 in 1972 to 480 persons per 100,000 in 2012. Indeed, despite the declining crime rates of recent years, prison and jail populations continue to climb, as the full impact of sentencing reforms from previous years becomes felt (Raphael & Stoll 2009).[51]

For economists, it is natural to wonder if more severe prison sentences deter crime by a sufficient amount to make paying for them worthwhile. Furthermore, can a lower level of crime be attained by an expenditure-neutral reallocation of funds from prisons to alternative uses, such as social programs (Donohue & Siegelman, 1998) or policing (Chalfin & McCrary, 2017; McCrary, 2010)? These questions have recently acquired a renewed policy relevance, as state and local governments scramble to lower costs in the wake of the financial crisis.

A key input to these considerations is the magnitude of the deterrence effect of prison sentences. In this chapter, we use a quasi-experimental approach to identify the deterrence effect of adult criminal sanctions, relative to juvenile criminal sanctions. Standard economic models of crime imply that participation in crime should drop discontinuously at 18, when sanctions become more punitive. Our central empirical finding is that the rate of criminal involvement of young offenders is generally a smooth function of age, with only a small change at 18. Our findings are based on a high frequency, longitudinal administrative data set on arrests in Florida.

Our focus on the pattern of criminal involvement around 18 years of age yields two key benefits. First, under mild smoothness assumptions, our estimated effects are unlikely to be tainted by omitted variables bias. Second, our estimated effect reflects purely a deterrence effect, rather than a conflation of deterrence and incapacitation – which may well be contributing to the larger estimated deterrence elasticities in the some of the existing literature. We document that incapacitation effects of prison are large and nearly immediate, at least for youthful offenders in our data.

The point estimates from our discontinuity analysis indicate an approximately 2 percent decline in the rate of criminal offending when a juvenile turns 18, when the expected incarceration length conditional on arrest jumps discontinuously by roughly 230 percent. This suggests a small "reduced-form" elasticity of −0.007.[52] This elasticity is directly relevant to three types of criminal justice policy reforms that impact youthful offenders when they cross the age of criminal majority: (1) reducing the age of criminal majority, (2) increasing the rate at which juveniles are tried in the adult criminal court, and (3) marginal increases in the juvenile-adult gap in punishments. These kinds of policy changes are not simply hypothetical, but are instead relevant to recent and ongoing criminal justice discussions: (1) four states have reduced the age of majority in recent years (Chalfin & Loeffler, 2017; Snyder & Sickmund, 2006), (2) all 50 states and the District of Columbia had by 1995 adopted provisions for juvenile transfer to the adult criminal court (Government Accounting Office, 1995), and (3) during the 1980s and 1990s, many states increased the punitiveness of the sentencing regime for adults by much more than that for juveniles (Levitt, 1998).[53] Our estimates imply that changes to these types of policy changes may affect the crime rate of youthful offenders through incapacitation channels, but have at best minor effects on deterrence.

Investigating the implications of our estimates for broader policies of interest requires imposing some structural assumptions on criminal behavior. Combining ideas from Becker (1968) and McCall (1970), we develop a stochastic dynamic model of crime and calibrate it to match key empirical quantities. We recognize that our forward-looking, rational-expectations, rational-agent model may indeed be too restrictive. For example, although aggregate statistics give us a reasonably objective estimate of both the average probability of apprehension and expected incarceration lengths, potential offenders may not accurately perceive, and could conceivably vastly underestimate, those risks and punishments (Hjalmarsson, 2009a; Lochner, 2007). Furthermore, it may well be that criminal offending is better described by a hyperbolic discounting model, or some alternative model of time preferences, rather than time-consistent exponential discounting.[54]

Nevertheless, with the caveat that our theoretical framework is one particular lens through which to view the results, we draw upon it to reach three key conclusions. First, the magnitudes of our point estimates are consistent with impatient or even myopic behavior on the part of criminal offenders. For example, combining our point estimate with the mild assumption of a decreasing density function of criminal benefits rules out annual discount factors of 0.022 or more.[55]

Second, within the range of discount factors consistent with both our empirical findings and our theoretical model, our predicted elasticities can be reconciled with previous estimates of the elasticity of crime with respect to the probability of apprehension (e.g., Evans and Owens (2007) or Klick and Tabarrok (2005)). Interestingly, the model indicates that large policing elasticities, such as the −0.75 estimate of Di Tella and Schargrodsky

(2004), imply a very small discount factor – so small that it would imply an elasticity with respect to sentence lengths no more negative than −0.06. Generally, there is a trade-off between these two elasticities, if one stipulates to both our model and our empirical results.

Third, no matter the discount factor, the most negative sentence length elasticity consistent with our data and model is −0.13. This is several times smaller than the larger elasticities from the literature, but is consistent with the magnitudes from Helland and Tabarrok (2007) and Iyengar (2008).

We conclude by noting that while our findings point to small deterrence effects of prison, this does not mean that prison is ineffective in reducing the overall incidence of crime. Indeed, we show that once arrested, the marginal juvenile transferred from the juvenile to adult criminal justice system is less likely to recidivate. We suspect that this is occurring through an incapacitation mechanism, which could be quite important, as suggested by the work of Jacob and Lefgren (2003). A fruitful avenue for future research would thus be an investigation of optimal sentencing structure and the optimal mix of police and prisons when incapacitation is the primary mechanism for crime reduction.

NOTES

1. While in principle the case may then be transferred to the adult criminal court, this is rare generally (Snyder & Sickmund, 1999). We examine juvenile transfer empirically below in Section 5.

2. In Florida, our study state, the age of criminal majority is 18. This is typical of U.S. states, but some states have legislated cutoffs at 16 and 17 (Bozynski & Szymanski, 2003).

3. Deterrence refers to the behavioral reduction in crime due to offender anticipation of punishment. Incapacitation refers to the mechanical reduction in crime that occurs when offenders are incarcerated and unavailable to commit additional crimes (against the unimprisoned).

4. See, for example, Nagin's (1978) criticisms of much of the older literature due to its failure to recognize problems with endogeneity. See also the discussion in Freeman (1999) and Levitt and Miles (2007).

5. Relative punitiveness is defined as the number of adult prisoners per adult violent crimes, relative to the number of delinquents per juvenile violent crimes.

6. On page 1181, it is noted that "[b]etween 1978 and 1993, punishment per crime fell 20% for juveniles but rose 60% for adults. Over that same time period, rates of juvenile violent and property crime rose 107 and 7 percent, respectively. For adults, the corresponding increases were 52 and 19 percent. On the basis of the estimates of Table 2, if juvenile punishments had increased proportionally with those of adults, then the predicted percentage changes in juvenile violent and

property crime over this period would have been 74 and 2 percent." In other words, if punishments rose 60% instead of falling 20% (a difference of 80 percent) crime would have risen by only 74% rather than 107% (a difference of 33%), implying an elasticity of about $-33/80 \approx -0.41$.

7. See Drago et al. (2009, pp. 273–274).

8. The statute details those offenses which are strikeable. Essentially, a strikeable offense is a felony that is serious or violent.

9. Helland and Tabarrok (2007, pp. 327–328) argue that three-strikes increased expected sentences at third strike eligibility by 330%, or from about 60 months to at least 260 months.

10. Iyengar (2008) estimates a 16% decline in the incidence of crime at second strike eligibility (associated with a statutorily required doubling of the sentence) and a 29% decline in crime at third strike eligibility.

11. See McCrary (2010) for a review of the literature on dynamic models of criminal behavior. An important early paper in this literature is Flinn (1986), and a recent paper is Sickles and Williams (2008).

12. Further, taking into account the conditions of confinement as well as the duration of confinement surely amplifies the deterrence effect of the adult system. As noted by Parsell (2012), "Congressional findings in the Prison Rape Elimination Act of 2003 posited that juveniles were five times as likely to be sexually assaulted in adult rather than in juvenile facilities–often within their first 48 hours of incarceration. Youth advocacy groups report that juveniles housed in adult facilities are 36 times more likely to commit suicide."

13. Fig. A1 plots the frequency distribution of age (measured in years), for those arrested for an index crimes (murder, rape, robbery, assault, burlgary, larceny, and motor vehicle theft), as computed using the Federal Bureau of Investigation's Uniform Crime Reports (UCR). Because our administrative data pertain to Florida during the period 1995–2002, we show the data for Florida and the United States in 1995 and for the United States in 2002.

14. The vector X_t can also include interactions of the polynomial with the indicator for adulthood. Practically, because the regressors only vary at the level of the group, we estimate the model at the group level and avoid the construction of the (large) micro data set.

15. To do so, we estimate a logit model with functional form $F(W_t'\pi)$, where W_t is a series of indicators for each week, and compare the fit to the (nested) parametric approximation.

16. Note that this approach to estimating incapacitation effects of prison is accurate to the extent that there are small to no deterrence effects of prison.

17. A more detailed description of the database and its construction is provided in Appendix A.

18. The baseline arrest is allowed to be any arrest. The first and second arrests since 17 are restricted to be index arrests unless otherwise specified.

19. Also, in our main analysis we focus on the incidence of Index felonies, and so at any post-17 age, the individual could be incarcerated from a non-Index felony arrest.

20. For example, responding to a question regarding how he knew that sanctions were more punitive after the age of criminal majority, one 12-year old interviewed

by those authors who was earlier arrested for stealing from cars responded that the police had told him so: "Police come in our school and a lot of stuff, and I get caught and they tell me that" (p. 220).

21. We classified counties according to total arrests in the FDLE data. Large counties are Miami-Dade, Broward, and Orange. Medium counties are Franklin, Palm Beach, Duval, Pinellas, Polk, Escambia, and Volusia. Remaining counties are classified as small.

22. Another example that leads to the same prediction of an increase at 18 in the probability of observing criminality is from Levitt (1998). There, a hypothetical officer desires to see an arrestee punished and deems that the punishment accorded a juvenile is not worth the paperwork required to complete the arrest.

23. Here, X_t is a cubic polynomial in t. These predictions correspond to estimation of the model in column (1) of Table 2.

24. We report in Table 2 standard errors based on the observed information matrix (OIM). These are accurate if the logit model is the correct parametric model within our neighborhood of one year. We note that the standard errors are highly similar to those obtained from the outer product of the gradient (OPG) or (consequently) the robust or "sandwich" approach of White (1982). For example, to four digits, the estimate in column (1) is −0.0180 with an OIM standard error of 0.0474; the OPG standard error is 0.0472; and the robust standard error is 0.0475. As another example, the corresponding four numbers for column (7) are −0.0172, 0.0474, 0.0472, and 0.0475. The similarity of these estimated standard errors is consistent with the implications of the information matrix equality. One interpretation of this similarity is that the data contain little evidence against our parametric logit approximation within one year of the 18th birthday.

25. Use of a one-sided test implies rejection of any θ more negative than −0.095.

26. Practically, we implement this using the `rdrobust` package using the default options unless otherwise specified. The package was downloaded from `https://sites.google.com/site/rdpackages/rdrobust`

27. This standard error is similar to that advocated by Hahn et al. (2001) and is described by `rdrobust` as "conventional."

28. If we change to local cubic estimation with fourth-order models for bias adjustment, the point estimate (standard error) analogous to the local linear framework above increases in magnitude to −0.0498 (0.1091). Bias adjustment now reduces the magnitude of the point estimate, to −0.0424, and standard errors robust to bias estimation increase to 0.1220, or about 12% larger than the standard error ignoring bias estimation.

29. In Florida, prosecutors have the discretion to try a juvenile arrestee in the adult criminal court. This typically results in a larger number of juveniles tried as adults than the other common form of juvenile transfer, in which the juvenile court judge retains the authority to transfer juveniles (Snyder & Sickmund, 2006).

30. Therefore, the sample for this figure is the same as that underlying the top panel of Fig. 1.

31. We note that, analogous to McCrary and Royer (2011), assessing the magnitude of the "treatment" discontinuity is possible here because there is little evidence of deterrence. Were there to be strong evidence of deterrence, the composition of the sample would potentially be quite different for those aged 17.98 and 18.02.

32. The implication of this Florida law was summarized by a state attorney general opinion in 1995: "Under Florida law, crime and police records regarding crime have been a matter of public record. With limited exceptions, however, the identity of a juvenile who committed a crime has been protected. With the enactment of Chapter 94-209, Laws of Florida, an omnibus juvenile justice reform measure, the Legislature has amended the confidentiality provisions relating to juvenile offenders to allow for greater public dissemination of information. The clear goal of the Legislature was to establish the public's right to obtain information about persons who commit serious offenses, regardless of age" (Butterworth, 1995, p. 274).

33. Florida creates an exception to this general rule when the individual has committed a serious offense as an adult. Juvenile records for "habitual offenders" are retained by the FDLE until the offender is 26 (Fla. Stat. 943.0585, 943.059).

34. The dark bars represent the values used in the upper panel of Fig. 1, except that Fig. 1 normalizes each value with the at-risk population at each point in time to provide a probability value.

35. By "imprecise sorting" we mean that for each individual, the density function of age at arrest is continuous at 18. See Lee (2008) for further discussion.

36. It is worth noting that by Florida law, juvenile pre-trial detention cannot be longer than 21 days. No such restriction applies to adult pre-trial detention. Investigation of the juvenile hazard of re-offense indicates a strong spike near 21 days (results unreported).

37. Note also that an incapacitation interpretation is not the only one consistent with the data. Alternatively, these data are also consistent with no difference between juvenile and adult lengths of incarceration. This could occur if being incarcerated in the adult criminal justice system had a negative causal effect on criminal propensities upon release. This notion is sometimes referred to as "specific deterrence," particularly in the criminology literature. See Cook (1980) for discussion of the concept and Hjalmarsson (2009b) for a recent empirical approach. This alternative interpretation requires a mechanism by which the quality of the experience of being incarcerated as an adult induces individuals to be more law-abiding upon release.

38. In an earlier draft, we considered an alternative model in which there was heterogeneity in p, the probability of apprehension (Lee & McCrary, 2005).

39. Thus, criminal opportunities cannot be acted upon while incarcerated.

40. This mimics the standard reservation wage property of a job search model. See, for example, the textbook treatments in Adda and Cooper (2003) and Ljungqvist and Sargent (2004).

41. Setting $a = 0$ and $c = 1$ can be viewed as a normalization. Maximizing $\mathbb{E}_t\left[\sum_{\tau=t}^{\infty}\delta^{\tau-t}u_\tau\right]$ is equivalent to maximizing $\mathbb{E}_t\left[\sum_{\tau=t}^{\infty}\delta^{\tau-t}(u_\tau - a)/c\right]$. Thus, no matter the values of a and c, the solution will be equivalent to considering a problem where the flow utility takes the values of -1, 0, or B/c.

42. As noted in Section 4, $\ln\left(\frac{h}{1-h}\right) - \ln\left(\frac{h_J}{1-h_J}\right) \leq \theta < \ln(h) - \ln(h_J) \leq 0$, where h and h_J are the true arrest hazards for a boundary adult and juvenile, respectively. In our calibrations, we report the smaller (in magnitude) of the two, $\ln(h) - \ln(h_J)$, but as noted there is very little difference between the bounds in practice.

43. According to our model, the average duration between arrests as an adult is $\mathbb{E}[S] + 1/(p(1 - F(b^*)))$; that is, once an adult is arrested, the time until the next

arrest is given by the period of incarceration S plus the time between being released and the next arrest, which is exponentially distributed with hazard $p(1 - F(b^*))$ and thus mean $1/(p(1 - F(b^*)))$. Similarly, the time until next arrest for a boundary juvenile is $\mathbb{E}[S_J] + 1/(p(1 - F(b^*)))$.

44. Specifically, we model the juvenile and adult durations as separate Weibull models, allowing both the shape and scale parameters to depend on a cubic in age at initial arrest.

45. Clearance rates are the fraction of offenses known to police for which an individual has been arrested and handed over to prosecutors. These rates are estimated using the UCR data. Reporting rates are the fraction of Index offenses reported to police and are based on the NCVS. The aggregate clearance rate is 20 percent, and the aggregate reporting rate is 42 percent, for an approximate $0.2 \times 0.42 = 0.08$ probability of apprehension.

46. This robustness to distributional assumptions on the sentencing side is due to the fact that the distribution of sentences only matters to the extent that it affects $\left[\sum_{s=1}^{\infty} \pi_s(\delta - \delta^S)/(1 - \delta)\right]$.

47. As another example, Caskey (1996) shows that in states without interest rate regulations, pawnbrokers charge interest rates implying annual discount factors of 0.3.

48. A slightly more accurate approximation is $(\mathbb{E}[S] - \mathbb{E}[S_J])/(1 + \mathbb{E}[S])$ but since we measure time in days, this distinction is of little practical difference.

49. See Theory Appendix for discussion.

50. As we show in the Theory Appendix, our model and a declining density for B imply that $\mathbb{E}[B - b^* | B > b^*]$ can be no lower than $\underline{E}$, which is strictly positive as long as adults commit crime. See Appendix for the formula for $\overline{E}$.

51. Since 1993, the most recent peak of criminal activity, through 2012, violent crime had fallen by 37 percent and property crime had fallen by 26 percent, whereas the prison population had increased 34 percent. See *Sourcebook of Criminal Justice Statistics Online*, Tables 3.106.2012, 6.1.2011, and 6.28.2012, and *Sourcebook of Criminal Justice Statistics*, 1975, Table 1.131.

52. When we consider the smallest negative effects consistent with our (two-sided) confidence interval, −0.11, this elasticity is instead −0.047.

53. For further background on these institutional changes, see National Research Council (2001).

54. For exploration of models of this type in this context, see Lee and McCrary (2005).

55. Note, however, that the outer edge of our (two-sided) confidence region does not allow us to rule out any particular discount factor.

56. This identifier is constant across the FDLE files.

57. Roughly speaking, a record of the arrest file corresponds to the triple of (1) person identifier, (2) arrest date, and (3) charge code. Conceptually, the named triple will not uniquely identify a record due to the possibility of multiple arrest events for the same crime on the same day. Practically, there also appear to be some minor errors with double-counting in the file (e.g., two such triples, one with a missing offense date and another with an offense date equal, as usual, to the arrest date). However, the number of unique triples in the data is 94.3% of the total record count. We conclude that neither the conceptual distinction nor the double-counting issue is important empirically.

58. Thus the average arrest event is associated with 1.36 charges.

59. In the juvenile courts, a serious criminal offender will be placed in secure "detention" (the rough equivalent of jail), where they await an adjudication by the juvenile court judge. If they are found to be guilty, they are committed to a residential placement facility (the rough equivalent of a prison). In the table, column (2) (labeled "jail") includes juveniles in secure detention awaiting an adjudication as well as adults who are unconvicted, but awaiting court proceedings. Note that in the United States, jails not only incarcerate those awaiting hearings and trials, but also those who have been convicted to short terms. Therefore, the prison population includes adults in a state correctional facility as well as those incarcerated in jail who are serving a sentence. For juveniles, the prison population (column (3)) includes those in juvenile residential placement.

60. See McCrary (2010) for details on some of the calculations described in this appendix.

61. If S is distributed geometric with mean $1/q$, then this formula quickly shows that $\nu = \delta(1-q)/(1-\delta(1-q))$.

62. Note, however, that in the juvenile analogue to Eq. (B.3), $\psi(b^*)$ retains its adult definition because the juvenile will be an adult in subsequent periods.

63. The first inequality corresponds to a uniform distribution over the range b^* to $b^* + \frac{1-F(b^*)}{f(b^*)}$, which on inspection is the distribution which yields the smallest possible value of E and yet is consistent with a boundary adult crime rate of $1 - F(b^*)$ and does not violate the assumption of a non-increasing density. The second inequality follows since if $f(\cdot)$ is non-increasing then the distribution function $F(\cdot)$ is concave, implying $f(b^*) \leq \frac{F(b^*)-F(b_J^*)}{b^*-b_J^*} \leq f(b_J^*)$ since $b_J^* \leq b^*$.

ACKNOWLEDGMENTS

Brett Allen, Diane Alexander, and Yan Lau provided excellent research assistance. We are grateful to Jeff Kling for detailed suggestions on a previous draft. For useful comments, we thank Josh Angrist, David Autor, David Card, Kerwin Charles, Ken Chay, Gordon Dahl, John DiNardo, Michael Greenstone, Vivian Hwa, Christina Lee, Steve Levitt, Robert McMillan, Enrico Moretti, Daniel Parent, Dan Silverman, and seminar participants at UC Berkeley, the Center for Advanced Study in the Behavioral Sciences, Chicago GSB, Columbia, London School of Economics, the NBER Summer Institute, MIT, Northwestern Law, the University of Illinois at Urbana-Champaign, the University of Rochester, the University of Toronto, University College London, and the University of Wisconsin. Sue Burton of the Florida Department of Law Enforcement was instrumental in obtaining the data we use here. McCrary acknowledges funding from the Center for Local, Urban, and State Policy at the University of Michigan.

REFERENCES

Abrams, D. S., & Rohlfs, C. (2011). Optimal bail and the value of freedom: Evidence from the Philadelphia bail experiment. *Economic Inquiry*, *49*(3), 750–770.

Adams, W., Einav, L., & Levin, J. (2009). Liquidity constraints and imperfect information in subprime lending. *American Economic Review*, *99*(1), 49–84.

Adda, J., & Cooper, R. W. (2003). *Dynamic economics: Quantitative methods and applications*. Cambridge, MA: MIT Press.

Angrist, J., & Pischke, J.-S. (2009). *Mostly harmless econometrics*. Princeton, NJ: Princeton University Press.

Ayres, I., & Donohue, J. J. (2003). The latest misfires in support of the more guns, less crime hypothesis. *Stanford Law Review*, *55*, 1371.

Bayer, P., Hjalmarsson, R., & Pozen, D. (2009). Building criminal capital behind bars: Peer effects in juvenile corrections. *Quarterly Journal of Economics*, *124*(1), 105–147.

Beccaria, C. (1764). *On crimes and punishments*. Oxford: Clarendon Press.

Becker, G. S. (1968). Crime and punishment: An economic approach. *Journal of Political Economy*, *76*(2), 169–217.

Bentham, J. (1789). *An introduction to the principles of morals and legislation*. Oxford: Clarendon Press.

Bozynski, M., & Szymanski, L. (2003). National overviews. *State Juvenile Justice Profiles*. Pittsburgh: National Center for Juvenile Justice.

Butterworth, B. (1995). *Annual report of the Attorney General of the State of Florida*. Tallahassee, FL: Attorney General's Office.

Calonico, S., Cattaneo, M., & Titiunik, R. (2014). Robust nonparametric confidence intervals for regression-discontinuity designs. *Econometrica*, *82*(6), 2295–2326.

Case, A. C., & Katz, L. F. (1991). The company you keep: The effects of family and neighborhood on disadvantaged youths. NBER Working Paper No. 3705.

Caskey, J. P. (1996). *Fringe banking*. New York, NY: Russell Sage.

Chaddock, G. R. (2003). U.S. Notches World's Highest Incarceration Rate. *Christian Science Monitor*, August 18.

Chalfin, A., & Loeffler, C. (2017, February). Estimating the Crime Effects of Raising the Age of Majority: Evidence from Connecticut. *Criminology and Public Policy*, *16*(1), 45–71

Chalfin, A., & McCrary, J. (2017). Are U.S. cities underpoliced? Theory and evidence. *Review of Economics and Statistics*, forthcoming.

Cook, P. J. (1980). Research in criminal deterrence: Laying the groundwork for the second decade. In M. Tonry (Ed.), *Crime and justice: An annual review of research* (Vol. 2, pp. 211–268). Chicago, IL: University of Chicago Press.

Cook, P. J., & Ludwig, J. (2000). *Gun violence: The real costs*. New York, NY: Oxford University Press.

Donohue, J. J. III, & Levitt, S. D. (2001). The impact of legalized abortion on crime. *Quarterly Journal of Economics*, *116*(2), 379–420.

Donohue, J. J. III, & Siegelman, P. (1998). Allocating resources among prisons and social programs in the battle against crime. *Journal of Legal Studies*, *27*(1), 1–43.

Donohue, J. J. III, & Wolfers, J. (2008). Uses and abuses of empirical data in the death penalty debate. In S. D. Levitt & T. J. Miles (Eds.), *Economics of criminal law*. Northampton, MA: Edward Elgar.

Drago, F., Galbiati, R., & Vertova, P. (2009). The deterrent effects of prison: Evidence from a natural experiment. *Journal of Political Economy*, *117*(2), 257–280.

Duggan, M. (2001). More guns, more crime. *Journal of Political Economy*, *109*(5), 1086–1114.

Eeckhout, J., Persico, N., & Todd, P. (2010). A theory of optimal random crackdowns. *American Economic Review*, *100*(3), 1104–1135.

Efron, B. (1988). Logistic regression, survival analysis, and the Kaplan-Meier curve. *Journal of the American Statistical Association*, *83*(402), 414–425.

Ehrlich, I. (1973). Participation in illegitimate activities: A theoretical and empirical investigation. *Journal of Political Economy*, *81*(3), 521–565.

Ehrlich, I. (1987). On the issue of causality in the economic model of crime and law enforcement: Some theoretical considerations and experimental evidence. *American Economic Review*, *77*(2), 99–106.

Evans, W. N., & Owens, E. G. (2007). Cops and crime. *Journal of Public Economics*, *91*(1), 181–201.

Federal Bureau of Investigation. (2004). *Uniform crime reporting handbook*. Washington, DC: Federal Bureau of Investigation.

Flinn, C. (1986). Dynamic models of criminal careers. In A. Blumstein, J. Cohen, J. A. Roth, & C. A. Visher (Eds.), *Criminal careers and "career criminals"* (Vol. 2, pp. 356–379). Washington, DC: National Academy Press.

Freeman, R. B. (1996). Why do so many young American men commit crimes and what might we do about it. *Journal of Economic Perspectives*, *10*(1), 24–42.

Freeman, R. B. (1999). The economics of crime. In O. Ashenfelter & D. E. Card (Eds.), *Handbook of labor economics* (Vol. 3C, pp. 3529–3571). New York, NY: Elsevier-North Holland.

Gaviria, A., & Raphael, S. (2001). School-based peer effects and juvenile behavior. *Review of Economics and Statistics*, *83*(2), 257–268.

Glaeser, E. L., & Sacerdote, B. (1999). Why is there more crime in cities? *Journal of Political Economy*, *107*(2, Pt. 2), S225–S258.

Glaeser, E. L., Sacerdote, B., & Scheinkman, J. A. (1996). Crime and social interactions. *Quarterly Journal of Economics*, *111*(2), 507–548.

Glassner, B., Ksander, M., Berg, B., & Johnson, B. (1983). A note on the deterrent effect of juvenile vs. adult jurisdiction. *Social Problems*, *31*(2), 219–221.

Government Accounting Office. (1995). Juvenile justice: Juveniles processed in criminal court and case dispositions. Washington, DC: U.S. General Accounting Office.

Grogger, J. (1995). The effect of arrests on the employment and earnings of young men. *Quarterly Journal of Economics*, *110*(1), 51–72.

Grogger, J. (1998). Market wages and youth crime. *Journal of Labor Economics*, *16*(4), 756–791.

Hahn, J., Todd, P., & van der Klaauw, W. (2001). Identification and estimation of treatment effects with a regression discontinuity design. *Econometrica*, *69*(1), 201–209.

Helland, E., & Tabarrok, A. (2007). Does three strikes deter: A non-parametric investigation. *Journal of Human Resources*, *42*(2), 309–330.

Hjalmarsson, R. (2009a). Crime and expected punishment: Changes in perceptions at the age of criminal majority. *American Law and Economics Review*, *11*(1), 209–248.

Hjalmarsson, R. (2009b). Juvenile jails: A path to the straight and narrow or hardened criminality? *Journal of Law and Economics*, *52*(4), 779–809.

Iyengar, R. (2008, February). *I'd rather be hanged for a sheep than a lamb: The unintended consequences of 'three-strikes' laws*. NBER Working Paper No. 13784.

Jacob, B. A., & Lefgren, L. (2003). Are idle hands the devil's workshop? Incapacitation, concentration, and juvenile crime. *American Economic Review*, *93*(5), 1560–1577.

Katz, L. F., Levitt, S. D., & Shustorovich, E. (2003). Prison conditions, capital punishment and deterrence. *American Law and Economics Review*, *5*(2), 318–343.

Klick, J., & Tabarrok, A. (2005). Using terror alert levels to estimate the effect of police on crime. *Journal of Law and Economics*, *48*(1), 267–280.

Kling, J. R. (2006). Incarceration length, employment, and earnings. *American Economic Review*, *96*(3), 863–876.

Kling, J. R., Ludwig, J., & Katz, L. F. (2005). Neighborhood Effects on Crime for Female and Male Youth: Evidence from a Randomized Housing Voucher Experiment. *Quarterly Journal of Economics*, *120*(1), 87–130.

Kuziemko, I. (2007, September). *Going off parole: How the elimination of discretionary prison release affects the social cost of crime*. NBER Working Paper No. 13380.

LEAA, U.S. (1972). *Expenditure and employment data for the criminal justice system: 1969–1970*. Washington, DC: U.S. Government Printing Office.

Lee, D. S. (2008). Randomized experiments from non-random selection in U.S. House elections. *Journal of Econometrics*, *142*(2), 675–697.

Lee, D. S., & Lemieux, T. (2010). Regression discontinuity designs in economics. *Journal of Economic Literature*, *48*(2), 281–355.

Lee, D. S., & McCrary, J. (2005, July). Crime, punishment, and myopia. NBER Working Paper No. 11491.

Levitt, S. D. (1996). The effect of prison population size on crime rates: Evidence from prison overcrowding litigation. *Quarterly Journal of Economics*, *111*(2), 319–351.

Levitt, S. D. (1997). Using electoral cycles in police hiring to estimate the effect of police on crime. *American Economic Review*, *87*(3), 270–290.

Levitt, S. D. (1998). Juvenile crime and punishment. *Journal of Political Economy*, *106*(6), 1156–1185.

Levitt, S. D. (2004). Understanding why crime fell in the 1990s: Four factors that explain the decline and six that do not. *Journal of Economic Perspectives*, *18*(1), 163–190.

Levitt, S. D., & Lochner, L. (2001). The determinants of juvenile crime. In J. Gruber (Ed.), *Risky behavior among youths: An economic analysis* (pp. 327–374). Chicago, IL: University of Chicago Press.

Levitt, S. D., & Miles, T. J. (2007). Empirical study of criminal punishment. In A. M. Polinsky & S. Shavell (Eds.), *Handbook of law and economics* (Vol. 1, pp. 455–498). Amsterdam: North-Holland.

Ljungqvist, L., & Sargent, T. J. (2004). *Recursive macroeconomic theory*. Cambridge, MA: MIT Press.

Lochner, L. (2004). Education, work, and crime: A human capital approach. *International Economic Review*, *45*(3), 811–843.

Lochner, L. (2007). Individual perceptions of the criminal justice system. *American Economic Review*, *97*(1), 444–460.

Lochner, L., & Moretti, E. (2004). The effect of education on crime: Evidence from prison inmates, arrests, and self-reports. *American Economic Review*, *94*(1), 155–189.

Lott, J. (1998). *More guns, less crime*. Chicago, IL: University of Chicago Press.

Maguire, K., & Pastore, A. L. (Eds.), (2000). *Sourcebook of criminal justice statistics*. Washington, DC: U.S. GPO.

McCall, J. J. (1970). Economics of information and job search. *Quarterly Journal of Economics*, *84*(1), 113–126.

McCrary, J. (2008). Manipulation of the running variable in the regression discontinuity design: A density test. *Journal of Econometrics*, *142*(2), 698–714.

McCrary, J. (2010). Dynamic perspectives on crime. In B. Benson & P. R. Zimmerman (Eds.), *Handbook on the economics of crime*. Northampton, MA: Edward Elgar.

McCrary, J., & Royer, H. N. (2011, February). The effect of female education on fertility and infant health: Evidence from school entry policies using exact date of birth. *American Economic Review*, *101*(1), 158–195.

Mueller-Smith, M. (2015). *The criminal and labor market impacts of incarceration*. Unpublished manuscript, University of Michigan.

Nagin, D. (1978). General deterrence: A review of the empirical evidence. In A. Blumstein, J. Cohen, & D. Nagin (Eds.), *Deterrence and incapacitation: Estimating the effects of criminal sanctions on crime rates* (pp. 95–139). Washington, DC: National Academy of Sciences.

Nagin, D. (1998). Criminal deterrence research at the outset of the twenty-first century. In M. Tonry (Ed.), *Crime and justice: A review of research* (Vol. 23, pp. 1–42). Chicago, IL: University of Chicago Press.

National Institute of Justice. (2003, April). *2000 arrestee drug abuse monitoring: Annual report*.

National Research Council. (2001). *Juvenile crime, juvenile justice*. Washington, DC: National Academy Press.

Parsell, T. J. (2012). In prison, teenagers become prey. *New York Times*, June 5.

Polinsky, A. M., & Shavell, S. (2000). The economic theory of public enforcement of law. *Journal of Economic Literature*, *38*(1), 45–76.

Rafael, D. T., & Schargrodsky, E. (2004). Do police reduce crime? Estimates using the allocation of police forces after a terrorist attack. *American Economic Review*, *94*(1), 115–133.

Raphael, S., & Winter-Ebmer, R. (2001). Identifying the effect of unemployment on crime. *Journal of Law and Economics*, *44*(1), 259–284.

Sickles, R. C., & Williams, J. (2008). Turning from crime: A dynamic perspective. *Journal of Econometrics*, *145*(1–2), 158–173.

Silverman, D. (2004). Street crime and street culture. *International Economic Review*, *45*(3), 761–786.

Snyder, H. N., & Sickmund, M. (1999). *Juvenile offenders and victims: 1999 national report*. Washington, DC: Office of Juvenile Justice and Delinquency Prevention.

Snyder, H. N., & Sickmund, M. (2006). *Juvenile offenders and victims: 2006 national report*. Washington, DC: Office of Juvenile Justice and Delinquency Prevention.

Stigler, G. J. (1970). The optimum enforcement of laws. *Journal of Political Economy*, *78*(3), 526–536.

U.S. DOJ (2006). Justice expenditure and employment statistical extracts, 2006. Retrieved from http://www.ojp.usdoj.gov/bjs/eande.htm.

Viscusi, W. K. (1986). The risks and rewards of criminal activity: A comprehensive test of criminal deterrence. *Journal of Labor Economics*, *4*(3), 317–340.

Western, B., Kling, J. R., & Weiman, D. F. (2001). *The labor market consequences of incarceration*. IR Section Working Paper No. 450. Princeton University.

White, H. (1982). Maximum likelihood estimation of misspecified models. *Econometrica*, *50*(1), 1–25.

Wilson, J. Q., & Herrnstein, R. J. (1985). *Crime and human nature*. New York, NY: Simon and Schuster.

APPENDIX A: DATA APPENDIX

A.1. Arrest-Level Database

Our data set is constructed using four electronic files maintained by the FDLE. The FDLE refers to these files as the arrest, date of birth, judicial, and identifier files. They constitute the key elements of Florida's Computerized Criminal History (CCH) system, which is maintained by the Criminal Justice Information System (CJIS) division of the FDLE. We obtained from the FDLE records on all felony arrests for the period 1989 to 2002.

A.1.1. Construction of Data Set

We construct our data set as follows. First, we begin with the arrests file, which contains a person identifier,[56] the offense date, the arrest date, the charge code, and the arresting agency. Each record in the arrest file pertains to a separate offense.[57] The total number of records in the arrest file we received is 4,498,139. Because a single arrest event may result in multiple records (due to multiple offenses), we collapse the data down to the level of the (1) person identifier and (2) arrest date, coding the offense as the most serious offense with which the individual was charged on that date, using the FBI hierarchy (Federal Bureau of Investigation, 2004, p. 10) There are 3,314,851 unique arrest-person observations.[58]

We similarly collapse the judicial file down to the person-arrest-date level. The judicial file represents all arrests that result in a formal prosecution. For each collapsed observation, if any of the potentially multiple arrests led to a conviction and prison or jail sentence, a prison or jail sentence was associated with the person-arrest-date. The collapsed judicial and arrest files were then merged on the unique person-identifier-arrest-date pair. We use the race variable in the identifier file and birthday from the birth date file.

A.1.2. Date Variables

The key variables we utilize from the arrest file are the person identifier, the arrest date, and the offense code. The offense date is missing for many observations, so we use the arrest date to proxy for the date of crime commission. This is due primarily to a reporting problem – officers do not always submit information on the offense date. On the other hand, there are no missing values for the arrest date. Among the 1,948,096 records with information on offense date, every one of those records has an associated arrest date. 90% of those arrest dates are equal to the offense date, and over 93% of those arrest dates occur within the first week subsequent to the stated offense date.

To further assess the validity of date of arrest as a proxy for date of offense, we obtained data from the Miami Police Department, which recorded arrest and offense dates for all charges pertaining to arrests made between July 1999 and December 2002. For the 272,494 arrests we obtained, 257,263 have a valid offense date, and 91.3% of those have offense and arrest dates that are identical, and with 95.8% of arrests occurring within the first week after the offense date. Focusing only on felony arrests, we find that of the 33,698 felony arrests, 32,033 have valid offense dates, and of these 78.9% have identical arrest and offense dates, and 90.6% have associated arrest dates that fall within a week of the offense date.

A.2. Average Incarceration Lengths: $\mathbb{E}[S]$ *and* $\mathbb{E}[S_J]$

To estimate the average number of weeks that an arrestee can expect to spend incarcerated, we obtain the cumulative number of person-weeks that are spent incarcerated for a given year, and divide by the total number of arrests that occur within the year. As long as jail/prison populations and arrest numbers are reasonably stable, this should provide the average number of weeks spent incarcerated per arrest.

Our estimates of the jail and prison populations are complied from the 1999 Census of Jails, the 2000 Census of State and Federal Correctional Facilities, and the 1999 Census of Juveniles in Residential Placement. Our estimates of the number of arrests come from the 1999 FBI Uniform Crime Reports.

The first two numbers in column (1) of Table A1 provide the number of arrests for juveniles (younger than 18 years old) and adults (18 and older). Column (2) provides the stock of juveniles and adults incarcerated in jail awaiting courtproceedings.[59] Provided that these population numbers are reasonable estimates of the average daily population throughout the year, this number multiplied by 52 gives the number of person-weeks spent incarcerated in "jail" throughout the year. The next column takes the ratio of the first two columns to produce an average duration of incarceration conditional on an arrest. For juveniles, it is about 0.59 weeks, and for adults it is about 2.21 weeks. The ratio of these is given in the same column (3.77).

Table A1. Juvenile and Adult Incarceration Length in Weeks.

	Arrests (1)	Jail (2)	Average Duration 52×(2)/(1)	Prison (3)	Average Duration 52×(3)/(1)	Total (2)+(3)	Average Duration 52×((2)+(3))/(1)	Average Duration, Adjusted for Juvenile Transfera
Panel A. Florida								
All offenses as denominator								
Juvenile	131,330	1,482	0.59	5,211	2.06	6.693	2.65	3.71
Adult	766,259	32,585	2.21	89,730	6.09	122,315	8.30	8.12
Adult/Juvenile Ratio			3.77		2.95		3.13	2.19
Index offenses as denominator								
Juvenile	53,967	1,482	1.43	5,211	5.02	6,693	6.45	9.03
Adult	133,853	32,585	12.66	89,730	34.86	122,315	47.52	46.48
Adult/Juvenile ratio			8.86		6.94		7.37	5.15
Panel B. United States								
All offenses as denominator								
Juvenile	1,588,839	26,439	0.87	76,926	2.52	103,365	3.38	4.74
Adult	7,552,362	354,379	2.44	1,546,456	10.65	1,900,835	13.09	12.80
Adult/Juvenile Ratio			2.82		4.23		3.87	2.70

Index offenses as denominator								
Juvenile	420,543	26,439	3.27	76,926	9.51	103,365	12.78	17.89
Adult	1,091,530	354,379	16.88	1,546,456	73.67	1,900,835	90.55	88.59
Adult/ Juvenile Ratio			5.16		7.75		7.09	4.95

Notes: Table gives estimates of average incarceration length based on stock-flow comparisons. Adult and juvenile arrest counts are for the year 1999, from the FBI's Uniform Crime Reports, as reported in the Sourcebook of Criminal Justice Statistics. Adult jail population counts come from the Census of Jails and pertain to December 1999. Adult prison population counts come from the Census of Correctional facilities and pertain to June 2000. Juvenile jail and prison population counts come from the Census of Juveniles in Residential Placement and pertain to October 1999.

[a]Adjustment for juvenile transfer assumes 20 percent of juvenile arrestees tried in adult criminal court and 50 percent tried in juvenile court; see Appendix for details of calculations.

Column (3) provides the prison populations of both juveniles and adults, and the subsequent column divides by the number of arrests to give the average length of incarceration in juvenile or adult prison (2.06 and 6.09, respectively) conditional on an arrest. It is important to note that this average will include many zeroes, for those who are not convicted/committed, or for whom formal charges are dropped. The penultimate column adds the two averages. Overall, the expected length of incarceration conditional on an arrest is about 2.65 weeks for a juvenile, and 8.30 for adults.

The final column of Table A1 adjusts the figures in the penultimate column for juvenile transfer, assuming that 20 percent of juveniles are tried as juveniles, 50 percent are tried as adults, and that the remainder have no charges brought against them (cf., Fig. 2). To understand the adjustment calculations we use, we introduce some new notation. We observe N_A individuals in adult prisons and jails, N_J individuals in juvenile facilities, M_A adult arrestees, and M_J juvenile arrestees. The only people in juvenile facilities are juveniles, but some of the individuals in adult prisons and jails are adults, and some are juveniles.

Decompose

$$N_A = N_A^A + N_A^J, \tag{A.1}$$

where N_A^A is the number of adults in adult facilities and N_A^J is the number of juveniles in adult facilities, both of which are unobserved.

Ideally, we would approximate expected sentences for adults and juveniles as

$$\widehat{E}[S] = \frac{P_A}{M_A}, \tag{A.2}$$

$$\widehat{E}[S_J] = \frac{P_J}{M_J}, \tag{A.3}$$

where $P_A = N_A^A$ is the number of adults incarcerated and $P_J = N_A^J + N_J$ is the number of juveniles incarcerated. Because N_A^A and N_A^J are unobserved, so are P_A and P_J.

To back out P_A and P_J, let the number of juveniles in adult facilities, relative to the number of incarcerated juveniles, be denoted by

$$\alpha = \frac{N_A^J}{P_J}. \tag{A.4}$$

For example, if 20 percent of juveniles are tried as adults, and 50 percent are tried as juveniles, and if juveniles tried as adults receive adult sentences, then we can approximate $\alpha = 20/(20 + 50) \doteq 0.286$.

With α in hand, we adjust the observed quantities N_J and N_A to estimate N_A^J and P_J. Specifically, since

$$\frac{1}{1-\alpha} = \frac{P_J}{N_J} \tag{A.5}$$

$$\frac{\alpha}{1-\alpha} = \frac{N_A^J}{N_J} \tag{A.6}$$

we obtain

$$\widehat{E}[S] = \frac{N_A - N_J \frac{\alpha}{1-\alpha}}{M_A}, \tag{A.7}$$

$$\widehat{E}[S_J] = \frac{N_J \frac{1}{1-\alpha}}{M_J}. \tag{A.8}$$

These are the quantities reported in the final column of Table A1.

Our analysis focuses on Index crimes. More minor crimes are likely to lead to very short periods of custody, and many offenders – particularly for misdemeanors – may be released almost immediately after a formal arrest. Thus, the numbers in the first set of rows are probably a lower bound on the incarceration lengths, conditional on an Index arrest.

To obtain an upper bound, the second set of rows re-computes the average durations using the number of Index Crime arrests as the denominator. This is an upper bound since surely some non-Index Arrests lead to a positive incarceration length.

A.3. Probability of Arrest

In our model, p is the expected probability of being apprehended, conditional on committing a crime. We provide an estimate of this quantity using so-called "clearance rates" from the FBI Uniform Crime Reports. A reported crime is "cleared by arrest" when an incident is followed up by law enforcement, and results in the arrest of an alleged offender who is then charged with the offense.

Column (2) of Table A2 reports clearance rates for all FBI Index Crimes, and the various sub-categories. The overall rate is 0.20. This is

Table A2. Arrest Probabilities.

Crime Category	Offenses Known to Police (1)	Fraction of Offenses Cleared by Arrest (2)	Fraction of Victimizations Reported to Police (3)	Probability of Apprehension (2)×(3)
All index crimes	10,121,721	0.20	0.42	0.08
Violent crime	1,184,453	0.47	0.48	0.23
Murder	13,561	0.64	0.77*	0.49
Forcible rape	80,515	0.45	0.54	0.24
Robbery	343,023	0.26	0.71	0.18
Aggravated assault	747,354	0.57	0.46	0.26
Property crime	8,937,268	0.17	0.40	0.07
Burglary	1,842,930	0.13	0.58	0.08
Larceny-theft	6,014,290	0.18	0.33	0.06
Motor vehicle theft	1,080,048	0.14	0.86	0.12
Source	UCR	UCR	NCVS, NCHS	Authors' calculations

Notes: Figures pertain to 2002 and are taken from the 30th Online Edition of the Sourcebook of Criminal Justice Statistics. Figures labelled "UCR" are from the FBI's Uniform Crime Reports system (Table 4.19). Figures labelled "NCVS" are from the Census Bureau's National Crime Victimization Survey (Table 3.36). Asterisk indicates that the fraction of victimizations reported to police was estimated by taking the ratio of offenses known to police to the number of 2002 murders reported to the National Center for Health Statistics (Table E, National Vital Statistics Reports, Vol. 53, No. 17, 2005).

Table A3. Robustness of Main Results.

	Imposing Equality of Derivatives at 18					Allowing for Different Derivatives at 18				
	(1)	(2)	(3)	(4)	(5)	(6)	(7)	(8)	(9)	(10)
Panel A. Unconditional Estimates										
Estimated Discontinuity	0.174	0.135	−0.018	0.002	0.029	0.147	−0.040	0.014	−0.054	−0.065
	(0.035)	(0.036)	(0.047)	(0.048)	(0.057)	(0.036)	(0.053)	(0.072)	(0.090)	(0.109)
Order of Polynomial	1	2	3	4	5	1	2	3	4	5
Test Against Saturated	175.8	144.2	119.8	113.8	113.0	153.5	119.5	112.6	110.3	109.3
Model degrees of freedom	[102]	[101]	[100]	[99]	[98]	[101]	[99]	[97]	[95]	[93]
p-Value	0.0000	0.0031	0.0865	0.1461	0.1421	0.0006	0.0789	0.1325	0.1351	0.1194
Panel B. Controlling for Race, Size of County, and Type of Baseline Crime										
Estimated discontinuity	0.173	0.134	−0.017	0.002	0.029	0.147	−0.039	0.014	−0.054	−0.065
	(0.035)	(0.036)	(0.047)	(0.048)	(0.057)	(0.036)	(0.053)	(0.072)	(0.090)	(0.109)
Black	0.534	0.533	0.533	0.533	0.533	0.534	0.533	0.533	0.533	0.533
	(0.018)	(0.018)	(0.018)	(0.018)	(0.018)	(0.018)	(0.018)	(0.018)	(0.018)	(0.018)
Small county	0.256	0.255	0.255	0.255	0.255	0.255	0.255	0.255	0.255	0.255
	(0.021)	(0.021)	(0.021)	(0.021)	(0.021)	(0.021)	(0.021)	(0.021)	(0.021)	(0.021)
Medium county	0.088	0.088	0.087	0.087	0.087	0.088	0.087	0.087	0.087	0.087
	(0.021)	(0.021)	(0.021)	(0.021)	(0.021)	(0.021)	(0.021)	(0.021)	(0.021)	(0.021)
Baseline crime violent	0.009	0.009	0.009	0.009	0.009	0.009	0.009	0.009	0.009	0.009
	(0.028)	(0.028)	(0.028)	(0.028)	(0.028)	(0.028)	(0.028)	(0.028)	(0.028)	(0.028)
Baseline Crime property	0.190	0.190	0.190	0.190	0.190	0.190	0.190	0.190	0.190	0.190
	(0.026)	(0.026)	(0.026)	(0.026)	(0.026)	(0.026)	(0.026)	(0.026)	(0.026)	(0.026)
	−0.153	−0.154	−0.154	−0.154	−0.154	−0.154	−0.154	−0.154	−0.154	−0.154

Table A3. *(Continued)*

	Imposing Equality of Derivatives at 18					Allowing for Different Derivatives at 18				
	(1)	(2)	(3)	(4)	(5)	(6)	(7)	(8)	(9)	(10)
Baseline crime non-index Dr	(0.042)	(0.042)	(0.042)	(0.042)	(0.042)	(0.042)	(0.042)	(0.042)	(0.042)	(0.042)
Order of polynomial	1	2	3	4	5	1	2	3	4	5
Test against saturated	173.8	143.8	119.7	113.9	113.1	152.7	119.5	112.7	110.3	109.3
model degrees of freedom	[102]	[101]	[100]	[99]	[98]	[101]	[99]	[97]	[95]	[93]
p-Value	0.0000	0.0033	0.0873	0.1455	0.1414	0.0007	0.0791	0.1323	0.1351	0.1192

Notes: Standard errors in parentheses. Table presents alternative parametrizations of the models given in Table II. Coefficients on the polynomial model are suppressed throughout. In each panel the bottom row tests the fit of the presented model against a saturated model which includes a series of exhaustive and mutually exclusive indicators for each possible age, in weeks.

Table A4. Discontinuity in Probability of Adult Incarceration.

	(1)	(2)	(3)	(4)	(5)	(6)	(7)	(8)
Panel A. Main Estimates								
Estimated discontinuity in probability of confinement	0.1475	0.1474	0.1459	0.1479	0.1464	0.1463	0.1459	0.1460
	(0.0155)	(0.0155)	(0.0154)	(0.0155)	(0.0154)	(0.0154)	(0.0154)	(0.0154)
Log Discontinuity	*1.7246*	*1.7238*	*1.7156*	*1.7264*	*1.7181*	*1.7176*	*1.7152*	*1.7162*
Non-white		−0.0053	0.0012	−0.0043	0.0020	0.0017	0.0017	0.0018
		(0.0037)	(0.0038)	(0.0038)	(0.0038)	(0.0039)	(0.0039)	(0.0039)
Size of County of Baseline Arrest (Relative to Large)								
Small			−0.0476		−0.0472	−0.0471	−0.0469	−0.0468
			(0.0047)		(0.0047)	(0.0047)	(0.0047)	(0.0047)
Medium			−0.0360		−0.0358	−0.0358	−0.0359	−0.0360
			(0.0045)		(0.0045)	(0.0045)	(0.0045)	(0.0045)
Type of Crime, Baseline Arrest (Relative to Non-index Non-drug)								
Index Crime, Violent				−0.0156	−0.0142	−0.0143	−0.0143	−0.0143
				(0.0062)	(0.0062)	(0.0062)	(0.0062)	(0.0062)
Index Crime, Property				−0.0041	−0.0039	−0.0041	−0.0040	−0.0040
				(0.0056)	(0.0056)	(0.0056)	(0.0056)	(0.0056)
Non-index Drug				0.0029	0.0033	0.0044	0.0039	0.0040
				(0.0079)	(0.0079)	(0.0079)	(0.0079)	(0.0079)
Controls for Age at Baseline Arrest?	N	N	N	N	N	Y	Y	Y
Order of Polynomial in Age at Baseline Arrest					1	3	5	
R^2	0.0649	0.0650	0.0695	0.0654	0.0699	0.0700	0.0702	0.0703

Table A4. *(Continued)*

	(1)	(2)	(3)	(4)	(5)	(6)	(7)	(8)
Panel B: Decomposition of probability of incarceration into prison and jail								
Estimated discontinuity in probability of prison	0.0401	0.0400	0.0394	0.0401	0.0395	0.0395	0.0392	0.0394
	(0.0107)	(0.0107)	(0.0107)	(0.0107)	(0.0107)	(0.0107)	(0.0107)	(0.0107)
Log discontinuity	*1.4066*	*1.4054*	*1.3943*	*1.4067*	*1.3954*	*1.3953*	*1.3907*	*1.3930*
Estimated discontinuity in probability of jail	0.1075	0.1074	0.1065	0.1078	0.1069	0.1068	0.1066	0.1067
	(0.0120)	(0.0120)	(0.0120)	(0.0120)	(0.0120)	(0.0120)	(0.0120)	(0.0120)
Log discontinuity	*1.9417*	*1.9410*	*1.9339*	*1.9442*	*1.9370*	*1.9364*	*1.9350*	*1.9353*

Notes: Standard errors in parentheses. Log discontinuity calculated from presented discontinuity estimate using a baseline rate of 0.032 for the marginal juvenile. Delta-method standard errors for the log discontinuity are in each case approximately 0.086. Bottom panel of table presents estimated discontinuity in probability of being sentenced to adult prison and to adult jail, respectively. Log discontinuities in bottom panel use a baseline marginal juvenile rate for prison and jail of 0.013 and 0.018, respectively. Delta-method standard errors for these log discontinuities are in each case approximately 0.15 and 0.086, respectively. Controls for regressions in bottom panel suppressed.

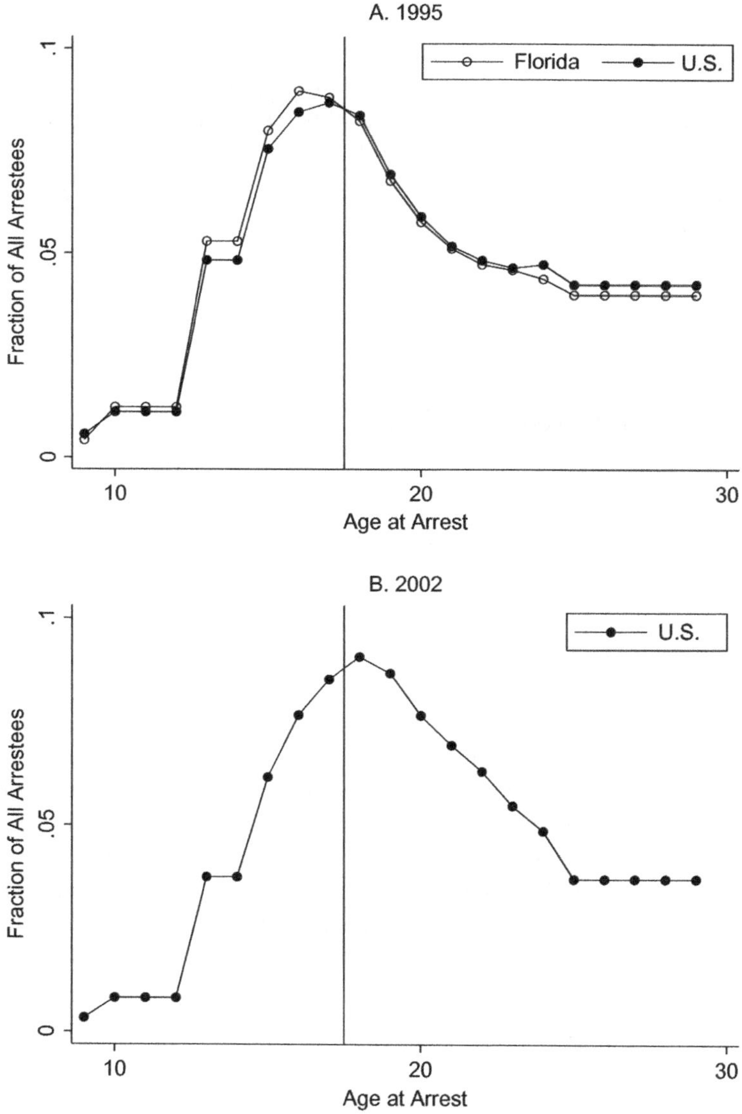

Fig. A1. Age Distribution of Crime: FBI Uniform Crime Reports. (a) 1995 and (b) 2002. *Notes*: Figure presents the histogram of age at arrest, for 1995 and 2002, for Florida and the U.S, for cities of 10,000 population or more. Data pertain to arrests for murder, rape, robbery, aggravated assault, burglary, larceny, and motor vehicle theft (index arrests) and are from the FBI Uniform Crime Reports. 1995 is the last year Florida provided the FBI with detailed age breakdowns for arrests.

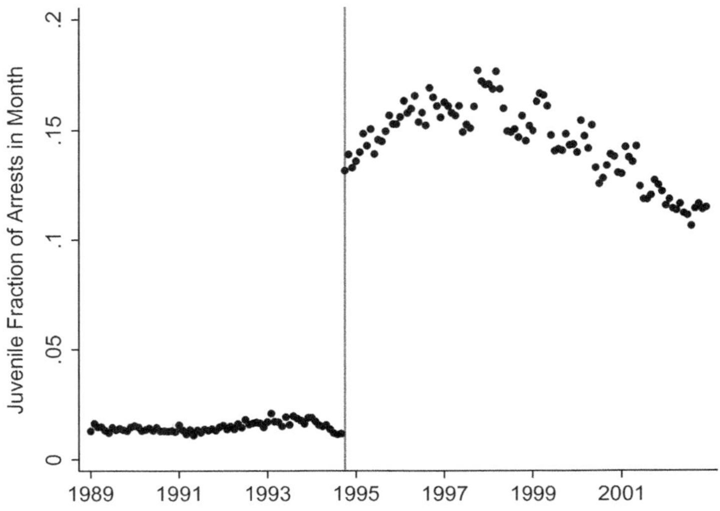

Fig. A2 Impact of Juvenile Justice Reform Act of 1994 on FDLE Coverage of Juvenile Arrests. *Notes*: Figure shows fraction of arrests in FDLE data set pertaining to juveniles, by month since 1989. Vertical line indicates effective date of Juvenile Justice Reform Act of 1994.

likely an upwardly biased estimate of p, because not all criminal incidents are reported to the police, and therefore the denominator of column (2) is likely too small. To address this, we obtained estimates of the rate of reporting victimizations to the police from the National Criminal Victimization Survey. By multiplying these rates (column 3) by column 2, we obtain an adjusted estimate of the probability of arrest conditional on committing an Index crime. The average for all Index crimes is about 0.08, with the lowest for larceny (0.06), and the highest for assault (0.26) and murder (0.49).

APPENDIX B: THEORY APPENDIX

B.1. Expressions for $\mathbb{E}[V(B)]$, b^*, Comparative Statics, and ν

In this subsection, we derive $\mathbb{E}[V(B)]$, define and then characterize b^*, and provide some results for ν.[60] First, to develop $\mathbb{E}[V(B)]$, note that for opportunities $b<b^*$, the value is fixed at $V(b) = a + \delta\mathbb{E}[V(B)]$. For opportunities $b>b^*$, higher value opportunities are worth more: $V(b)$ is linear in b with slope $1 - p$. These two segments of $V(b)$ connect at $b = b^*$. We thus have

$$V(b) = a + \delta\mathbb{E}[V(B)] + (1 - p)(b - b^*)\mathbf{1}(b>b^*). \tag{B.1}$$

Taking expectations with respect to the distribution of B and rearranging yields

$$(1 - \delta)\mathbb{E}[V(b)] = a + (1 - F(b^*))(1 - p)\mathbb{E}[B - b^*|B>b^*]. \tag{B.2}$$

Second, to obtain b^*, substitute Eq. (B.2) into Eq. (3) from the main text. Doing so shows that b^* solves

$$G(p,\delta,c,b^*) \equiv c\frac{p}{1-p}\left[1 + \nu\left(1 + \frac{1-p}{c}\psi(b^*)\right)\right] - b^* = 0, \tag{B.3}$$

where $\psi(b) = (1 - F(b))\mathbb{E}[B - b|B>b]$ and we recall from the main text that $\nu = \sum_{s=1}^{\infty}\pi_s\frac{\delta-\delta^s}{1-\delta}$.

Third, to characterize how b^* changes with changes to δ, c, p, and $\mathbb{E}[S]$, we use the implicit function theorem. Specifically:

- $\frac{\partial b^*}{\partial \delta} = -\frac{\partial G/\partial \delta}{\partial G/\partial b^*} > 0$, because $\frac{\partial G}{\partial \delta} = c\frac{p}{1-p}\frac{\partial \nu}{\partial \delta}\left(1 + \frac{1-p}{c}\psi(b^*)\right)$ is positive and because $\frac{\partial G}{\partial b^*} = -p\nu(1 - F(b^*)) - 1$ is negative. To obtain the sign for $\frac{\partial \nu}{\partial \delta}$, rewrite $\nu = \sum_{s=1}^{\infty}\pi_s\left(\delta + \cdots + \delta^{s-1}\right)$ and the result follows on inspection. To obtain the result for $\partial G/\partial b^*$, note that integration by parts shows $\psi(b) = \int_b^{\infty}(1 - F(t))\,\mathrm{dt}$. The Leibniz integral rule then shows that $\psi'(b) = -(1 - F(b))$.
- $\frac{\partial b^*}{\partial c} = -\frac{\partial G/\partial c}{\partial G/\partial b^*} > 0$ because as shown above $\frac{\partial G}{\partial b^*}$ is negative, and because $\frac{\partial G}{\partial c} = \frac{p}{1-p}(1 + \nu)$ is positive.
- $\frac{\partial b^*}{\partial p} = -\frac{\partial G/\partial p}{\partial G/\partial b^*} > 0$ because $\frac{\partial G}{\partial b^*}$ is negative, and because $\frac{\partial G}{\partial p} = c\frac{1}{(1-p)^2}(1 + \nu) + \nu\psi(b^*)$ is positive.

- $\frac{\partial b^*}{\partial \mathbb{E}[S]} = \frac{\partial b^*}{\partial \nu}\frac{\partial \nu}{\partial \mathbb{E}[S]}$ because changes to $\mathbb{E}[S]$ affect b^* only through ν. Then note that $\frac{\partial b^*}{\partial \nu} = -\frac{\partial G/\partial \nu}{\partial G/\partial b^*} > 0$ because $\frac{\partial G}{\partial b^*}$ is negative and $\frac{\partial G}{\partial \nu} = c\frac{p}{1-p} + p\psi(b^*)$ is positive. Consequently, $\frac{\partial b^*}{\partial \mathbb{E}[S]}$ is positive since $\frac{\partial \nu}{\partial \mathbb{E}[S]}$ is positive. This follows for any first-order stochastically dominating rightward shift in the distribution of S, which results in an increase in $\mathbb{E}[S]$ and an increase in ν (since $\frac{\delta - \delta^s}{1-\delta}$ is strictly increasing in s).

Fourth, to develop results for ν, observe that it is a shifted geometric series and thus both absolutely and uniformly convergent, allowing rearrangements and differentiation, respectively. Rearrangement then implies $\nu = \pi_1 \times 0 + \pi_2 \times \delta + \pi_3 \times (\delta + \delta^2) + \cdots = \sum_{s=1}^{\infty} \delta^s \mathbb{P}(S > s)$.[61] Differentiation then provides confirmation that $\partial \nu / \partial \delta > 0$ because the derivative can be written as $\partial \nu / \partial \delta = \sum_{s=1}^{\infty} s\delta^{s-1}\mathbb{P}(S > s) > 0$.

B.2. Elasticities of Crime with Respect to p and $\mathbb{E}[S]$

In this subsection, we give expressions for the policing elasticity and the sentencing elasticity. In addition to the notation above, define the expected surplus associated with crimes worth doing, $E \equiv \mathbb{E}[B - b^* | B > b^*]$, the elasticity $\eta_\nu \equiv \frac{\partial \nu}{\partial \mathbb{E}[S]}\frac{\mathbb{E}[S]}{\nu}$, and the adult and juvenile arrest rates, $h \equiv p(1 - F(b^*))$ and $h_J \equiv p\left(1 - F\left(b_J^*\right)\right)$, respectively, where b_J^* is defined analogously to Eq. (B.3) with ν_J replacing ν.[62]

The *policing elasticity* is given by

$$\eta_p \equiv \frac{\partial(1 - F(b^*))}{\partial p}\frac{p}{1 - F(b^*)} = -\frac{f(b^*)p}{1 - F(b^*)}\frac{\partial b^*}{\partial p}. \tag{B.4}$$

By the mean value theorem, there is a $\bar{b}$ between b_J^* and b^* such that

$$f\left(\bar{b}\right) = \frac{(1 - F(b^*)) - \left(1 - F\left(b_J^*\right)\right)}{b_J^* - b^*}. \tag{B.5}$$

Our model implies that the denominator in this expression is proportional to $\nu_J - \nu$. Too see this, note that Eqs. (6) and (7) in the main text imply that

$$b_J^* - b^* = \left(c\frac{p}{1-p} + p\psi(b^*)\right)(\nu_J - \nu) = \left(c\frac{p}{1-p} + hE\right)(\nu_J - \nu), \quad \text{(B.6)}$$

which in turn means that

$$f(\overline{b}) = \frac{h_J - h}{\nu - \nu_J}\frac{1}{c\frac{p}{1-p} + hE}\frac{1}{p}. \quad \text{(B.7)}$$

Combining Eq. (B.7) with Eq. (B.4), we have

$$\eta_p = -\frac{f(b^*)}{f(\overline{b})}\frac{f(\overline{b})p}{1-F(b^*)}\frac{\partial b^*}{\partial p} = -\frac{f(b^*)}{f(\overline{b})}\frac{h_J - h}{\nu - \nu_J}\frac{1}{c\frac{p}{1-p} + hE}\frac{p}{h}\frac{\partial b^*}{\partial p} \quad \text{(B.8)}$$

$$= -\frac{f(b^*)}{f(\overline{b})}\frac{h_J - h}{h}\frac{\nu}{\nu - \nu_J}\frac{p}{c\frac{p}{1-p}\nu + h\nu E}\frac{\partial b^*}{\partial p}. \quad \text{(B.9)}$$

Using the results from the prior section, we have

$$\frac{\partial b^*}{\partial p} = -\frac{\partial G/\partial p}{\partial G/\partial b^*} = -\frac{c\frac{1}{(1-p)^2}(1+\nu) + \nu\psi(b^*)}{-p\nu(1-F(b^*)) - 1} = \frac{c\frac{p}{(1-p)^2}(1+\nu) + \nu hE}{1+\nu h}\frac{1}{p}. \quad \text{(B.10)}$$

Finally, combining Eqs. (B.9) and (B.10), we have

$$\eta_p = -\frac{f(b^*)}{f(\overline{b})}\left(\frac{h_J - h}{h} \bigg/ \frac{\nu - \nu_J}{\nu}\right)\frac{c\frac{p}{(1-p)^2}(1+\nu) + \nu hE}{c\frac{p}{1-p}\nu + h\nu E}\frac{1}{1+\nu h}. \quad \text{(B.11)}$$

Turning now to the *sentencing elasticity*, note that since changes in $\mathbb{E}[S]$ affect ν but no other parameter,

$$\eta_{\mathbb{E}[S]} = \frac{\partial(1-F(b^*))}{\partial \nu}\frac{\nu}{1-F(b^*)}\eta_\nu = -\frac{f(b^*)\nu}{1-F(b^*)}\frac{\partial b^*}{\partial \nu}\eta_\nu. \quad \text{(B.12)}$$

Using the results from the prior section, we have

$$\frac{\partial b^*}{\partial \nu} = -\frac{\partial G/\partial \nu}{\partial G/\partial b^*} = -\frac{c\frac{p}{1-p} + p\psi(b^*)}{-p\nu(1-F(b^*)) - 1} = \frac{c\frac{p}{1-p} + hE}{1+\nu h}. \tag{B.13}$$

Parallel to above, we now substitute in Eq. (B.7) and then Eq. (B.13) into Eq. (B.12) to obtain

$$\eta_{\mathbb{E}[S]} = -\frac{f(b^*)\nu}{1-F(b^*)}\frac{\partial b^*}{\partial \nu}\eta_\nu \tag{B.14}$$

$$= -\frac{f(b^*)}{f(\bar{b})}\frac{f(\bar{b})\nu}{1-F(b^*)}\frac{\partial b^*}{\partial \nu}\eta_\nu \tag{B.15}$$

$$= -\frac{f(b^*)}{f(\bar{b})}\frac{h_J - h}{\nu - \nu_J}\frac{\nu}{c\frac{p}{1-p} + hE}\frac{1}{h}\frac{\partial b^*}{\partial \nu}\eta_\nu \tag{B.16}$$

$$= -\frac{f(b^*)}{f(\bar{b})}\left(\frac{h_J - h}{h}\Big/\frac{\nu - \nu_J}{\nu}\right)\frac{1}{c\frac{p}{1-p} + hE}\frac{\partial b^*}{\partial \nu}\eta_\nu \tag{B.17}$$

$$= -\frac{f(b^*)}{f(\bar{b})}\left(\frac{h_J - h}{h}\Big/\frac{\nu - \nu_J}{\nu}\right)\frac{1}{1+\nu h}\eta_\nu. \tag{B.18}$$

B.3. Implications of Non-Increasing Density $f(\cdot)$

Suppose now that the density of B is such that $f(\cdot)$ is non-increasing past b_J^*. The elasticity η_p is the most negative when the product $\frac{f(b^*)}{f(\bar{b})}\frac{c\frac{p}{(1-p)^2}(1+\nu)+\nu hE}{c\frac{p}{1-p}\nu+\nu hE}$ is as large as possible. Consider then the smallest possible E (since the second factor is positive and decreasing in E) consistent with the model and the assumption that $f(\cdot)$ be non-increasing. We have the following inequalities: (1) $E \geq \frac{1-F(b^*)}{2f(b^*)}$, and (2) $f(b^*) \leq \frac{F(b^*)-F(b_J^*)}{b^*-b_J^*}$.[63] Inequalities (1) and (2), together with Eq. (B.4), yield

$$E \geq \frac{1-F(b^*)}{2f(b^*)} \geq \frac{(1-F(b^*))(b^*-b_J^*)}{2\left(F(b^*)-F\left(b_J^*\right)\right)} = \frac{(1-F(b^*))(\nu-\nu_J)\left(c\frac{p}{1-p}+hE\right)}{2\left(F(b^*)-F\left(b_J^*\right)\right)}. \tag{B.19}$$

Minor rearrangement allows us to isolate E, and we obtain the bound

$$E \geq \frac{(1-F(b^*))(\nu-\nu_J)c\frac{p}{1-p}}{2\left(F(b^*)-F\left(b_J^*\right)\right)-h(1-F(b^*))(\nu-\nu_J)} \equiv \underline{E}. \tag{B.20}$$

Since $E > 0$, we must have $\underline{E} > 0$, which is true if and only if the denominator is positive, which in turn is true if and only if

$$\nu-\nu_J < \frac{2}{h}\frac{h_J-h}{h}. \tag{B.21}$$

This means that since $\nu-\nu_J$ is increasing in δ, certain values of δ may be strictly inconsistent with the model and measured rates of arrest h and h_J.

When Inequality (B.21) is satisfied, the smallest E consistent with the model is given by the $\underline{E}$ defined in (B.20). When $E=\underline{E}$, Inequalities (1) and (2) are both binding, and if (2) is binding then the density between b_J^* and b^* is flat, which implies $f(b^*)/f\left(\overline{b}\right)=1$. Thus, the most negative elasticity consistent with our modeling assumptions is given by

$$\overline{\eta}_p = -\left(\frac{h_J-h}{h}\Big/\frac{\nu-\nu_J}{\nu}\right)\frac{c\frac{p}{(1-p)^2}(1+\nu)+\nu h\underline{E}}{c\frac{p}{1-p}\nu+\nu h\underline{E}}\frac{1}{1+\nu h}. \tag{B.22}$$

Similarly, under a non-increasing $f(\cdot)$, by setting $f(b^*)=f\left(\overline{b}\right)$, we obtain the most negative sentencing elasticity consistent with our estimates and model as

$$\overline{\eta}_{\mathbb{E}[S]} = -\left(\frac{h_J-h}{h}\Big/\frac{\nu-\nu_J}{\nu}\right)\frac{1}{1+\nu h}\eta_\nu \tag{B.23}$$

B.4. Intertemporal Elasticity

The sentencing elasticities discussed above incorporate two distinct mechanisms: (1) an increase in punishments reduces the immediate attractiveness of crime, and (2) an increase in punishments reduces the expected benefit from a future crime, and hence reduces the opportunity cost of being incarcerated (and therefore unable to commit crime). An elasticity that isolates the first mechanism would immediately change punishments today, while keeping future punishments constant. Call this notion of elasticity an "intertemporal elasticity."

Consider a generic response to a 1 percent increase in the juvenile's punishment $\mathbb{E}[S_J]$. Because the adult's punishment does not change, $\mathbb{E}[B - b^* | B > b^*]$ is held constant as $\mathbb{E}[S_J]$ changes. This elasticity is

$$\eta_{\mathbb{E}[S_J]} = -\frac{pf\left(b_J^*\right)}{h_J}\left[c\frac{p}{1-p}\nu_J + \nu_J hE\right]\eta_{\nu_J}. \tag{B.24}$$

To isolate the intertemporal elasticity described above, we evaluate this elasticity assuming that the distribution of sentence lengths is equal for adults and juveniles. When this is the case, we have

$$\eta_{\mathbb{E}[S_J]} = \eta_{\mathbb{E}[S]}(1 + \nu h). \tag{B.25}$$

The resulting analogue to $\bar{\eta}_{\mathbb{E}[S]}$ becomes

$$\bar{\eta}_{\mathbb{E}[S_J]} = -\left(\frac{h_J - h}{h} \Big/ \frac{\nu - \nu_J}{\nu}\right)\eta_{\nu}. \tag{B.26}$$

Both $\eta_{\mathbb{E}[S_J]}$ and $\bar{\eta}_{\mathbb{E}[S_J]}$ differ from the adult analogue only by the term $1/(1 + \nu h)$. This discussion clarifies that this term reflects the fact that an increase in punishments reduces the expected benefit from a future crime, and hence reduces the opportunity cost of being incarcerated.

AN OVERVIEW OF GEOGRAPHICALLY DISCONTINUOUS TREATMENT ASSIGNMENTS WITH AN APPLICATION TO CHILDREN'S HEALTH INSURANCE☆

Luke Keele[a], Scott Lorch[b], Molly Passarella[b], Dylan Small[c] and Rocío Titiunik[d]

[a]*Georgetown University, Washington, DC, USA*
[b]*The Children's Hospital of Philadelphia and School of Medicine, University of Pennsylvania, Philadelphia, PA, USA*
[c]*Department of Statistics, University of Pennsylvania, Philadelphia, PA, USA*
[d]*Department of Political Science, University of Michigan, Ann Arbor, MI, USA*

☆Authors are in alphabetical order.

Regression Discontinuity Designs: Theory and Applications
Advances in Econometrics, Volume 38, 147–194

ISSN: 0731-9053/doi:10.1108/S0731-905320170000038007

ABSTRACT

We study research designs where a binary treatment changes discontinuously at the border between administrative units such as states, counties, or municipalities, creating a treated and a control area. This type of geographically discontinuous treatment assignment can be analyzed in a standard regression discontinuity (RD) framework if the exact geographic location of each unit in the dataset is known. Such data, however, is often unavailable due to privacy considerations or measurement limitations. In the absence of geo-referenced individual-level data, two scenarios can arise depending on what kind of geographic information is available. If researchers have information about each observation's location within aggregate but small geographic units, a modified RD framework can be applied, where the running variable is treated as discrete instead of continuous. If researchers lack this type of information and instead only have access to the location of units within coarse aggregate geographic units that are too large to be considered in an RD framework, the available coarse geographic information can be used to create a band or buffer around the border, only including in the analysis observations that fall within this band. We characterize each scenario, and also discuss several methodological challenges that are common to all research designs based on geographically discontinuous treatment assignments. We illustrate these issues with an original geographic application that studies the effect of introducing copayments for the use of the Children's Health Insurance Program in the United States, focusing on the border between Illinois and Wisconsin.

Keywords: Geographic discontinuity; natural experiment

1. INTRODUCTION

We study a form of research designs based on geography, where a treatment changes discontinuously at the border between administrative units such as states, counties, or municipalities. The opportunity to use designs of this type is frequent given that policies often vary with the borders of government or administrative units that are themselves based on geography. Indeed, the extant literature contains numerous examples studying the effect of pollution (Chen, Ebenstein, Greenstone, & Li, 2013), foreclosure laws (Pence, 2006), collective bargaining (Magruder, 2012), nation building, governance and ethnic relations in Africa (Asiwaju, 1985; Berger, 2009; Dell, 2010; Laitin, 1986; Michalopoulos & Papaioannou, 2014; Miguel, 2004; Miles, 1994; Miles & Rochefort, 1991; Posner, 2004), media effects in Europe and the United States (Huber & Arceneaux, 2007; Kearney & Levine, 2014; Kern & Hainmueller, 2008; Krasno & Green, 2008), local policies in U.S. cities (Gerber, Kessler, & Meredith, 2011), mosquito eradication (Salazar, Maffioli, Aramburu, & Agurto Adrianzen, 2016), population shocks (Schumann, 2014), the effects of tax rates on residential mobility (Young, Varner, Lurie, & Prisinzano, 2014), the effect of private police forces (MacDonald, Klick, & Grunwald, 2016), and mobilization and polarization in the American electorate (Middleton & Green, 2008; Nall, 2015.).

We discuss the general features of research designs based on a treatment assignment that is geographically discontinuous, in particular how the implementation of such designs is closely linked to the availability of geo-referenced information at the appropriate level. When the exact geographic location of each unit of analysis is available, the geographically discontinuous treatment assignment can be analyzed in a standard regression discontinuity (RD) setup (Hahn, Todd, & van der Klaauw, 2001; Imbens & Lemieux, 2008; Lee & Lemieux, 2010) with a two-dimensional running variable. However, the availability of this kind of information is limited in many applications, often because of confidentiality or measurement reasons.

When geo-referenced data at the individual level is not available, the standard RD framework cannot be readily applied. When this happens, there are at least two ways to proceed with the analysis. If researchers have information about each observation's location within aggregate but small geographic units, a modified RD framework can be applied, where the running variable is treated as discrete instead of continuous. If instead researchers only have access to the location of units within coarse aggregate geographic units that are too large to be considered in an RD framework, the coarse geographic information can be used to create a band or buffer around the border that contains observations within a maximum distance from the border; the analysis then only includes observations within the buffer. In this scenario, the identification assumptions must be modified accordingly.

In what follows, we discuss these issues in detail, and also review other challenges that arise in the study of geographically discontinuous treatment assignments such as multiple treatments that coincide at the border of interest; the enhanced ability of subjects to sort very precisely around the border; the possibility of interference between treated and control units due to their spatial proximity; and the potential heterogeneity in treatment effects along the border. Our discussion draws on Keele and Titiunik (2015b), Keele and Titiunik (2016), and Keele, Titiunik, and Zubizarreta (2015), where we explored in detail several types of geographic designs. The geographic regression discontinuity (GRD) is a special case of an RD design with multiple running variables (e.g., Dell, 2010), which is discussed in general by Imbens and Zajonc (2011), Papay, Willett, and Murnane (2011), Reardon and Robinson (2012), and Wong, Steiner, and Cook (2013).

We illustrate with an original geographic application that studies the effect of introducing copayments for the use of the Children's Health Insurance Program (CHIP), a public health insurance program in the United States that covers children in families with modest incomes. Specifically, we study whether the introduction of copayments in Wisconsin's CHIP in 2008 led to a decrease in the usage of health services, relative to a control group of Illinois residents who live just across the state border and whose CHIP program did not introduce copayments in the period under study.

2. EMPIRICAL APPLICATION: COPAYMENTS IN THE CHIP

Approximately, 8.1 million or 11% of all children in the United States have health insurance through the CHIP (Kaiser Family Foundation 2016). This program provides health coverage for uninsured children in families with modest incomes that are not low enough to qualify for Medicaid coverage. Faced with a combination of serious budgetary crises and rapid growth in health care spending over the years, many states have adopted cost-containment strategies used in the private insurance sector, most predominantly the implementation of copayments. Between 2003 and 2012, only 1 state had instituted deductibles, but 20 states had made 65 copayment changes, with 12 states instituting 23 new copayment policies for one or more medical services such as inpatient stays, emergency department visits, nonpreventive outpatient visits, and prescription drugs (Heberlein, Brooks, Alker, Artiga, & Stephens, 2013).

Extant research shows that copayments tend to reduce overall health spending in a limited number of studies in pediatrics (Haggerty, 1985; Leibowitz et al., 1985; Lohr et al., 1986; Valdez et al., 1985), but may also reduce potentially beneficial items such as medications (Campbell, Allen-Ramey, Sajjan, Maiese, & Sullivan, 2011). For example, in Alabama, the increase in copayments in 2004 was associated with a 2.4% reduction in the use of brand name drugs, a 1.4% decrease in generic drugs, and a 2.5% decrease in outpatient visits to physicians (Sen et al., 2012). Preventive health visits and inpatient visits temporarily declined after these copayment changes, but the effects did not persist over time, unlike the changes to other more discretionary services. Investigators found that demand for inpatient services was less price sensitive compared to demand for other services such as medications and outpatient visits (Sen et al., 2012). However, it is not known whether copayments for one type of service will change the use of other health care services which are substitutes or complements. Moreover, this study relied on a single-state, before-and-after design. The lack of a robust control group raises questions about the validity of its estimated effects of copayment increases.

In February 2008, Wisconsin expanded its CHIP program, known as BadgerCare, to create a new program known as BadgerCare Plus (BC+). BC+ operates as a single program with two insurance products. The first, known as the Standard Plan, operates as Wisconsin's traditional Medicaid plan for enrollees with incomes less than the Federal Poverty Level. The second insurance product, known as the Benchmark Plan, is for enrollees with incomes above 200% of the Federal Poverty Level. Premiums are subsidized until incomes exceed 300% of the Federal Poverty Level. Under the Benchmark Plan, enrollees cannot have been offered employer-sponsored insurance in the last 12 months or have the opportunity to gain employer-based coverage in the next 3 months. BC+ simplified eligibility rules and enrollment processes and included a marketing and outreach program. Finally, BC+ also added payments for many services for some enrollees, specifically those children under age 18 whose family income was at or below 100% of the Federal Poverty Line plus all enrollees in the Benchmark Plan (Department of Health and Family Services 2008b).

The new copayment amounts ranged from \$1 for acute care visits; \$3 for inpatient services; \$1–5 for prescription drugs, typically \$1–5 for generic and compound drugs, \$3 for brand name drugs, and \$0.50 for over-the-counter drugs; to \$3 for emergency department visits that did not result in a hospital admission (Department of Health and Family Services, 2008a; Ross & Marks, 2009). Copays were not added to well child visits, which are designed to administer preventative care. Copayments have continued to be modified since the enactment of this legislation, such that by January of 2011, the maximum copayment was increased to $15 for a nonpreventive outpatient visit and $100 for an inpatient visit, but had dropped to $0 for emergency department visits (Heberlein, Brooks, Guyer, Artiga, & Stephens, 2011).

The BC+ program is emblematic of many changes in U.S. health policy. These policy changes occur at the state level and may have significant effects on target populations. Many such policy changes do not have a randomized component to aid evaluation – but see Baicker et al. (2013) for an exception. The standard research strategy for studying such policy changes is differences-in-differences (DID), but geography-based designs are a natural alternative.

We use this policy change as a case study. Since Wisconsin did not enact other legislation that may affect the use of health care, we seek to understand whether the addition of copayments to the Wisconsin CHIP program changed health care utilization. During this period, states that border Wisconsin left their CHIP programs intact and did not add copayments. While these states are relatively similar to Wisconsin, one can readily imagine reasons why health care utilization may differ across these states other than the addition of copays in Wisconsin. Differences in health care usage may arise from differences in health care training that may influence the type and quality of care delivered by practitioners; differential access to inpatient or outpatient care, particularly in the large urban centers in the two states; and differences in the care quality available to children in different regions of each state (Asch, Nicholson, Srinivas, Herrin, & Epstein, 2009).

We treat the change in insurance copays as a treatment that changes discontinuously at the Wisconsin state border. In our application, we use residents from Illinois as the control group. With a geographic design, we can compare families living close to the state border. This provides a more similar control population than if we used larger state populations, as we will demonstrate through a comparison of Wisconsin and Illinois children receiving CHIP insurance.

3. DISCONTINUOUS ASSIGNMENT OF TREATMENT AT A GEOGRAPHIC BOUNDARY

We focus on the general problem of studying the effect of a binary intervention or treatment that (i) is given to all units in a geographic area, and

(ii) is withheld from all units who are located on the other side of this area's geographic boundary. In other words, the border between the treated and control areas marks the boundary where the treatment assignment changes discontinuously from zero to one. Under certain circumstances, this setup can be analyzed by directly applying a generalization of the standard RD framework. In other cases, typically when geo-located data is not available, the standard RD machinery cannot be applied, and a natural experimental framework may be used instead.

We denote the binary treatment of interest by T, and assume we have a random sample of n subjects or units, indexed by $i = 1, 2, \ldots, n$, from a larger population. The treatment is assigned based on geography, so that all units who are located in area $\mathcal{A}^t$ have $T_i = 1$, and all units located in area $\mathcal{A}^c$ have $T_i = 0$. For example, in our health care usage application the treated area $\mathcal{A}^t$ is Wisconsin, while the control area $\mathcal{A}^c$ is Illinois. Thus, $T_i = 1$ if i resides in Wisconsin and is subject to insurance copays, and $T_i = 0$ if i resides in Illinois and does not have copays. Note that the setup assumes that there are no compliance problems, so that the treatment assignment and the actual treatment received are identical for every unit.

We adopt the potential outcomes framework and let each individual have two potential outcomes, Y_{i1} and Y_{i0}, which correspond to levels of treatment $T_i = 1$ and $T_i = 0$, respectively. The observed outcome is $Y_i = T_i Y_{i1} + (1 - T_i) Y_{i0}$. We also assume that the Stable Unit Treatment Value Assumption or SUTVA holds (Cox, 1958; Rubin, 1986). SUTVA is comprised of two parts: there are no hidden forms of treatment, which implies that for unit i under $T_i = t$, we have $Y_{it} = Y_i$; and the potential outcomes of one unit do not depend on the treatment of other units. As we outline later, the validity of both parts of SUTVA may be questionable in geographic designs.

Under this framework, the treatment is a deterministic function of the unit's geographic location. The analysis of this type of designs will therefore depend on whether the exact location of each individual in the sample is known. We now consider the different types of parameters that can be defined and estimated with and without this form of geo-located data.

3.1. GRD Designs When Data Is Geo-Located

If researchers have access to the exact geographic location of each unit, the discontinuous treatment assignment based on geography can be analyzed in a standard RD setup, with the only modification that the running variable or score that determines treatment has two dimensions instead of one (see Imbens & Zajonc, 2011; Papay et al., 2011; Wong et al., 2013).

To consider this case, which we discussed in more detail in Keele and Titiunik (2015b), we assume that each unit's geographic location is known and given by a pair of geographic coordinates such as longitude and latitude. We define the two-dimensional score $\mathbf{S_i} = (S_{i1}, S_{i2})$, which records the geographic location of individual i given by the two geographic coordinates. We call the set that collects the locations of all boundary points $\mathcal{B}$, and denote a single point on the boundary by $\mathbf{b}$, with $\mathbf{b} = (S_1, S_2) \in \mathcal{B}$. Thus, $\mathcal{A}^t$ and $\mathcal{A}^c$ are sets that collect all the locations that receive treatment and control, respectively. The treatment assignment is $T_i = T(\mathbf{S}_i)$, with $T(\mathbf{s}) = 1$ for $\mathbf{s} \in A^t$ and with $T(\mathbf{s}) = 0$ for $\mathbf{s} \in A^c$. This assignment has a discontinuity at the known boundary $\mathcal{B}$. We assume that the density of $\mathbf{S}_i$, $f(\mathbf{s})$, is positive and continuous in a neighborhood of the boundary $\mathcal{B}$ – an assumption that is often particularly restrictive in geographic contexts.

Under this setup, a natural parameter of interest is $\tau(\mathbf{b}) \equiv \mathbb{E}[Y_{i1} - Y_{i0}|\mathbf{S}_i = \mathbf{b}]$, for $\mathbf{b} \in \mathcal{B}$. Since there is a (possibly different) treatment effect $\tau(\mathbf{b})$ for every point $\mathbf{b}$ on the boundary, this defines a treatment effect curve. Alternatively, these effects can be averaged across all boundary points, leading to the parameter $\tau = \mathbb{E}[\tau(\mathbf{b})|\mathbf{b} \in \mathcal{B}]$.

Identification of $\tau(\mathbf{b})$ follows from generalizing the standard RD identification results in Hahn et al. (2001) to a two-dimensional running variable. Thus, the main assumption required for identification is continuity of conditional regression functions $\mathbb{E}[Y_{i1}|\mathbf{s}]$ and $\mathbb{E}[Y_{i0}|\mathbf{s}]$, at all points on the boundary. In the context of our application, this assumption implies that the average potential health care utilization under a copayment regime for a unit located near a point $\mathbf{b}$ on the Wisconsin–Illinois boundary is very similar to the average utilization that would be observed exactly at this boundary point, regardless of the direction in which we approach the boundary. Thus, when data is geo-located, this design may be deemed a GRD design, and it is a particular case of an RD design with two running variables (Keele & Titiunik, 2015b). For example, the GRD design is mathematically equivalent to an RD design where students take two exams and receive a treatment only if each of their two exam scores exceeds a known (and possibly different) cutoff.

When the relevant continuity conditions hold, the average treatment effect at the cutoff can be identified as the limit of two regression functions on the observed outcomes. In other words, letting superscripts t and c denote locations in the treated and control areas, respectively, we have

$\tau(\mathbf{b}) = \lim_{\mathbf{s}^t \to \mathbf{b}} \mathbb{E}[Y_i|\mathbf{S}_i = \mathbf{s}^t] - \lim_{\mathbf{s}^c \to \mathbf{b}} E[Y_i|\mathbf{S}_i = \mathbf{s}^c]$ for all $\mathbf{b} \in \mathcal{B}$, which is analogous to the single-dimensional standard RD result.

The availability of geo-located data together with a continuous two-dimensional running variable and the assumption of continuity of the conditional regression functions means that standard smoothness-based RD methods for estimation and inference can be applied directly to this problem. In essence, geo-located data allows us to approximate the regression functions arbitrarily close to any boundary point (assuming data density is positive everywhere).

In particular, local polynomial methods (Fan & Gijbels, 1996) are now standard in the analysis of RD designs, and can be applied with appropriate modifications to the geographic setup we are considering here. For a given point $\mathbf{b}$ on the boundary, we calculate a distance measure between the location $\mathbf{S}_i$ of unit i and the boundary point $\mathbf{b}$. For every unit i in the sample, we define this distance as $d_{i\mathbf{b}} := d(\mathbf{b}, \mathbf{S}_i)$. For example, if Euclidean distance is used, $d_{i\mathbf{b}} = \sqrt{(b_1 - S_{i1})^2 + (b_2 - S_{i2})^2}$. Note that $d(\mathbf{b}, \mathbf{b}) = 0$ by definition.

Letting

$$\mu(\mathbf{b})^c \equiv \lim_{\mathbf{s}^c \to \mathbf{b}} E[Y_{i0}|d_{i\mathbf{b}} = d(\mathbf{b}, \mathbf{s}^c)],$$
$$\mu(\mathbf{b})^t \equiv \lim_{\mathbf{s}^t \to \mathbf{b}} E[Y_{i1}|d_{i\mathbf{b}} = d(\mathbf{b}, \mathbf{s}^t)],$$

we can estimate these functions by local linear regression. In order to do so, we solve

$$\left(\widehat{\alpha}_{\mathbf{b}}^c, \widehat{\beta}_{\mathbf{b}}^c\right) = \arg\min_{\alpha_{\mathbf{b}}^c, \beta_{\mathbf{b}}^c} \sum_{i \in A^c} \left\{Y_i - \alpha_{\mathbf{b}}^c - \beta_{\mathbf{b}}^c d_{i\mathbf{b}}\right\}^2 w_{i\mathbf{b}},$$
$$\left(\widehat{\alpha}_{\mathbf{b}}^t, \widehat{\beta}_{\mathbf{b}}^t\right) = \arg\min_{\alpha_{\mathbf{b}}^t, \beta_{\mathbf{b}}^t} \sum_{i \in A^t} \left\{Y_i - \alpha_{\mathbf{b}}^t - \beta_{\mathbf{b}}^t d_{i\mathbf{b}}\right\}^2 w_{i\mathbf{b}},$$

where

$$w_{i\mathbf{b}} = \frac{1}{h_{\mathbf{b}}} K\left(\frac{d_{i\mathbf{b}}}{h_{\mathbf{b}}}\right)$$

are spatial weights with $K(\cdot)$ representing a kernel weighting function and $h_{\mathbf{b}}$ a bandwidth. Note that the bandwidth is specific to each boundary point $\mathbf{b}$;

thus, for implementation, a different bandwidth should be chosen at every **b**. Given these solutions, the GRD treatment effect is estimated as

$$\widehat{\tau}(\mathbf{b}) = \widehat{\mu^t(\mathbf{b})} - \widehat{\mu^c(\mathbf{b})} = \widehat{\alpha}^t_{\mathbf{b}} - \widehat{\alpha}^c_{\mathbf{b}}.$$

Inference procedures must be implemented with care, as the standard asymptotic distribution of the least-squares estimator and robust standard errors ignore the asymptotic bias of the nonparametric local polynomial estimator and lead to invalid inferences in general. The most common procedure of bandwidth selection is based on asymptotic mean-squared error (MSE) minimization (see, e.g., Imbens & Kalyanaraman, 2012), a method which leads to bandwidth choices that are too large for conventional confidence intervals to be valid. In order to obtain valid inferences, researchers may select a smaller bandwidth to undersmooth, a procedure that is ad hoc and leads to power loss. An automatic, data-driven alternative is to estimate the asymptotic bias ignored by conventional inference, and correct the standard errors appropriately to produce robust confidence intervals that are valid even for large bandwidths, including those selected by MSE minimization (Calonico, Cattaneo, & Titiunik, 2014b). These methods are implemented in the `rdrobust` software – see Calonico, Cattaneo, and Titiunik (2014a) and Calonico, Cattaneo, Farrell, and Titiunik (2017) for details on the STATA implementation, and Calonico, Cattaneo, and Titiunik (2015) for details on the `R` implementation.[1]

In practice, since the boundary $\mathcal{B}$ is an infinite collection of points, we can select a grid of G points along the boundary for estimation, $\mathbf{b}^1, \mathbf{b}^2, \ldots, \mathbf{b}^G$. For this grid of points, we define a series of treatment effects $\tau(\mathbf{b}^g)$ for $g = 1, 2, \ldots, G$. In this case, the estimation procedure leads to a collection of G treatment effects that can vary along the boundary that separates the treatment and control areas, and in fact leads to a treatment effect curve, where each effect can then be mapped in its specific location, $\mathbf{b}^g$.

3.2. Geographic Treatment Assignments in the Absence of Geo-Located Individual Data

When geo-located data is not available, the smoothing methods described above cannot be applied, as there is no way to estimate the relevant regression functions arbitrarily close to the boundary. The unavailability of geocoded data is common in applications, typically caused by measurement limitations or confidentiality restrictions. For example, in our application, any information that allows the precise identification of individual patients is removed from the data, and this naturally includes the exact address of their residence.

We now discuss two scenarios that may arise when the geo-location of each individual observation is absent. In the first, the dataset contains geographic information for a sufficiently small unit, and an RD analysis can proceed with some modifications. In the second, the geographic information is too coarse; in this case, the analysis loses some of the distinctive RD features.

3.2.1. Scenario 1: Geo-Location of Small Aggregate Units

In the first scenario, the researcher has information about each observation's location within aggregate geographic units that are still sufficiently small. For example, the data may contain information on each observation's census block, the geo-location of which is often readily available. Armed with this information, the researcher can then assign the census block coordinates to each observation in the dataset, and treat those coordinates as the RD running variable. The aggregation step in this strategy, however, creates some complications because it causes all units in the same aggregate unit to share the same coordinates and leads to mass points in the running variable. This renders the standard RD methods discussed above inapplicable, since such methods rely on the assumption of a continuous score.

An appropriate analysis of this scenario would involve RD methods that allow for discrete running variables. One alternative is to use a randomization-based RD framework, where instead of continuity one assumes that there is a neighborhood or window around the cutoff where the treatment is as-if randomly assigned. Implementation requires choosing the window where this assumption plausibly holds, which can be done based on observable characteristics of the units. These methods have been developed for standard RD designs with a single score (see Cattaneo, Frandsen, & Titiunik, 2015; Cattaneo, Titiunik, & Vazquez-Bare, 2017), but can be extended straightforwardly to accommodate scores with two dimensions, and in particular to geographic RD applications where the window would be a geographic region. This randomization-based method requires assumptions that are stronger than the usual continuity assumptions invoked in the standard RD framework – see Sekhon and Titiunik (2017) for an in-depth discussion of the local randomization interpretation of RD designs.

An alternative strategy to deal with a discretized score is to approximate the unknown regression function connecting the mass points in the running variable, and model the deviation between the expected and predicted outcome as a random specification error, as proposed by Lee and Card (2008). In practice, this strategy involves fitting a polynomial model of the outcome on the score and clustering the standard errors by the discrete values of the score. This can be adapted to the geographic case with a two-dimensional score; the implementation would include a polynomial on both dimensions of the score and would cluster the standard errors by indicators corresponding to the aggregate geographic units where the mass points occur. Unlike the randomization-based method described above, this method requires a global fit and relies on the strong assumption that the specification errors are orthogonal to the score values.

3.2.2. Scenario 2: Geo-Location of Few Coarse Geographic Units

The second scenario occurs when the data contains information about aggregate geographic units that are simply too large to be considered in an RD framework. This occurs when researchers have coarse geographic information, and can only classify observations into a few categories according to their maximum distance to the boundary. Our application falls into this category. For confidentiality reasons, we are unable to access patients' addresses or census block locations; instead, the smallest geographic unit contained in the data is zip code. Aggregating the information to the zip code level, however, is undesirable for various reasons. First, such aggregation would dramatically reduce the sample size, as there are only 71 treated zip codes contiguous to the segment of the Illinois–Wisconsin border we analyze. Second, it would force us to conduct the analysis at an arbitrary level of aggregation, introducing the possibility of seeing a modifiable areal unit problem (MAUP) (Openshaw, 1984). MAUP refers to the fact that areal units such as zip codes have borders that are relatively arbitrary with respect to the spatial variation of the units measured. In this case, aggregate measures will not accurately reflect individual level phenomena unless those phenomena are spatially constant with respect to the areal unit. Naturally, the biases caused by the MAUP would be avoided if we could geo-locate every observation in the dataset and use more spatially informative measures.

What, then, is the appropriate strategy for analysis in this scenario of coarse geographic information? One possibility is to treat the geographically discontinuous treatment assignment as a natural experiment, and use the coarse geographic information available to focus the analysis on treated and control areas along the border that are sufficiently close to each other. In some applications, geographically proximate treated and control areas are similar to each other in other relevant dimensions, increasing the plausibility of the analysis relative to a treated–control comparison that does not rely on geographic information.

In this second scenario, the available geographic information can be used in at least two ways. First, it can be used to select segments along the border where the treated and control populations are comparable. This is most relevant in cases where the boundary is long and contains some segments that are, for example, unpopulated or overlapping with rivers or mountains that drastically separate and differentiate the treated and control populations. For example, in our application, our treated area of Wisconsin borders four states that do not have CHIP copayments, but only the border with Illinois has enough population density to conduct the analysis.

The other way in which coarse geographic information can be used is by only including in the analysis observations that are within a maximum distance from the border. For example, we might only use units in zip codes that are within 5 miles from the Wisconsin–Illinois border. Researchers can then invoke the assumption that, conditional on being in this buffer around the boundary, potential outcomes and treatment assignment are unrelated to each other, as we also discussed in Keele and Titiunik (2016). We can formalize this assumption by defining, for each unit i in the dataset, the point on the border that is closest to i's location $\mathbf{S}_i$; we call this point $\mathbf{b}_i^\star$. We denote the distance between $\mathbf{b}_i^\star$ and $\mathbf{S}_i$ by $d_i^\star := d(\mathbf{b}_i^\star, \mathbf{S}_i)$; thus, $d_i^\star$ is the perpendicular distance between i's location and the border. The assumption that the comparison of treated and control observations close to the border leads to valid inferences can be formalized as follows:

Assumption 1 (Geographic Mean Independence). The potential outcomes Y_{i0}, Y_{i1} are mean independent of the treatment assignment T_i within a buffer of length $D > 0$ around the border:

$$\mathbb{E}[Y_{i1}|d_i^\star < D, T_i] = \mathbb{E}[Y_{i1}|d_i^\star < D],$$
$$\mathbb{E}[Y_{i0}|d_i^\star < D, T_i] = \mathbb{E}[Y_{i0}|d_i^\star < D].$$

In other words, by focusing on units that are close together, the preexisting confounding differences between treated and control units can be eliminated. This assumption is invoked often in applications that study the effects of geographically discontinuous interventions, either formally or informally (e.g., Card & Krueger, 1994; Lavy, 2010; Posner, 2004).[2] Naturally, Assumption 1 is untestable, but researchers can nonetheless provide some indirect empirical evidence consistent with its plausibility. Analogous to experimental settings, this assumption suggests that treated and control units within the selected band should be similar in those observable characteristics that are likely to be related to the potential outcomes. Thus, researchers should provide evidence that treated and control units within the buffer are comparable in terms of relevant observable characteristics that are determined before the treatment is assigned.

It is not uncommon, however, to encounter applications where, even in a small band around the border, treated and control units differ significantly in observable characteristics, questioning the plausibility of Assumption 1. In this case, researchers must decide how to interpret these differences. One alternative is to view the observable differences as a symptom of unsolvable differences between the groups, differences that are due to "endogeneity" or "sorting" around the border and are likely to be present not only in observable but also in unobservable characteristics. In this case, the treatment effects estimated based on the geographically discontinuous treatment assignment would lack credibility.

The other alternative is to assume that the predetermined observable characteristics available to the investigator capture enough of the treatment assignment mechanism, so that conditioning on them would suffice to make valid treated–control comparisons. Collecting in the vector $\mathbf{X}_i$ the available observable characteristics for each unit, this interpretation invokes a weaker version of Assumption 1:

> **Assumption 2 (Conditional Geographic Mean Independence).** The potential outcomes Y_{i0}, Y_{i1} are conditionally mean independent of the treatment assignment T_i within a buffer of length $D > 0$ around the border:

$$\mathbb{E}\left[Y_{i1} | d_i^\star < D, \mathbf{X}_i, T_i\right] = \mathbb{E}\left[Y_{i1} | d_i^\star < D, \mathbf{X}_i\right],$$

$$\mathbb{E}\left[Y_{i0} | d_i^\star < D, \mathbf{X}_i, T_i\right] = \mathbb{E}\left[Y_{i0} | d_i^\star < D, \mathbf{X}_i\right].$$

By invoking Assumption 2, which we also discussed in Keele et al. (2015), the researcher admits that focusing on a narrow band around the border is not enough to create comparable groups, but she assumes that a valid comparison can be made after one conditions on observable characteristics within this band. Following the terminology introduced by Galiani, McEwan, and Quistorff (2017), we refer to research designs based on assumptions such as 1 or 2 as geographic quasi-experiments (GQE).

We use a GQE design in our health care utilization application, where we find persistent observable differences between the populations on either side of the Illinois–Wisconsin border. That is, we assume that the treatment is as-if randomly assigned for those who live near the Illinois–Wisconsin border, after conditioning on a set of pretreatment covariates. We address the need to condition on covariates in more detail in the next section.

4. PARTICULARITIES OF TREATMENT ASSIGNMENTS BASED ON GEOGRAPHY

In the previous section, we discussed the different scenarios that can arise when geographically discontinuous treatment assignments are studied, focusing on the availability of geo-referenced information and how such availability affects researchers' ability to implement a pure RD framework. In this section, we discuss some common challenges that arise in the analysis of geographically discontinuous treatments that are common to all the scenarios discussed above. As we note, many of these challenges are specific to geographically discontinuous treatments, as they rarely arise in nongeographic RD designs with two running variables.

4.1. Compound Treatments

When studying treatment assignments that change discontinuously at a geographic border, it is common for multiple administrative or political borders to perfectly overlap. When each of the overlapping borders induces a change that can separately affect the outcome of interest, we face the problem of "compound" treatments – a situation where two or more treatments affect the outcome of interest simultaneously. Although this phenomenon can also occur in standard RD designs (as when a person who turns 65 becomes simultaneously eligible to multiple social programs), it is more frequent in geographic treatment assignments because the border that induces the change in the intervention is often an administrative border that serves as a border for multiple units. For example, county borders tend to coincide with the border of other relevant units such as school districts, congressional districts, media markets, and cities. In our application, the discontinuity of interest is the state border, which overlaps perfectly with a city and county border.

Since the researcher is typically interested in the effect of a single intervention, compound treatments often pose a serious challenge and constitute a violation of the consistency component of SUTVA. When multiple borders overlap, absent any restrictions or assumptions, it will not be possible to separate the effect of the treatment of interest on the outcome from the effect of all other simultaneous "irrelevant" treatments. In our current application, we are unable to separate the effect of the new copays in Wisconsin from any other treatments that change at the state border and also affect health care utilization.

Keele and Titiunik (2015b) introduce the assumption of compound treatment irrelevance to address applications with compound treatments. To restate this assumption, we assume there are K binary treatments that coincide at the same geographic border. We denote these treatments as T_{ij}, $j = 1, 2, \ldots, K$, for each individual i, with $T_{ij} = \{0, 1\}$. Only the kth treatment, T_{ik}, is of interest. The potential outcomes notation can be generalized to allow all K versions of treatment to possibly affect the potential outcomes of each individual: we let $Y_{i\mathbf{T}_i}$ be i's potential outcome, with $\mathbf{T}_i = (T_{i1}, T_{i2}, \ldots, T_{ik}, \ldots, T_{iK})'$. In order to isolate the effect of T_{ik} on the outcome of interest, we can invoke the following assumption:

> **Assumption 3** (Compound Treatment Irrelevance). Assume the treatment of interest is the kth treatment. For each i and for all possible pairs of treatment vectors $\mathbf{T}_i$ and $\mathbf{T}'_i$, $Y_{i\mathbf{T}_i} = Y_{i\mathbf{T}'_i}$ if $T_{ik} = T'_{ik}$.

When Assumption 3 holds, the potential outcomes are only a function of the treatment of interest, so $Y_{i\mathbf{T}_i} = Y_{iT_{ik}}$ and we can go back to the original notation, with Y_{i1} and Y_{i0} the potential outcomes corresponding, respectively, to $T_{ik} = 1$ and $T_{ik} = 0$. In many cases, potential outcomes will be affected by each of these simultaneously occurring treatments, and isolating the effect of T_{ik} will not be possible. The ideal situation occurs when Assumption 3 can be avoided altogether because only the treatment of interest changes at the border. In some instances, analysts may also be able to exploit variation in other dimensions such as time to disentangle the compounded effects.

In our example, we must assume that there is no separate county effect on health care utilization, so that the county treatment can be exactly reduced to the state treatment. Another alternative is to define the estimand as a compound treatment effect that includes both a state effect and a county effect – but this is unsatisfactory, because our substantive interest is on isolating the effect of copays. That is, a compound treatment effect will be of little use to a policymaker who wishes to isolate the effect of one treatment in particular.

4.2. Geographically Discontinuous Treatment Assignments and Internal Validity

RD designs are generally assumed to have high levels of internal validity (Lee, 2008). One indication that an RD design may be internally valid is when the design passes a series of falsification tests. Under one form of falsification test, the investigator examines whether treated and control units are similar on predetermined covariates near the cutoff, and tests the hypothesis that there are no RD effects on these covariates. The falsification test is "passed" if these hypotheses cannot be rejected. The same type of falsification test can be applied to GRD and GQE designs. The units close to either side of the boundary are expected to be similar, which suggests testing for differences in observable covariates at the border.

Unfortunately, it is not uncommon to see applications where treated and control units differ in observable characteristics even very close to the border. In practice, we have often found that while covariate imbalances decrease as we move closer to the border, such imbalances are not entirely eliminated even when units are very close to the border (Keele & Titiunik, 2015a, 2016; Keele et al., 2015). Galiani, McEwan, and Quistorff (2017) note the same problem: balance improves as distance to the border decreases, but a few key imbalances remain. In our application, we find that urban areas along the Wisconsin–Illinois border are more comparable than the two states are in general, but we find that significant imbalances remain even restricting our comparison to residents close to the border. In our experience, such imbalances are also found when the data is precisely geo-located (Keele & Titiunik, 2015a, 2016). As such, it is likely that these differences are not simply a consequence of our inability to geo-locate individual observations, but are instead symptomatic of an "endogenous" or "confounding" selection process.

We believe the threats to the internal validity of geographic research designs result from the special nature of their treatment assignment rule. As noted by Lee and Lemieux (2010), RD designs work best when a known treatment assignment rule is imposed on participants – a rule over which participants have no precise control. This generally does not happen in geographically discontinuous treatment assignments. For most geographically assigned treatments, the treatment is assigned based on an existing border around which residents may have been sorting for years, decades, or even longer.[3] In all likelihood, geographic designs would have higher internal validity if a border was drawn for the purpose of treatment assignment rather than treatment assignment being based on an already-existing border.

Most importantly, in evaluating the plausibility of research designs based on geographically discontinuous treatments, it is crucial to remember that most units of interest in the social and biomedical sciences are often able to very precisely select which side of a border will affect them. In a GQE, researchers focus on treated and control units that are close to one another because proximity reduces differences in important observed variables, and there is some reason to believe that proximity also reduces differences in unobserved variables, possibly after conditioning on observables. However, as also discussed by Galiani, McEwan, and Quistorff (2017), the ability of agents to choose the location of their residences means that the required assumptions are less plausible in the typical GQE than in the typical nongeographic RD design. The reason is simply that, in most nongeographic RD designs, precise manipulation of the score (also known as "sorting around the threshold") is considerably more difficult and constitutes aberrant behavior rather than the norm. For example, score manipulation in RD designs based on vote shares or test scores requires engaging in fraud or a post-treatment appeal process. In contrast, many firms and households routinely choose the precise location of their residence to optimize access to education, transportation, tax rates, etc. In natural science applications, such sorting may be less prevalent (see, e.g., Wonkka, Rogers, & Kreuter, 2015), making the GQE potentially more promising.[4]

Given these difficulties, whenever faced with geographically discontinuous treatments, researchers might be tempted to adopt a simple selection-on-observables strategy that ignores distance to the border altogether. We would argue against such a strategy. If balance on observables is improving as the distance to the border decreases, it is possible that geographic proximity is also capturing some unobservable differences as well. In other words, failure to include geography in the conditioning set may, in some applications, make the selection-on-observables assumption less plausible.

The difficulty with invoking an assumption such as Assumption 2 is that, like for any other selection-on-observables assumption, a falsification test is less readily available. Since covariates must be conditioned on before units can be compared, falsification tests cannot rest on covariate balance tests but must instead rely on other forms of evidence. For example, analysts may wish to use other types of falsification tests such as outcomes "known to be unaffected by treatment" or negative controls (Angrist & Krueger, 1999; Lipsitch, Tchetgen, & Cohen, 2010; Rosenbaum, 2002). In addition, investigators might apply a sensitivity analysis for bias from unobserved confounders (Rosenbaum, 2005). Below, we evaluate the design by testing the null hypothesis that, after conditioning on the relevant covariates, there is no treatment effect on pretreatment outcomes. For designs where time-varying outcomes are available, placebo tests on past outcomes can be a very fruitful falsification test. Importantly, the effect on past outcomes must be evaluated in the same way as the actual outcome is analyzed – that is, within the same band and conditioning on the same covariates.

In sum, researchers employing geographically discontinuous treatment assignments must scrutinize their assumptions with extra care. Most designs based on geographically discontinuous treatment assignments in the social sciences will be more accurately characterized and analyzed as GQE designs than as pure RD designs with two-dimensional (geographic) scores.

4.3. Interference

Thus far, we have assumed that the no interference component of SUTVA holds, which implies that treated units cannot interfere with control units in a way that causes the treatment to spill over and affect the control units. In the nongeographic RD with two scores, SUTVA violations of this type would seem rare. For students taking two exams, it is relatively harder to imagine how a student who barely passes the two exams might interfere with other students who barely fail both exams. In geographically discontinuous treatment assignments, however, interference may be more common. The reason is that the analysis relies on the comparison of spatially proximate subjects who may be likely to interact in various ways. Thus, the evaluation of possible forms of interference is a key part of any research design based on a geographically discontinuous treatment.

How might interference arise in our current application? If we expect that the addition of copays reduces health care utilization, the question we must ask is whether less care among residents in Wisconsin can make residents in Illinois sicker – which could occur, for example, if sicker Wisconsin residents started spreading contagious diseases. There are a few reasons why this may not be a serious concern in this case. First, the main mechanism of interference in our application is contagious illnesses, and vaccinations are included in preventative care which still does not require a copay in Wisconsin. To minimize concerns of interference, we could test that preventative visits do not decrease after the addition of copays. Moreover, school districts do not overlap at the border, so a main form of illness transmission among children is precluded. We could also restrict our analysis to chronic conditions where transmission from treated to controls is unlikely or impossible.

Of course, the likelihood of interference varies from application to application. In our application, we would argue that interference is not impossible, but is probably not a first-order concern. In contrast, consider the study by Salazar et al. (2016), who examine the effects of a fruit fly eradication program in coastal areas of Peru on agricultural outcomes. The treatment consisted of both the release of sterile male fruit flies as well as the application of insecticides. The treatment was applied to some geographic regions but not others, and the analysis rests on the comparison of units that are spatially proximate. In this case, it is easy to imagine that fruit fly eradication solutions such as spraying insecticide could affect the control areas.

If interference occurs, analysts need not (and should not) ignore it. Progress can be made if analysts make assumptions about the spatial contagion (Gerber & Green, 2012, chapter 8). For example, Keele and Titiunik (2015a) discuss a framework for thinking about interference with geographically discontinuous treatments. Their approach amounts to adoption of a "doughnut hole" design, where the most spatially proximate units are dropped and less spatially proximate, but comparable, units are used. The validity of this method depends heavily on the underlying treatment assignment mechanism.

4.4. Local Nature of Effects

Treatment effect estimates in the standard RD design are often said to be local because they capture the effect of treatment at the cutoff. The same is true in most designs based on geographically discontinuous treatments, because such designs focus on treatment effects for units that reside within some short distance from the border of interest. In our application, the estimates are local since they apply to residents of both states that live 3–6 miles from the border. This population may or may not be representative of the larger population of either state. In their Honduras study, Galiani, McEwan, and Quistorff (2017) find that the population that resides near the border is systematically different from the population located at the geographic center.

The effect estimates in geographically discontinuous treatment assignments tend be local in nature on another dimension as well. It is often the case that we do not estimate treatment effects using the entire length of the border of interest. For example, we will be unable to estimate treatment effects along the entire Wisconsin–Illinois border. This is true for two reasons. First, most of the border is composed of rural areas with little to any data density. Second, it is often the case that the design is validated by falsification tests along only some parts of the border. For example, in our application, our falsification analysis is successful in Area 3 but not in Area 4. If our estimates are confined to only Area 3, our treatment effect estimates may not generalize to even other parts of the border, much less the entire state.

We note, however, that in applications where a GQE design seems valid and the border is long, the heterogeneity in the population along the border can allow us to estimate treatment effects for relevant subgroups based on demographic, socioeconomic, or other characteristics. This can prove valuable in understanding and possibly predicting the likely effect of the policy in new populations.

4.5. Spatial Treatment Effects

We note one final important feature of designs based on geographically discontinuous treatments. In both the GRD and GQE design, the estimated effects can be spatially located. This is most evident in the geographic RD, where the analysis leads to a curve or set of treatment effects along the border that separates the treated and control areas. This leads to estimated effects that are spatially located, and these treatment effects can be heterogeneous. Multiple spatially located treatment effects can also arise in designs that focus on a band around the border, if the border is divided into different segments that are analyzed separately, as we do in our application.

Thus, using the type of geographic designs we are considering, treatment effects can be mapped to their specific geographic locations to observe whether the treatment effect varies along the geographic border of interest. In other words, we can uncover interesting patterns of geographic treatment effect heterogeneity that may have, for example, important policy implications. Analysts should either outline whether they can identify a pattern in the treatment effects, or treat such heterogeneity as an exploratory analysis.

5. APPLICATION

We now turn to our application and study whether the addition of copays altered health care utilization in Wisconsin. While Wisconsin borders four different states that did not add copays for services, only the border with Illinois has enough population density to carry out the analysis. Even along this border, there were only a few areas with adequate population density. The first area is the border between Lake County in Illinois and Kenosha county in Wisconsin. The second area is the border between McHenry county in Illinois and Walworth county in Wisconsin. The third area is the border between Winnebago county in Illinois and Rock County, Wisconsin. Fig. 1 highlights the counties of interest along the Wisconsin–Illinois border. We refer to the area surrounding the border between Lake and Kenosha counties as Area 1; the area surrounding the border between McHenry and Walworth counties as Area 2; and the area surrounding the border between Winnebago and Rock counties as Area 3. Areas 1 and 2 are each comprised of seven separate zip codes, with four zip codes in the treated area of Wisconsin and three zip codes in the control area of Illinois. In Area 3, there is just a single zip code on each side of the state border.

This third area is, we think, the most promising. Areas 1 and 2 are at the edges of the Chicago metropolitan area. As such, residents in Wisconsin may be wealthier as they reside on the fringes of ex-urban Chicago. However, in Area 3, the state border splits the small urban area of Beloit. Fig. 2 displays the Beloit area, which is partially split by the Wisconsin–Illinois state border.

5.1. Data

As we noted above, due to privacy constraints, we are unable to obtain geographic information about respondents below the zip code level. This poses two challenges. First, we are unable to rely on a geographic RD, and instead have to rely on the analysis of a band around the border. Second, the need to rely on aggregate data also affects our ability to perform falsification tests. Since we do not have individual geographic information, observed covariate imbalances in a small band around the border may

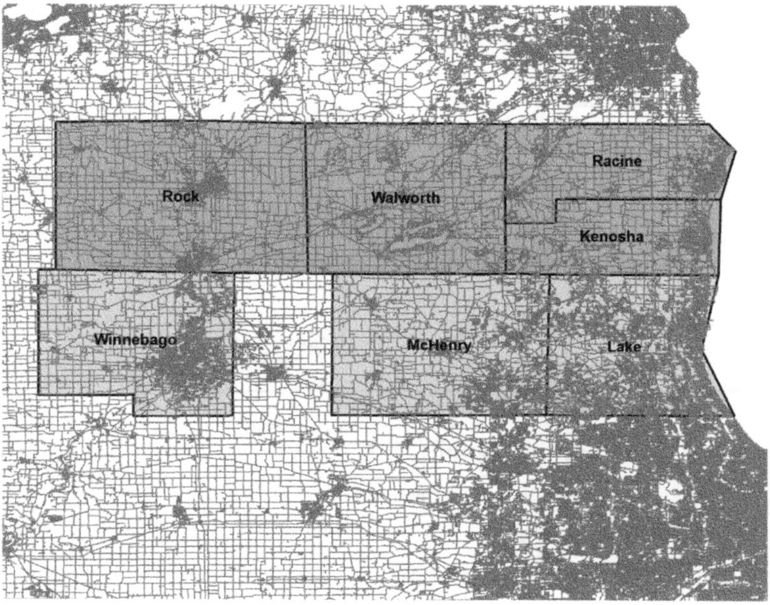

Fig. 1. Map of Counties along the Wisconsin and Illinois Border. *Note*: Counties in darker grey are in Wisconsin, and counties in lighter grey in Illinois.

either be a function of differences in the populations on either side or differences induced by the MAUP discussed above. That is, differences in zip code level means may reflect either actual differences in covariate distributions or bias introduced by aggregation.

While we are restricted by zip code aggregates in the main dataset, we can use other data with more precise geographic information to avoid the issues of aggregation outlined above. One such source of data in the United States is property sales records. Housing prices are often very important in geographic applications. While housing prices may not reflect all neighborhood characteristics (there is some evidence that racial differences are not reflected in house prices, see Bayer, Ferreira, & McMillan, 2007), in general house prices should capture many aspects of local geography. This is because, under hedonic pricing theory, housing prices reflect a wide variety of neighborhood characteristics, including the quality of local services, and school quality (Malpezzi, 2002; Sheppard, 1999). Thus, among all pretreatment covariates, property prices are often one of

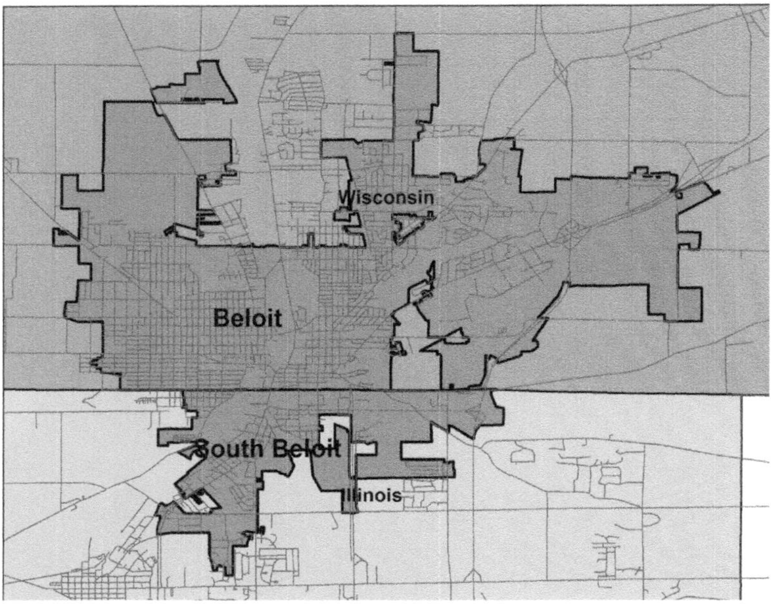

Fig. 2. The Beloit Municipal Area Split by the Wisconsin and Illinois State Border.

the most useful to falsify designs based on geographically discontinuous treatments. Moreover, these data are almost always available in an unaggregated form. Individual property records can be geocoded to allow the analyst to understand whether there is variation in property prices as a function of geographic distance to the border. As such, in our application, we can avoid the bias caused by MAUP by using housing price data. Naturally, in some applications, property sales data may be of limited use. In rural areas, property sales may be too sparse. In some counties, sales records may be unreliable or unavailable. However, when such data is available, it provides an important summary.

Next, we provide details on our primary data source on health care utilization. Our data is based on the Medicaid Analytics Extracts. For individual level patients, we have covariates on sex, three age categories (1–5, 6–14, 15–20), and race. Furthermore, we applied a validated algorithm to the billing codes to classify whether patients have any nonchronic conditions, noncomplex chronic conditions, or complex chronic conditions (Simon et al., 2014). For outcomes, we measure whether patient visited the emergency department, was hospitalized, required acute care, had a well child visit, and usage and type of medications. Since well child visits do not have copayment because they fall under the category of preventive case, they serve as a placebo outcome because they should not change in the post-treatment period. As we noted above, the only geographic information we have for each respondent is his or her zip code.

5.2. Analysis Plan

Next, we outline the steps we implemented to first evaluate our design, and then later estimate the treatment effect for copayments. We begin our analysis using the housing data to evaluate the design, comparing housing prices along the border in all three areas. Since this analysis reveals important differences between treated and control areas, we decide to invoke Assumption 2 – that is, we assume that a comparison of treated and control units is valid only after we restrict our analysis to a narrow band around the border and we condition on a set of observable characteristics. Moreover, we split the long Illinois–Wisconsin border into several areas, each of which we analyze separately – one way to characterize this strategy is to consider border segment indicators as covariates included in the conditioning set.

For areas where the housing data validate the design, we further adjust for patient level covariates. We adjust for differences in patient level covariates using matching. We implemented matching via an integer programming using the `R` package `designmatch` (Zubizarreta, 2012; Zubizarreta & Kilcioglu, 2016). Matching based on integer programming achieves covariate balance directly by minimizing the total sum of distances while constraining the measures of imbalance to be less than or equal to certain tolerances. This allows us to directly set a target level of imbalance before matching. This type of matching also allows us to impose constraints for exact and near-exact matching, and near and near-fine balance for nominal covariates. We perform separate matches for each area along the border. Statistical inferences are based on conventional least-squares methods applied to the matched dataset.

As we noted above, for both treated and control subjects, we have outcome data from before the intervention was put into place in Wisconsin. We could use these covariates in two ways. We could treat them as baseline covariates and match on them along with the other baseline covariates. Alternatively, we could use them as outcomes in a falsification test. That is, we could perform the match on a more limited set of covariates that excludes past outcomes, and then use this past or pretreatment outcomes as outcomes in a placebo analysis. We should find that treated and control subjects do not differ on these placebo outcomes. We opt for this second approach because it allows us to further assess the similarity of Wisconsin and Illinois patients prior to the policy change. That is, using these measures as placebo outcomes allows us to understand whether balancing observable covariates is enough to remove differences in pretreatment outcomes. As we discussed above, this can be an effective falsification strategy when the design relies on Assumption 1. Finally, using the matched data, we analyzed true outcomes. Given that we have pretreatment outcomes, we estimate treatment effects using the method of DID rather than simply report average differences from the post-treatment time period. We report DID estimates based on linear regression models with standard errors clustered at the individual level.

5.3. Falsification Test: Housing Prices

We start by conducting a falsification test using housing sales data. We expect to find no differences in housing prices at the Wisconsin–Illinois border. It is important to note, however, that the expectation that house prices should be equal on either side of the border rests on the assumption that property tax rates are not considerably different. If, for example, property tax rates were higher in the treated than in the control area, a finding that average house prices are similar in both areas would mask a difference in "effective" prices. State-level property tax rates are very similar in Illinois and Wisconsin and therefore we decided not to adjust the raw house prices.[5] But such adjustment may be necessary in other applications.

Falsification tests for geographic treatments vary according to whether continuity or local mean independence assumptions are invoked. When geo-referenced data is available for individual observations or small units and we can implement a pure RD framework, there are two different forms of falsification tests that investigators can employ. First, we can test the hypothesis that there is no treatment effect on observed pretreatment characteristics at each boundary point **b**, using the same local polynomial (or other smoothing methods) used to estimate treatment effects on the outcome.

Alternatively, analysts can apply a geographic balance-test approach to falsification tests. Keele and Titiunik (2015b) outline an algorithm for assessing covariate balance in a geographic context as follows:

- For treated unit *i*, calculate the geographic distance between it and all control units.
- Match unit *i* to the nearest control unit (or set of control units) in terms of this geographic distance.
- Break ties randomly, so that each treated unit *i* is matched to a single control unit.
- Repeat for all treated units.
- Apply standard balance tests such as KS tests or *t*-tests to the spatially matched data.

First, note that the algorithm above is simply a greedy nearest neighbor matching algorithm; however, it is only applied to distance, so the end result is a set of spatially proximate of pairs. Once the spatially proximate pairs have been formed, standard balance tests can be applied to the data. The advantage of this geographic balance-test approach is that one need not select specific points on the border and need not select a bandwidth.

When data availability prevents the implementation of RD methods and researchers invoke local mean independence assumptions instead, the falsification tests must be modified accordingly. When the design is based on Assumption 1, treated and control units within the chosen band D around the border – that is, units with $d_i^\star < D$ – should have indistinguishable average observable characteristics. This serves as an indirect falsification test on the assumption of mean independence within D.

We acquired property records for all houses sold in these counties in both states from January 2007 to January 2010. Using this data, we performed three different comparisons for each of the areas. First, we estimated the difference in house prices for the counties in each area. Next, we restricted our comparison to the zip codes in each area that are contiguous with the state border. Finally, we performed a third comparison where we performed balance tests using the matching algorithm outlined above. Table 1 contains the results for all three areas.

Based on the balance tests, the evidence for the validity of the design is good in one case, mixed in another, and poor in the last one. First, in Area 1, we see that when we compare the counties, home prices in the treated area are, on average, \$150,000 more expensive than in Illinois. However, when we compare adjacent zip codes, the difference shrinks to just under \$8,000. However, once we match, the difference increases to just under \$100,000.

Table 1. Covariate Balance across Wisconsin and Illinois Housing Markets as a Function of Distance.

	County Comparison	Border Zip Codes	Geographic Match
Area 1			
Average price difference	$154,864	$7,624	$92,647
Median price difference	−$88,400	$550	$54,600
Standardized difference	−0.65	0.05	0.54
KS-test *p*-value	0.000	0.378	0.000
Tr sample size	8,387	1,276	1,276
Co sample size	19,302	1,694	1,276
Area 2			
Average price difference	$38,620	$115,163	−$28,518
Median price difference	−$48,000	$55,000	−$51,500
Standardized difference	−0.25	0.52	−0.21
KS-test *p*-value	0.000	0.002	0.00
Tr sample size	74	50	50
Co sample size	9,240	528	50
Area 3			
Average price difference	−$15,231	−$11,197	$2,811
Median price difference	$15,000	−$14,000	$8,950
Standardized difference	0.19	−0.16	−0.04
KS-test *p*-value	0.000	0.027	0.146
Tr sample size	836	71	71
Co sample size	8,587	349	71

Note: The standardized difference is the difference-in-means divided by the pooled standard deviation.

Why does this difference shrink and then grow again after we match? The difference is driven by the fact that we restricted the matched analysis to only those zip codes that are contiguous with the state border. For these two zip codes, the treated area tends to contain homes that are more expensive than the control area. This is even true when one excludes house prices that are more than $500,000. However, when we include homes in all the zip codes around the border, prices in Illinois housing prices actually increase such that the imbalance is removed.

Given this mixed evidence, how should our analysis proceed? Since we are treating this application as a GQE, the results are consistent with what we should expect there: balance holds within a band of zip codes adjacent to the border. However, while balance holds in this band, it clearly does not hold when we pair the most proximate units as we would expect in a GRD design. The balance results for housing in Area 1 illustrate the challenges and complications that may arise in geographic-based identification strategies. Falsification test results are rarely as clean as might be expected under a standard nongeographic RD design.

In Area 2, houses also tended to be more expensive for the first two comparisons and less expensive after matching. However, in every comparison there is little reason to believe that these two areas are particularly comparable with differences in housing prices that range from \$25,000 to \$115,000. In Area 3, we find the best results. While the difference is statistically significant at the county level, the difference is a much smaller \$15,000. Once we match houses, we find the average difference is less than \$3,000 and not statistically significant, while the median difference is less than \$9,000. This difference might be explained by differential property tax rates – see the appendix for more details on property tax rates. Based on this evidence, we discard Area 2 as a possible area of analysis.

5.4. Balance Results and Placebo Tests

Next, we adjust for differences in the subject level covariates via matching. The results before and after matching for Area 1 are in Table 2. After matching, we have 224 matched pairs in Area 1. We measure the discrepancy between treated and control areas using a measure known as the standardized difference, which is the difference-in-means divided by the pooled standard deviation before matching. Ideally, the standardized difference after matching should be less than 0.10. In Area 1, we find that in general the treated and control populations are fairly similar with one exception. The treated area contains a substantially larger fraction of Hispanics than the control area: this part of Illinois is 51% Hispanic, while this area of Wisconsin is 22% Hispanic. We cannot rule out that this is an important confounder. But we note that, in terms of medical conditions, the two populations are quite similar, with standardized differences of less than 0.10. While racial disparities in pediatric care are well documented (Bindman, Chattopadhyay, Osmond, Huen, & Bacchetti, 2005; Eberly, Davidoff, & Miller, 2010; Fisher-Owens, Isong, Soobader, Gansky,

Table 2. Balance Table for Area 1.

	Before Matching				After Matching			
	Mean T	Mean C	Std Diff.	p-value	Mean T	Mean C	Std Diff.	p-value
% White	0.51	0.29	0.47	0.00	0.51	0.35	0.34	0.00
% African-American	0.16	0.12	0.11	0.21	0.16	0.14	0.04	0.69
% Hispanic	0.22	0.51	−0.61	0.00	0.22	0.40	−0.39	0.00
% Other	0.11	0.09	0.06	0.48	0.11	0.10	0.02	0.88
Age 1−5	0.00	0.01	−0.04	0.68	0.00	0.00	0.00	1.00
Age 6−14	0.80	0.85	−0.11	0.22	0.80	0.83	−0.07	0.46
Age 15−20	0.19	0.15	0.12	0.18	0.19	0.17	0.07	0.46
Male	0.54	0.48	0.11	0.22	0.54	0.49	0.10	0.30
Nonchronic condition	0.78	0.78	−0.01	0.93	0.78	0.79	−0.04	0.65
Noncomplex chronic condition	0.19	0.17	0.04	0.66	0.19	0.16	0.07	0.46
Complex chronic condition	0.04	0.05	−0.06	0.51	0.04	0.04	−0.04	0.63

Notes: "*T*" denotes treated observations in Wisconsin, "*C*" denotes control observations in Illinois, and "Std Diff." denotes standardized difference, that is, the treated−control difference-in-means divided by the prematching pooled standard deviation. The p-value columns report the p-value associated with the test of the hypothesis that the treated−control difference-in-means is zero. Sample size before matching: IL = 273, WI = 224, analysis above is based on 224 matched pairs.

Weintraub, Platt, & Newacheck, 2013; Hakmeh, Barker, Szpunar, Fox, & Irvin, 2010; Rose, Parish, Yoo, Grady, Powell, & Hicks-Sangster, 2010), race or ethnicity is often used a proxy for socioeconomic status in the study of health disparities. Here, both treated and control have similar incomes by design. As such, the differences in the racial make-up across areas might be less consequential if the health status is similar across the two areas.

Table 3 contains the same information for Area 3, where after matching we have 52 matched pairs. Here, we observe a similar disparity in racial and ethnic makeup, but in reverse. In Area 3, the treated area tends to contain more Hispanic and African-American residents. Once we match those disparities are reduced. However, even after we match, there are notable discrepancies in the standardized differences for several covariates. For example, there are notable differences in medical conditions. Next, we further evaluate both areas using placebo tests.

Table 3. Balance Table for Area 3.

	Before Matching				After Matching			
	Mean *T*	Mean *C*	Std Diff.	*p*-value	Mean *T*	Mean *C*	Std Diff.	*p*-value
% White	0.48	0.72	−0.49	0.01	0.48	0.60	−0.24	0.24
% African-American	0.23	0.14	0.25	0.18	0.23	0.19	0.10	0.64
% Hispanic	0.19	0.07	0.35	0.06	0.19	0.12	0.23	0.28
% Other	0.10	0.07	0.08	0.66	0.10	0.10	0.00	1.00
Age 1−5	0.02	0.01	0.05	0.76	0.02	0.02	0.00	1.00
Age 6−14	0.79	0.84	−0.13	0.47	0.79	0.83	−0.10	0.62
Age 15−20	0.19	0.15	0.12	0.52	0.19	0.15	0.10	0.61
Male	0.67	0.54	0.27	0.13	0.67	0.63	0.08	0.68
Nonchronic condition	0.69	0.77	−0.16	0.36	0.69	0.75	−0.13	0.52
Noncomplex chronic condition	0.23	0.17	0.14	0.43	0.23	0.15	0.19	0.32
Complex chronic condition	0.08	0.06	0.06	0.74	0.08	0.10	−0.08	0.73

Notes: "*T*" denotes treated observations in Wisconsin, "*C*" denotes control observations in Illinois, and "Std Diff." denotes standardized difference, that is, the treated−control difference-in-means divided by the prematching pooled standard deviation. The *p*-value columns report the p-value associated with the test of the hypothesis that the treated−control difference-in-means is zero. Sample size before matching: IL = 81, WI = 52, analysis above is based on 52 matched pairs.

For all children enrolled in CHIP, we observe several measures of health care utilization in 2007 before the copayments were added to BC+. As we noted above, we chose not to match on these outcomes. Instead, knowing these measures must be unaffected by the treatment, we use them as placebo outcomes after matching to evaluate whether our treated and control populations are comparable. Ideally, we would like to see, conditional on the matched covariates, that levels of health care utilization are highly comparable before the addition of copays. The outcomes are observed as raw counts of health care utilization. For example, we observe the number of hospitalizations for each enrollee. However, not all patients have a full year of data, since people may enroll at any time in the program. Therefore, we express the outcome as a rate: the average level of usage per month. All outcomes in Tables 4 and 5 are expressed as average monthly rates of usage.

Table 4 contains these placebo test results for Area 1. Unfortunately, the placebo test results demonstrate that treated and control enrollees have different patterns of health care utilization in the pretreatment period. Children in Wisconsin tend to use the emergency room at higher rates, and use a larger number of medications. However, one pattern in the table is reassuring. The pattern of differences switches for each outcome. In some cases, the treated use more health care, in others, the controls use more. This, at least, does not suggest that one area displays a pattern of systematically higher or lower health care usage. Given the results from the placebo tests, we rematched units from Area 1, this time including the outcomes from 2007 in the conditioning set. This helps further remove differences in the treated–control populations – but it is only a "solution" to the lack of comparability between treatment and control units under the assumption that, after these placebo outcomes are conditioned on, no systematic differences remain.

Table 4. Effect of Copays on "Placebo" Health Care Utilization (Outcomes Measured before Copay Adoption), Area 1.

	Wisconsin	Illinois	Difference	95% CI	*p*-value
Emergency Department Visit	0.38	0.30	0.07	[−0.01, 0.16]	0.110
Hospitalization	0.004	0.04	−0.04	[−0.06, −0.003]	0.032
Acute Visit	1.68	1.41	0.27	[−0.007, 0.54]	0.056
Well Child Visit	0.36	0.54	−0.17	[−0.24, −0.11]	0.000
Any Outpatient Medication	0.43	0.27	0.16	[0.061, 0.26]	0.002

Notes: Sample size before matching: IL = 273, WI = 224, analysis above is based on 224 matched pairs. Outcome is scaled as average counts per month.

Table 5. Effect of Copays on "Placebo" Health Care Utilization (Outcomes Measured before Copay Adoption), Area 3.

	Wisconsin	Illinois	Difference	95% CI	*p*-value
Emergency department visit	0.29	0.08	0.21	[0.012, 0.41]	0.033
Hospitalization	0.04	0.02	0.02	[−0.05, 0.09]	0.569
Acute visit	2.0	1.6	0.48	[−0.46, 1.42]	0.311
Well child visit	0.35	0.5	−0.15	[−0.40, 0.09]	0.209
Any outpatient medication	0.54	0.47	0.07	[−0.21, 0.35]	0.628

Notes: Sample size before matching: IL = 273, WI = 224, analysis above is based on 224 matched pairs. Outcome is scaled as average counts per month.

Table 5 contains the same results for Area 3. In Area 3, the treated and control have much more similar patterns of pretreatment health care utilization. Here, only the difference in emergency room visits is statistically significant. For the other variables, while the treated area does tend to have higher usage rates, the differences tend to be small. While the validation of the design has serious issues in Area 1, Area 3 shows much better comparability around the state border. The additional balance results are in the appendix.

5.5. Outcome Estimates

We now turn to outcome estimates. As we noted in the analysis plan, we use DID for impact estimates. Of course, under DID, we must assume that without the increase in copayments, the average level of health care utilization in Wisconsin would have followed the same path from 2007 to 2010 as the path of health care utilization in Illinois. To the best of our knowledge, no other policy changes relating to health care occurred during this period. However, we plot the average monthly usage by quarter from 2007 to 2008 for each outcome in the appendix.

Table 6 contains the DID estimates for Area 1. If we examine the overall pattern of effects, the evidence in favor of a causal effect is weak. First, while the sign of the effect is in the expected direction for ER visits, hospitalizations, and acute care visits, none of these estimates are statistically significant. Moreover, the magnitude of the estimates for ER visits and

Table 6. Effect of Copays on Health Care Utilization: Differences-in-Differences Estimates for Area 1.

	DID Estimate	*p*-value
Emergency department visit	−0.003	0.638
Hospitalization	−0.0003	0.929
Acute visit	−0.04	0.062
Well child visit	0.02	0.005
Any outpatient medication	0.14	0.075

Notes: Standard errors adjusted for individual level clustering. Sample size before matching: IL = 273, WI = 224, analysis above is based on 224 matched pairs. Outcome is scaled as average counts per month.

Table 7. Effect of Copays on Health Care Utilization: Differences-in-Differences Estimates for Area 3.

	DID Estimate	*p*-value
Emergency department visit	−0.005	0.830
Hospitalization	−0.002	0.492
Acute visit	−0.12	0.014
Well child visit	0.07	0.001
Any outpatient medication	0.04	0.796

Notes: Standard errors adjusted for individual level clustering. Sample size before matching: IL = 81, WI = 52, analysis above is based on 52 matched pairs. Outcome is scaled as average counts per month.

hospitalizations is quite small, −0.003 and −0.0003, respectively. Second, well child visits are a placebo outcome, since no additional copay was added for these visits. However, the point estimate for well child visits shows a statistically significant increase. For medications, the sign of the estimate is in the unexpected direction.

Table 7 contains the DID estimates for Area 3. The pattern is remarkably similar to Area 1. Again, the point estimates for ER visits and hospitalizations are very small, −0.005 and −0.002, respectively. The estimate for acute care visits is in the expected negative direction and is statistically significant at the 0.05 level, but again the treatment effect estimate for well child visits is in the unexpected positive direction and clearly statistically significant. Finally, the point estimate for medications is positive but statistically indistinguishable from zero. In sum, in both areas, there is little evidence that the addition of copays changed health care usage for CHIP enrollees. Importantly, the copay "effects" on an outcome that is known not to be affected by the treatment – well child visits – raises doubts about the credibility of the design.

6. DISCUSSION

The credibility revolution in the social sciences has been led by an emphasis on research design (Angrist & Pischke, 2010), and the RD design has been an important part of that revolution. A key advantage of RD designs over other research strategies is that a known treatment assignment rule is designed and enforced, giving researchers an objective basis to at least partly assess the plausibility of the assumptions invoked.

As we have discussed, however, research designs based on geographically discontinuous treatment assignments often fall short of the promise of the best RD designs. This is true for several reasons. First, the lack of availability of geo-referenced data at the individual level prevents researchers from using the RD framework because it is impossible to identify the observations arbitrarily close to the boundary. This is why in many applications, including ours, researchers are forced to rely on ignorability assumptions in some band around the border. Second, borders tend to be created temporally prior to treatment assignment. That is, while treatments change at borders, it is rare for policymakers to draw borders and then assign treatments based on these newly drawn borders. Instead, borders tend to be divisions of long-standing importance. This, together with people's and organizations' ability to sort very precisely around existing borders, can lead to systematic differences in the populations on either side of them.

We do not believe these obstacles invalidate all research designs based on discontinuous geographic treatment assignments, but the existence of such challenges does imply that these designs need careful evaluation. On this aspect, geographic designs are like many other natural experimental designs. For example, good instrumental variable designs can provide convincing evidence for causal effects, especially in the context of randomized designs with noncompliance. Many instrumental variables designs fall well short of that ideal, but can still provide important evidence about causal effects. Geographic designs are similar in this regard: many (perhaps most) fall short of the ideal design, but some of them can provide plausible and useful evidence about causal effects.

Our application illustrated many of the challenges that are common in research designs based on geographically discontinuous treatment assignments. In Area 1, significant demographic differences between treated and control units near the border remain even after matching, and the "effects" of copays on placebo outcomes – various measures of health care utilization before copays were adopted – are also statistically significant. In Area 3, the preexisting treated–control differences are less severe, and the effects on most placebo outcomes are indistinguishable from zero. But in both Area 1 and Area 3, copays seem to significantly affect well child visits, an outcome that is known not to be affected by the treatment. Taken together, the results from these falsification tests raise doubts about the credibility of the design, and suggest that we should interpret the effects on the actual outcomes with caution.

NOTES

1. Software available at https://sites.google.com/site/rdpackages/home

2. Note that the notation we have used so far suggests that the potential outcomes do not depend on the unit's specific geographic coordinates other than through the treatment indicator. This exclusion restriction may be plausible within a narrow band, but it is important to note that it is not implied by mean independence assumptions such as Assumptions 1 and 2. See Sekhon and Titiunik (2017) for a discussion of the important role of exclusion restriction assumptions in the local randomization interpretation of RD designs.

3. For an exception, see MacDonald et al. (2016), where the treatment is assigned based on a newly drawn border.

4. In the study by Wonkka et al. (2015), researchers use the geographic variation in statutory reforms to study the impact of less stringent liability standards on the use of prescribed fires. In this case, focusing on spatially proximate units is likely to be effective, because proximate areas have similar weather, vegetation, topography, etc., all of which are crucial determinants of both spontaneous and prescribed burning.

5. A report by the Tax Foundation based on U.S. Census Bureau data estimated a median 2009 property tax rate (as a percentage of home value) of 1.73% for Illinois and 1.76% for Wisconsin (The Tax Foundation, 2010), placing both states in the top ten states in the country with highest property tax rates (Wisconsin ranks number 4 and Illinois ranks number 6). However, we note that, looking at county-level tax rates in the 2006–2008 period, we find somewhat higher property tax rates in the Illinois counties than in the Wisconsin counties in our sample; see the appendix for details.

ACKNOWLEDGMENTS

For helpful comments and suggestions we thank Matias Cattaneo, Sebastian Galiani, Patrick McEwan, two anonymous reviewers, and participants at the Advances in Econometrics conference held in May 2016 at the Department of Economics at the University of Michigan. Titiunik gratefully acknowledges financial support from the National Science Foundation (SES 1357561).

REFERENCES

Angrist, J. D., & Krueger, A. B. (1999). Empirical strategies in labor economics. In O. Ashenfelter & D. Card (Eds.), *Handbook of labor economics* (Vol. 3, pp. 1277–1366). Amsterdam: Elsevier Science.

Angrist, J. D., & Pischke, J.-S. (2010). The credibility revolution in empirical economics: How better research design is taking the con out of econometrics. *Journal of Economic Perspectives*, *24*(2), 3–30.

Asch, D. A., Nicholson, S., Srinivas, S., Herrin, J., & Epstein, A. J. (2009). Evaluating obstetrical residency programs using patient outcomes. *Jama*, *302*(12), 1277–1283.

Asiwaju, A. I. (1985). *Partitioned Africa: Ethnic relations across Africa's International Boundaries, 1884-1984*. London: C. Hurst.

Baicker, K., Taubman, S. L., Allen, H. L., Bernstein, M., Gruber, J. H., Newhouse, J. P., ... Finkelstein, A. N. (2013). The Oregon experiment – Effects of Medicaid on clinical outcomes. *New England Journal of Medicine*, *368*(18), 1713–1722. PMID: 23635051.

Bayer, P., Ferreira, F., & McMillan, R. (2007). A unified framework for measuring preferences for schools and neighborhoods. *Journal of Political Economy*, *115*(4), 588–638.

Berger, D. (2009). *Taxes, Institutions and Local Governance: Evidence from a Natural Experiment in Colonial Nigeria*. Unpublished manuscript.

Bindman, A. B., Chattopadhyay, A., Osmond, D. H., Huen, W., & Bacchetti, P. (2005). The impact of Medicaid managed care on hospitalizations for ambulatory care sensitive conditions. *Health Services Research*, *40*(1), 19–38.

Calonico, S., Cattaneo, M. D., & Titiunik, R. (2014a). Robust data-driven inference in the regression-discontinuity design. *Stata Journal*, *14*(4), 909–946.

Calonico, S., Cattaneo, M. D., & Titiunik, R. (2014b). Robust nonparametric confidence intervals for regression-discontinuity designs. *Econometrica*, *82*(6), 2295–2326.

Calonico, S., Cattaneo, M. D., & Titiunik, R. (2015). Rdrobust: An R package for robust nonparametric inference in regression-discontinuity designs. *R Journal*, *7*(1), 38–51.

Calonico, S., Cattaneo, M. D., Farrell, M. H., & Titiunik, R. (2017). rdrobust: Software for regression discontinuity designs. *Stata Journal*, forthcoming.

Campbell, J. D., Allen-Ramey, F., Sajjan, S. G., Maiese, E. M., & Sullivan, S. D. (2011). Increasing pharmaceutical copayments: Impact on asthma medication utilization and outcomes. *American Journal of Managed Care*, *17*(10), 703–710. Retrieved from http://www.ncbi.nlm.nih.gov/entrez/query.fcgi?cmd=Retrieve&db = PubMed&dopt = Citation list_uids = 22106463

Card, D., & Krueger, A. B. (1994). Minimum wages and employment: A case study of the fast-food industry in New Jersey and Pennsylvania. *The American Economic Review*, *84*(4), 772–793.

Cattaneo, M. D., Frandsen, B., & Titiunik, R. (2015). Randomization inference in the regression discontinuity design: An application to party advantages in the U.S. Senate. *Journal of Causal Inference*, *3*(1), 1–24.

Cattaneo, M. D., Titiunik, R., & Vazquez-Bare, G., (2017). Comparing inference approaches for RD designs: A reexamination of the effect of head start on child mortality. *Journal of Policy Analysis and Management*, forthcoming.

Chen, Y., Ebenstein, A., Greenstone, M., & Li, H. (2013). Evidence on the impact of sustained exposure to air pollution on life expectancy from China's Huai River policy. *Proceedings of the National Academy of Sciences*, *110*(32), 12936–12941.

Cox, D. R. (1958). *Planning of experiments*. New York, NY: Wiley.

Dell, M. (2010). The persistent effects of Peru's mining Mita. *Ecometrica*, *78*(6), 1863–1903.

Department of Health and Family Services. (2008a). BadgerCare plus information for members. Retrieved from https://www.dhs.wisconsin.gov/forwardhealth/customerhelp/memberupdate-01-2008-pharmacy.pdf. Accessed on April 2016.

Department of Health and Family Services. (2008b). Wisconsin Medicaid and Badgercare recipient update. Retrieved from https://www.dhs.wisconsin.gov/forwardhealth/customerhelp/memberupdate-01-2008-bcplus.pdf. Accessed on April 2016.

Eberly, T., Davidoff, A., & Miller, C. (2010). Managing the gap: Evaluating the impact of Medicaid managed care on preventive care receipt by child and adolescent minority populations. *Journal of Health Care for the Poor and Underserved*, *21*(1), 92–111.

Fan, J., & Gijbels, I. (1996). *Local polynomial modelling and its applications*. Boca Raton, FL: Chapman & Hall.

Fisher-Owens, S. A., Isong, I. A., Soobader, M.-J., Gansky, S. A., Weintraub, J. A., Platt, L. J., & Newacheck, P. W. (2013). An examination of racial/ethnic disparities in children's oral health in the United States. *Journal of Public Health Dentistry*, *73*(2), 166–174.

Galiani, S., McEwan, P. J., & Quistorff, B. (2017). External and internal validity of a geographic quasi-experiment embedded in a cluster-randomized experiment. In M. D. Cattaneo & J. C. Escanciano (Eds.), *Regression discontinuity designs: Theory and applications* (Vol. 38). Advances in Econometrics. Bingley: Emerald Publishing Limited.

Gerber, A. S., & Green, D. P. (2012). *Field experiments: Design, analysis, and interpretation*. New York, NY: Norton.

Gerber, A. S., Kessler, D. P., & Meredith, M. (2011). The persuasive effects of direct mail: A regression discontinuity based approach. *Journal of Politics*, *73*(1), 140–155.

Haggerty, R. J. (1985). The Rand Health Insurance Experiment for children. *Pediatrics*, *75*(5), 969–971.

Hahn, J., Todd, P., & van der Klaauw, W. (2001). Identification and estimation of treatments effects with a regression-discontinuity design. *Econometrica*, *69*(1), 201–209.

Hakmeh, W., Barker, J., Szpunar, S. M., Fox, J. M., & Irvin, C. B. (2010). Effect of race and insurance on outcome of pediatric trauma. *Academic Emergency Medicine*, *17*(8), 809–812.

Heberlein, M., Brooks, T., Alker, J., Artiga, S., & Stephens, J. (2013). Getting into Gear for 2014: Findings from a 50-State Survey of Eligibility, Enrollment, Renewal, and Cost-Sharing Policies in Medicaid and CHIP, 2012–2013. Retrieved from https://kaiserfamilyfoundation.files.wordpress.com/2013/01/8130.pdf. Accessed on April 2016.

Heberlein, M., Brooks, T., Guyer, J., Artiga, S., & Stephens, J. (2011). Holding steady, looking ahead: Annual findings of a 50-State Survey of Eligibility Rules, Enrollment and Renewal Procedures, and Cost-Sharing Practices in Medicaid and CHIP, 2010-2011. Retrieved from https://kaiserfamilyfoundation.files.wordpress.com/2013/01/8130.pdf. Accessed on April 2016.

Huber, G. A., & Arceneaux, K. (2007). Identifying the persuasive effects of presidential advertising. *American Journal of Political Science*, *51*(4), 957–977.

Imbens, G. W., & Kalyanaraman, K. (2012). Optimal bandwidth choice for the regression discontinuity estimator. *Review of Economic Studies*, *79*(3), 933–959.

Imbens, G. W., & Zajonc, T. (2011). *Regression discontinuity design with multiple forcing variables*. Working Paper. Unpublished manuscript.

Imbens, G., & Lemieux, T. (2008). Regression discontinuity designs: A guide to practice. *Journal of Econometrics*, *142*(2), 615–635.

Kaiser Family Foundation. (2016). *Total number of children ever enrolled in CHIP annually*. Retrieved from http://kff.org/other/state-indicator/annual-chip-enrollment/

Kearney, M. S., & Levine, P. B. (2014). Media influences on social outcomes: The impact of MTV's 16 and pregnant on teen childbearing. *American Economic Review*, *105*(12), 3597–3632.

Keele, L. J., & Titiunik, R. (2015a). *Bounding treatment effects under interference in geographic natural experiments: An application to all-mail voting in Colorado*. Working Paper. Unpublished manuscript.

Keele, L. J., & Titiunik, R. (2015b). Geographic boundaries as regression discontinuities. *Political Analysis*, *23*(1), 127–155.

Keele, L. J., & Titiunik, R. (2016). Natural experiments based on Geography. *Political Science Research and Methods*, *4*(1), 65–95.

Keele, L. J., Titiunik, R., & Zubizarreta, J. (2015). Enhancing a geographic regression discontinuity design through matching to estimate the effect of ballot initiatives on voter turnout. *Journal of the Royal Statistical Society: Series A*, *178*(1), 223–239.

Kern, H. L., & Hainmueller, J. (2008). Opium for the masses: How foreign media can stabilize authoritarian regimes. *Political Analysis*, *17*(2), 377–399.

Krasno, J. S., & Green, D. P. (2008). Do televised presidential ads increase voter turnout? Evidence from a natural experiment. *Journal of Politics*, *70*(1), 245–261.

Laitin, D. D. (1986). *Hegemony and culture: Politics and religious change among the Yoruba*. Chicago, IL: University of Chicago Press.

Lavy, V. (2010). Effects of free choice among public schools. *The Review of Economic Studies*, *77*(3), 1164–1191.

Lee, D. S. (2008). Randomized experiments from non-random selection in U.S. house elections. *Journal of Econometrics*, *142*(2), 675–697.

Lee, D. S., & Card, D. (2008). Regression discontinuity inference with specification error. *Journal of Econometrics*, *142*(2), 655–674.

Lee, D. S., & Lemieux, T. (2010). Regression discontinuity designs in economics. *Journal of Economic Literature*, *48*(2), 281–355.

Leibowitz, A., Jr., Manning, W. G., Keeler, E. B., Duan, N., Lohr, K. N., & Newhouse, J. P. (1985). Effect of cost-sharing on the use of medical services by children: Interim results from a randomized controlled trial. *Pediatrics*, *75*(5), 942–951.

Lipsitch, M., Tchetgen, E. T., & Cohen, T. (2010). Negative controls: A tool for detecting confounding and bias in observational studies. *Epidemiology (Cambridge, Mass.)*, *21*(3), 383–388.

Lohr, K. N., Brook, R. H., Kamberg, C. J., Goldberg, G. A., Leibowitz, A., Keesey, J., ... Newhouse, J. P. (1986). Use of medical care in the Rand Health Insurance Experiment. Diagnosis- and service-specific analyses in a randomized controlled trial. *Medical Care*, *24*(9 Suppl), S1–S87. Retrieved from http://www.ncbi.nlm.nih.gov/entrez/query.fcgi?cmd=Retrieve&db=PubMed&dopt=Citation&list_uids=3093785

MacDonald, J. M., Klick, J., & Grunwald, B. (2016). The effect of private police on crime: evidence from a geographic regression discontinuity design. *Journal of the Royal Statistical Society: Series A (Statistics in Society)*, *3*, 831–846.

Magruder, J. R. (2012). High unemployment yet few small firms: The role of centralized bargaining in South Africa. *American Economic Journal: Applied Economics*, *4*(3), 138–166.

Malpezzi, S. (2002). Hedonic pricing models and house price indexes: A select review. In K. Gibb & A. O'Sullivan (Eds.), *Housing economics and public policy: Essays in Honour of Duncan Maclennan* (pp. 67–89). Oxford: Blackwell Publishing.

Michalopoulos, S., & Papaioannou, E. (2014). National institutions and subnational development in Africa. *The Quarterly Journal of Economics*, *129*(1), 151–213.

Middleton, J. A., & Green, D. P. (2008). Do community-based voter mobilization campaigns work even in battleground states? Evaluating the effectiveness of MoveOn's 2004 Outreach Campaign. *Quarterly Journal of Political Science*, *3*(1), 63–82.

Miguel, E. (2004). Tribe or nation? Nation building and public goods in Kenya versus Tanzania. *World Politics*, *56*(3), 327–362.

Miles, W. F. S. (1994). *Hausaland divided: Colonialism and independence in Nigeria and Niger*. Ithaca, NY: Cornell University Press.

Miles, W. F. S., & Rochefort, D. (1991). Nationalism versus ethnic identity in Sub-Saharan Africa. *American Political Science Review*, *85*(2), 393–403.

Nall, C. (2015). The political consequences of spatial policies: How interstate highways caused geographic polarization. *Journal of Politics*, *77*(2), 394–406.

Openshaw, S. (1984). *The modifiable areal unit problem*. Norwich, CT: Geo Books.

Papay, J. P., Willett, J. B., & Murnane, R. J. (2011). Extending the regression-discontinuity approach to multiple assignment variables. *Journal of Econometrics*, *161*(2), 203–207.

Pence, K. M. (2006). Foreclosing on opportunity: State laws and mortgage credit. *The Review of Economics and Statistics*, *88*(1), 177–182.

Posner, D. N. (2004). The political salience of cultural difference: Why Chewas and Tumbukas are allies in Zambia and Adversaries in Malawi. *The American Political Science Review*, *98*(4), 529–545.

Reardon, S. F., & Robinson, J. P. (2012). Regression discontinuity designs with multiple rating-score variables. *Journal of Research on Educational Effectiveness*, *5*(1), 83–104.

Rose, R. A., Parish, S. L., Yoo, J., Grady, M. D., Powell, S. E., & Hicks-Sangster, T. K. (2010). Suppression of racial disparities for children with special health care needs among families receiving Medicaid. *Social Science & Medicine*, *70*(9), 1263–1270.

Rosenbaum, P. R. (2002). *Observational studies* (2nd ed.). New York, NY: Springer.

Rosenbaum, P. R. (2005). Sensitivity analysis in observational studies. In B. S. Everitt & D. C. Howell (Eds.), *Encyclopedia of statistics in behavioral science* (Vol. 4, pp. 1809–1814). Chichester: John Wiley and Sons.

Ross, D. C., & Marks, C. (2009). *Challenges of providing health coverage for children and parents in a recession: A 50 state update on eligibility rules, enrollment and renewal procedures, and cost-sharing practices in Medicaid and SCHIP in 2009*. Retrieved from https://kaiserfamilyfoundation.files.wordpress.com/2013/01/7855.pdf. Accessed on April 2016.

Rubin, D. B. (1986). Which ifs have causal answers. *Journal of the American Statistical Association*, *81*(396), 961–962.

Salazar, L., Maffioli, A., Aramburu, J., & Agurto Adrianzen, M. (2016). *Estimating the impacts of a fruit fly eradication program in Peru: A geographical regression discontinuity approach*. IDB Working Paper Series, IDB-WP-677. Infrastructure and Environment Sector. Environment Rural Development Disaster Risk Management Division.

Schumann, A. (2014). Persistence of population shocks: Evidence from the occupation of West Germany after World War II. *American Economic Journal: Applied Economics*, *6*(3), 189–205.

Sekhon, J. S., & Titiunik, R. (2017). On interpreting the regression discontinuity design as a local experiment. In M. Cattaneo & J. C. Escanciano (Eds.), *Regression discontinuity designs: Theory and applications* (Vol. 38). Advances in Econometrics. Bingley: Emerald Publishing Limited.

Sen, B., Blackburn, J., Morrisey, M. A., Kilgore, M. L., Becker, D. J., Caldwell, C., & Menachemi, N. (2012). Did copayment changes reduce health service utilization among CHIP enrollees? Evidence from Alabama. *Health Services Research*, *47*(4), 1603–1620.

Sheppard, S. (1999). Hedonic analysis of housing markets. In P. Cheshire & E. S. Mills (Eds.), *Handbook of regional and urban economics* (Vol. 3, pp. 1595–1635). Springer: Elsevier, Applied Urban Economic.

Simon, T. D., Cawthon, M. L., Stanford, S., Popalisky, J., Lyons, D., Woodcox, P., ... Mangione-Smith, R. (2014). Pediatric Medical complexity algorithm: A new method to stratify children by medical complexity. *Pediatrics*, *133*(6), e1647–e1654.

The Tax Foundation. (2010). *Property taxes on owner-occupied housing, By State, 2009*. Technical report The Tax Foundation.

The Tax Foundation. (2012). *Median effective property tax rates by county, ranked by total taxes paid, 3-year average, 2008-2010*. Technical report The Tax Foundation.

Valdez, R. B., Brook, R. H., Rogers, W. H., Jr., Ware, J. E., Keeler, E. B., Sherbourne, C. A., ... Newhouse, J. P. (1985). Consequences of cost-sharing for children's health. *Pediatrics*, *75*(5), 952–961.

Wong, V. C., Steiner, P. M., & Cook, T. D. (2013). Analyzing regression-discontinuity designs with multiple assignment variables: A comparative study of four estimation methods. *Journal of Educational and Behavioral Statistics*, *38*(2), 107–141.

Wonkka, C. L., Rogers, W. E., & Kreuter, U. P. (2015). Legal barriers to effective ecosystem management: Exploring linkages between liability, regulations, and prescribed fire. *Ecological Applications*, *25*(8), 2382–2393.

Young, C., Varner, C., Lurie, I., & Prisinzano, R. (2014). Millionaire migration and the taxation of the elite: Evidence from administrative data. *American Sociological Review*, *81*(3), 421–446.

Zubizarreta, J. R. (2012). Using mixed integer programming for matching in an observational study of kidney failure after surgery. *Journal of the American Statistical Association*, *107*(500), 1360–1371.

Zubizarreta, J. R., & Kilcioglu, C. (2016). *Designmatch: Construction of optimally matched samples for randomized experiments and observational studies that are balanced by design*. R package version 0.1.1.

APPENDIX A: ADDITIONAL BALANCE RESULTS

Table A1. Balance Table for Area 1 that Matches on Past Outcomes.

	Before Matching				After Matching			
	Mean *T*	Mean *C*	Std Diff.	*p*-Value	Mean *T*	Mean *C*	Std Diff.	*p*-Value
% White	0.51	0.29	0.47	0.00	0.51	0.35	0.34	0.00
% African-American	0.16	0.12	0.11	0.21	0.16	0.14	0.04	0.69
% Hispanic	0.22	0.51	−0.61	0.00	0.22	0.40	−0.39	0.00
% Other	0.11	0.09	0.06	0.48	0.11	0.10	0.02	0.88
Age 1–5	0.00	0.01	−0.04	0.68	0.00	0.00	0.00	1.00
Age 6–14	0.80	0.85	−0.11	0.22	0.80	0.83	−0.07	0.46
Age 15–20	0.19	0.15	0.12	0.18	0.19	0.17	0.07	0.46
Male	0.54	0.48	0.11	0.22	0.54	0.46	0.14	0.13
Non-chronic condition	0.78	0.78	−0.01	0.93	0.78	0.79	−0.03	0.73
Noncomplex chronic condition	0.19	0.17	0.04	0.66	0.19	0.17	0.06	0.54
Complex chronic condition	0.04	0.05	−0.06	0.51	0.04	0.04	−0.04	0.63
ED visit	0.38	0.29	0.12	0.21	0.38	0.30	0.10	0.31
IP visit	0.00	0.04	−0.16	0.06	0.00	0.04	−0.18	0.07
Sick visit	1.68	1.32	0.16	0.07	1.68	1.41	0.12	0.19
Well child visit	0.36	0.60	−0.36	0.00	0.36	0.54	−0.26	0.01
Total medications	4.90	2.79	0.27	0.00	4.90	3.10	0.23	0.02

Notes: "*T*" denotes treated observations in Wisconsin, "*C*" denotes control observations in Illinois, and "Std Diff." denotes standardized difference, that is, the treated–control difference-in-means divided by the prematching pooled standard deviation. The *p*-value columns report the *p*-value associated with the test of the hypothesis that the treated–control difference-in-means is zero. Sample size before matching: IL = 273, WI = 224, analysis above is based on 224 matched pairs.

APPENDIX B: TRENDS IN HEALTH CARE UTILIZATION 2007–2008

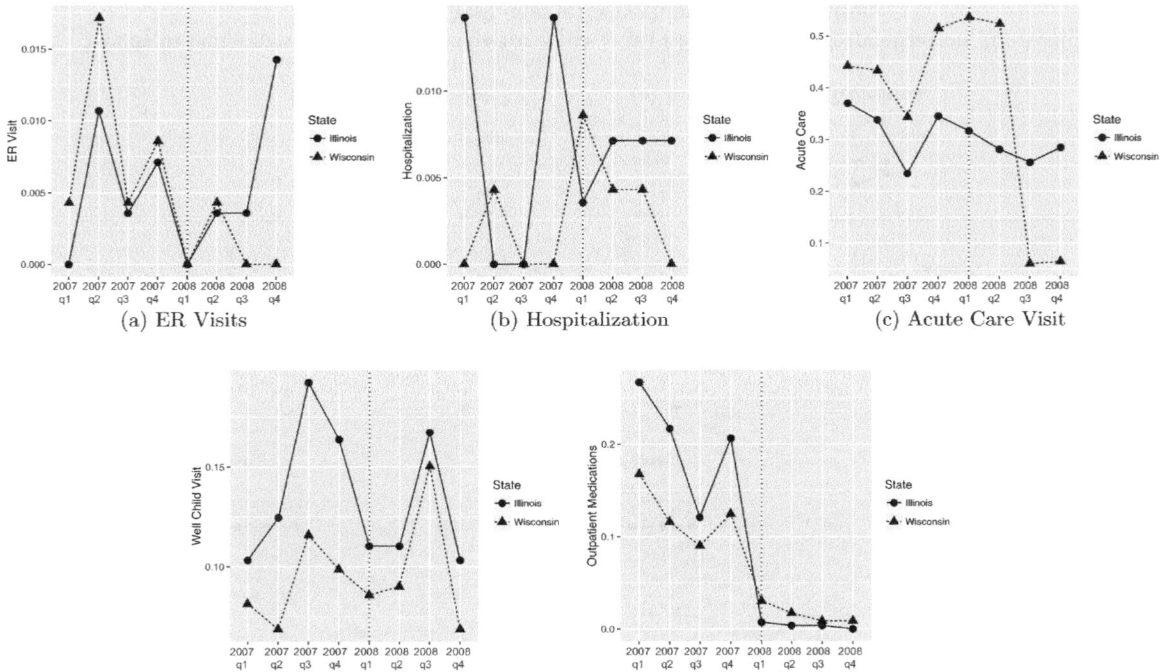

(a) ER Visits (b) Hospitalization (c) Acute Care Visit

(d) Well Child Visit (e) Any Outpatient Medication

Fig. B1. Trends in Outcomes from 2007 to 2008 for Area 1. (a) ER Visits, (b) Hospitalization, (c) Acute Care Visit, (d) Well Child Visit, and (e) Any Outpatient Medication.

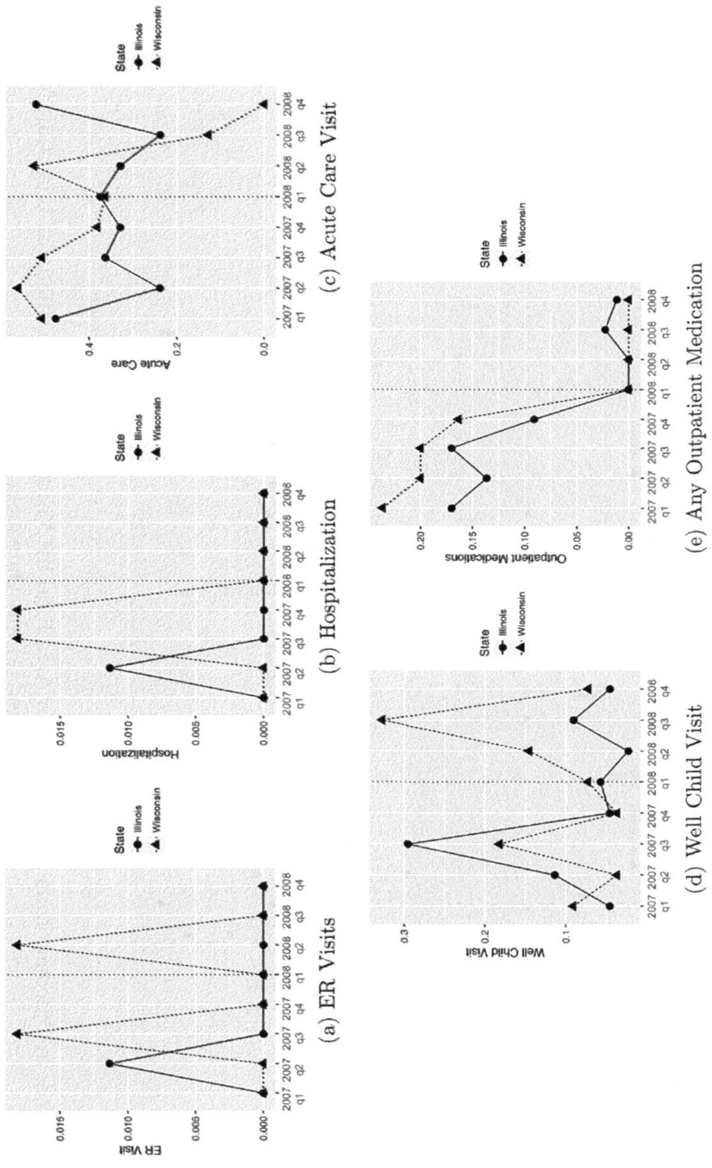

(a) ER Visits

(b) Hospitalization

(c) Acute Care Visit

(d) Well Child Visit

(e) Any Outpatient Medication

Fig. B2. Trends in Outcomes from 2007 to 2008 for Area 3. (a) ER Visits, (b) Hospitalization, (c) Acute Care Visit, (d) Well Child Visit, and (e) Any Outpatient Medication.

APPENDIX C: PLACEBO TESTS

The parallel trends assumption can be probed via a set of placebo tests. This is the primary alternative to visual inspection. For each area, we observe pretreatment data from 2007. We treat the first quarter of 2007 as the baseline time period. We then estimated the treatment effect via DID using our matched data, using the second quarter of 2007 as the posttreatment period. We then repeated this analysis using the third and fourth quarters as the posttreatment periods.

Table C1. Effect of Copays on "Placebo" Health Care Utilization (Outcomes Measured before Copay Adoption), Differences-in-Differences Estimates for Area 1.

	Q2	Q3	Q4
Emergency department visit	0.005	−0.005	−0.005
Hospitalization	0.018*	0.013	−0.004
Acute visit	0.027	0.054	0.098
Any outpatient medication	−0.112	0.268	0.005

Notes: Standard errors adjusted for individual-level clustering. In each case, Q1 is the baseline untreated period, and each subsequent quarter is treated as the posttreatment time period. Each cell represents the DID point estimate. Sample size before matching: IL = 273, WI = 224, analysis above is based on 224 matched pairs.
*Indicate whether the point estimate was significant at the 0.05 level.

Table C2. Effect of Copays on "Placebo" Health Care Utilization (Outcomes Measured before Copay Adoption), Differences-in-Differences Estimates for Area 3.

	Q2	Q3	Q4
Emergency department visit	−0.019	0.019	0
Hospitalization	−0.019	0.019	0.019
Acute visit	0.365	0.212	0.058
Any outpatient medication	0.211	0.096	0.077

Notes: Standard errors adjusted for individual-level clustering. In each case, Q1 is the baseline untreated period, and each subsequent quarter is treated as the posttreatment time period. Each cell represents the DID point estimate. Sample size before matching: IL = 81, WI = 52, analysis above is based on 52 matched pairs.
*Indicate whether the point estimate was significant at the 0.05 level.

APPENDIX D: ADDITIONAL OUTCOME RESULTS

In the main text, we report DID estimates. Here, we report the estimated treated-control mean differences for outcomes in the posttreatment period. We also include the full results for medication usage by type of medication.

Table D1. Effect of Copays on Health Care Utilization: Mean Differences, 2008–2010 Outcomes, Area 1.

	Wisconsin	Illinois	Difference	95% CI	*p*-value
Emergency department visit	0.38	0.41	−0.03	[−0.15, 0.09]	0.647
Hospitalization	0.02	0.06	−0.04	[−0.12, 0.03]	0.323
Acute visit	0.52	1.29	−0.77	[−1.04, −0.51]	0.000
Well child visit	0.38	0.47	−0.10	[−0.17, −0.02]	0.014
Any outpatient medication	0.58	0.28	0.29	[0.12, 0.46]	0.000

Notes: Sample size before matching: IL = 273, WI = 224, analysis above is based on 224 matched pairs. Outcome is scaled as average counts per month.

Table D2. Effect of Copays on Health Care Utilization: Mean Differences, 2008–2010 Medications by Type, Area 1.

	Wisconsin	Illinois	Difference	95% CI	*p*-value
Preventive medications	0.03	0.01	0.02	[−0.03, 0.07]	0.468
Acne medications	0.01	0.006	0.006	[−0.004, 0.01]	0.227
Allergy medications	0.04	0.02	0.01	[−0.009, 0.04]	0.191
Topical antibiotics	0.005	0.005	−0.0004	[−0.004, 0.003]	0.816
ADHD medications	0.15	0.02	0.13	[0.01, 0.24]	0.025
Antibiotics	0.06	0.06	−0.005	[−0.02, 0.01]	0.674
Eczema medications	0.01	0.01	0.0007	[−0.012, 0.014]	0.918
Gastrointestinal medications	0.009	0.004	0.005	[−0.003, 0.01]	0.253
Hypertension medications	0.008	0	0.008	[0.001, 0.01]	0.019
Neuro medications	0.01	0.002	0.01	[0.002, 0.023]	0.016
Pain medications	0.02	0.01	0.007	[−0.008, 0.02]	0.352
Psychiatric medications	0.04	0.01	0.03	[0.004, 0.06]	0.023
Reflux medications	0.009	0.005	0.003	[−0.004, 0.01]	0.406
Respiratory medications	0.06	0.04	0.01	[−0.03, 0.05]	0.544
Steroids	0.01	0.003	0.01	[−0.004, 0.02]	0.165
Optional medications	0.06	0.03	0.03	[−0.01, 0.07]	0.188

Notes: Sample size before matching: IL = 273, WI = 224, analysis above is based on 224 matched pairs. Outcome is scaled as average counts per month.

Table D3. Effect of Copays on Health Care Utilization: Mean Differences, 2008–2010 Outcomes, Area 3.

	Wisconsin	Illinois	Difference	95% CI	p-value
Emergency department visit	0.27	0.29	−0.02	[−0.29, 0.24]	0.847
Hospitalization	0.006	0.006	0	[−0.01, 0.01]	1
Acute visit	0.36	1.91	−1.54	[−2.47, −0.61]	0.001
Well child visit	0.47	0.34	0.12	[−0.01, 0.27]	0.084
Any outpatient medication	0.62	0.51	0.11	[−0.42, 0.64]	0.677

Notes: Sample size before matching: IL = 81, WI = 52, analysis above is based on 52 matched pairs. Outcome is scaled as average counts per month.

Table D4. Effect of Copays on Health Care Utilization: Mean Differences, 2008–2010 Medications by Type, Area 3.

	Wisconsin	Illinois	Difference	95% CI	p-value
Preventive medications	0.01	0.001	0.01	[−0.01, 0.03]	0.330
Acne medications	0.02	0.001	0.019	[0.0006, 0.04]	0.044
Allergy medications	0.02	0.04	−0.02	[−0.06, 0.02]	0.360
Topical antibiotics	0.003	0.004	−0.001	[−0.005, 0.003]	0.589
ADHD medications	0.28	0.10	0.18	[−0.15, 0.52]	0.277
Antibiotics	0.08	0.06	0.02	[−0.02, 0.07]	0.329
Eczema medications	0.009	0.007	0.002	[−0.01, 0.02]	0.823
Gastrointestinal medications	0	0.008	−0.008	[−0.01, 0.003]	0.168
Hypertension medications	0	0.046	−0.04	[−0.11, 0.01]	0.122
Neuro medications	0.04	0.075	−0.02	[−0.18, 0.12]	0.721
Pain medications	0.01	0.034	−0.01	[−0.04, 0.008]	0.167
Psychiatric medications	0.01	0.038	−0.02	[−0.07, 0.02]	0.254
Reflux medications	0.002	0.002	0.0001	[−0.004, 0.004]	0.934
Respiratory medications	0.07	0.02	0.04	[−0.01, 0.11]	0.121
Steroids	0.003	0.0005	0.002	[−0.002, 0.008]	0.246
Optional medications	0.026	0.05	−0.03	[−0.08, 0.008]	0.112

Notes: Sample size before matching: IL = 81, WI = 52, analysis above is based on 52 matched pairs. Outcome is scaled as average counts per month.

APPENDIX E: COUNTY LEVEL TAX RATES

In the 2006–2008 period, county-level tax rates were higher in the Illinois counties in our sample than in the Wisconsin counties in our sample. According to a report by the Tax Foundation (The Tax Foundation 2012), property tax rates (calculated as median taxes paid as a percent of median home value) were as follows in our three areas. Area 1: Lake County (IL) 2.19% and Kenosha County (WI) 1.93%; Area 2: McHenry county (IL) 2.09% and Walworth (WI) 1.68%; Area 3: Winnebago County (IL) 2.39% and Rock County (WI) 1.96%.

The highest difference occurs in Area 3, where Winnebago County (IL) has a 2.39% rate and Rock County (WI) a 1.96% rate, a 0.43 percentage point difference (in the 2006–2008 period). This means that a house that sold for $100,000 in Wisconsin would pay roughly the same amount in property tax as a house that sold for $82,000 in Illinois. Thus, if house prices fully incorporated the different tax rates in these these two areas, we would expect to observe lower house prices in Illinois than in Wisconsin in equilibrium (assuming there is free mobility). This is what we observe when we compare median prices in our County Comparison column in Areas 1 and 2 (Table 1), although the magnitude of most of the differences reported for Areas 1 and 2 are too large to be explained solely by difference in tax rates, which suggests that the imbalances likely reflect other, perhaps more problematic, differences between the adjacent counties. In contrast, Area 3 is the only area where the Geographic Match column shows a treated-control difference in house prices that is small, and the magnitude of the negative differences in the other two columns are consistent with the county-level property tax differences we report in this area; this reinforces our decision to focus on Area 3.

EXTERNAL AND INTERNAL VALIDITY OF A GEOGRAPHIC QUASI-EXPERIMENT EMBEDDED IN A CLUSTER-RANDOMIZED EXPERIMENT

Sebastian Galiani[a,b], Patrick J. McEwan[c] and Brian Quistorff[d]

[a]*University of Maryland, College Park, MD, USA*
[b]*NBER, Cambridge, MA, USA*
[c]*Wellesley College, Wellesley, MA, USA*
[d]*Microsoft AI and Research, Redmond, WA, USA*

ABSTRACT

This chapter analyzes a geographic quasi-experiment embedded in a cluster-randomized experiment in Honduras. In the experiment, average treatment effects of conditional cash transfers on school enrollment and child labor were large – especially in the poorest experimental blocks – and could be generalized to a policy-relevant population given the original sample selection criteria. In contrast, the geographic quasi-experiment

Regression Discontinuity Designs: Theory and Applications
Advances in Econometrics, Volume 38, 195–236

ISSN: 0731-9053/doi:10.1108/S0731-905320170000038009

yielded point estimates that, for two of three dependent variables, were attenuated. A judicious policy analyst without access to the experimental results might have provided misleading advice based on the magnitude of point estimates. We assessed two main explanations for the difference in point estimates, related to external and internal validity.

Keywords: Geographical natural experiment; internal validity and external validity

1. INTRODUCTION

In a typical regression-discontinuity design, treatments are assigned on the basis of a single, continuous covariate and a cutoff. The identification of treatment effects relies on the assumption that the relation between potential outcomes and the assignment variable is continuous at the cutoff (Hahn, Todd, & van der Klaauw, 2001; Lee & Lemieux, 2010). The assumption is particularly credible if a stochastic component in the assignment variable (e.g., a noisy test score) ensures that the agents cannot precisely manipulate their values of the assignment variable. That is, agents are subject to "local" random assignment (Lee, 2008; Lee & Lemieux, 2010).[1]

Researchers have extended the continuity results to regression-discontinuity designs in which assignment is based on a vector of variables (Imbens & Zajonc, 2009; Keele & Titiunik, 2015). A special case is the geographic discontinuity design (GDD), in which exposure to a treatment depends on the latitude and longitude of agents with respect to an administrative or territorial boundary. Researchers compare treated and untreated agents residing near boundaries, using parametric and/or non-parametric methods (Black, 1999; Dell, 2010; Keele & Titiunik, 2015). In the words of Lee and Lemieux (2010), these are often "nonrandomized" discontinuity designs since agents are usually aware of boundaries (and associated treatments) and can precisely choose their locations.[2] This places a special burden on researchers to rule out location-based sorting on observed or unobserved variables as a threat to the internal validity of treatment effects (Keele, Lorch, Passarella, Small, & Titiunik, 2017; Keele & Titiunik, 2015, 2016).

In addition to internal validity, one might assess the external validity of estimates obtained from geographic designs.[3] Suppose that a treatment is (non-randomly) assigned to 10 states, and that policy-makers are interested in the average treatment effect in this population. Yet, for identification purposes, geographic designs must exclude treated individuals that are (1) far from a state border or (2) near a state border with no cross-state variation in the treatment (as often occurs when treated states are contiguous). If excluded individuals have a different distribution of variables that moderate treatment effects – such as income, race, or even distance-to-border – then the geographic design will not recover the average treatment effect in the policy-relevant population. The challenge is well-understood in the context of non-representative convenience samples often used in randomized experiments (Cole & Stuart, 2010; Hotz, Imbens, & Mortimer, 2005; Muller, 2015). We note that it also applies to geographic designs.

This chapter assesses both validity concerns in a geographic quasi-experiment (GQE) that is embedded in a cluster-randomized experiment conducted in Honduras.[4] In the original experiment, 70 malnourished municipalities were identified, and 40 were randomly awarded conditional cash transfers (International Food Policy Research Institute [IFPRI], 2000). Using the 2001 census, Galiani and McEwan (2013) found that the treatment increased the probability that children enrolled in school and reduced their probabilities of working outside and inside the home. The effects were especially large in two of five strata (or blocks) with the highest rates of malnutrition. In this subsample, the effects on enrollment, work outside the home, and work inside the home were, respectively, 15, −6.8, and −6.2 percentage points. Relative to the control group, these represented changes of 25%, −54%, and −41%.

We compare these results to those of a geographic quasi-experiment using the same census data. Specifically, we identify a sample of treated children that are close to municipal borders shared with untreated, non-experimental municipalities. Children on the opposite side of the border constitute the quasi-experimental control group. Using the same covariates as Galiani and McEwan (2013), we show that treatment-control balance for nearly all covariates improves in samples that are increasingly close to the border (our preferred distance buffer is 2 kilometers). We can also rule out that households sorted across municipal borders in direct response to the treatment, addressing a common internal validity concern in geographic designs. Nevertheless, we find that treated children are more likely to self-identify as Lenca – an indigenous group – even very close to municipal borders. Ultimately, our analysis of the GQE sample finds that point estimates for two of the three dependent variables are attenuated relative to the experimental benchmarks.

Is this because of imbalance in unobserved variables (i.e., a threat to internal validity) or simply because the GQE sample has a different distribution of observed or unobserved variables that moderate treatment effects? We separately assess each explanation using subsamples of the randomized experiment. First, we re-estimate experimental effects in the subsample of treatment and control children within 2 kilometers of *any* municipal border (we refer to this as experiment 1). These estimates are slightly (but consistently) *stronger* than full-sample results. There are no mean differences in census covariates between the two samples, suggesting that distance-to-border proxies unobserved moderators of treatment effects.

Second, we further restrict the sample in experiment 1 to treatment and control children residing near the border of an untreated, non-experimental municipality (we refer to this sample as experiment 2). Note that it includes exactly the same treated children as the GQE. However, it uses experimental rather than the quasi-experimental controls. This permits us to (momentarily) abstract from internal validity. The point estimates for school enrollment and work-in-home are attenuated relative to experiment 1. Descriptive statistics suggests that the sample in experiment 2 is better-off than that of experiment 1, given higher rates of electricity use, asset ownership, and other income proxies. This plausibly explains the attenuated effects, since the literature on conditional cash transfers finds smaller effects in when children are less poor (Fiszbein & Schady, 2009; Galiani & McEwan, 2013).

Third, we assess internal validity by comparing the unbiased estimates from experiment 2 to those of the GQE (noting again that both include the same treatment group but different control groups). Particularly for school enrollment, the GQE estimates are attenuated relative to those of experiment 2. It suggests that imbalance in unobserved variables results in downward biases in the GQE enrollment estimates. This is perhaps consistent with the higher proportion of indigenous children in the GQE treatment group, relative to its quasi-experimental control group.

In summary, we find that the GQE cannot fully replicate the policy-relevant experimental benchmark in Galiani and McEwan (2013) for reasons related to both validity concerns. Based on these results, we make two concrete recommendations. First, it is essential that researchers using a geographic design carefully assess treatment-control balance on a wide range of observed covariates that are plausibly correlated with dependent variables (echoing the recommendations of Keele et al., 2017). Our GQE is an especially cautionary tale, since it had very good (but not perfect) balance in observed variables, but still could not replicate school enrollment estimates using the same treatment group and an experimental control group.

Second, we recommend that researchers assess the external validity of their geographic design by comparing the distributions of observed moderators of treatment effects – such as household income – to those of a well-defined, policy-relevant population. Aided by theory or prior empirical evidence on the relevance of moderators, this can be used to speculate about the generalizability of a GQE. More concretely, Cole and Stuart (2010) describe how one might construct inverse-probability weights and re-weight a convenience sample – whether experimental or quasi-experimental – to resemble a well-defined population. We conduct and report a similar analysis, re-weighting the sample of experiment 2 to resemble that of experiment 1. The weights are estimated using a wide range of covariates that are plausible moderators of treatment effects. Ultimately, however, the weighted estimates in experiment 2 are still attenuated relative to experiment 1, suggesting that some relevant moderators are unobserved.

Our results contribute to a growing literature that compares regression-discontinuity designs with a single assignment variable to experimental benchmarks (see Shadish, Galindo, Wong, Steiner, & Cook, 2011 and the citations therein). In this literature, several papers analyze conditional cash transfer experiments in which eligibility was determined by a poverty proxy. Oosterbeek, Ponce, and Schady (2008) found that experimental enrollment effects in Ecuador were large for the poorest households, but that RDD effects were zero for less-poor households in the vicinity of the eligibility cutoff. Similarly, Galiani and McEwan (2013) found no effects on enrollment and child labor in the vicinity of the cutoff used to determine assignment to the experimental sample, but large experimental effects among households residing in municipalities with the lowest levels of the assignment variable. In the absence of an experiment, both papers caution against generalizing "away" from cutoffs when the assignment variable is a plausible or well-documented moderator of treatment effects.[5] A recent strand of methodological literature has considered situations in which such generalizations might still be possible.[6]

We first revisit the results from the original experiment in Galiani and McEwan (2013), and then analyze the GQE sample and show that it fails to recover the experimental benchmark. Finally, we analyze experiments 1 and 2 to assess the external and internal validity of the GQE to offer an explanation for the divergence between the original experimental results and the results from the GQE.

2. THE PRAF-II EXPERIMENT

2.1. Design and Treatment

In the late 1990s, the International Food Policy Research Institute (IFPRI) designed a cluster-randomized experiment to estimate the impact of conditional cash transfers (CCTs) on the poverty, education, and health outcomes of households in poor Honduran municipalities. In the absence of a national poverty map, researchers identified poor municipalities with a nutrition-related proxy from a 1997 census of first-graders' heights (Secretaría de Educación, 1997). IFPRI (2000) ordered 298 municipalities by their mean municipal height-for-age *z*-scores. Seventy-three municipalities with the lowest scores were eligible (the implied cutoff was −2.3, highlighting the extremely high rates of stunting). Three were excluded due to accessibility, leaving an experimental sample of 70.

In 1999, IFPRI divided the sample into five quintiles of mean municipal height-for-age. Within quintiles, municipalities were randomly assigned to three treatment arms and a control group (in a ratio of 4:4:2:4). Arms 1 and 2 received CCTs, while arms 2 and 3 were to receive grants to schools and health centers. Moore (2008) suggests the grants were sparsely implemented as late as 2002. Using this paper's census data, Galiani and McEwan (2013) failed to reject the null that average treatment effects in arms 1 and 2 were equal (relative to the control). Arm 3 had small and statistically insignificant effects relative to the control (but its effect was statistically different from arm 2). Following Galiani and McEwan (2013), we compare 40 municipalities in a pooled CCT treatment arm and 30 in a pooled control group.

In the CCT treatment, households were eligible for an annual per-child cash transfer of L 800 (about US$50) if a child between 6 and 12 enrolled in primary school grades 1 to 4.[7] Children with higher attainment were not eligible, and households could receive up to 3 per-child transfers. During the experiment, transfers were distributed in November 2000, May–June 2001, October 2001, and late 2002 (Galiani & McEwan, 2013; Morris, Flores, Olinto, & Medina, 2004). The average household in experimental municipalities would have been eligible for transfers equal to about 5% of median per capita expenditure (Galiani & McEwan, 2013). This is smaller than most Latin American CCTs such as Progresa (Fiszbein & Schady, 2009). Indeed, payments were only intended to cover the out-of-pocket and opportunity costs of enrolling a child in school (IFPRI, 2000).

2.2. Data and Replication

Galiani and McEwan (2013) used the 2001 Honduran census – collected in late July 2001 – to estimate the short-run effects of offering CCTs to eligible children.[8] Their sample contained 120,411 six to twelve year olds eligible for the education transfer, residing in the 70 experimental municipalities. The census includes three dummy dependent variables: (1) whether a child was enrolled on the census date, (2) whether a child worked outside the home in the week preceding the census, and (3) whether a child worked exclusively in the home during the preceding week (Table A1 provides variable definitions).[9]

Their preferred specification regressed each dependent variable on a treatment dummy, dummy variables indicating randomization blocks, and a set of individual and household covariates unlikely to have been affected by the treatment. Table 1 replicates the main results. The regressions in this and subsequent tables control for block dummy variables, the 21 covariates described in Table A1, squared terms for continuous variables, and dummy variables indicating missing values of any covariate. In addition to analytic standard errors clustered by municipality, we report symmetric p-values from the wild cluster-bootstrap percentile-t that imposes the null hypothesis (Cameron, Gelbach, & Miller, 2008).

In the full sample, the treatment increases the probability of enrollment by 8.1 percentage points (a 13% increase relative to the control group). The treatment reduces the probability of work outside the home by 3.1 p.p. (32%) and work only inside the home by 4.1 p.p. (30%). The effects are larger in the two poorest blocks (1 and 2), and closer to zero and not statistically significant in three less-poor blocks (3, 4, and 5). In blocks 1 and 2, the effect on enrollment, work outside the home, and work inside the home are, respectively, 15 p.p. (25%), −6.8 p.p. (−54%), and −6.2 p.p. (−41%). The magnitude of these effects is notable given the comparatively small size of the transfer.[10]

3. A GEOGRAPHIC QUASI-EXPERIMENT

3.1. Sample

The Honduran census does not record the precise location of dwellings. To proxy location, we use the latitude and longitude of caseríos. In Honduras,

Table 1. Replication of Experimental Estimates in Galiani and McEwan (2013).

	All Blocks	Blocks 1–2	Blocks 3–5
Enrolled in school	0.081***	0.150***	0.035
	(0.024)	(0.035)	(0.025)
N	120,411	44,358	76,053
Control mean	0.646	0.600	0.680
BS *p*(sym)	0.007	0.005	0.198
Works outside home	−0.031***	−0.068***	−0.008
	(0.012)	(0.017)	(0.013)
N	98,783	36,261	62,522
Control mean	0.097	0.126	0.075
BS *p*(sym)	0.013	0.005	0.585
Works in home	−0.041***	−0.062***	−0.027
	(0.013)	(0.018)	(0.017)
N	98,783	36,261	62,522
Control mean	0.136	0.150	0.126
BS *p*(sym)	0.010	0.005	0.145

Notes: Robust standard errors are in parentheses, clustered by municipality (70, 28, and 42, respectively, in all blocks, blocks 1–2, and blocks 3–5). BS *p*(sym) is the symmetric *p*-value from a wild cluster-bootstrap percentile-*t* procedure with 399 replications. The regressions in this and all subsequent tables control for block dummy variables, the 21 covariates in Table A1, squared terms for six continuous variables, and dummy variables indicating missing values of variables. Sample sizes are lower for the work-related variables because the census restricts the question to children 7 and older.
*** indicates statistical significance at 1%.

18 departments contain 298 municipalities and over 3,700 aldeas (villages). Within the boundaries of villages, points identify the center of over 24,000 caseríos ("hamlets") that are contiguous groups of dwellings. We calculated the straight-line distance between each caserío and its nearest municipal border.[11]

We then identified a sample of 801 experimental, treated caseríos (with 23,974 children) that share a municipal border with 794 non-experimental, untreated caseríos (with 25,025 children). In the pooled sample of children, the 10th, 50th, and 90th percentiles of the distance-to-border distribution are 0.37, 2.02, and 4.22 kilometers, respectively. We henceforth refer to this as the GQE (or geographic quasi-experiment) sample.

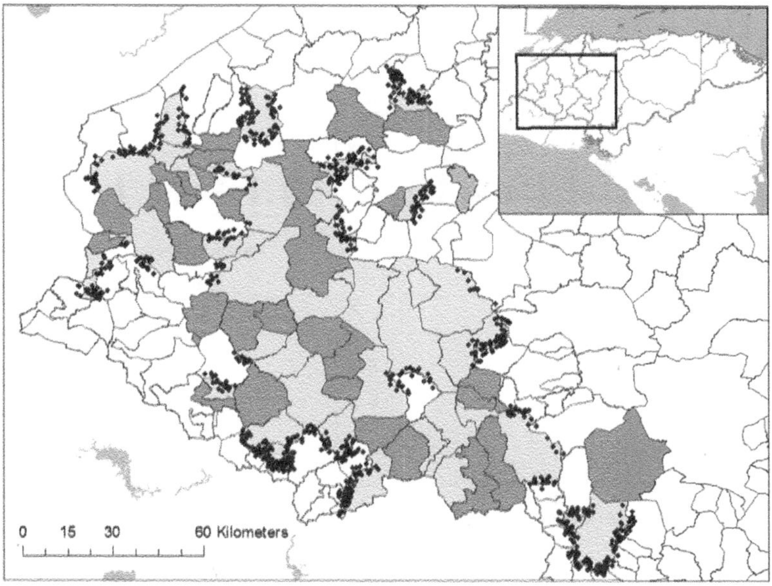

Fig. 1. Caseríos in the Geographic Quasi-Experiment. *Notes*: Experimental treatment municipalities are lightly shaded; experimental control municipalities are darkly shaded. Unshaded areas are untreated, non-experimental municipalities. Dots indicate caseríos within 2 kilometers of municipal borders shared by experimental treatment municipalities and untreated non-experimental municipalities. The inset map indicates department borders.

The map in Fig. 1 illustrates the subsample of GQE caseríos that fall within 2 kilometers of a municipal border (in the next section, we provide a rationale for using this distance buffer). It highlights that treated caseríos are a non-random sample of all treated caseríos. In particular, treated caseríos are excluded when their municipalities are fully circumscribed by other treatment or control municipalities in the experimental sample.

3.2. Covariate Balance Near Municipal Borders

By focusing on treated and untreated children that reside near municipal borders, there may be fewer differences in observed and, perhaps, unobserved variables that affect child outcomes. We assess this in the top-left panel of Fig. 2, using 21 covariates from the experimental analysis. Dots

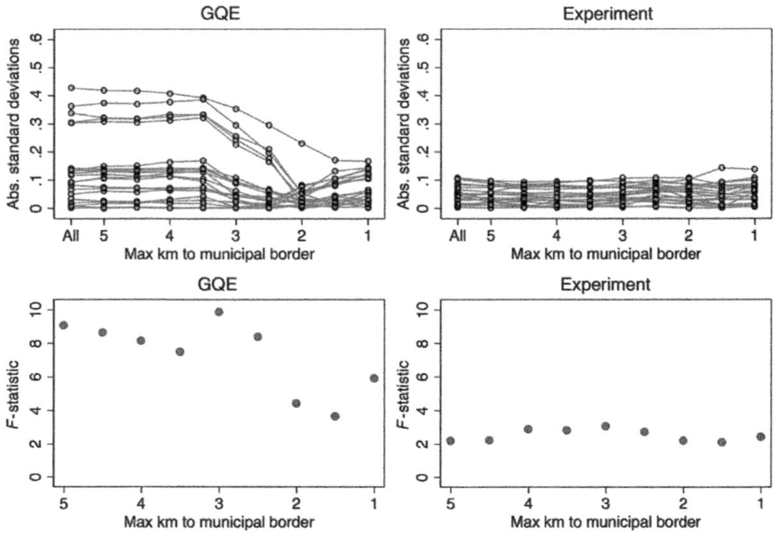

Fig. 2. Covariate Balance Increasingly Close to Municipal Borders. *Notes*: In the upper panels, dots indicate the absolute value of the standardized mean difference (using the full-sample standard deviation) between the treatment and comparison group for 21 covariates in Table A1, within the specified distance-to-border (Appendix B reports unstandardized differences, standardized differences, and cluster-adjusted *p*-values). In the lower panels, dots indicate *F*-statistics from regressions of the treatment dummy on the 21 covariates, squared terms for six continuous covariates, and dummy variables indicating missing values.

indicate the absolute values of standardized treatment-control differences for each covariate. The left-most dots pertain to the full GQE sample just described, while others refer to GQE samples restricted by increasingly narrow distance buffers.

Balance in the GQE sample is sensitive to distance-to-border, markedly improving when caseríos fall within 2 kilometers of the border. In the 2-kilometer subsample, it is notable that 20 of 21 covariates show treatment-control differences of less than 8% of a standard deviation, and none are statistically different from zero at conventional levels (Table 2). This is despite the fact that there are large mechanical differences in mean municipal height-for-age *z*-scores due to the selection rule for the experimental sample (Fig. 3). The bottom-left panel of Fig. 2 shows *F*-statistics from omnibus tests of covariate balance.[12] In each subsample, we regress the treatment dummy on the 21 covariates, squared terms for continuous variables, and

Table 2. Balance in the Geographic Quasi-Experiment (≤2 kilometers from border).

	T/*C* Differences	*p*-Value
Age	−0.002/−0.001	0.979
Female	0.007/0.013	0.383
Born in municipality	−0.004/−0.013	0.849
Lenca	0.094/0.231	0.166
Moved	0.006/0.032	0.505
Father is literate	0.040/0.081	0.470
Mother is literate	0.030/0.060	0.531
Father's schooling	0.231/0.076	0.689
Mother's schooling	0.147/0.050	0.778
Dirt floor	0.008/0.017	0.915
Piped water	0.004/0.009	0.949
Electricity	0.022/0.053	0.827
Rooms in dwelling	0.041/0.054	0.747
Sewer/septic	0.030/0.063	0.683
Auto	0.006/0.029	0.780
Refrigerator	−0.014/−0.051	0.791
Computer	−0.001/−0.020	0.746
Television	0.012/0.035	0.890
Mitch	−0.002/−0.009	0.890
Household members	0.054/0.022	0.692
Household members, 0–17	0.013/0.007	0.920
Predicted mean municipal child height-for-age *z*-score	−0.464/−1.538	0.001

Note: In the difference column, the first number is the mean difference and the second number is mean difference divided by the full-sample standard deviation. *p*-values account for clustering by municipality.

dummies indicating missing values. The *F*-statistic declines as distance-to-border restrictions are applied, consistent with prior evidence.

In the right-hand panels of Fig. 2, we can compare balance in GQE samples to balance in experimental samples with similar distance-to-border restrictions. As anticipated, given randomized assignment, covariate balance in the experiment does not depend on the distance of caseríos to municipal borders. The top-right panel shows that absolute values of treatment-control

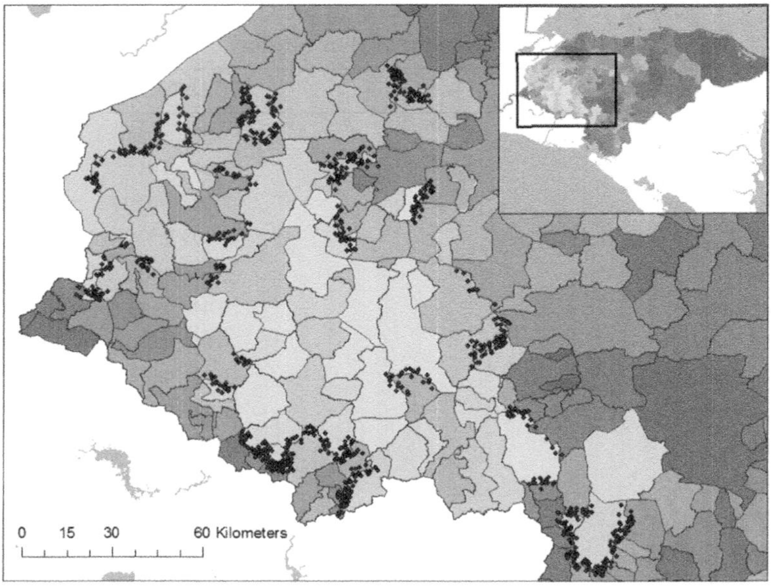

Fig. 3. Mean Municipal Height-for-Age *Z*-Scores in 1997. *Notes*: Lighter shades indicate increasingly negative values of the predicted mean municipal height-for-age *Z*-scores (Galiani & McEwan, 2013), using 20 quantiles of the municipal distribution. Municipal borders are outlined. Dots indicate caserios in the geographic quasi-experimental sample within 2 kilometers of municipal borders.

differences for 21 covariates are rarely larger than 10% of a standard deviation, regardless of distance. The bottom-right panel of Fig. 2 shows relatively stable *F*-statistics of around 2 from the omnibus test.

In fact, the smallest *F*-statistics in the GQE are approximately twice as large as those in the experimental sample. In the 2-kilometer GQE sample, this is driven by imbalance in a single covariate (Lenca). Children on the treated side of borders are about 10 percentage points more likely to self-identify as a member of the indigenous Lenca group (Table 2). National poverty headcounts are higher among Lenca than non-indigenous Hondurans.[13] Nevertheless, this does not necessarily imply imbalance in unobserved socioeconomic outcomes such as poverty, because there is demonstrable balance in many poverty proxies such as schooling and household assets.

3.3. Potential Threats to Internal Validity

Why might imbalance persist close to borders? One explanation is that Lenca households manipulated treatment status by moving after experimental assignment, but prior to the census. We regard this as unlikely for three reasons. First, the cash transfer for a typical household was extremely small (less than 5% of a household's consumption) and unlikely to provide sufficient liquidity for poor households to move. Second, 91% of children in the 2-kilometer GQE sample were born in their municipality of residence, and only 4% lived in a different caserío, aldea, or city in 1996 (five years before the census date). Both variables are among those with the smallest cross-border differences in the GQE, and neither is statistically different from zero (Table 2).[14] Third, there is a similar pattern of imbalance in the 2-kilometer sample of untreated, experimental caseríos that share a border with untreated, non-experimental caseríos (see Table B2; we later use this sample to conduct a placebo test). In other words, imbalance persists even when there is no CCT treatment to create incentives for cross-border sorting.

Thus, a second explanation is that some households chose dwellings years or decades before treatment assignment, but were not indifferent to attributes of communities on opposite sides of the border. Fig. 4 illustrates the proportion of Lenca children in aldeas (a sub-territory of municipalities), along with the GQE caseríos. Cross-border imbalance is most evident on the eastern-most municipal borders of the experimental sample, but is certainly not a feature of all border segments. Most notably, one interior segment cleaves the "twin cities" of La Esperanza and Intibucá, both rich centers of Lenca culture. Though in separately governed municipalities, the towns are commonly referred to by a single name (and treated as such by locals).

In summary, there is evidence of balance in the GQE sample (with a buffer of 2 kilometers) on 20 of 21 covariates that are typically correlated with child education and labor outcomes. From a design perspective, it is notable that covariate selection was imposed by an earlier paper (Galiani & McEwan, 2013). However, there is persistent imbalance in one covariate (Lenca) that is plausibly correlated with unobserved determinants of child outcomes. In this volume, Keele et al. (2017) report similar covariate imbalances close to borders, leading them to invoke an assumption of conditional geographic mean independence. That is, potential outcomes are assumed to be mean independent of treatment assignment within a specified buffer, conditional on observed covariates.[15] We make a similar ignorability

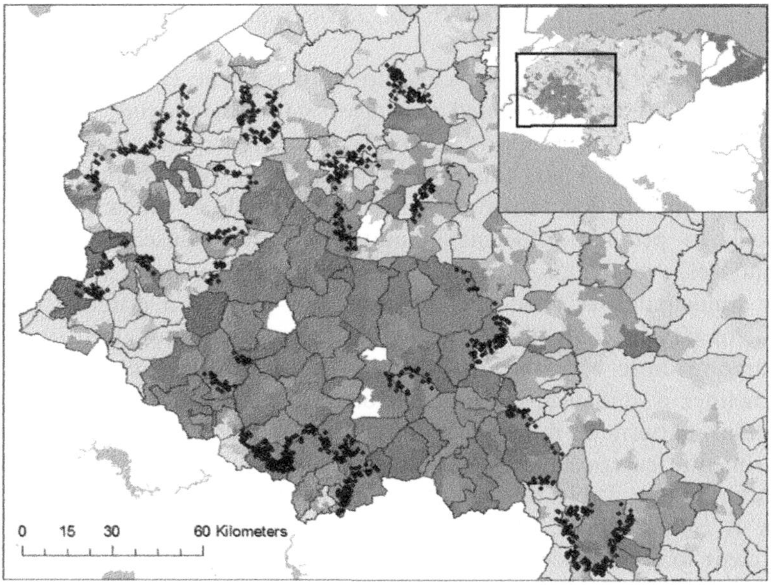

Fig. 4. Proportion of Eligible Children Self-Identifying as Lenca in 2001. *Notes*: Darker shades indicate higher proportions of children self-identify as Lenca (or another racial or ethnic minority), using 20 quantiles of the aldea (village) distribution. Municipal borders are outlined. Dots indicate caserios in the geographic quasi-experimental sample within 2 kilometers of municipal borders.

assumption within a 2-kilometer buffer and refer to the design as a geographic quasi-experiment.

We do not assume local continuity in potential outcomes at municipal borders, as one might in a "pure" GDD (Imbens & Zajonc, 2009; Keele & Titiunik, 2015). First, our evidence suggests that assignment is not locally randomized, given long-standing municipal borders and households' ability to sort around them. Second, we are forced to proxy the location of dwellings using the latitude and longitude of caseríos. This leads to mass points in the putative assignment variables,[16] even though standard methods for analyzing discontinuity designs rely on continuous assignment variables (Calonico, Cattaneo, & Titiunik, 2014; Keele et al., 2017).

3.4. GQE Estimates

Using the 2-kilometer GQE sample, Table 3 reports estimates from ordinary least squares regressions that control for the same covariates as Table 1, but excluding experimental block dummy variables. The analytic standard errors apply multi-way clustering on both municipality and border segments, given the spatial proximity of treatment and control caseríos (Cameron, Gelbach, & Miller, 2011). A separate border segments exists for every unique combination of bordering municipalities. Overall, there are 81 municipalities and 65 non-nested border segments.[17] As in Table 1, we also report symmetric p-values from the wild cluster-bootstrap percentile-t that

Table 3. Estimates of the Geographic Quasi-Experiment (≤2 kilometers from border).

	All Blocks	Blocks 1–2	Blocks 3–5
Enrolled in school	0.057*	0.106**	0.035
	(0.032)	(0.054)	(0.041)
N	24,360/81/65	6,888/26/19	17,472/61/46
BS p(sym)	0.095	0.072	0.463
Diff RCT p(sym)	0.537	0.580	0.998
Works outside home	−0.024	−0.086*	−0.005
	(0.015)	(0.044)	(0.016)
N	20,009/81/65	5,591/26/19	14,418/61/46
BS p(sym)	0.122	0.025	0.743
Diff RCT p(sym)	0.677	0.715	0.885
Works only inside home	0.014	−0.013	0.026
	(0.017)	(0.026)	(0.022)
N	20,009/81/65	5,591/26/19	14,418/61/46
BS p(sym)	0.468	0.613	0.330
Diff RCT p(sym)	0.010	0.105	0.048

Notes: Robust standard errors are in parentheses, clustered by municipality and border segment (see text for details). *N* indicates the number of eligible children, municipalities, and border segments. BS p(sym) is the symmetric p-value from a wild cluster-bootstrap percentile-t procedure (clustering on municipalities) with 399 replications. Diff RCT p(sym) is the p-value from the test of equality between a GQE estimate and the corresponding experimental estimate from Table 1, computed using 399 replications. All regressions control for the variables described in the note to Table 1 (but excluding block dummy variables).
** indicates statistical significance at 5%, and * at 10%.

imposes the null hypothesis, clustering by municipality (Cameron et al., 2008).

In the 2-kilometer GQE sample, the treatment increases enrollment by a marginally significant 5.7 percentage points, or 2.4 p.p. smaller than the experimental estimate in Table 1. In blocks 1–2, the enrollment effect is 10.6 p.p., or 4.4 p.p. lower than the experimental estimate. In blocks 3–5, both the GQE and the experiment find similarly small and statistically insignificant estimates.

The results are mixed for the two work-related variables. In the GQE sample, the treatment reduces work outside the home by 2.4 p.p. in all blocks and 8.6 p.p. in blocks 1–2 (only the latter is marginally statistically significant). The experimental estimates are roughly similar. Neither the GQE nor the experiment suggest any effects in blocks 3–5. In contrast, the GQE estimates are attenuated for work inside the home, relative to the experimental estimates. There is never a large or statistically significant effect for this dependent variable in the GQE sample. Yet, in the experiment, there were reductions of 4.1 and 6.2 p.p., respectively, in all blocks and blocks 1–2.

Despite these differences between the GQE and the experiment, bootstrapped *p*-values in Table 3 suggest that the estimates are not statistically distinguishable from one another. Even so, one can pose a practical question: would a reasonable policy analyst – relying on the GQE point estimates and blinded to the experimental ones – have reached conclusions as optimistic as those of Galiani and McEwan (2013)? Most likely the attenuated GQE estimates would have yielded more guarded conclusions.

4. EMPIRICAL STRATEGY

4.1. External and Internal Validity of the GQE

We consider two explanations for the divergence in point estimates of the GQE and the experiment, related to external and internal validity. Recall that treated caseríos in the GQE are a non-random subset of all treated caseríos. First, they are close to municipal borders. Second, they share a border with untreated, non-experimental caseríos. This naturally excludes caseríos in the spatial core of the experimental sample. As Figs. 3 and 4 suggest, excluded caseríos might exhibit higher rates of child stunting or greater concentrations of indigenous children.

Thus, treated children in the GQE are plausibly different in variables observed by the econometrician – such as distance-to-border, height-for-age, and ethnicity – and perhaps in unobserved variables, such as income. If these variables moderate treatment effects, then the GQE estimates – even internally valid ones – will differ from the experimental benchmarks in Table 1. In the present application, it is plausible that GQE caseríos are "less poor" than other treated ones. Since treatment effects are much larger in the poorest municipalities (Table 1), this provides a plausible explanation for the attenuated GQE treatment effects.

An alternative explanation for the divergence of point estimates is related to internal validity. Suppose that treated children in the GQE differ in unobserved ways from their bordering control group, even after conditioning on a rich set of covariates (e.g., they are more likely to be poor, an unmeasured variable). This too could explain attenuated treatment effects, assuming that poorer children are less likely to enroll in school and more likely to work.

4.2. Experimental Samples Used to Assess Validity

Table 4 summarizes the experimental samples that we use to assess the external and, then, the internal validity of the GQE. The full experimental sample was already used, in Table 1, to obtain estimates of the average treatment effect (ATE). Given the design of the experiment, the estimates are generalizable to a well-defined, policy-relevant population of Honduran children residing in malnourished municipalities.

Table 4. Experimental Samples Used to Assess External and Internal Validity of the GQE.

	Sample Restriction on Full Experimental Sample	Parameter(s)
Full experimental sample	–	ATE
Experiment 1	2 km from any municipal border	ATE_1
Experiment 2	2 km from municipal borders shared with untreated, non-experimental caseríos	Unweighted: ATE_2 [a]Weighted: ATE_1

Note: ATE is the average treatment effect in the full experimental sample (Galiani & McEwan, 2013), and subscripts indicate average treatment effects in subsamples of the experiment.
[a]Indicates that identification relies on a selection-on-observables assumption described in the text.

We next limit the experimental sample to children residing in caseríos no more than 2 kilometers from *any* municipal border. This sample – denoted experiment 1 – is used to obtain estimates of ATE_1. If distance-to-border does not moderate treatment effects, then ATE and ATE_1 (and estimates thereof) should be similar.

We further limit the sample of experiment 1 to children residing in caseríos no more than 2 kilometers from a municipal border shared with untreated, non-experimental caseríos. This sample – denoted experiment 2 – includes exactly the same sample of treated caseríos (and children) as the GQE. However, its control group consists of the experimental control group subject to the same sample restriction (illustrated in Fig. 5). By using an experimental control group instead of a quasi-experimental one, we can abstract from the internal validity of the GQE and focus on external validity. If observed and unobserved moderators are similarly distributed across the

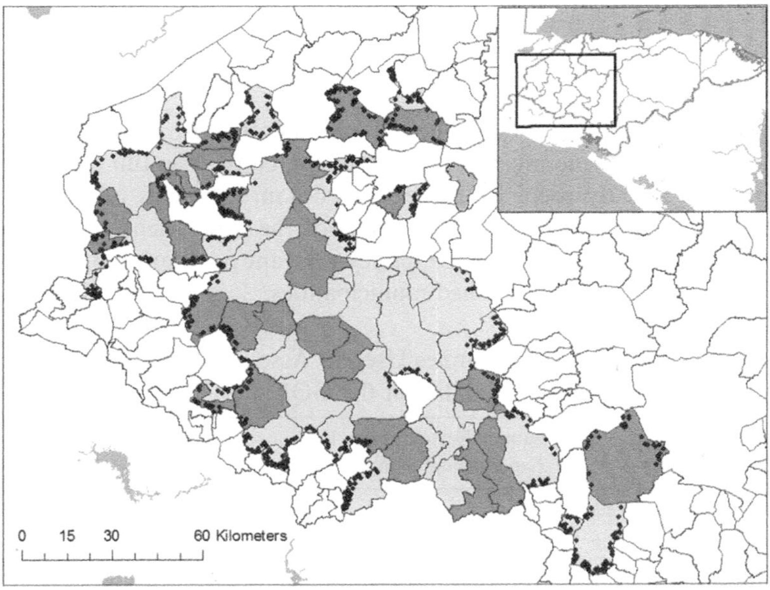

Fig. 5. Caseríos in Experiment 2. *Notes*: Experimental treatment municipalities are lightly shaded; experimental control municipalities are darkly shaded. Unshaded areas are untreated, non-experimental municipalities. Dots indicate caseríos within 2 kilometers of a municipal border with untreated non-experimental municipalities. The inset map indicates department borders.

samples of experiments 1 and 2, then ATE_1 and ATE_2 (and estimates thereof) should be similar.

If they differ, then it weakens the GQE's claim on external validity. More constructively, one can further diagnose whether the samples of experiment 1 and 2 differ in their distributions of observed moderators of treatment effects. If they do, then one can re-weight the sample in experiment 2 to resemble that of experiment 1, and re-estimate effects (Cole & Stuart, 2010). To the extent that relevant moderators are observed and contribute to the estimation of the weights, then weighted estimates should be similar to estimates of ATE_1. If they still diverge, then it suggests that a relevant moderator is unobserved. The next subsection will further elaborate the assumptions and method.

Lastly, we assess the internal validity of the GQE in two ways. We first compare GQE estimates from Table 3 to those of ATE_2, which uses an experimental rather than quasi-experimental control group for the same sample of treated children. Any divergence is indicative of bias in the GQE. Second, we implement the placebo test alluded to in an earlier section. We construct a placebo sample of untreated, experimental caseríos no more than 2 kilometers from a municipal border shared with untreated, non-experimental caseríos. We anticipate finding zero effects in the "GQE" placebo sample, conditional on covariates. A positive or negative effect is likely the result of imbalance in unobserved variables.

4.3. Inverse-Probability Weighting and External Validity

Using the potential outcomes framework, let the outcome Y for individual i be a function of a randomly assigned treatment T_i. The difference in potential outcomes under treated (1) and untreated (0) conditions is $Y_i(1) - Y_i(0)$. Table 1 reported estimates of the ATE in a well-defined, policy-relevant sample, denoted S.

We will further report estimates of ATE_2 in the non-random subsample of experiment 2, denoted S_2. If the treatment has heterogeneous effects on individuals, moderated by a set of variables X, then effects may differ across samples depending on the distribution of X. An intuitive method of correcting for this difference is to re-weight S_2 so that its distribution of moderators is similar to S (Cole & Stuart, 2010; Stuart, Bradshaw, & Leaf, 2015).

Closely following Hotz et al. (2005), we specify three assumptions under which the procedure can recover ATE from S_2. Assumption 1 is that the treatment is randomly assigned in S_2:

$$T_i \perp [Y_i(1), Y_i(0)] | S_2.$$

The assumption is satisfied because there was random assignment in S, and S_2 is a subsample of treatment and control groups obtained by imposing exogenous sample restrictions.

Assumption 2 asserts that one's presence in the subsample does not depend on potential outcomes, given the moderators:

$$(i \in S_2) \perp [Y_i(1), Y_i(0)] | X_i.$$

Hotz et al. (2005) refer to this as unconfounded location. Stuart, Cole, Bradshaw, and Leaf (2011) invoke a similar assumption and refer to it as unconfounded sample selection. Both papers highlight the need to measure all relevant moderators in order to satisfy the assumption. Lastly, Assumption 3 imposes a requirement of common support for the moderators between the two samples. For each moderator, it must be the case that:

$$0 < P(i \in S_2 | X_i) < 1.$$

In the present context, assumption 3 does not hold for one moderator, namely distance-to-border (since observations more than 2 kilometers from the border have zero probability of contributing to S_2). However, we can reframe the task as generalizing from the sample S_2 to S_1, which also imposes the 2-kilometer distance restriction (Stuart et al., 2011). Common support holds for all other covariates.

Assumption 2 may not hold if there are unobserved moderators. A typical study cannot verify this, just as a typical observational study cannot directly test for selection-on-unobservables into a treatment or control group. In contrast, we can compare weighted estimates in experiment 2 to estimates of ATE_1. Any difference suggests that relevant moderators are unobserved.

To implement the method, we estimate inverse-probability weights (Cole & Stuart, 2010; Stuart et al., 2015).[18] In S_1 – which imposes the 2-kilometer distance buffer – we estimate a logit regression in which the dependent variable indicates observations in S_2. The regressors include

21 covariates, 6 squared terms, and dummy variables indicating missing values.[19] They further include block dummy variables, mean municipal height-for-age, distance-to-border, and squared terms for the latter two. Lastly, we calculate inverse-probability weights for observations in S_2 as $w_i(X_i) = 1/\hat{p}(X_i)$, where $\hat{p}$ is the estimated probability that an observation is selected for S_2.

5. RESULTS

5.1. External Validity: Experiment 1

Tables 5–7 report results for the three dependent variables. Experiment 1 includes children residing in caseríos within 2 kilometers of *any* municipal border. Imposing this restriction slightly increases the positive enrollment estimates and makes the work-related estimates slightly more negative.

Table 5. Estimates for Experiments 1 and 2 (Dependent Variable: Enrolled in School).

	All Blocks	Blocks 1–2	Blocks 3–5
Experiment 1	0.089***	0.163***	0.041
	(0.025)	(0.042)	(0.025)
N	65,310/70	26,122/28	39,188/42
BS *p*(sym)	0.013	0.010	0.180
Experiment 2	0.063*	0.133**	0.035
	(0.035)	(0.059)	(0.040)
N	21703/52	6996/17	14707/35
BS *p*(sym)	0.133	0.095	0.455
Experiment 2 (weighted)	0.064*	0.134**	0.022
	(0.035)	(0.054)	(0.037)
N	21703/52	6996/17	14707/35
BS *p*(sym)	0.107	0.055	0.560

Notes: Robust standard errors are in parentheses, clustered by municipality. *N* indicates the number of eligible children and the number of municipalities. BS *p*(sym) is the symmetric *p*-value from a wild cluster-bootstrap percentile-*t* procedure with 399 replications. All regressions control for the variables described in the note to Table 1.

*** indicates statistical significance at 1%, ** at 5%, and * at 10%.

Table 6. Estimates for Experiments 1 and 2 (Dependent Variable: Works Outside Home).

	All Blocks	Blocks 1–2	Blocks 3–5
Experiment 1	−0.042***	−0.082***	−0.017
	(0.012)	(0.022)	(0.012)
N	53,703/70	21,387/28	32,316/42
BS *p*(sym)	0.003	0.003	0.217
Experiment 2	−0.032*	−0.090**	−0.009
	(0.017)	(0.034)	(0.017)
N	17,883/52	5,691/17	12,192/35
BS *p*(sym)	0.095	0.013	0.632
Experiment 2 (weighted)	−0.034*	−0.087**	−0.009
	(0.017)	(0.033)	(0.018)
N	17,883/52	5,691/17	12,192/35
BS *p*(sym)	0.070	0.007	0.637

Notes: Robust standard errors are in parentheses, clustered by municipality. *N* indicates the number of eligible children and the number of municipalities. BS *p*(sym) is the symmetric *p*-value from a wild cluster-bootstrap percentile-*t* procedure with 399 replications. All regressions control for the variables described in the note to Table 1.
*** indicates statistical significance at 1%, ** at 5%, and * at 10%.

In Table 5, for example, the enrollment estimate is 8.9 percentage points inside the buffer (vs. 8.1 in Table 1). Further limiting the sample to blocks 1 and 2, the estimate in experiment 1 is 16.3 p.p. (vs. 15 p.p. in Table 1). There is no ready explanation for the slight increases in enrollment effects in Table 5 (and slightly more negative work effects reported in Tables 6 and 7). The mean covariate differences between the full experimental sample and experiment 1 are small and statistically insignificant (full results are available from the authors). This suggests that distance-to-border is a proxy for other, unobserved moderators.

5.2. *External Validity: Experiment 2*

Tables 5–7 also report estimates for experiment 2. Recall that it includes the same treated observations as the GQE sample, but with an experimental control group. Imposing this sample restriction reduces the enrollment estimates by 2.6 p.p. relative to experiment 1, apparently driven by a 3 p.p.

Table 7. Estimates for Experiments 1 and 2 (Dependent Variable: Works Only in Home).

	All Blocks	Blocks 1–2	Blocks 3–5
Experiment 1	−0.041***	−0.067***	−0.024
	(0.013)	(0.021)	(0.016)
N	53,703/70	21,387/28	32,316/42
BS *p*(sym)	0.007	0.025	0.188
Experiment 2	−0.024	−0.026	−0.023
	(0.019)	(0.025)	(0.023)
N	17,883/52	5,691/17	12,192/35
BS *p*(sym)	0.320	0.352	0.448
Experiment 2 (weighted)	−0.018	−0.020	−0.013
	(0.019)	(0.025)	(0.022)
N	17,883/52	5,691/17	12,192/35
BS *p*(sym)	0.438	0.515	0.630

Notes: Robust standard errors are in parentheses, clustered by municipality. *N* indicates the number of eligible children and the number of municipalities. BS *p*(sym) is the symmetric *p*-value from a wild cluster-bootstrap percentile-*t* procedure with 399 replications. All regressions control for the variables described in the note to Table 1.
*** indicates statistical significance at 1%.

decline in the blocks 1–2 subsample. A similar pattern of attenuation is evident for work-in-home estimates (Table 7), but not for work-outside-home (Table 6). For the latter variable, the coefficient in the blocks 1–2 sample is slightly *more* negative.

Did the sample restriction in experiment 2 change the distribution of plausible moderating variables? In fact, observations in the experiment 2 sample are 8 percentage points less likely to belong to blocks 1–2 (Table C1). There are also substantial differences for specific covariates, especially within blocks 1–2 (Fig. 6). For example, households of children in experiment 2 are 12 p.p. more likely to have electric light, 10 p.p. more likely to own a television, and their mothers have two-thirds of a year more schooling, on average. In addition to Galiani and McEwan (2013), the literature on Latin American CCTs usually finds that effects on school enrollment are larger among poorer households (Fiszbein & Schady, 2009). This implies that sample selection on observed (and perhaps unobserved) moderating variables – all common proxies for poverty – is responsible for the pattern of attenuated estimates in experiment 2.

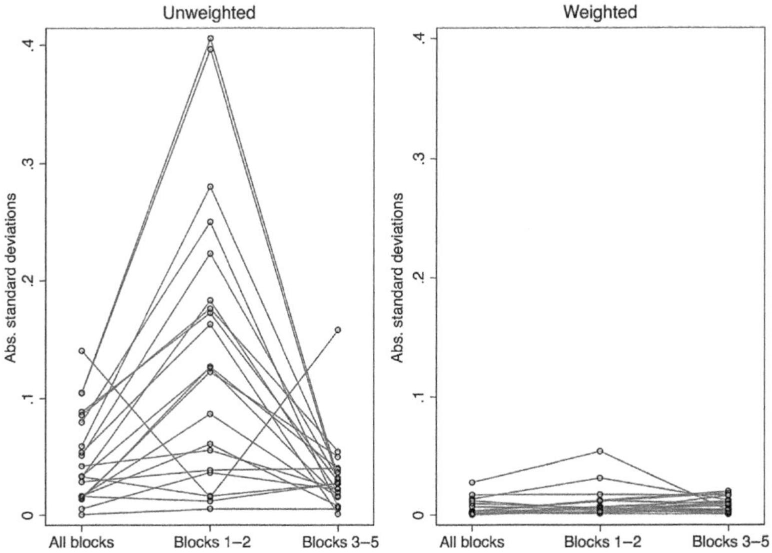

Fig. 6. Comparing Covariate Means between Samples in Experiments 1 and 2. *Notes*: Dots indicate the absolute value of the standardized mean difference (using the full-sample standard deviation) between the pooled experiment 2 sample and the pooled experiment 1 sample, for the 21 covariates in Table A1. In all panels, the sample includes caseríos within 2 kilometers of municipal borders. In the right panel, the mean of the experiment 2 sample is weighted, as described in the text.

5.3. External Validity: Weighted Estimates in Experiment 2

To further examine this issue, we estimated the probability that each observation in experiment 1 was selected for experiment 2 (using the logit specification described earlier). The mean difference in the estimated propensity score between the samples of experiment 2 and experiment 1 is 0.045 (or 37% of the standard deviation in the experiment 1 sample). For each of the 21 covariates, we then estimated the standardized difference between the weighted mean in experiment 2 – applying the inverse-probability weights described earlier – and the unweighted mean in experiment 1. As Fig. 6 illustrates, re-weighting nearly eliminates observed differences between the two samples.

Finally, Tables 5–7 report weighted estimates in the experiment 2 sample. We anticipate that the weighted estimates will more closely resemble those from experiment 1. On the contrary, we find that the point estimates from unweighted and weighted specifications in experiment 2 are quite similar (and both exhibit similar patterns of attenuation relative to experiment 1). One interpretation is that sample selection into experiment 2 altered the distribution of unobserved variables that moderate treatment effects, leading to a violation of assumption 2.

What else might be done? One possibility is to implement a two-stage correction for sample selection, à la Heckman (1979). In the sample of experiment 1, one estimates a first-stage probit with a dependent variable indicating selection into experiment 2. It includes the same independent variables as the logit used to estimate the inverse-probability weights, in addition to variables that affects selection into experiment 1, but not child outcomes. Of course, compelling exclusion restrictions are usually hard to come by (and no obvious candidates exist in our application). Finally, in the second-stage regression, one includes the inverse Mills ratio as a regressor (along with other covariates) and examines its sign and significance for evidence of sample selection bias.

5.4. Internal Validity

Recall the estimates in experiment 2 use the same group of treated children as the GQE, but with an experimental control group. How do they compare to the quasi-experimental GQE estimates reported in Table 3? For enrollment, the GQE estimates are attenuated relative to those of experiment 2 (which themselves were attenuated relative to those of experiment 1). This is especially evident in the blocks 1−2 subsample. The enrollment effect is 13.3 p.p. in experiment 2 and 10.6 in the GQE. The work-related variables provide less obvious conclusions because the estimates – in the experiment 2 and GQE samples – are small and not significant at 5% in blocks 1−5. In blocks 1−2, however, the point estimates for work-outside-home have a similar magnitude in both samples. In summary, the evidence is suggestive the GQE enrollment estimates are downward-biased relative to the unbiased estimates from experiment 2.

Finally, Table 8 reports the placebo test described earlier. In the first column, the coefficients are small and statistically insignificant, providing some evidence that the GQE estimates are not explained by selection-on-unobservables. The pattern is not as clear in blocks 1−2, likely due to the much smaller samples of municipalities (recalling the experimental control group contained fewer municipalities than the treatment group). Overall,

Table 8. Placebo Estimates ($\leq$ 2 kilometers from border).

	All Blocks	Blocks 1–2	Blocks 3–5
Enrolled in school	−0.025	0.073	−0.052
	(0.036)	(0.063)	(0.035)
N	13,980/47/40	4,064/17/12	9,916/33/28
BS *p*(sym)	0.525	0.415	0.165
Works outside home	−0.014	−0.035	−0.008
	(0.018)	(0.054)	(0.015)
N	11,573/47/40	3,365/17/12	8,208/33/28
BS *p*(sym)	0.542	0.705	0.623
Works only inside home	−0.001	−0.069**	0.024
	(0.019)	(0.032)	(0.018)
N	11,573/47/40	3,365/17/12	8,208/33/28
BS *p*(sym)	0.930	0.135	0.188

Notes: Robust standard errors are in parentheses, clustered by municipality and border segment (see text for details). *N* indicates the number of eligible children, municipalities, and border segments. BS *p*(sym) is the symmetric *p*-value from a wild cluster-bootstrap percentile-*t* procedure (clustering on municipalities) with 399 replications. All regressions control for the variables described in the note to Table 1 (but excluding block dummy variables).
**indicates statistical significance at 5%.

the imprecision prevents us from ruling out some bias in the GQE estimates.

5.5. *Compound Treatment Irrelevance*

The GQE must assume that the CCT is the only treatment that varies across borders (or, if there is another that does not affect potential outcomes). Keele and Titiunik (2015, 2016) describe this assumption as compound treatment irrelevance. In the Honduran context, the most likely violation occurs when a municipal border is also a department border. Although the management and financing of Honduran public schools is still highly centralized, each department controls some functions, especially related to personnel management. This leaves open the possibility that the assumption is violated, and so we assess robustness to the dropping of observations near municipal border segments that also happen to be department borders.

Of course, these occasions further non-random sample restrictions, which may affect external validity. Thus, Tables D1–D3 repeat all experimental analyses from Tables 5–7 after excluding municipal border segments that are also a department border. The immediate result is a reduction in the number of municipalities in experiment 2 (from 52 to 43). Despite the reduced precision, the substantive findings are similar to earlier ones, focusing especially on blocks 1–2. The large effects for enrollment and work-at-home (in experiment 1) are attenuated in experiment 2, while the effects are more robust across samples for work-outside-home. Applying inverse-probability weights to experiment 2 again has little effect on the point estimates. Table D4 then replicates the GQE estimates. Here too, the substantive conclusions are similar. The only obvious difference is an attenuation of the enrollment estimate in blocks 1–2 (from 10.6 to 7.3 percentage points), though the full-sample estimates are within 0.2 percentage points of one another.

6. CONCLUSIONS

This chapter analyzed a geographic quasi-experiment embedded in a cluster-randomized experiment in Honduras. In the experiment, ATEs on school enrollment and child labor were large – especially in the poorest blocks – and could be generalized to a policy-relevant population given the original sample selection criteria (Galiani & McEwan, 2013; IFPRI, 2000). In contrast, the geographic quasi-experiment yielded point estimates that, for two of three dependent variables, were attenuated. A judicious policy analyst without access to the experimental results might have provided misleading advice based on the magnitude of point estimates.

We assessed two main explanations for the difference in point estimates, related to external and internal validity. The GQE sample is necessarily restricted to children residing close to a municipal border with cross-border variation in the treatment. Sample selection modifies the distribution of some observed and (perhaps) unobserved variables that moderate treatment effects, relative to the original experiment. We find that this explains some, but not all of the attenuation, especially for school enrollment effects. The remainder is plausibly explained by imbalance in unobserved variables between treatment and control groups in the 2-kilometer GQE sample. While there is treatment-control balance along a wide range of pre-specified covariates, the GQE enrollment estimates are still attenuated relative to the benchmark estimates of experiment 2.

Both findings suggest that researchers using geographic designs should carefully describe how their geographically imposed convenience sample differs from that of a well-defined, policy-relevant population. If feasible, they might further apply inverse-probability weighting as a robustness check (following Cole & Stuart, 2010 and the analysis herein). Moreover, they should carefully assess treatment-control balance in the geographic sample. In this volume, Keele et al. (2017) discuss related consideration when units of analysis (such as households) cannot be precisely geo-located.

The findings on external validity have broader implications for the design and interpretation of randomized field experiments, which often rely on convenience samples defined by observed and unobserved moderators of treatment effects (such as poverty, distance to urban centers, agents' willingness to submit to randomization). At a minimum, experiments should specifically describe the criteria for sample selection (Campbell, Piaggio, Elbourne, & Altman, 2012), and whether these variables are plausible moderators of treatment effects. Our chapter suggests that authors can push this exercise further and assess robustness after re-weighting the experimental convenience sample to resemble a policy-relevant population, with appropriate caveats about selection-on-unobservables into the convenience sample (Cole & Stuart, 2010; Hotz et al., 2005; Stuart et al., 2015).

NOTES

1. Cattaneo, Frandsen, and Titiunik (2015) push this interpretation further by implementing randomization inference in samples near the cutoff.

2. In rare cases, boundaries might be suddenly (and quasi-randomly) redrawn, leading to a more credible "re-randomization" of households before endogenous sorting begins anew (Billings, Deming, & Rockoff, 2013).

3. External validity exists when causal relationships "[hold] over variations in persons, settings, treatment variables, and measurement variables" (Shadish, Cook, & Campbell, 2002, p. 507). Some authors assess the importance of variation in treatments (particularly implementer characteristics) and measurement variables (Allcott, 2015; Bold et al., 2013; Lucas, McEwan, Ngware, & Oketch, 2014). This paper focuses on the potentially moderating role of independent variables related to persons and settings.

4. We refer to it as a geographic quasi-experiment because assignment is not transparently random (even local to borders), and because we must proxy the location of children using the coordinates of their caserío (a cluster of dwellings). The latter introduces mass points in the putative assignment variables of latitude and longitude in a geographic discontinuity design. For related explanations, see Keele et al. (2016) and later sections of this paper.

5. Buddelmeyer and Skoufias (2004) analyzed Mexico's well-known Progresa experiment (and a proxy means test and cutoff used to determine eligibility). In contrast to other results, they found that experimental enrollment estimates in samples "close" to the eligibility cutoff were roughly similar or slightly larger than full-sample estimates.

6. Angrist and Rokkanen (2015) show how RDD effects might be estimated "away" from the cutoff if the assignment variable is ignorable, conditional on a set of covariates unaffected by the treatment. Dong and Lewbel (2015) note that the relative slopes of lines fit to data within bandwidths on either side of the cutoff provide insights into how modest changes in the assignment cutoff could affect the magnitude of estimated effects.

7. A school attendance condition was apparently not enforced (Glewwe & Olinto, 2004).

8. A related literature uses a panel household survey – collected in 2000 and 2002 – to estimate effects on child health and nutrition (Morris et al., 2004), education (Glewwe & Olinto, 2004), and adult labor supply (Alzúa, Cruces, & Ripani, 2013).

9. The interpretation of work-only-inside-home variable is governed by the flow of survey questions.

10. Benedetti, Ibarrarán, and McEwan (2016) analyze a later Honduran CCT experiment – also conducted in a sample of poor municipalities – which offered much larger transfers. They found smaller effects on both enrollment and child labor, which they attributed to a weaker application of the education enrollment condition.

11. We identified the caserío coordinates for 93% of all 6–12 year olds in the census (and 95% of the full experimental sample). The missing coordinates are due to incomplete geocoding of caseríos in an ArcGIS file obtained from the Infotecnología unit of the Secretaría de Educación in 2008.

12. A simulation in Hansen and Bowers (2008) shows that a similar test using logistic regression leads to over-rejection of the null with a modest number of assigned units (100). In the present case, we are concerned with comparing balance across subsamples.

13. In a national sample from 2004, the poverty headcount is 49% among non-indigenous individuals and 71% among ethnic and racial minorities (World Bank, 2006).

14. It is possible that households somehow misreported their answers, but seems unlikely given the fact that census data collection was independent of PRAF, IFPRI, and the original impact evaluation's data collection schedule.

15. The standardized differences are within thresholds beyond which regression adjustment is particularly sensitive to specification (Imbens & Wooldridge, 2009; Rubin, 2001).

16. In a histogram of distance-to-border distribution in the GQE sample – available from the authors – there is a puzzlingly large spike on the untreated side of the border, between 3 and 4 kilometers away. In fact, this is the town of Santa Bárbara (identified as a single caserío in Honduran data). It stretches about 2 kilometers across at its widest point and contains 1,178 dwellings with eligible children. If distance-to-border had been measured without error for each dwelling, it might have

"filled" an apparent notch in the histogram. This is the most severe example in the GQE sample of mis-measurement of the assignment variable, given the use of caserío rather than dwelling location. In the full GQE sample of caseríos, the mean (median) number of dwellings is 16.6 (8), and the 90th and 95th percentiles are 33 and 47, respectively.

17. When GQE estimates are reported within subsamples of blocks 1–2 and 3–5, control observations in untreated, non-experimental municipalities are assigned to the block corresponding to the treated observations on the opposite side of the border segment.

18. Cole and Stuart (2011) prove that the method yields consistent estimates – in this paper, of ATE_1 – under assumptions similar to those just described.

19. When weighted estimates are reported in subsamples (e.g., blocks 1–2), we separately estimate weights in that subsample.

ACKNOWLEDGMENTS

We are grateful to Matias Cattaneo, Juan Carlos Escanciano, Luke Keele, Rocío Titiunik, the anonymous referees, and participants of the Advances in Econometrics conference at the University of Michigan for their helpful comments, without implicating them for errors or interpretations.

REFERENCES

Allcott, H. (2015). Site selection bias in program evaluation. *The Quarterly Journal of Economics*, *130*(3), 1117–1165. doi:10.1093/qje/qjv015

Alzúa, M. L., Cruces, G., & Ripani, L. (2013). Welfare programs and labor supply in developing countries: Experimental evidence from Latin America. *Journal of Population Economics*, *26*(4), 1255–1284. doi:10.1007/s00148-012-0458-0

Angrist, J. D., & Rokkanen, M. (2015). Wanna get away? Regression discontinuity estimation of exam school effects away from the cutoff. *Journal of the American Statistical Association*, *110*, 1331–1344.

Benedetti, F., Ibarrarán, P., & McEwan, P. J. (2016). Do education and health conditions matter in a large cash transfer? Evidence from a Honduran experiment. *Economic Development and Cultural Change*. doi:10.1086/686583

Billings, S. B., Deming, D. J., & Rockoff, J. (2013). School segregation, educational attainment, and crime: Evidence from the end of busing in Charlotte-Mecklenburg. *The Quarterly Journal of Economics*, *129*(1), 435–476. doi:10.1093/qje/qjt026

Black, S. (1999). Do better schools matter? Parental valuation of elementary education. *Quarterly Journal of Economics*, *114*, 577–599.

Bold, T., Kimenyi, M., Mwabu, G., Ng'ang'a, A., & Sandefur, J. (2013). *Scaling-up what works: Experimental evidence on external validity in Kenyan Education*. Centre for the Study of African Economies WPS/2013-04.

Buddelmeyer, H., & Skoufias, E. (2004). An evaluation of the performance of regression discontinuity design on PROGRESA. Policy Research Working Paper No. 3386. World Bank, Washington, DC.

Calonico, S., Cattaneo, M. D., & Titiunik, R. (2014). Robust nonparametric confidence intervals for regression-discontinuity designs. *Econometrica*, *82*(6), 2295–2326. doi:10.3982/ecta11757

Cameron, A. C., Gelbach, J. B., & Miller, D. L. (2008). Bootstrap-based improvements for inference with clustered errors. *Review of Economics and Statistics*, *90*(3), 414–427. doi:10.1162/rest.90.3.414

Cameron, A. C., Gelbach, J. B., & Miller, D. L. (2011). Robust inference with multiway clustering. *Journal of Business & Economic Statistics*, *29*(2), 238–249. doi:10.1198/jbes.2010.07136

Campbell, M. K., Piaggio, G., Elbourne, D. R., & Altman, D. G. (2012). Consort 2010 statement: Extension to cluster randomised trials. *BMJ*, *345*, e5661. doi:10.1136/bmj.e5661

Cattaneo, M. D., Frandsen, B. R., & Titiunik, R. (2015). Randomization inference in the regression discontinuity design: An application to party advantages in the U.S. Senate. *Journal of Causal Inference*, *3*(1), 1–24. doi:10.1515/jci-2013-0010

Cole, S. R., & Stuart, E. A. (2010). Generalizing evidence from randomized clinical trials to target populations: The ACTG 320 trial. *American Journal of Epidemiology*, *172*(1), 107–115. doi:10.1093/aje/kwq084

Dell, M. (2010). The persistent effects of Peru's mining Mita. *Econometrica*, *78*(6), 1863–1903.

Dong, Y., & Lewbel, A. (2015). Identifying the effect of changing the policy threshold in regression discontinuity models. *Review of Economics and Statistics*, *97*(5), 1081–1092.

Fiszbein, A., & Schady, N. (2009). *Conditional cash transfers: Reducing present and future poverty*. Washington, DC: World Bank.

Galiani, S., & McEwan, P. J. (2013). The heterogeneous impact of conditional cash transfers. *Journal of Public Economics*, *103*, 85–96. doi:10.1016/j.jpubeco.2013.04.004

Glewwe, P., & Olinto, P. (2004). *Evaluating of the impact of conditional cash transfers on schooling: An experimental analysis of Honduras' PRAF program*. Final Report for USAID.

Hahn, J., Todd, P., & Van der Klaauw, W. (2001). Identification and estimation of treatment effects with a regression-discontinuity design. *Econometrica*, *69*(1), 201–209.

Hansen, B. B., & Bowers, J. (2008). Covariate balance in simple, stratified and clustered comparative studies. *Statistical Science*, *23*(2), 219–236. doi:10.1214/08-sts254.

Heckman, J. J. (1979). Sample selection bias as a specification error. *Econometrica*, *47*, 153–161.

Hotz, V. J., Imbens, G. W., & Mortimer, J. H. (2005). Predicting the efficacy of future training programs using past experiences at other locations. *Journal of Econometrics*, *125*(1–2), 241–270. doi:10.1016/j.jeconom.2004.04.009

Imbens, G., & Zajonc, T. (2009). Regression discontinuity design with vector-argument assignment rules. *Mimeo*.

Imbens, G. W., & Wooldridge, J. M. (2009). Recent developments in the econometrics of program evaluation. *Journal of Economic Literature*, *47*(1), 5–86. doi:10.1257/jel.47.1.5

International Food Policy Research Institute (IFPRI). (2000). *Second report: Implementation proposal for the PRAF/IDB Project—Phase II*. Washington, DC: International Food Policy Research Institute.

Keele, L., & Titiunik, R. (2015). Geographic boundaries as regression discontinuities. *Political Analysis*, *23*(1), 127–155. doi:10.1093/pan/mpu014

Keele, L., & Titiunik, R. (2016). Natural experiments based on geography. *Political Science Research and Methods*, *4*(1), 65–95. doi:10.1017/psrm.2015.4

Keele, L., Lorch, S., Passarella, M., Small, D., & Titiunik, R. (2017). An overview of geographically discontinuous treatment assignments with an application to children's health insurance. In M. D. Cattaneo & J. C. Escanciano (Eds.), *Advances in econometrics* (p. 38). Bingley: Emerald Group Publishing Limited.

Lee, D. S. (2008). Randomized experiments from non-random selection in U.S. House elections. *Journal of Econometrics*, *142*(2), 675–697. doi:10.1016/j.jeconom.2007.05.004

Lee, D. S., & Lemieux, T. (2010). Regression discontinuity designs in economics. *Journal of Economic Literature*, *48*(2), 281–355. doi:10.1257/jel.48.2.281

Lucas, A. M., McEwan, P. J., Ngware, M., & Oketch, M. (2014). Improving early-grade literacy in East Africa: Experimental evidence from Kenya and Uganda. *Journal of Policy Analysis and Management*, *33*(4), 950–976. doi:10.1002/pam.21782

Moore, C. (2008). Assessing Honduras' CCT Programme PRAF, Programa de Asignación Familiar: Expected and unexpected realities. Country Study No. 15. International Poverty Center.

Morris, S. S., Flores, R., Olinto, P., & Medina, J. M. (2004). Monetary incentives in primary health care and effects on use and coverage of preventive health care interventions in rural honduras: Cluster randomised trial. *The Lancet*, *364*(9450), 2030–2037. doi:10.1016/s0140-6736(04)17515-6

Muller, S. M. (2015). Causal interaction and external validity: Obstacles to the policy relevance of randomized evaluations. *The World Bank Economic Review*, *29*, S217–S225. doi:10.1093/wber/lhv027

Oosterbeek, H., Ponce, J., & Schady, N. (2008). *The impact of cash transfers on school enrollment: Evidence from Ecuador*. Policy Research Working Paper No. 4645. World Bank, Washington, DC.

Rubin, D. B. (2001). Using propensity scores to help design observational studies: Application to the tobacco litigation. *Health Services and Outcomes Research Methodology*, *2*(3–4), 169–188. doi:10.1023/a:1020363010465

Secretaría de Educación (1997). *VII Censo Nacional de Talla, Informe 1997*. Tegucigalpa: Secretaría de Educación, Programa de Asignación Familiar.

Shadish, J. W., Cook, T. D., & Campbell, D. T. (2002). *Experimental and quasi-experimental designs for generalized causal inference*. Boston: Houghton Mifflin.

Shadish, W. R., Galindo, R., Wong, V. C., Steiner, P. M., & Cook, T. D. (2011). A randomized experiment comparing random and cutoff-based assignment. *Psychological Methods*, *16*(2), 179–191.

Stuart, E. A., Bradshaw, C. P., & Leaf, P. J. (2015). Assessing the generalizability of randomized trial results to target populations. *Prevention Science*, *16*(3), 475–485. doi:10.1007/s11121-014-0513-z

Stuart, E. A., Cole, S. R., Bradshaw, C. P., & Leaf, P. J. (2011). The use of propensity scores to assess the generalizability of results from randomized trials. *Journal of the Royal Statistical Society: Series A (Statistics in Society)*, *174*(2), 369–386. doi:10.1111/j.1467-985x.2010.00673.x

World Bank. (2006). *Honduras poverty assessment: Attaining poverty reductions*. Report No. 35622-HN. Washington, DC: World Bank.

APPENDIX A

Table A1. Census Variable Definitions.

	Variable Definition and Census Question(s) Used to Construct Variable
Dependent variables	
Enrolled in school	1 = Enrolled in school on census date; 0 = not (F8).
Works outside home	1 = Worked during past week, including self-employment, family business, and agricultural work; 0 = not (F12, F13A01-04); only reported for ages 7 and up.
Works only in home	1 = Worked during past week, exclusively on household chores; 0 = not (F13B10). Variable only reported for ages 7 and up.
Independent variables	
Age	Integer age on census date (F3).
Female	1 = Female; 0 = Male (F2).
Born in municipality	1 = Born in present municipality; 0 = not (F4A).
Lenca	1 = Lenca or other non-mestizo ethnicity/race; 0 = not (F5).
Moved	1 = Resided in a different caserío, aldea, or city in 1996; 0 = resided in current caserío, aldea, or city in 1996 (F6).
Father is literate	1 = Father is literate; 0 = not (F7, F1, F2).
Mother is literate	1 = Mother is literate; 0 = not (F7, F1, F2).
Father's schooling	Years of father's schooling (F9, F1, F2).
Mother's schooling	Years of mother's schooling (F9, F1, F2).
Dirt floor	1 = Dwelling has dirt floor; 0 = not (B5).
Piped water	1 = Dwelling has piped water from public or private source; 0 = not (B6).
Electricity	1 = Electric light from private or public source; 0 = light from another source (ocote, etc.) (B8).
Rooms in dwelling	Number of bedrooms used by household (C1).
Sewer/septic	1 = Household has toilet connected to sewer or septic system; 0 = not (C5).
Auto	1 = Household has at least one auto; 0 = not (C7).
Refrigerator	1 = Household has refrigerator; 0 = not (C8a).
Computer	1 = Household has computer; 0 = not (C8g).
Television	1 = Household has television; 0 = not (C8e).
Mitch	1 = at least 1 household member emigrated after Hurricane Mitch in October 1998; 0 = not (E1).
Household members	Total individuals residing in household.
Household members, 0–17	Total individuals, ages 0–17, residing in household.

Note: The 2001 Honduran census form is available at http://unstats.un.org/unsd/demographic/sources/census/quest/HND2001es.pdf

APPENDIX B: TREATMENT-CONTROL BALANCE

Table B1. Balance in Experiments 1 and 2 (≤2 kilometers from border).

	Experiment 1 Sample		Experiment 2 Sample	
	T/*C* differences	*p*-value	*T*/*C* differences	*p*-value
Age	−0.055/−0.029	0.124	−0.066/−0.036	0.333
Female	0.000/0.000	0.961	0.005/0.011	0.452
Born in municipality	−0.016/−0.059	0.325	−0.015/−0.055	0.482
Lenca	−0.007/−0.014	0.925	−0.057/−0.129	0.632
Moved	0.004/0.026	0.419	0.015/0.082	0.088
Father is literate	0.030/0.062	0.305	0.029/0.059	0.539
Mother is literate	0.027/0.055	0.277	0.030/0.061	0.467
Father's schooling	0.228/0.086	0.285	0.432/0.150	0.382
Mother's schooling	0.215/0.083	0.234	0.407/0.144	0.350
Dirt floor	0.046/0.102	0.312	0.078/0.167	0.347
Piped water	−0.000/−0.001	0.994	−0.001/−0.003	0.981
Electricity	0.012/0.032	0.802	0.085/0.217	0.322
Rooms in dwelling	0.050/0.071	0.308	0.104/0.139	0.342
Sewer/septic	0.049/0.107	0.171	0.050/0.105	0.434
Auto	−0.008/−0.040	0.455	0.012/0.058	0.512
Refrigerator	0.004/0.019	0.831	0.035/0.142	0.377
Computer	0.001/0.022	0.404	0.003/0.055	0.211
Television	0.009/0.033	0.779	0.068/0.219	0.341
Mitch	0.008/0.067	0.088	0.024/0.170	0.006
Household members	0.186/0.078	0.087	0.310/0.128	0.067
Household members, 0–17	0.146/0.076	0.115	0.274/0.141	0.075
Predicted mean municipal child height-for-age *z*-score	0.001/0.004	0.989	−0.029/−0.129	0.735

Notes: See Table 4 for sample definitions of experiments 1 and 2. In the difference columns, the first number is the mean difference and the second number is the mean difference divided by the full-sample standard deviation. *p*-values account for clustering by municipality.

Table B2. Balance in the Placebo Sample (≤2 kilometers from border).

	T/*C* Differences	*p*-Value
Age	0.057/0.030	0.191
Female	0.005/0.010	0.513
Born in municipality	0.012/0.045	0.537
Lenca	0.200/0.485	0.085
Moved	−0.013/−0.076	0.063
Father is literate	0.009/0.019	0.750
Mother is literate	−0.043/−0.087	0.111
Father's schooling	−0.069/−0.026	0.771
Mother's schooling	−0.239/−0.095	0.261
Dirt floor	−0.020/−0.041	0.800
Piped water	−0.037/−0.077	0.439
Electricity	−0.046/−0.127	0.382
Rooms in dwelling	−0.007/−0.010	0.867
Sewer/septic	0.023/0.050	0.632
Auto	−0.015/−0.075	0.177
Refrigerator	−0.008/−0.037	0.691
Computer	−0.000/−0.014	0.587
Television	−0.019/−0.073	0.495
Mitch	−0.008/−0.084	0.066
Household members	−1.036/−0.156	0.266
Household members, 0 – 17	−0.977/−0.161	0.254
Predicted mean municipal child height-for-age *z*-score	−0.495/−1.423	0.001

Notes: See the text for definition of the placebo sample. In the difference column, the first number is the mean difference and the second number is the mean difference divided by the full-sample standard deviation. *p*-values account for clustering by municipality.

APPENDIX C: COVARIATE DIFFERENCES BETWEEN EXPERIMENTAL SUBSAMPLES (≤2 KILOMETERS FROM BORDER)

Table C1. All Blocks.

	Experiment 2 − Experiment 1	
	Unweighted	Weighted
Age	0.009/0.005	−0.002/−0.001
Female	0.000/0.000	−0.001/−0.001
Born in municipality	−0.004/−0.016	0.008/0.028
Lenca	−0.066/−0.141	0.002/0.003
Moved	0.003/0.016	0.000/0.001
Father is literate	−0.014/−0.029	−0.001/−0.003
Mother is literate	0.017/0.034	0.000/0.000
Father's schooling	0.087/0.033	−0.007/−0.003
Mother's schooling	0.208/0.080	−0.003/−0.001
Dirt floor	−0.040/−0.089	0.001/0.003
Piped water	0.007/0.015	−0.007/−0.014
Electricity	0.037/0.105	0.003/0.007
Rooms in dwelling	0.037/0.052	−0.001/−0.001
Sewer/septic	0.039/0.086	−0.008/−0.017
Auto	0.003/0.013	0.000/0.001
Refrigerator	0.013/0.060	0.002/0.010
Computer	0.001/0.016	0.000/0.002
Television	0.028/0.105	0.003/0.012
Mitch	0.006/0.054	0.000/0.001
Household members	−0.079/−0.033	0.003/0.001
Household members, 0–17	−0.082/−0.043	0.000/0.000
Predicted mean municipal child height-for-age z-score	0.052/0.204	0.015/0.058
Block 1 or 2 (proportion)	−0.078/−0.158	−0.007/−0.015
Estimated propensity score	0.045/0.369	0.005/0.039

Notes: In each cell, the first number is the mean difference between the experiment 2 sample and experiment 1 sample; the second number is the mean difference divided by the standard deviation in the experiment 1 sample. In the weighted column, the means for experiment 2 apply inverse-probability weights described in the text.

Table C2. Blocks 1–2.

	Experiment 2 – Experiment 1	
	Unweighted	Weighted
Age	−0.069/−0.037	−0.003/−0.002
Female	−0.002/−0.005	−0.003/−0.005
Born in municipality	0.003/0.012	0.015/0.055
Lenca	−0.008/−0.015	0.002/0.004
Moved	0.010/0.061	−0.001/−0.006
Father is literate	0.019/0.039	0.006/0.012
Mother is literate	0.063/0.126	0.001/0.003
Father's schooling	0.486/0.183	0.031/0.012
Mother's schooling	0.658/0.250	0.006/0.002
Dirt floor	−0.073/−0.173	−0.005/−0.012
Piped water	0.060/0.123	−0.016/−0.032
Electricity	0.122/0.396	0.001/0.004
Rooms in dwelling	0.166/0.223	−0.005/−0.007
Sewer/septic	0.083/0.176	−0.008/−0.018
Auto	0.023/0.127	−0.001/−0.007
Refrigerator	0.055/0.280	0.001/0.006
Computer	0.003/0.087	0.000/0.001
Television	0.103/0.405	0.001/0.004
Mitch	0.020/0.163	0.001/0.005
Household members	−0.038/−0.016	−0.006/−0.002
Household members, 0–17	−0.107/−0.056	−0.005/−0.003
Predicted mean municipal child height-for-age *z*-score	0.041/0.213	0.026/0.134
Estimated propensity score	0.077/0.535	0.005/0.035

Notes: In each cell, the first number is the mean difference between the experiment 2 sample and experiment 1 sample; the second number is the mean difference divided by the standard deviation in the experiment 1 sample. In the weighted column, the means for experiment 2 apply inverse-probability weights described in the text.

Table C3. Blocks 3–5.

	Experiment 2 – Experiment 1	
	Unweighted	Weighted
Age	0.041/0.022	0.008/0.004
Female	0.003/0.005	−0.000/−0.001
Born in municipality	−0.008/−0.027	0.001/0.004
Lenca	−0.068/−0.158	−0.003/−0.008
Moved	−0.001/−0.007	−0.001/−0.004
Father is literate	−0.020/−0.041	−0.010/−0.020
Mother is literate	−0.008/−0.016	−0.005/−0.011
Father's schooling	−0.052/−0.020	−0.049/−0.019
Mother's schooling	−0.016/−0.006	−0.044/−0.017
Dirt floor	−0.014/−0.029	0.004/0.010
Piped water	−0.024/−0.051	−0.006/−0.013
Electricity	−0.011/−0.029	0.001/0.002
Rooms in dwelling	−0.022/−0.031	−0.008/−0.012
Sewer/septic	0.025/0.055	−0.008/−0.017
Auto	−0.008/−0.040	−0.002/−0.008
Refrigerator	−0.009/−0.037	0.001/0.004
Computer	−0.001/−0.015	0.000/0.001
Television	−0.009/−0.032	0.003/0.009
Mitch	0.000/0.001	−0.000/−0.002
Household members	−0.065/−0.027	0.007/0.003
Household members, 0–17	−0.043/−0.023	0.009/0.005
Predicted mean municipal child height-for-age z-score	0.007/0.084	0.001/0.008
Estimated propensity score	0.048/0.358	0.001/0.007

Notes: In each cell, the first number is the mean difference between the experiment 2 sample and experiment 1 sample; the second number is the mean difference divided by the standard deviation in the experiment 1 sample. In the weighted column, the means for experiment 2 apply inverse-probability weights described in the text.

APPENDIX D: EXPERIMENTAL AND QUASI-EXPERIMENTAL ESTIMATES THAT EXCLUDE DEPARTMENT BORDERS (≤2 KILOMETERS FROM BORDER)

Table D1. Estimates for Experiments 1 and 2 (Dependent Variable: Enrolled in School).

	All Blocks	Blocks 1–2	Blocks 3–5
Experiment 1	0.094***	0.167***	0.047*
	(0.026)	(0.045)	(0.024)
N	54,089/70	22,344/28	31,745/42
BS *p*(sym)	0.007	0.013	0.092
Experiment 2	0.056	0.141*	0.029
	(0.036)	(0.074)	(0.039)
N	17,156/43	5,736/14	11,420/29
BS *p*(sym)	0.210	0.150	0.575
Experiment 2, weighted	0.056	0.139*	0.012
	(0.039)	(0.071)	(0.039)
N	17,156/43	5,736/14	11,420/29
BS *p*(sym)	0.253	0.135	0.777

Notes: Robust standard errors are in parentheses, clustered by municipality. *N* indicates the number of eligible children and the number of municipalities. BS *p*(sym) is the symmetric *p*-value from a wild cluster-bootstrap percentile-*t* procedure with 399 replications. All regressions control for the variables described in the note to Table 1.

*** indicates statistical significance at 1%, and * at 10%.

Table D2. Estimates for Experiments 1 and 2 (Dependent Variable: Works Outside Home).

	All Blocks	Blocks 1–2	Blocks 3–5
Experiment 1	−0.047***	−0.086***	−0.020*
	(0.013)	(0.024)	(0.012)
N	44,433/70	18,277/28	26,156/42
BS *p*(sym)	0.003	0.003	0.152
Experiment 2	−0.033	−0.107**	−0.004
	(0.020)	(0.043)	(0.017)
N	14,139/43	4,656/14	9,483/29
BS *p*(sym)	0.182	0.058	0.858
Experiment 2, weighted	−0.040*	−0.104**	−0.004
	(0.022)	(0.041)	(0.017)
N	14,139/43	4,656/14	9,483/29
BS *p*(sym)	0.120	0.043	0.855

Notes: Robust standard errors are in parentheses, clustered by municipality. *N* indicates the number of eligible children and the number of municipalities. BS *p*(sym) is the symmetric *p*-value from a wild cluster-bootstrap percentile-*t* procedure with 399 replications. All regressions control for the variables described in the note to Table 1.

*** indicates statistical significance at 1%, ** at 5%, and * at 10%.

Table D3. Estimates for Experiments 1 and 2 (Dependent Variable: Works Only in Home).

	All Blocks	Blocks 1–2	Blocks 3–5
Experiment 1	−0.042***	−0.061**	−0.029*
	(0.014)	(0.024)	(0.015)
N	44,433/70	18,277/28	26,156/42
BS *p*(sym)	0.018	0.075	0.105
Experiment 2	−0.016	−0.023	−0.015
	(0.020)	(0.030)	(0.024)
N	14,139/43	4,656/14	9,483/29
BS *p*(sym)	0.545	0.502	0.578
Experiment 2, weighted	−0.008	−0.022	−0.001
	(0.020)	(0.030)	(0.024)
N	14,139/43	4,656/14	9,483/29
BS *p*(sym)	0.790	0.545	0.958

Notes: Robust standard errors are in parentheses, clustered by municipality. *N* indicates the number of eligible children and the number of municipalities. BS *p*(sym) is the symmetric *p*-value from a wild cluster-bootstrap percentile-*t* procedure with 399 replications. All regressions control for the variables described in the note to Table 1.

*** indicates statistical significance at 1%, ** at 5%, and * at 10%.

Table D4. Estimates for Geographic Quasi-Experiment (All Dependent Variables).

	All Blocks	Blocks 1–2	Blocks 3–5
Enrolled in school	0.059*	0.073	0.047
	(0.030)	(0.055)	(0.041)
N	17,288/56/43	5,835/17/13	11,453/41/30
BS *p*(sym)	0.062	0.205	0.300
Works outside home	−0.025	−0.085*	−0.001
	(0.017)	(0.052)	(0.014)
N	14,200/56/43	4,724/17/13	9,476/41/30
BS *p*(sym)	0.133	0.058	0.968
Works only inside home	0.014	−0.005	0.023
	(0.017)	(0.028)	(0.024)
N	14,200/56/43	4,724/17/13	9,476/41/30
BS *p*(sym)	0.468	0.887	0.398

Notes: Robust standard errors are in parentheses, clustered by municipality and border segment (see text for details). *N* indicates the number of eligible children, municipalities, and border segments. BS *p*(sym) is the symmetric *p*-value from a wild cluster-bootstrap percentile-*t* procedure (clustering on municipalities) with 399 replications. All regressions control for the variables described in the note to Table 1 (but excluding block dummy variables).
* indicates statistical significance at 10%.

THE COMPARATIVE REGRESSION DISCONTINUITY (CRD) DESIGN: AN OVERVIEW AND DEMONSTRATION OF ITS PERFORMANCE RELATIVE TO BASIC RD AND THE RANDOMIZED EXPERIMENT

Yang Tang[a], Thomas D. Cook[b,c], Yasemin Kisbu-Sakarya[d], Heinrich Hock[c] and Hanley Chiang[e]

[a]*Northwestern University, Evanston, IL, USA*
[b]*Northwestern University, Evanston, IL, USA*
[c]*Mathematica Policy Research, Washington, DC, USA*
[d]*Koç University, Istanbul, Turkey*
[e]*Mathematica Policy Research, Cambridge, MA, USA*

ABSTRACT

Relative to the randomized controlled trial (RCT), the basic regression discontinuity (RD) design suffers from lower statistical power and lesser

Regression Discontinuity Designs: Theory and Applications
Advances in Econometrics, Volume 38, 237–279

ISSN: 0731-9053/doi:10.1108/S0731-905320170000038011

ability to generalize causal estimates away from the treatment eligibility cutoff. This chapter seeks to mitigate these limitations by adding an untreated outcome comparison function that is measured along all or most of the assignment variable. When added to the usual treated and untreated outcomes observed in the basic RD, a comparative RD (CRD) design results. One version of CRD adds a pretest measure of the study outcome (CRD-Pre); another adds posttest outcomes from a nonequivalent comparison group (CRD-CG). We describe how these designs can be used to identify unbiased causal effects away from the cutoff under the assumption that a common, stable functional form describes how untreated outcomes vary with the assignment variable, both in the basic RD and in the added outcomes data (pretests or a comparison group's posttest). We then create the two CRD designs using data from the National Head Start Impact Study, a large-scale RCT. For both designs, we find that all untreated outcome functions are parallel, which lends support to CRD's identifying assumptions. Our results also indicate that CRD-Pre and CRD-CG both yield impact estimates at the cutoff that have a similarly small bias as, but are more precise than, the basic RD's impact estimates. In addition, both CRD designs produce estimates of impacts away from the cutoff that have relatively little bias compared to estimates of the same parameter from the RCT design. This common finding appears to be driven by two different mechanisms. In this instance of CRD-CG, potential untreated outcomes were likely independent of the assignment variable from the start. This was not the case with CRD-Pre. However, fitting a model using the observed pretests and untreated posttests to account for the initial dependence generated an accurate prediction of the missing counterfactual. The result was an unbiased causal estimate away from the cutoff, conditional on this successful prediction of the untreated outcomes of the treated.

Keywords: Comparative regression discontinuity design; causal estimates away from the cutoff; comparison with experimental estimates away from the cutoff; within-study comparison

1. INTRODUCTION

The regression discontinuity (RD) design has been widely used to produce internally valid causal impact estimates. It requires just three basic elements: an outcome assessed at posttest, a continuous assignment variable, and a dummy treatment indicator defined by a cutoff value on the assignment variable. In the "sharp" version of RD, all subjects on one side of cutoff are eligible for treatment and take it up, whereas those on the other side are ineligible and cannot access the treatment. In the "fuzzy" version, there is some noncompliance with treatment eligibility status. This chapter assumes a sharp design.

As commonly used today, the basic RD design has two major disadvantages relative to the randomized controlled trial (RCT). First, the basic RD provides valid estimates only of the local average treatment effect (LATE) at the cutoff, whereas RCTs typically estimate the average treatment effect (ATE) for the entire population or the treatment-on-treated (TOT) impact for those who receive the treatment. Since the RD assignment rule deterministically separates the treatment group from the comparison group, there is no area of common support where both treated and untreated outcomes can be observed at the same points on the assignment variable. Fig. 1 illustrates this limitation using the potential outcomes framework of Rubin (1974). The treated outcomes of treatment-eligible individuals can only be observed on one side of the cutoff and untreated outcomes of ineligibles on the other side. Missing are the untreated outcomes of eligible individuals, the dashed line in Fig. 1, which indicates our ignorance of what would have happened to the treated cases had they not been treated. In the basic posttest-only RD design, the LATE is based on the difference between the observed solid lines at the cutoff of the assignment variable

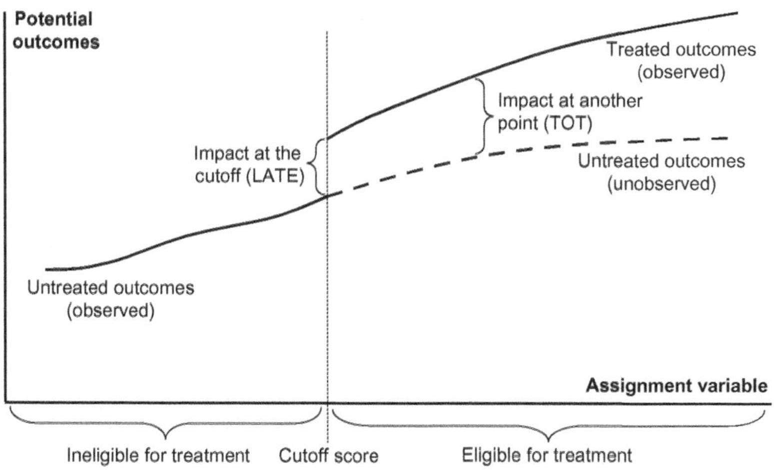

Fig. 1. LATE and TOT Estimates in the Basic RD Design.

only. Additional assumptions about the unobserved dashed lines are required to identify causal effects elsewhere.

Although impact estimates in RD were historically obtained based on functional forms assumed to hold globally, current practice generally favors nonparametric estimation in which observed outcome functions only need to be approximated locally near the cutoff (Hahn, Todd, & Van der Klaauw, 2001). This local approximation can be extrapolated away from the cutoff by taking into account the curvature of the outcome-assignment variable relationship observed near the cutoff (Dong & Lewbel, 2015), and creative research is now taking place to obtain more robust inference using local polynomial approximations (Calonico, Cattaneo, & Titiunik, 2014; Cattaneo, Titiunik, & Vazquez-Bare, in press). Nonetheless, modern estimation techniques for the posttest-only RD design must base extrapolation away from the cutoff on the information contained in a relatively narrow window around that cutoff. The farther away from this point causal effects are estimated, the less likely it is that such an extrapolation is valid. That is why RD analysts generally restrict their causal inference to the cutoff LATE and rarely venture more general estimates for other regions of the assignment variable – where the majority of cases will actually be located.

Another limitation of basic RD is its statistical power for estimating the LATE at the cutoff. Goldberger (1972) found that RD sample sizes have to be at least 2.75 times larger than the RCTs in order to produce impacts estimates with the same degree of statistical precision. Schochet (2009) has generalized this and has shown that the variance of an RD impact estimate is $1/\left(1-\rho_{TS}^2\right)$ times that of an RCT impact estimate, where ρ_{TS} is the correlation between the treatment indicator and the assignment variable. Since the treatment indicator corresponds to a point on the assignment variable, the correlation ρ_{TS} is nonzero (and can often be large in absolute value), thus reducing the RD design's power relative to the RCT design. Improving power in RD studies requires attenuating ρ_{TS}.

The ultimate goal of this chapter and related research is to provide researchers with a broader and better range of RD design options than the basic RD that currently dominates the literature. In this chapter, we show how the two main limitations to the basic RD design might be addressed by adding data to create a "comparative regression discontinuity" (CRD) design. We describe two versions of this design. One adds a pretest measure of the study outcome (CRD-Pre), while the other adds posttest outcomes from a nonequivalent comparison group (CRD-CG). We describe the key assumptions needed for the CRD strategy to produce valid impact estimates away from the RD cutoff and propose a diagnostic test to lend support to those assumptions. For each version of CRD we then use data from a prior RCT, the National Head Start Impact Study, to illustrate (1) use of this diagnostic in practice, (2) the gains in precision that result from using CRD instead of basic RD to estimate impacts at the cutoff, and (3) the extent of bias in CRD's causal estimates both at and away from the cutoff. Our assessments of bias and precision are made relative to RCT benchmarks using a within-study comparison (WSC) approach, also known as a design experiment (LaLonde, 1986; Cook, Shadish, & Wong, 2008). They reveal that each version of CRD increases power for estimating the basic RD's traditional LATE and also generates TOT effect estimates away from the cutoff that correspond closely to the RCT estimates in the same area.

The rest of this chapter is organized as follows. In Section 2, we provide an overview of the CRD design and how it might be used to improve upon the basic RD design. We also discuss the within-study comparison approach that we use to assess CRD's possible gains in precision and/or generalizability. In Section 3, we describe the National Head Start Impact Study data and the specific analysis methods we use to estimate impacts. In Section 4, we present the results of our analysis for each CRD design separately, including comparisons of precision and bias between the CRD estimates and RCT benchmark estimates. Section 5 includes a synthesis of results across the two CRD designs, a discussion of the different reasons why CRD-Pre and CRD-CG eliminated bias with these data, and what our findings suggest about the potential strengths and limitations of the CRD designs.

2. OVERVIEW OF CRD DESIGNS AND THE WITHIN-STUDY COMPARISON APPROACH

2.1. Improving the Basic RD Design by Adding a Pretest or a Nonequivalent Comparison Group

2.1.1. Main Features of the CRD Design

It is well recognized that controlling for baseline covariates other than the assignment variable – particularly pretest measures of the outcome – may improve the precision of the basic RD design's estimate of the LATE (Lee & Lemieux, 2010). An additional feature of the CRD-Pre approach, introduced by Wing and Cook (2013), is that it also leverages the pretest to improve causal inference elsewhere along the assignment variable. It does this by including untreated pretest data on both sides of the cutoff to provide some empirical support for imputing what is crucially missing in basic RD – the untreated outcomes of individuals assigned to the treatment group. That information is still missing in CRD; but, as Fig. 2 shows, CRD now provides three sources of data on untreated outcomes, compared to the single source of such data in basic RD. Both basic RD and CRD share the same untreated posttest-only cases on the ineligible side of the cutoff. But CRD now also includes untreated pretest data not only within that same region but also in the region of the assignment variable where only treated cases are located in basic RD. It is these added pretest data that support causal inference away from the cutoff in CRD-Pre, in addition to yielding improvements in in statistical power. Although CRD-Pre requires a parametric extrapolation, as with posttest-only methods, it can be made based on data available along the full length of the assignment variable (unlike posttest-only methods). Estimates of the untreated outcomes of the treatment group can be formed by combining all observed pretest and posttest data on the ineligible side of the cutoff with the pretest observations on the eligible side.

The CRD-Pre framework also provides checks on the plausibility of extrapolation based on the coherence of the observed relationships between outcomes and assignment variable. For example, we may assess how closely the pretest and posttest functions track one another below the cutoff

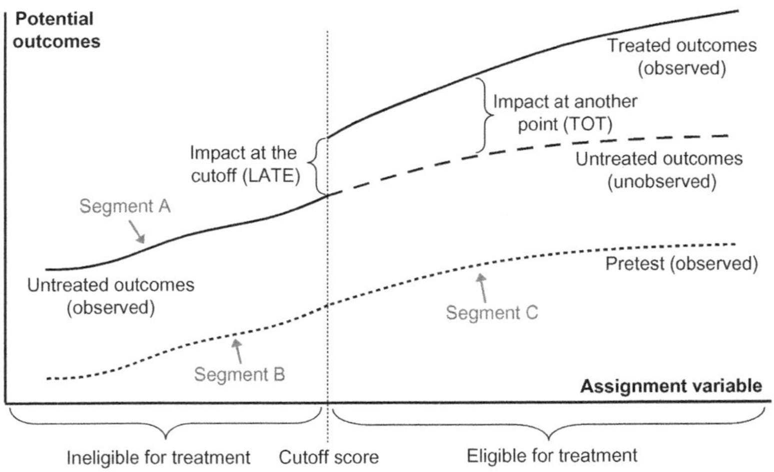

Fig. 2. Adding a Pretest Function in the CRD Design.

(Segments A and B in Fig. 2), how stable the pretest function is across the cutoff (Segment C compared to Segment B), and how parallel all three untreated outcome functions are to one another. This insight dates to Lohr (1972), although he did not make use of the pretest data available in his RD study when fitting a model to estimate impacts.

CRD-CG is an alternative approach to supplementing the basic RD with untreated outcomes data described by Kisbu-Sakarya, Cook, and Tang (2016). This design adds the posttests of a nonequivalent comparison group instead of adding pretests to cases already in the basic RD. Being completely ineligible for treatment, the posttest observations of a comparison group can again be used to trace out an untreated outcomes function along the length of the assignment variable, thus serving the same purposes as the pretest in CRD-Pre. That is, the supplementary posttest data provide the basis for increasing statistical power and support a parametric prediction of the untreated outcomes of the treated. Moreover, these data may help assess the likely validity of that prediction based on the similarity and stability of the observed relationships between untreated outcomes and the assignment variable along its length for the comparison group and on the ineligible side of the cutoff for the basic RD group.

CRD is similar in spirit to the covariate-based extrapolation method of Angrist and Rokkanen (2015), but there are also important differences between the two. Both draw on additional data elements to render treatment eligibility status conditionally ignorable for potential outcomes. Both also ignore the theoretically banal but practically important circumstance in which the assignment variable choice results in no initial relationship with untreated potential outcomes. (In this case, the independence assumption is unconditionally met, which obviates the need to draw in additional data to make these outcomes conditionally independent of treatment eligibility.) Angrist and Rokkanen (2015) use covariates other than the assignment variable to impute untreated outcomes based on techniques grounded in the classic matching assumptions of conditional independence and common support. That is, the covariates must collectively make the assignment variable irrelevant for predicting potential outcomes and there must also be sufficient overlap to allow for matching across the cutoff. As discussed in detail below, CRD differs in two major ways. First, it relies on the assignment variable to produce a reliable model of untreated outcomes that is de-confounded from actual assignment status, which achieves conditional independence in the prognostic sense described by Hansen (2008). Second, CRD does not require the assumption of common covariate-based support across the cutoff of assignment variable; in

CRD-Pre, for example, the pretest scores between individuals need not overlap across each side of the cutoff. However, in practice, the comparison cases added by CRD-CG are likely to result in an improved model of untreated outcomes only when they overlap with the basic RD cases in their values of the assignment variable.

2.1.2. Identification of Causal Effects Using CRD

We focus on using the CRD design to estimate the TOT because, as formulated in this chapter, CRD cannot identify treatment effects on the ineligible side of the RD cutoff. The aim of CRD is to develop a valid estimate of the average potential untreated outcomes of the treated. Given that their treated outcomes can be observed, this is sufficient to identify the TOT. For all observations i:

- Let K_i be an indicator equal to 1 if the observation belongs to the basic RD design. Both versions of CRD bring in a supplemental set of observations ($K_i = 0$) – from a pretreatment period in the case of CRD-Pre and from a nonequivalent comparison group in the case of CRD-CG.
- Let AV_i denote the assignment variable, and $Z_i = Z(\text{AV}_i) \in \{0, 1\}$ be an indicator that the individual's assignment variable would result in eligibility for treatment based on the RD cutoff rule (even if the observation is not part of the basic RD design). For example, in Fig. 2, $Z_i = 1$ denotes all observations to the right of the cutoff (regardless of K_i).

Thus, using Y_i^0 to denote the untreated potential outcome of observation i, the objective is to identify the counterfactual $E(Y_i^0|K_i = 1, Z_i = 1)$. For simplicity, the remainder of this section focuses on CRD-Pre, but the same framework applies to CRD-CG.

CRD relies on a variant of the differences-in-differences framework and uses information about the RD assignment variable to improve inference about the average untreated posttests of the treated. Expressing the conditional expectation of untreated outcomes as a function of the assignment variable, we can write:

$$E\left(Y_i^0|K_i, \text{AV}_i,\ Z_i\right) = [(1 - Z_i)F(\text{AV}_i) + Z_iG(\text{AV}_i)] \\ + K_i[(1 - Z_i)H(\text{AV}_i) + Z_iI(\text{AV}_i)],$$

where $F(\text{AV}_i)$ describes the relationship between pretests and the assignment variable on the ineligible side of the cutoff and $G(\text{AV}_i)$ describes this relationship on the eligible side. In this setup, time-period effects (i.e., differences between untreated posttests and pretests) are described by $H(\text{AV}_i)$ on the ineligible side of the cutoff and by $I(\text{AV}_i)$ on the eligible side. This general specification allows period effects to differ along the assignment variable, potentially in different ways on each side of the cutoff. The functions $F(\cdot)$, $G(\cdot)$, and $H(\cdot)$ can all be estimated with the available data, but additional assumptions must be made to determine $I(\text{AV}_i)$ because no untreated posttest data are available on the eligible side of the cutoff.

The standard differences-in-differences assumption in this setting would be that period effects are independent of treatment eligibility: $E[H(\text{AV}_i)|Z_i = 0] = E[I(\text{AV}_i)|Z_i = 1]$. In this case, we could first estimate $H(\text{AV}_i)$ using pretest and posttest data among treatment ineligibles (with $Z_i = 0$) and take the average of predicted values across the assignment variable on the ineligible side of the cutoff. Similarly, we could then estimate $G(\text{AV}_i)$ using pretests among treatment eligibles (with $Z_i = 1$) and take the average of predicted values across the assignment variable on the eligible side of the cutoff.[1] Because the expected period effect is transferrable across the cutoff, adding the two averages together would yield an unbiased estimate of $E(Y_i^0|K_i = 1, Z_i = 1)$.

CRD strengthens the standard differences-in-differences assumption in a way that provides testable implications in the RD setting.

> ***Assumption CRD1.*** *Period effects are independent of the assignment variable.*

Because treatment eligibility is a deterministic function of the assignment variable in an RD design, this condition implies that period effects are also independent of treatment eligibility. Formally, this condition is that $H(\text{AV}_i) = I(\text{AV}_i) = \delta$; that is, that there is a constant period effect. In this case, we can write $E\left(Y_i^0|K_i, \text{AV}_i, Z_i\right) = [(1 - Z_i)F(\text{AV}_i) + Z_iG(\text{AV}_i)] + \delta K_i$. Based on this assumption, we can estimate the counterfactual untreated outcomes of treatment eligibles using a similar process as above, but with δ estimated as the average difference between posttests and pretests on the ineligible side of the cutoff.

Although it is stronger than the usual differences-in-differences assumption, Assumption CRD1 has the advantage of being partially verifiable using data on the untreated side of the cutoff. On that side, the function $H(\text{AV}_i)$ can be directly estimated because posttests and pretests can be observed along the length of the assignment variable, and so the claim that $H(\text{AV}_i) = \delta$ is testable on the treatment-ineligible side of the cutoff. Finding that posttest-pretest differences are related to AV_i would falsify Assumption CRD1. Meanwhile, finding that $H(\cdot)$ does not depend on AV_i – that is, finding that the pretest and posttest functions are parallel – would provide supporting evidence for the plausibility of Assumption CRD1. However, this assumption would remain unverifiable on the treatment-eligible side where $Z_i = 1$, as well as when moving across the cutoff.

The spirit of CRD is that untreated potential outcomes have a stable, predictable relationship with the assignment variable – so stable and predictable that there is constant period effect. However, taken alone, Assumption CRD1 does not require that this relationship be the same on either side of the cutoff; the only requirement is that the period effect stays the same. Yet, if the relationship between pretest outcomes and the assignment variable were to change substantially across the cutoff, this would belie the spirit of CRD. In this case, Assumption CRD1 would also be highly suspect. For this reason, CRD also postulates an additional assumption:

> ***Assumption CRD2.*** *The relationships between untreated outcomes and the assignment variable do not change across the cutoff.*

Formally, this assumption states that, after specifying a parametric form for each function, $F(\mathrm{AV}_i) = G(\mathrm{AV}_i)$ and $H(\mathrm{AV}_i) = I(\mathrm{AV}_i)$. The latter equality is already implied by Assumption CRD1. Hence, Assumption CRD2 adds the requirement that $F(\mathrm{AV}_i) = G(\mathrm{AV}_i)$, which can be verified for observations with $K_i = 0$ by examining whether the pretest has the same relationship with the assignment variable on either side of the cutoff.

The combination of Assumptions CRD1 and CRD2 imply that untreated outcomes can be characterized using a single, common relationship with the assignment variable plus a period effect:

$$E(Y_i^0 | K_i, \mathrm{AV}_i, Z_i) = E(Y_i^0 | K_i, \mathrm{AV}_i) = F(\mathrm{AV}_i) + \delta K_i.$$

Hence, the two assumptions yield a model of untreated outcomes that does not depend on treatment eligibility. Using the terminology of Hansen (2008), under these assumptions $\hat{F}(\mathrm{AV}_i) + \hat{\delta} K_i$ provides an estimate of Y_i^0 that is de-confounded from treatment eligibility – so long as the model sufficiently reduces the information contained in (AV_i, K_i). An advantage of this formulation is that the function $F(\cdot)$ can be fit using data from both sides of the cutoff, which results in greater precision. A failure of Assumption CRD2 could be accommodated, but with less empirical support for the model; in this case $F(\cdot)$ would be fit using data only from the treatment-eligible side of the cutoff. However, if Assumption CRD1 does not hold, then the period effect cannot be transferred across the cutoff without additional assumptions; in this case, the counterfactual untreated posttests of the eligibles cannot be identified.

2.1.3. Implementation and Estimation

Wing and Cook (2013) and Kisbu-Sakarya et al. (2016) obtained predicted values for the untreated outcomes of the treated using variations of the following model:

$$Y_i = \boldsymbol{\theta}'\boldsymbol{p}(\mathrm{AV}_i) + \delta K_i + e_i,$$

where Y_i denotes *observed* untreated outcomes, $\boldsymbol{p}(\mathrm{AV}_i)$ is a polynomial in the assignment variable, K_i is as defined before, and e_i is an error term. This model was estimated using all data on untreated outcomes: (1) posttests on the ineligible side of the cutoff for $K_i = 1$, and (2) pretests or the comparison-group posttests on both sides of the cutoff for $K_i = 0$. In the CRD-Pre case, this amounts to pooling together the data defining Segments A, B, and C in Fig. 2; a similar set of three untreated posttest outcome function segments are pooled together for the CRD-CG design.

One of the main tasks in estimation is to ascertain the correct functional form for the outcome-assignment variable relationship. Both Wing and Cook (2013) and Kisbu-Sakarya et al. (2016) applied CRD to settings in which untreated outcomes were judged to depend *linearly* on the assignment variable. As discussed below, linearity cannot be rejected in the application presented in this chapter. With linear outcome functions, tests of the two CRD assumptions can be stated more simply. In particular, the combination of Assumptions CRD1 and CRD2 implies that all untreated outcome functions have the same slopes and that there is no discontinuity in pretests at the cutoff. The latter is already a common falsification test used to check the validity of the basic RD design.

Hence, the main diagnostic innovation of CRD, as implemented to date, is to determine whether the three observed untreated outcome function segments in Fig. 2 are parallel lines. Parallelicity of the pretest and posttest lines on the ineligible side of the cutoff (Segments A and B) provides partial support for Assumption CRD1, and finding the same slope for the pretest function on the eligible side of the cutoff could help bolster the credibility of Assumption CRD2. Wing and Cook (2013) and Kisbu-Sakarya et al. (2016) conducted informal versions of this test for CRD-Pre and CRD-CG, respectively, comparing the results of this diagnostic assessment to an estimate of the bias in the impacts derived from CRD. They found that settings in which the untreated outcomes functions were plausibly parallel tended to produce CRD impact estimates that corresponded relatively closely to RCT estimates of the same treatment effect.

2.1.4. Gains in Statistical Power from Using CRD

Under almost all conditions, power will increase for estimating the LATE at the cutoff when data on untreated outcomes are brought into to the basic RD structure. For these added data points, the treatment indicator is fixed at zero. Hence, including them in a pooled model of untreated outcomes along with the data from the basic RD design will attenuate the ordinarily high correlation between the assignment status and the treatment indicator, ρ_{TS}. Tang, Cook, and Kisbu-Sakarya (2016) demonstrated these tendencies more formally for CRD-Pre design. They showed that CRD-Pre can be almost as powerful as an RCT of the same sample size when there is a high partial correlation between pretests and posttests after conditioning on the assignment variable. In this case, power is also less dependent on the location of the cutoff and the distribution of the assignment variable than is the case in basic RD. In addition, Tang, Cook, and Kisbu-Sakarya (2017) examined the statistical power of CRD-CG, contrasting it with the power of both the RCT and basic RD designs. One result showed that, holding the total number of cases constant, CRD-CG can achieve the same power as basic RD while requiring that fewer cases actually receive the treatment, thus saving resources if the cost of treated cases exceeds that of untreated cases. But this assumption of equal sample sizes belies the very advantage of CRD-CG – adding untreated cases to a basic RD. Thus, Tang et al. (2017) also showed that accounting for CRD-CG's increased data on untreated outcomes results in improvements in power that parallel those found in the CRD-Pre design. Relative to basic RD, the untreated cases in CRD-CG attenuate the correlation between the assignment variable and treatment eligibility and reduce the extent to which precision depends on the distribution of the assignment variable and the location of the cutoff.

2.2. Testing CRD Using a Within-Study Comparison Approach

Our empirical focus is on comparing estimates of the impacts and standard errors produced by CRD-Pre, CRD-CG, and an RCT away from the cutoff.

However, for completeness, we will also compare CRD estimates of the LATE to estimates of the same quantity obtained using a basic RD design and an RCT, thereby adding to the evidence on precision gains at the cutoff previously provided by Kisbu-Sakarya et al. (2016, 2017). We base these comparisons on a setup in which each CRD design has exactly the same treatment group as the RCT, so as to concentrate on variation in how the comparison data are formed. The underlying RCT data we use come from the National Head Start Impact Study, described below, and we use them to develop estimate impacts and standard errors for all four designs (RCT, basic RD, CRD-CG, and CRD-Pre) across three different outcomes measuring performance in math, performance in language arts, and social behavior.

These study goals require a within-study comparison approach, in which estimates from a quasi-experiment are judged against a benchmark that is usually provided by an RCT. LaLonde (1986) was the first to use this approach to examine whether the various selection-bias adjustments of his day would generate results closely approximating those from an RCT. Within-study comparisons are now used more broadly, especially to assess which among many different quasi-experimental designs and analyses can produce causal results similar to those from an RCT benchmark. The quasi-experimental designs examined to date include (1) basic RD, for which the earliest comparisons with RCT results are in Cook et al. (2008); (2) comparative interrupted time series designs, most recently reviewed in St.Clair, Cook, and Hallberg (2014); and (3) various nonequivalent control group designs, most recently reviewed in Wong, Valentine, and Miller-Bains (2017).

Quality standards for within-study comparisons have evolved over time. One standard is that the research designs being compared should estimate the same treatment effect. Traditionally, RCTs yield an ATE estimate, while basic RD yields a LATE estimate. Yet within-study comparisons require the same estimand to avoid confounding the effects of the design and the choice of the estimand. In comparisons with basic RD it is therefore important to estimate RCT at the cutoff, even if fewer cases are located there and analysts would not normally compute such an estimate. However, each CRD variant aspires to estimate valid treatment effects for all those eligible for treatment, and not just those at the cutoff. Given this study's context of a sharp RD, we seek to estimate TOT using the CRD, and so we must also produce an RCT estimate impact of treatment effects that is also limited to those on the eligible side of the cutoff. We compute three separate RCT estimates, therefore: one at the cutoff (a LATE);

another away from it (a TOT for those treated in the RD); and the third is the usual ATE, which we use for general background purposes only.

Another requirement for high quality within-study comparisons is a credible defense of the criteria used to claim that the RCT and nonexperimental estimates are functionally exchangeable. It is unrealistic to expect identical RCT and CRD estimates, given the sampling error in the impact estimates produced by each design. Hence, in this chapter, we assess whether each design's point estimates fall within a region of practical equivalence. Somewhat arbitrarily, we use the criterion that treatment effect estimates are similar if they are within 0.10 standard deviations of each other. We supplement this by visually depicting the overlap between RCT and CRD estimates, recognizing that this is not independent of the 0.10 criterion.

3. DATA AND ANALYSIS METHODS

3.1. Study Data

The data we use come from the National Head Start Impact Study (Puma, Bell, Cook, & Heid, 2010). Mandated by Congress, this random-assignment evaluation collected data from 84 randomly selected grantee agencies located in 23 states, 383 randomly selected Head Start centers, and 4,442 children ages 3 to 4 seeking to enroll in those centers. Within this nationally representative sampling frame, the study randomly assigned new applicants in one year to either (1) a treatment group that was allowed to enroll immediately in Head Start or (2) a control group that was embargoed from Head Start participation for one year but otherwise free to receive services elsewhere. Because the control group could enroll in Head Start during the second year of the study, we focus on outcomes measured one year after random assignment.

The analyses we present are limited to 3 year olds except in CRD-CG where we add a nonequivalent comparison group of untreated 4 year olds. As already noted, we examine three preschool outcomes. Two are measures of cognitive development derived from Woodcock-Johnson III assessments of applied problems in math and letter word identification in literacy. The third is a measure of social emotional development based on parents' ratings of their children's problem behavior (including aggressiveness, hyperactivity, and withdrawal). In our analysis, all of these outcomes are expressed in standard deviation units; therefore estimated means, differences, and impacts are effect sizes.

Some observations had missing data on these outcomes and the associated pretests. Of the 2,449 three year olds in the Head Start RCT, 568 children were missing their pretest for applied problems in math and 414 their posttest. For letter word identification, 51 children had no pretest and 395 no posttest. For total problem behavior, no pretests were missing but 387 posttests were. We imputed all the missing data using the EM algorithm and conducted our analyses based on the full set of 2,449 observations, including both observed and imputed data. (When presenting results below, we include an assessment of how sensitive our findings are to missing data patterns.) The imputation included pretest and posttest measures of the outcomes, the assignment variables, and other covariates.

Of the 17 socioeconomic covariates available for the analysis, two were removed because of perfect collinearity with another measure. The remaining 15 covariates assess three main domains: (1) *caregiver characteristics*, including the baseline caregiver's age, whether both biological parents lived with child, the biological mother's recent immigration status, the race/ethnicity of the biological mother/caregiver, the mother's educational attainment and marital status, and whether she was a teenage mother; (2) *child characteristics*, including the child's gender, race, preacademic skills, depression level, and whether he or she had special needs; (3) *living environment*, including the language spoken at home, a household risk index, and whether the child lived in an urban environment. Up to 458 children's records had missing data on or more of these covariates; in such cases we imputed values based on the EM algorithm already noted.

3.2. Creating the Basic RD, CRD-Pre and CRD-CG Designs

For the RCT, we rely on the 3-year-old treatment and control groups only, omitting the 4-year-old data. To create the basic RD structure, we defined an assignment variable (see below) and removed all the 3 year olds in the RCT treatment group below the cutoff and all the untreated cases above it. To create the CRD-Pre design, we again restricted the analysis to the 3 year olds but utilized their pretest scores as the untreated comparison data. To create the CRD-CG design, we focus on the 3 year olds for the basic RD but, instead of pretests, draw in posttest data from all 4 year olds in the RCT control group who, for present purposes, thus become the nonequivalent comparison group.

Every variant of RD requires an assignment variable and cutoff value. For CRD-Pre, the assignment variable was a child's score on the Peabody Vocabulary Test (PPVT), and the cutoff was chosen as the mean score in the sample. For evaluating CRD-CG, the assignment variable was the date that pretest assessments were given, and the cutoff was again the mean. (We did not use PPVT scores as the assignment variable for CRD-CG because children learn over time, and so the 3- and 4-year-old scores had little overlap on this particular test. This precluded the chance to observe both groups along the same range of the assignment variable.) In both cases, we assumed that only values above the cutoff value rendered children eligible for treatment and that children with values below the cutoff were ineligible.

Using the date of pretesting as the assignment variable in CRD-CG has an interesting repercussion: we will see later that it is not meaningfully correlated with any of the posttest outcomes we studied. This should limit the role of CRD-CG in reducing bias away from the cutoff since each outcome is likely to be already *unconditionally* independent of treatment status. That is, when outcomes are uncorrelated with the assignment variable they are also (by definition) uncorrelated with treatment eligibility. This makes the basic RD assignment rule in CRD-CG functionally equivalent to a random-assignment rule. Hence, the RCT and CRD-CG estimates should not differ whether at or away from the cutoff. We will test this, of course; but the result might be preordained by virtue of the choice of the assignment variable. Fortunately, the same is not true for CRD-Pre, for we will see that all three outcomes are related to the assignment variable – reading and math especially. This means there is more room to improve inferences about causal effects away from the cutoff, whether via the use of covariates in general (Angrist & Rokkanen, 2015) or via the use of pretest data as described above. Thus, we test here whether the added pretest data render treatment eligibility ignorable for untreated outcomes and also increase power.

Tables 1 and 2 summarize sample sizes for the four study designs – RCT, basic RD, CRD-Pre, and CRD-CG. Sample sizes obviously vary by design. The RCT includes 2,449 children; CRD-Pre and its corresponding basic RD each have 1,163; CRD-CG's basic RD includes 1,045 children, but after the 4 year olds are added, the full CRD-CG sample includes 1,856 children. These differences in sample sizes confound comparisons of efficiency, and we discuss below our approach to adjusting the standard errors obtained from each design to produce a fairer comparison across them.

Table 1. Sample Sizes for the CRD-Pre Design.

Outcome	RCT Sample	Basic RD Sample	CRD-Pre Sample
Applied problems in math	2,449	1,163	1,163
Letter word identification	2,449	1,165	1,165
Total behavior problem	2,449	1,162	1,162

Table 2. Sample Sizes for the CRD-CG Design.

Outcome	RCT Sample	Basic RD Sample	Nonequivalent Comparison Group Sample	CRD-CG Sample
Applied problems in math	2,449	1,045	811	1,856
Letter word identification	2,449	1,025	811	1,836
Total behavior problem	2,449	1,029	811	1,840

3.3. Checking the Parallel Assumptions of CRD

Given linearity of the outcome data, the main diagnostic check on the plausibility of CRD's key assumptions is that the untreated outcome function segments are parallel. Otherwise, as discussed previously, there would be a greater potential for bias in our counterfactual estimates of the outcomes of the treatment group had they not been treated. A potential dilemma is that a diagnostic test has to have high power to be useful, yet very high power will detect deviations from parallelicity that are so small they might be conceptually and computationally insignificant. Consequently, we use both the informal graphical assessment undertaken in past studies (Kisbu-Sakarya et al., 2016; Wing & Cook, 2013) and conventional statistical tests to ascertain if the untreated slopes differ from each other.

In our graphical assessment, we use local linear regression to plot (1) the untreated posttest data on the eligible side of the cutoff, which is part of the basic RD; and (2) the untreated outcomes data (from the pretest or from the independent comparison group) on both sides of the cutoff. For context, we also include in these graphs the treated posttests of children on the eligible side of the cutoff. These graphs use a triangular kernel with the bandwidth set to the optimal value obtained for the basic RD data using the procedure described by Imbens and Kalyanaraman (2012). We display the resulting data later and all regression segments seem to be basically linear and parallel for all three outcomes in each of the two CRD designs.

To check this visual impression statistically, for each outcome we conducted a chi-squared test of whether the slopes differ from each other (within the power limits imposed by the sample size and data structure). The model is:

$$Y_i = \beta_0 + \beta_1 K_i + \beta_2 Z_i + \beta_3 \mathrm{AV}_i + \beta_4 K_i \mathrm{AV}_i + \beta_5 Z_i \mathrm{AV}_i + u_i,$$

where all of the variables are as described in Section 2.1 and u_i is an error term with mean zero and variance σ^2. For CRD-Pre, Y_i is the pretest measure of the outcome variable or the posttest measure of the untreated cases in the basic RD, and K_i is the indicator for the post period. The slope of the pretest below the cutoff is β_3; the slope of the untreated posttest below the cutoff is $\beta_3 + \beta_4$; and the slope of the pretest above the cutoff is $\beta_3 + \beta_5$; and. For CRD-CG, Y_i is the posttest of untreated cases in the basic RD, and $K_i = 0$ indicates that the observation comes from the comparison group. In this case, the slope of the comparison group's posttest below the cutoff is β_3; the slope of the untreated posttests from the group included in the basic RD is $\beta_3 + \beta_4$; and the slope of the comparison group's posttests above the cutoff is $\beta_3 + \beta_5$.

The null hypothesis is that the three observed segments of untreated outcomes have the same slopes, which would imply that $\beta_4 = \beta_5 = 0$. However, if $\beta_4 \neq 0$, then Assumption CRD1 fails because the two segments below the cutoff are not parallel, indicating that the period or group effect is related to the assignment variable. In this case, it is unlikely that the period/group effect estimated below the cutoff can be used without modification to generate counterfactual estimates of the untreated outcomes of treatment-eligible children above the cutoff. And, if $\beta_5 \neq 0$, then Assumption CRD2 fails because the pretest or comparison-group's outcome function has different slopes on either side of the cutoff. This also suggests caution in extrapolating across the cutoff because untreated outcomes do not appear to have a stable, linear relationship with the assignment variable that spans the cutoff.

3.4. Estimating Causal Parameters for the RCT and CRD Designs

3.4.1. Causal Estimates for the RCT Design

Because children were nested within centers in the Head Start study, we estimate the ATE from RCT using a two-level hierarchical model:

$$Y_{ij} = \alpha + \beta T_i + \upsilon_j + \varepsilon_{ij},$$

where Y_i is the posttest measure of the outcome variable for child i in center j; T_i is the treatment assignment indicator for child i (with $T_i = 1$ indicating assignment to the treatment group); υ_j is a center-level error term with mean zero and variance κ^2; and ε_{ij} is an individual-level error term with mean zero and variance σ^2. The estimated ATE is given by β.

Rubin (2008) advocates including all available covariates in the RCT estimation in order to adjust for any small-sample imbalances. Hence, we also estimate a covariate-adjusted version of ATE based on:

$$Y_{ij} = \alpha + \beta T_i + \gamma' \boldsymbol{X}_i + \upsilon_j + \varepsilon_{ij},$$

where the vector $\boldsymbol{X}_i$ includes the pretest measure of the given outcome and all of the covariates described previously in Section 3.1.

For the RCT, we also need to estimate the LATE at the cutoff. This need not be the same as the ATE if treatment effects are heterogeneous along the assignment variable. To account for this possibility, we estimate for each RCT group separate linear regression functions of the outcome variable on the assignment variable, the covariates, and (in specifications corresponding to the CRD-CG design) the applicable pretest. Our estimate of the LATE is then the point-wise difference between the treatment- and control-group regression functions at the cutoff.

Finally, we need to estimate RCT treatment effects for all those who were on the eligible side of the RD cutoff – that is, the TOT based on the sharp RD assignment rule. To estimate this in a way that is consistent with our implementation of CRD, we follow Wing and Cook (2013) by taking a weighted average of estimated treatment effects across values of the assignment variable above the cutoff. Letting TE(m) denote the treatment effect for $\text{AV}_i = m$, the ATE above the cutoff, c, is:

$$TE(\text{AV} \geq c) = \sum_{m=c}^{\max(\text{AV})} TE(m) \times \frac{\Pr(\text{AV}_i = m)}{\Pr(\text{AV}_i \geq c)}.$$

We estimate each TE(m) using the same technique described above for estimating effects at the cutoff, and we calculate $\Pr(\text{AV}_i = m)/\Pr(\text{AV}_i \geq c)$ based on the distribution of the sample along the assignment variable on the eligible side of the cutoff.

Because there is no closed-form expression for the variance of this TOT estimate, we use bootstrap methods to estimate standard errors, resampling program centers with replacement and using data from all children within centers sampled for a given replication. We resample 1,000 times, using the standard deviation of the resulting estimates as the standard error of original estimated treatment effect. For consistency, we use the same bootstrap-based approach to calculate the standard errors of the other estimates based on the RCT (i.e., the ATE for the whole sample and the LATE estimated for children at the cutoff).

3.4.2. Causal Estimates for the CRD and Basic RD Designs

We follow Wing and Cook (2013) in estimating causal effects based on the CRD-Pre design in three steps, adapting their framework to account for the nesting of children within centers in the Head Start study. In the first step, we combine all data on untreated outcomes to fit a pooled model of how they are related to the assignment variable. The covariate-adjusted specification of this model with assumed linearity in the assignment variable is:

$$Y_{ij}(K_i = k,\ Z_i = 0) = r_{00} + r_{01}AV_i + r_{02}\boldsymbol{X}_i + r_{03}k + \varsigma_j + \varphi_{ij} + \xi_{ijk}, \quad (1)$$

where Y_{ij} is the pretest or posttest (depending on the value of K_i) of child i in center j; AV_i is her or his value of the assignment variable; ς_j is a center-level error term with mean zero and variance σ^2_ς; φ_{ij} is a child-level error term that has a mean of zero, a variance of σ^2_φ, and is correlated across the child's pretest and posttest observations; and ξ_{ijk} is a test-level error term with mean zero and variance σ^2_ξ. In the second step, we estimate a similar model separately to obtain adjusted estimates of posttest scores above the cutoff for the treatment-eligible cases in the RD. For example, with a linear form for the relationship between posttests and the assignment variables, this is given by:

$$Y_{ij}(K_i = 1,\ Z_i = 1) = r_{10} + r_{11}\mathrm{AV}_i + r_{12}\boldsymbol{X}_i + \varpi_j + \vartheta_{1ij}. \quad (2)$$

Then, in the third step, we take the difference between the predicted values of treated outcomes from Eq. (2) and the predicted values of untreated outcomes from Eq. (1) to estimate the casual effect at a given value of the assignment variable.

We use a basically similar three-step approach to estimating causal effects at the each value of the assignment variable based on the CRD-CG design. The main difference is that the indicator K_i denotes whether children were 3 year olds, assumed to be subject to the RD assignment rule, or part of the comparison group of 4 year olds. Because the added outcomes data come from a different group, Eq. (1) does not contain the child-level error term when fit for CRD-CG.

Based on this approach, we estimate the LATE at the cutoff for each CRD design using predicted values with AV_i set to the cutoff value. We also use the technique of Wing and Cook (2013) described previously to estimate the TOT for all children eligible for treatment based on the RD assignment rule for the given CRD design. This amounts to taking a weighted average of the point-wise differences in predicted values across all values of AV_i above the cutoff, as described in the RCT section above.

Finally, we also obtain estimates of the LATE from the basic RD design using a small variation of this approach. When doing so for the CRD-Pre design, we estimate Eq. (1) using only data points corresponding to each child's posttest observation, with pretests included as covariates. When doing this for the basic RD design corresponding to the CRD-CG design, data on the comparison group's outcomes are excluded entirely. In both cases, we used a parametric model because the functional forms are clear and our assessment of the basic RD estimates is secondary to our assessment of the CRD estimates. Visual and statistical evidence we report later indicate that the outcome data have a dominant linear component and no sources of obvious and systematic nonlinearity that would require a nonparametric analysis.

For each LATE and TOT causal estimand, we also estimate variations of the models described by Eqs. (1) and (2) in which higher-order polynomial terms in the assignment variable, AV_i, are included. We specifically consider quadratic and cubic functions as alternatives to the linear function. However, the outcomes data appear to be close to linear (and we cannot formally reject linearity based on statistical tests), and so the quadratic and cubic polynomials seem likely to over-fit the data. Hence, while the tables present the results of all polynomial specifications, our discussion in the text focuses on results from the models that are linear in the assignment variable.

We calculate bootstrapped standard errors for both LATE estimates at the cutoff and TOT estimates above the cutoff. As with the RCT, in each bootstrap replication, we randomly select with replacement the same number of centers as in the original data and, for each center so selected, we include all children. We draw 1,000 bootstrap samples and estimate the standard error of the original impact estimate using the standard deviation of replicate estimates across bootstrap samples.

3.5. Comparing Efficiency across Designs

Given our WSC approach, comparing the standard errors of each design on an "as is" basis would potentially confound differences in precision that arise from the unique features of each design with differences in precision arising simply from the sample size differentials shown previously in Tables 1 and 2. To provide a fairer comparison, we divide the standard errors of LATEs obtained from the basic RD and CRD samples by the following factor:

$$f_{\mathrm{LATE}} = \sqrt{N_{\mathrm{RCT}}/N_{\mathrm{B}}},$$

where N_{RCT} is the number included in the RCT sample for the given CRD design and N_{B} is the number of cases in the corresponding basic RD sample. Applying this factor to the standard errors of the basic RD's LATE estimate equates its effective sample size with that of the RCT to isolate the RD design effect. We also apply this factor to the CRD estimates of the cutoff LATE. This allows us to assess the gains from using pretests as separate data points on untreated outcomes (rather than including them as covariates) in the CRD-Pre design and from drawing in the added data on untreated outcomes provided by the comparison group in the CRD-CG design.

For similar reasons, we divide the standard errors of CRD TOT estimates by:

$$f_{\mathrm{TOT}} = \sqrt{N_{\mathrm{RCT}}/N_{\mathrm{C}}},$$

where N_{RCT} is as described above and N_{C} is the number of children included in the CRD sample. For CRD-Pre, this adjustment allows us to compare the efficiency of CRD-Pre with the efficiency of the RCT based on a combination of how the sample is allocated and how pretest data are used. For CRD-CG, the adjusted standard errors allow for an assessment of efficiency, relative to the RCT, based on a combination of how the sample is allocated and the nature of the group (experimental control vs. nonequivalent comparison) providing additional data on untreated outcomes.

4. RESULTS

4.1. Results for the CRD-Pre Design

4.1.1. RCT Results

Table 3 shows the RCT ATE results for Head Start's effects on the three outcomes for 3 year olds. For the math and literacy outcomes, ATEs

Table 3. ATE Estimates from the RCT.

Outcome	RCT, No Covariates		RCT, with Covariates	
	ATE	Std. Err.	ATE	Std. Err.
Applied problems in math	0.08	(0.04)	0.10	(0.04)
Letter word identification	0.24	(0.05)	0.26	(0.04)
Total child behavior problems	−0.11	(0.04)	−0.11	(0.04)

Notes: The ATE is the average RCT treatment effect across the length of the assignment variable and is estimated using the approach described in Section 3.4.1. Standard errors are adjusted for clustering across children within Head Start centers. Outcomes are normalized so that all estimates are expressed in standard deviation units.

without covariates are 0.08 and 0.24, respectively.[2] Including covariates marginally increases the effects to 0.10 and 0.26 respectively. For the social emotional outcome, the ATE is −0.11 standard deviations, irrespective of whether covariates are included. Though the differences attributable to covariates are small, we will use as our benchmark the ATE estimates with covariates to reduce the potential role of sampling error (Rubin, 2008).

Table 4 lists estimated impacts using RCT design both at and above the cutoff, and both with and without covariates. For the literacy and social emotional outcomes, the local effects at the cutoff broadly agree with the TOT estimated for all children above the cutoff, and covariate adjustment results in very little change in either impact estimate. At the cutoff, the LATE for the letter word identification test is 0.27 without covariates and 0.25 with them, and the LATEs for behavior problems are −0.12 and −0.10, respectively. Above the cutoff, TOTs for letter word identification are 0.31 without covariates and 0.27 with them, and TOTs for behavior problems are −0.13 without covariates and −0.12 with them. All of these numbers are also similar to the corresponding ATEs reported above.

Although covariates do not appear to affect the LATE estimate for math, there are greater differences between the estimates away from the cutoff – the ATE is 0.10 and the TOT above the cutoff is 0.05. This anomaly may be due to chance or an imbalance that post-hoc investigation revealed between children with assignment variable values above the cutoff. Among such children on the eligible side of the cutoff, the standardized math pretest means were 0.08 in the treatment group and

Table 4. LATE Estimates at the Cutoff and TOT Estimates Above the Cutoff from the RCT Using the CRD-Pre Assignment Variable.

Functional Form	RCT at the Cutoff, No Covariates		RCT at the Cutoff, with Covariates		RCT away from the Cutoff, No Covariates		RCT away from the Cutoff, with Covariates	
	LATE	Std. Err.	LATE	Std. Err.	TOT	Std. Err.	TOT	Std. Err.
Applied problems in math								
Linear	0.12	(0.04)	0.10	(0.04)	0.08	(0.03)	0.05	(0.04)
Quadratic	0.12	(0.05)	0.09	(0.04)	0.08	(0.04)	0.04	(0.04)
Cubic	0.13	(0.05)	0.09	(0.04)	0.06	(0.04)	0.04	(0.04)
Letter word identification								
Linear	0.27	(0.04)	0.25	(0.03)	0.31	(0.05)	0.27	(0.04)
Quadratic	0.29	(0.05)	0.29	(0.04)	0.31	(0.05)	0.28	(0.04)
Cubic	0.30	(0.05)	0.30	(0.04)	0.29	(0.05)	0.26	(0.04)
Total child behavior problems								
Linear	−0.12	(0.03)	−0.10	(0.03)	−0.13	(0.04)	−0.12	(0.04)
Quadratic	−0.09	(0.05)	−0.08	(0.04)	−0.13	(0.04)	−0.11	(0.04)
Cubic	−0.08	(0.04)	−0.07	(0.04)	−0.13	(0.05)	−0.12	(0.04)

Notes: The LATE is the RCT treatment effect at the assignment variable cutoff, and the TOT is the average of RCT treatment effects across the length of the assignment variable above the cutoff; both are estimated using the approach described in Section 3.4.1. Standard errors are adjusted for clustering across children within Head Start centers. Outcomes are normalized so that all estimates are expressed in standard deviation units.

−0.09 in the control group – a difference that could becloud covariate-adjusted estimates of how well the RCT and CRD estimates coincide above the cutoff for the math outcome. Nonetheless, we still use the covariate-adjusted TOT estimate as the RCT benchmark away from the cutoff for all outcomes.

4.1.2. Support for Functional Form Assumptions

Fig. 3 displays local linear regressions of relationships with the assignment variable for the three untreated segments of the math, literacy, and social emotional outcomes in the CRD-Pre design. All three segments are plausibly linear and parallel for each outcome, although not perfectly so. Under assumed linearity, statistical tests do not reject the null hypotheses of equal

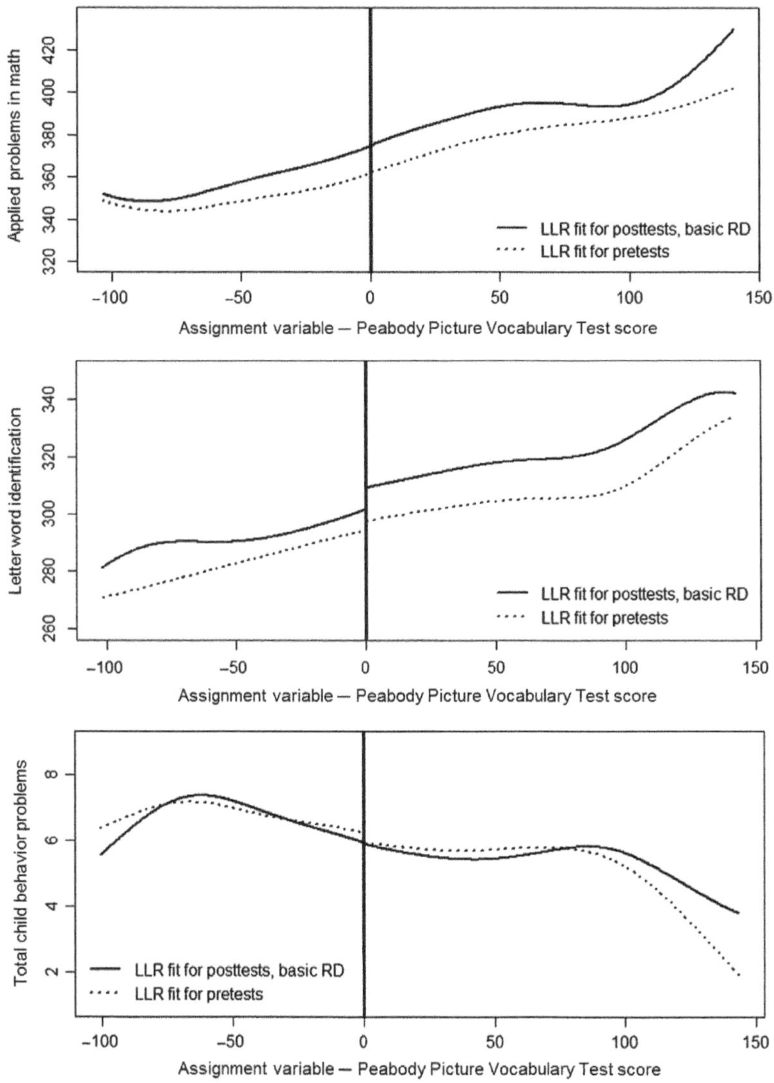

Fig. 3. Local Linear Regression of Each Outcome on the CRD-Pre Assignment Variable. (a) Applied Problems in Math. (b) Letter Word Identification. (c) Total Child Behavior Problems. *Notes*: Local linear regressions of posttests and pretests on the assignment variable were fit separately on either side of the cutoff for each outcome. In all cases, we used a triangle kernel with the bandwidth determined using the Imbens and Kalyaraman (2012) selection procedure.

slopes between all three untreated outcome function segments. (For the math, literacy, and social emotional outcomes, p-values from this test are 0.15, 0.10, and 0.59, respectively.) This is consistent with the key assumptions of CRD and suggests that the observed untreated outcomes might reasonably be used to produce a counterfactual estimate of the outcomes of treatment-eligible children away from the cutoff. The plausibly parallel trends do not prove this, of course; but they provide some empirical support for CRD's assumptions.

Striking is that the slopes for math (0.027) and reading/literacy (0.019) are visibly positive and statistically significant (p-values < 0.001). The slope for the social emotional outcomes is, as expected, in the opposite direction; students with high PPVT scores have fewer behavioral problems. The coefficient is statistically significant (p-value < 0.001), albeit smaller in magnitude (−0.009). In sum, then, the observed values of untreated outcomes are strongly related to the assignment variable and thus to treatment eligibility. For CRD-Pre, the task is to see if model-based predictions based on this apparently stable relationship are sufficient to render untreated posttests conditionally independent of treatment eligibility status. If so, then CRD-Pre could yield causal estimates away from the cutoff that are unbiased and, therefore, similar to the estimates in the RCT design.

4.1.3. Estimated Bias

Table 5 lists the CRD-Pre impact estimates for the three outcomes, including estimates based on linear, quadratic, and cubic functional forms for the elationship between outcomes and the assignment variable. Limiting ourselves to the most plausible linear form and to estimates at the cutoff, Fig. 4 plots bias for the RD and CRD-Pre estimates relative to the appropriate RCT benchmark. Considering the LATE at the cutoff, the biases for math, literacy, and social emotional outcomes are 0.02, 0.05, and 0.01 for CRD-Pre, while they are −0.01, 0.06, and 0.13 for basic RD. Overall, then, CRD-Pre provides estimates at the cutoff that are as bias-free as those of RD (if not more so). Fig. 4 also plots the estimated bias in the CRD-Pre estimates of the TOT for each outcome. They are 0.05, −0.01, and 0.07 for applied math, letter word identification, and social problem behavior, respectively – all smaller in absolute value than 0.10 and evidence that there is little if any bias in the CRD-Pre above the cutoff, too.

Table 5. LATE Estimates at the Cutoff and TOT Estimates Above the Cutoff from the CRD-Pre Design.

Functional Form	CRD-Pre at the Cutoff, with Covariates		CRD-Pre away from the Cutoff, with Covariates	
	LATE	Std. Err.	TOT	Std. Err.
Applied problems in math				
Linear	0.12	(0.07)	0.10	(0.05)
Quadratic	0.05	(0.08)	0.10	(0.05)
Cubic	0.00	(0.10)	0.08	(0.05)
Letter word identification				
Linear	0.30	(0.06)	0.27	(0.05)
Quadratic	0.30	(0.08)	0.27	(0.05)
Cubic	0.23	(0.10)	0.26	(0.06)
Total child behavior problems				
Linear	−0.09	(0.06)	−0.05	(0.06)
Quadratic	−0.06	(0.08)	−0.05	(0.06)
Cubic	0.07	(0.09)	−0.04	(0.06)

Notes: The LATE is the CRD-Pretreatment effect at the assignment variable cutoff, and the TOT is the average of CRD-Pretreatment effects across the length of the assignment variable above the cutoff; both are estimated using the approach described in Section 3.4.2. Standard errors are adjusted for clustering across children within Head Start centers and across test dates for each child. Outcomes are normalized so that all estimates are expressed in standard deviation units.

4.1.4. Statistical Efficiency

Table 5 includes the CRD-Pre standard errors for each outcome variable and impact estimand. Applying the adjustment factor described in Section 3.5, Fig. 5 plots ratios of (1) the adjusted standard errors from basic RD and CRD, and (2) the actual standard error from the RCT. The smaller these ratios are, the more efficient is the non-RCT design. At the cutoff, the relative size of CRD-Pre standard errors for the LATE values range from 1.12 to 1.48, compared to the RCT, while these ratios for basic RD are larger and range from 1.37 to 1.89. So CRD-Pre standard errors are smaller at the cutoff. Moreover, the ratios of standard errors for the TOT range from 0.87 to 0.99, indicating that CRD-Pre yields at least the same precision as the RCT above the cutoff.

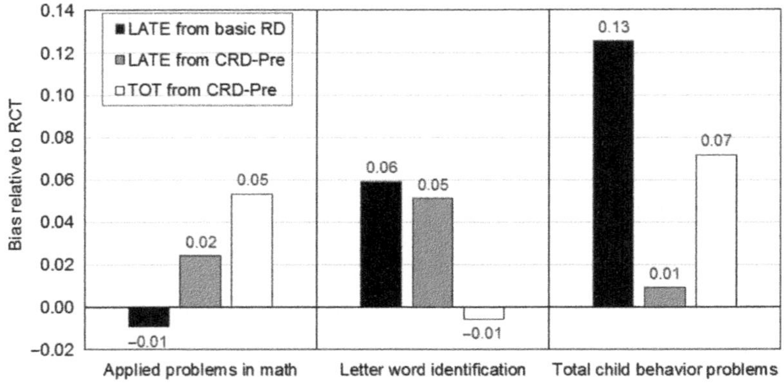

Fig. 4. Bias in Basic RD and CRD-Pre Estimates, by Outcome and Estimand. *Notes*: LATE impacts are estimated at the RD cutoff, and TOT impacts are estimated for all treated children with scores above the cutoff. Estimated bias is calculated as the difference between the basic RD or CRD-Pre estimate and the RCT estimate using models that specify a linear outcome-assignment variable relationship. Outcomes are normalized so that all estimates are expressed in standard deviation units.

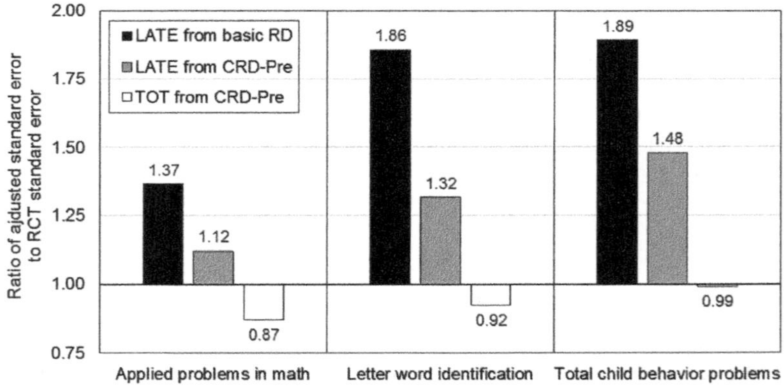

Fig. 5. Relative Efficiency of Basic RD and CRD-Pre Estimates, by Outcome and Estimand. *Notes*: LATE impacts are estimated at the RD cutoff, and TOT impacts are estimated for all treated children with scores above the cutoff. Efficiency is measured using the ratio of the adjusted standard error of the basic RD or CRD-Pre estimate to the actual standard error of the RCT estimate based on models that specify a linear outcome-assignment variable relationship. The adjustment factor applied to the original standard errors of the basic RD and CRD-Pre estimates is described in Section 3.5. Outcomes are normalized so that all estimates are expressed in standard deviation units.

4.2. Results for the CRD-CG Design

4.2.1. RCT Results

The CRD-CG design uses a different assignment variable than CRD-Pre and supplements the basic RD design by adding a comparison group of 4 year olds rather than a pretest measure of the outcome for the focal group. Because both designs examine the pool of 3 year olds, the ATE estimates for CRD-CG and CRD-Pre are the same as those reported earlier: 0.10, 0.26, and −0.11 for math, literacy, and social emotional outcomes, respectively. However, the new assignment variable entails different RCT estimates at and above the cutoff. Table 6 list the estimates obtained from this RCT design variant for each outcome and estimand. The LATE and TOT estimates are similar to one another and to the ATE estimates. LATE estimates at the cutoff are 0.10, 0.25, and −0.10, respectively, for

Table 6. LATE Estimates at the Cutoff and TOT Estimates Above the Cutoff for the RCT Using the CRD-CG Assignment Variable.

Functional Form	RCT at the Cutoff, No Covariates		RCT at the Cutoff, with Covariates		RCT away from the Cutoff, No Covariates		RCT away from the Cutoff, with Covariates	
	LATE	Std. Err.	LATE	Std. Err.	LATE	Std. Err.	LATE	Std. Err.
Applied problems in math								
Linear	0.08	(0.04)	0.10	(0.03)	0.13	(0.05)	0.14	(0.05)
Quadratic	0.09	(0.05)	0.08	(0.05)	0.13	(0.05)	0.13	(0.05)
Cubic	0.07	(0.06)	0.07	(0.06)	0.13	(0.05)	0.14	(0.05)
Letter word identification								
Linear	0.24	(0.04)	0.25	(0.03)	0.21	(0.06)	0.22	(0.05)
Quadratic	0.30	(0.05)	0.29	(0.05)	0.24	(0.06)	0.23	(0.05)
Cubic	0.25	(0.07)	0.25	(0.05)	0.24	(0.06)	0.24	(0.05)
Total child behavior problems								
Linear	−0.10	(0.04)	−0.10	(0.03)	−0.09	(0.05)	−0.10	(0.05)
Quadratic	−0.06	(0.05)	−0.06	(0.04)	−0.07	(0.05)	−0.08	(0.05)
Cubic	−0.09	(0.05)	−0.09	(0.05)	−0.07	(0.05)	−0.08	(0.05)

Notes: The LATE is RCT treatment effect at the assignment variable cutoff, and the TOT is the average of RCT treatment effects across the length of the assignment variable above the cutoff; both are estimated using the approach described in Section 3.4.1. Standard errors are adjusted for clustering within Head Start centers. Outcomes are normalized so that all estimates are expressed in standard deviation units.

each outcome, while the corresponding estimates of the TOT above the cutoff are 0.14, 0.22, and –0.10. Again, the covariates make little difference, and so we use the covariate-adjusted RCT estimates as benchmarks.

4.2.2. Support for Functional Form Assumptions

Fig. 6 portrays local linear regressions against the assignment variable for the three untreated segment of each outcome in the CRD-CG design. As with CRD-Pre, the assumptions of linear and parallel slopes might not be satisfied exactly in the CRD-CG design, but the assumption seems plausible for each outcome. In addition, when fitting globally linear models in the assignment variable, we cannot reject the null hypothesis of equal slopes between all three segments of observed outcomes. (The *p*-values from tests of this hypothesis are 0.68, 0.32, and 0.33 for the math, literacy, and social emotional outcomes, respectively.) Hence, these results support (but do not prove) the assumptions under which CRD-CG is expected to produce valid causal estimates away from the cutoff.

However, it is not clear that CRD-CG's untreated comparison data is actually required to yield unbiased estimates away from the cutoff – the slopes in the basic RD portion of the CRD-CG data seem flat, and the untreated outcomes of the comparison group corroborate this. So where untreated outcomes can be observed, they appear to be *unconditionally* independent of the assignment variable and, therefore, of treatment eligibility status. For both the math and social behavior outcomes, there is neither visual nor statistical evidence of any slope different from zero. The slope of the untreated literacy outcome functions appears to be relatively flat, but the statistical test rejects the null hypothesis of a zero slope in the two segments below the cutoff ($p < 0.05$) but not in the third segment above the cutoff. As a whole, the evidence suggests that untreated outcomes are more or less unconditionally independent of treatment eligibility in CRD-CG, in which case there should be little to no bias when making inference away from the cutoff.

4.2.3. Estimated Bias

Table 7 displays the CRD-CG estimates of the LATE and TOT for math, literacy, and social emotional outcomes based on each assumed polynomial form of the relationship between untreated outcomes and the assignment variable. Focusing on the linear model, Fig. 7 plots the bias of the CRD-CG estimate, relative to the RCT, both at and away from the cutoff. Bias in the LATE for the three outcomes based on the full CRD-CG design is estimated to be –0.03, 0.02, and –0.08, respectively, compared to –0.02,

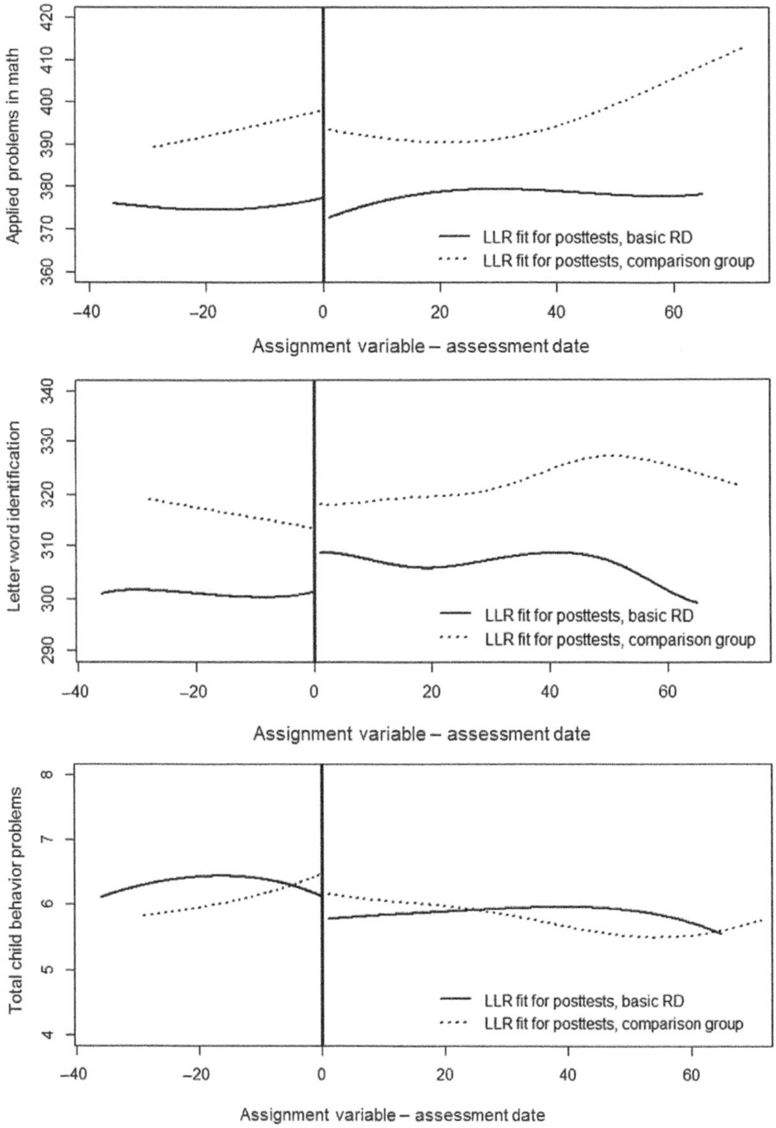

Fig. 6. Local Linear Regression of Each Outcome on the CRD-CG Assignment Variable. (a) Applied Problems in Math. (b) Letter Word Identification. (c) Total Child Behavior Problems. *Notes*: Local linear regressions of posttests and pretests on the assignment variable were fit separately on either side of the cutoff for each outcome. In all cases, we used a triangle kernel with the bandwidth determined using the Imbens and Kalyaraman (2012) selection procedure.

Table 7. LATE Estimates at the Cutoff and TOT Estimates Above the Cutoff for CRD-CG.

Functional Form	CRD-CG at the Cutoff, with Covariates		CRD-CG away from the Cutoff, with Covariates	
	LATE	Std. Err.	TOT	Std. Err.
Applied problems in math				
Linear	0.07	(0.07)	0.13	(0.07)
Quadratic	−0.04	(0.09)	0.14	(0.07)
Cubic	−0.11	(0.11)	0.16	(0.07)
Letter word identification				
Linear	0.27	(0.07)	0.19	(0.07)
Quadratic	0.38	(0.09)	0.20	(0.07)
Cubic	0.45	(0.10)	0.20	(0.07)
Total child behavior problems				
Linear	−0.18	(0.07)	−0.14	(0.06)
Quadratic	−0.20	(0.10)	−0.14	(0.07)
Cubic	−0.17	(0.12)	−0.12	(0.07)

Notes: The LATE is CRD-CG treatment effect at the assignment variable cutoff, and the TOT is the average of CRD-CG treatment effects across the length of the assignment variable above the cutoff; both are estimated using the approach described in Section 3.4.2. Standard errors are adjusted for clustering across children within Head Start centers. Outcomes are normalized so that all estimates are expressed in standard deviation units.

0.05, and −0.05 in the associated basic RD. This suggest that that CRD-CG produces estimates of the LATE that tend to be as valid as those of the basic RD at the cutoff.

Fig. 7 also shows that CRD-CG's estimates of the TOT display a relatively small amount of bias relative to the RCT. Bias estimates are −0.01, −0.03, and −0.04 across the three outcomes, indicating that CRD-CG provides valid causal estimates above the cutoff. However, this was expected given that the date-of-testing assignment variable had little or no relationship with the untreated outcomes in either the basic RD group or the comparison group.

4.2.4. Statistical Efficiency

Table 7 lists the estimated standard errors for each CRD-CG estimand by outcome. To describe relative efficiency at the cutoff, Fig. 8 shows

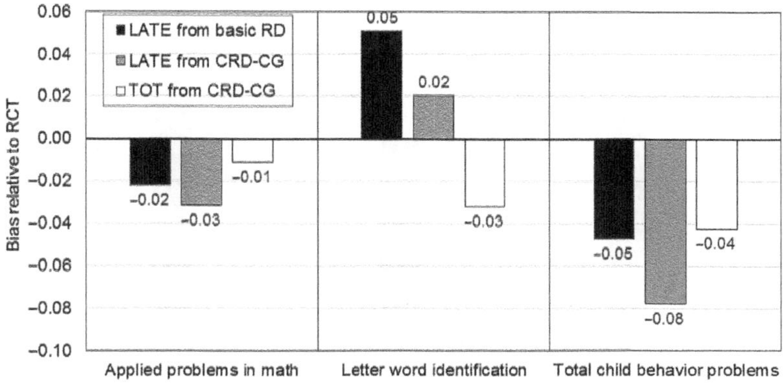

Fig. 7. Bias in Basic RD and CRD-CG Estimates, by Outcome and Estimand. *Notes*: LATE impacts are estimated at the RD cutoff, and TOT impacts are estimated for all treated children with scores above the cutoff. Estimated bias is calculated as the difference between the basic RD or CRD-CG estimate and the RCT estimate using models that specify a linear outcome-assignment variable relationship. Outcomes are normalized so that all estimates are expressed in standard deviation units.

the ratios of these standard errors to those of the corresponding RCT estimand and outcome, after adjusting the basic RD and CRD-CG standard errors using the method described in Section 3.5. For LATE estimates at the cutoff, these ratios range from 1.31 to 1.43 using the full CRD-CG data while they vary from 1.80 to 1.96 when using just the basic RD data. Hence, the added comparison group substantially improves the precision of LATE estimates at the cutoff, compared to basic RD. Fig. 8 also plots the ratio of adjusted CRD-CG standard errors and the actual RCT standard errors for the TOT above the cutoff. The ratios range from 1.14 to 1.23, indicating that the TOT estimates produced using the CRD-CG design are somewhat less precisely estimated than the TOT estimates obtained from the RCT design.

4.3. Sensitivity Test

As described earlier, there was a nontrivial amount of missing data in the Head Start study. Of the 2,449 3 year olds, 23% were missing the pretest for math and 17% did not have math posttest data. For letter word

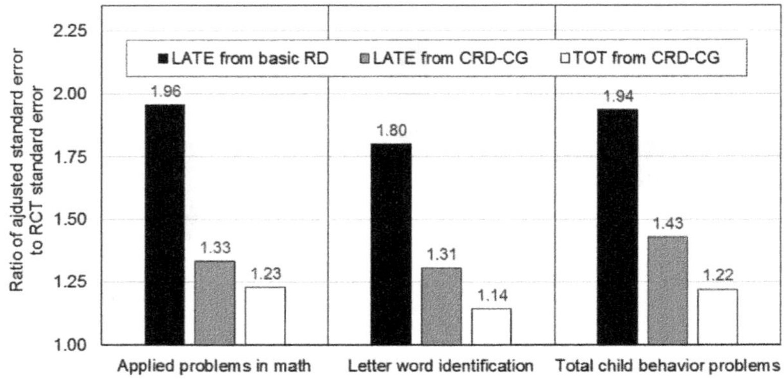

Fig. 8. Relative Efficiency of Basic RD and CRD-CG Estimates, by Outcome and Estimand. *Notes*: LATE impacts are estimated at the RD cutoff, and TOT impacts are estimated for all treated children with scores above the cutoff. Efficiency is measured using the ratio of the adjusted standard error of the basic RD or CRD-CG estimate to the actual standard error of the RCT estimate based on models that specify a linear outcome-assignment variable relationship. The adjustment factors applied to the original standard errors of the basic RD and CRD-CG estimates are described in Section 3.5. Outcomes are normalized so that all estimates are expressed in standard deviation units.

identification, 2% were missing the pretest and the posttest was missing for 16% of children. For total problem behavior, no children had missing pretest data but 16% were without a posttest. For the other covariates, up to 19% of the children's records had incomplete data. Although we imputed missing values using all available pretests, posttests, and covariates, this imputation could prove problematic if missingness is nonignorable with respect to key outcomes or, given the RD design, to the assignment variable. Although most outcomes had a fairly even pattern of missingness along the assignment variable, the same was not true for the math pretest. As shown in Fig. 9, more math pretest data were missing below the cutoff than above it.

To check the sensitivity of math results to this pattern of missing data, we removed all cases with missing data at pretest or posttest, leaving 1,569 observations. The RCT ATEs for math were 0.09 without covariates and 0.08 with them, compared to 0.08 and 0.10 when data were imputed. Without imputation, the estimated LATEs at the cutoff based on the RCT design were 0.10 without covariates and 0.08 with them; and after imputation they were 0.12 and 0.10. Above the cutoff and without imputation,

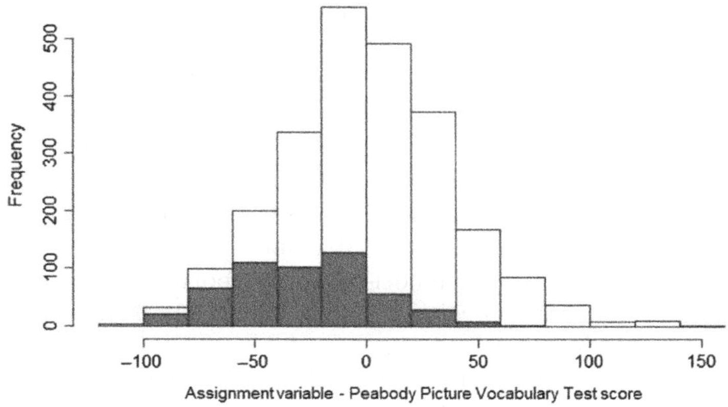

Fig. 9. Distribution of Missing Data for the Math Outcome along the CRD-Pre Assignment Variable.

TOTs estimated from the RCT were 0.08 without covariates and 0.04 with them, as compared to 0.08 and 0.05 with imputation. Given this pattern of results from the RCT design, imputation likely had little effect on the math results for the CRD designs.

5. DISCUSSION

Over the last 20 years, most research on RD has focused on how to analyze data so as to produce more valid and robust causal inference at the cutoff. This chapter is predicated on a different (and complementary) philosophy of progress that seeks to strengthen RD by adding to the design. In this application, we added data on untreated outcomes to create CRD-Pre and CRD-CG designs that might increase power and extend causal generalization away from the cutoff. We assessed how well these CRD designs performed, as judged by two main criteria: (1) the extent to which each design replicated the basic RD's unbiased inference at the cutoff but out-performed it there with respect to statistical power; and (2) the extent to which each CRD variant came close to RCT estimates of treatment effects and their standard errors away from the cutoff. Our test took place in the context of evaluating Head Start's effects on 3 year olds' math, language arts, and social behavior outcomes. RCT estimates served as the empirical benchmarks against which precision and bias of the basic RD, CRD-Pre, and CRD-CG designs were gauged. Unique is the study's preschool context that differs from the health policy topic in Wing and Cook (2013) and the university education topic in Kisbu-Sakarya et al. (2016).

Based on our visual and statistical inspection of the data, we used linear models of outcome-assignment variable relationships to assess the plausibility of the key assumptions of CRD. For both CRD-Pre and CRD-CG, we found that the observed segments of the untreated outcomes functions were reasonably parallel and that statistical tests could not reject the null hypothesis that they were so. These stable, linear relationships provide the support we sought for using a model-based extrapolation along the assignment variable to obtain predicted values of the untreated outcomes of treatment-eligible individuals – the critical piece of missing information needed to estimate impacts. If the key CRD assumptions hold in truth, this approach yields valid causal effects because these predicted values are conditionally independent of treatment eligibility.

Our results indicate that both CRD designs yielded LATE estimates at the cutoff that were more precise than the LATE estimates obtained from the corresponding basic RD designs. This can be seen in Figs. 10 and 11, which show LATE estimates from the basic RD and CRD designs alongside LATE estimates from the RCT samples corresponding to each CRD

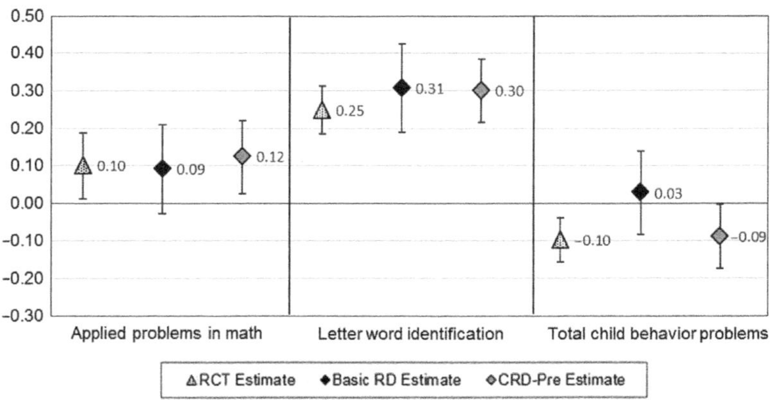

Fig. 10. Estimates of the LATE and Adjusted 95 percent Confidence Intervals, CRD-Pre Assignment Variable. *Notes*: LATE impacts are estimated at the RD cutoff based on models that assume a linear outcome-assignment variable relationship. Confidence intervals for the basic RD and CRD-Pre estimates are based on standard errors that have been adjusted to account for differences in sample sizes across designs, as described in Section 3.5. Outcomes are normalized so that all estimates are expressed in standard deviation units.

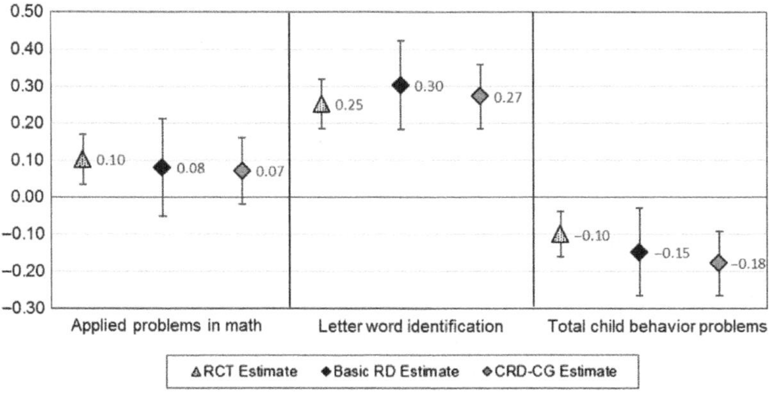

Fig. 11. Estimates of the LATE and Adjusted 95 percent Confidence Intervals, CRD-CG Assignment Variable. *Notes*: LATE impacts are estimated at the RD cutoff based on models that assume a linear outcome-assignment variable relationship. Confidence intervals for the basic RD and CRD-CG estimates are based on standard errors that have been adjusted to account for differences in sample sizes across designs, as described in Section 3.5. Outcomes are normalized so that all estimates are expressed in standard deviation units.

variant's assignment variable. The figure also includes 95 percent confidence intervals based on standard errors that have been adjusted to account for differences in sample sizes by design, as described in Section 3.5. The figures shows that basic RD and RCT samples for each assignment variable produce LATE estimates at the cutoff that are generally similar, but (as expected) the confidence intervals are consistently and markedly larger for estimates obtained using the basic RD design.

In contrast, Fig. 10 shows that CRD-Pre produces LATE estimates that are also similar to the other two estimates at the cutoff, but with confidence intervals that are tighter there than with basic RD. The same is true of the LATE estimates and confidence intervals for CRD-CG, as shown in Fig. 11. For example, confidence intervals for the LATE at the cutoff based on CRD-Pre were 0.71 to 0.82 times as large as those obtained using the basic RD design. Similarly, CRD-CG produced confidence intervals for the LATE that were 0.68 to 0.74 times as large as those of the basic RD. Hence, relative to basic RD, CRD introduces no novel biases at the cutoff and reduces standard errors there.

Our results also indicate a close correspondence between the causal estimates away from the cutoff produced by each CRD design and the RCT estimate of the same effect. Figs. 12 and 13 show estimates of TOT impacts for all children eligible for the treatment based on the cutoff rule associated with the assignment variables for CRD-Pre and CRD-CG. It includes point estimates and confidence intervals based on sample-size-adjusted standard errors. CRD-Pre and CRD-CG each produce TOT estimates that are quite similar in both magnitude and precision to the RCT estimates of the same quantities. Little indicates that one CRD design performed markedly better than the other in this application, although CRD-Pre yielded TOT estimates that were somewhat more efficient, relative to the RCT, than those of CRG-CG. Of course, the comparison is not exact, given the unique parameters influencing the power of each design (Tang et al., 2016, 2017) and also the fact that the differences across the two CRD designs are potentially confounded with the choice of assignment variable in this Head Start application, as discussed below.

Our conclusions about bias depend, in part, on the standard we used for considering differences between CRD and RCT impacts as tolerable or functionally equivalent. Our standard was that the two impact estimates should be less than 0.10 standard deviation apart. Based on this criterion, all six CRD estimates (3 outcomes × 2 assignment variables) reproduced

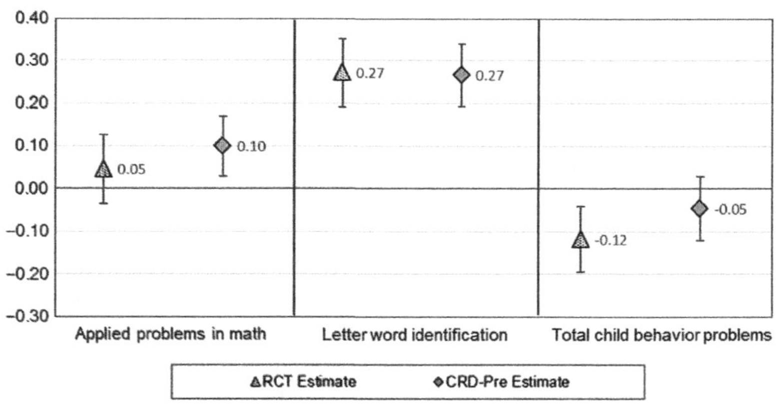

Fig. 12. Estimates of the TOT and Adjusted 95 percent Confidence Intervals, CRD-Pre Assignment Variable. *Notes*: TOT impacts are estimated for all treated children with scores above the cutoff based on models that assume a linear outcome-assignment variable relationship. Confidence intervals for the basic CRD-Pre estimates are based on standard errors that have been adjusted to account for differences in sample sizes across designs, as described in Section 3.5. Outcomes are normalized so that all estimates are expressed in standard deviation units.

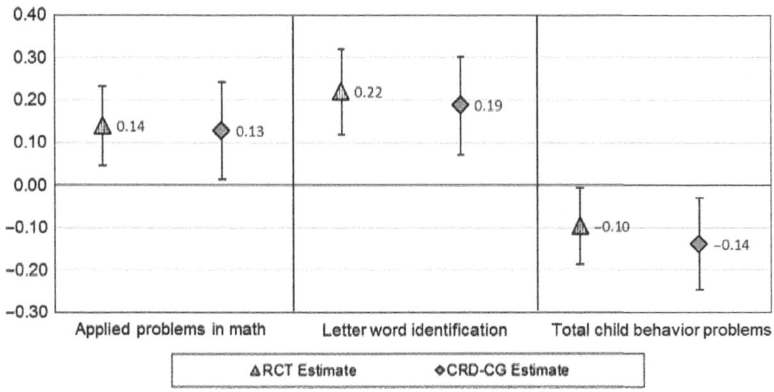

Fig. 13. Estimates of the TOT and Adjusted 95 percent Confidence Intervals, CRD-CG Assignment Variable. *Notes*: TOT impacts are estimated for all treated children with scores above the cutoff based on models that assume a linear outcome-assignment variable relationship. Confidence intervals for the basic CRD-CG estimates are based on standard errors that have been adjusted to account for differences in sample sizes across designs, as described in Section 3.5. Outcomes are normalized so that all estimates are expressed in standard deviation units.

their RCT benchmarks both at the cutoff and above the cutoff. As is evident from Figs. 11–13, varying this criterion to use a smaller tolerance would not substantially alter our conclusions.

The CRD-CG and CRD-Pre results were likely close to the RCTs as a result of two different routes to achieving independence between untreated outcomes and treatment eligibility. With CRD-CG, the observed untreated outcomes were uncorrelated with the assignment variable to start with. Hence, the CRD-CG assignment rule itself likely made treatment eligibility essentially random without the need for further analytic steps. In contrast, the baseline PPVT assignment variable was related to observable untreated outcomes in the CRD-Pre design, which likely made it necessary to achieve conditional independence by drawing on other data (pretests, in this case). Based on the relatively small bias found in CRD-Pre's estimates of the TOT, the model fit using the observed data on untreated outcomes appears to have successfully predicted the counterfactual outcomes of the treated. Moreover, this success was signaled by the results of the diagnostic tests, which indicated that the three untreated outcome function segments were parallel, as required by the assumptions of CRD given a linear outcome-assignment variable relationship.

Thus, it seems that CRD-CG closely approximated the RCT results due to the unconditional independence brought about by the assignment variable choice, whereas CRD-Pre achieved conditional independence due to the untreated pretest and posttest outcomes being essentially parallel along the assignment variable. This difference speaks to a broader point of importance. Although conditional independence might be achieved under certain assumptions by drawing on additional data to index untreated outcomes, it can also be achieved through a well-chosen assignment variable. Like Wing and Cook (2013) and Angrist and Rokkanen (2015), this chapter set out to induce strong ignorability by the use of more data and ignored the most basic prevention step of all – selecting assignment variables that are unrelated to outcomes, or are as weakly related as possible. We chanced upon fulfilling this requirement; yet it has to be the first step in any manual of how to justify using RD to make inference about causal effects away from the cutoff.

However, the fact that the RCT and CRD designs produced similar causal estimates and standard errors does not mean that RCT and CRD designs are interchangeable. For one, the TOT effect in CRD applies only to those treated who score above the cutoff. In contrast, the RCT's usual ATE estimate applies to all treated and untreated cases irrespective of their assignment variable scores. The RCT is therefore broader. Moreover, RCTs do not need the strong assumption that presently undergirds CRD – that each set of untreated outcomes follow a stable, parallel functional form along all or most of the assignment variable. Being less general and more assumption-dependent than the RCT, CRD is indisputably riskier. Nonetheless, genuine dispute is possible about how large and meaningful this risk differential is. Even more fundamental, though, is that RCTs and CRDs will rarely be contending design options. Each could have value in different circumstances.

The present chapter has several limitations. First is the fundamental confound between CRD design variants and choice of the assignment variable, although this did shine a light on the utility of assignment variables that are irrelevant for outcomes. Second, our results cannot clarify the degree to which the linear version of CRD can deviate from the assumption of parallel trends without introducing substantively important bias. We used conventional statistical tests to determine the significance of any observed deviations from parallelicity, and our working assumption is that the available sample sizes would permit the reliable detection of meaningful differences should they exist. But the extent to which we would have concluded that the untreated slopes are parallel with more powerful tests is not clear; a simulation study could help quantify the sensitivity of CRD bias to the extent to which the assumption of parallel trends is violated. Third, the RCT, CRD-Pre, and CRD-CG designs are based on different samples. The comparisons of standard errors we reported included adjustments for sample size differences by design and also used the same set of pretest and other covariates for each design. However, we did not account for differences by design in the allocation of the sample between treated and untreated cases, a topic that is addressed in Tang et al. (2017) for the case of CRD-CG. Subsequent within-study comparison could add analytic adjustments for how the sample allocation affects the relative efficiency of each CRD design in order to better isolate the gains in precision that arise from the added data on untreated outcomes.

A final limitation is that we treat the RCT benchmarks as if they were true causal parameters although they are in fact estimated with error. This problem is exacerbated when computing RCT estimates at the cutoff and on the treatment-eligible side only – these estimates include fewer cases than conventional ATE estimates. The within-study comparison literature contains many criteria for judging the similarity between RCT and nonexperimental estimates, given sampling error in each (Cook & Steiner, 2010; Cook et al., 2008; Shadish, Galindo, Wong, Steiner, & Cook, 2011; St.Clair et al., 2014). None is perfect, and "how close is close enough" (Wilde & Hollister, 2007) remains a critical question for within-study comparisons. The criterion we used – whether an estimated difference in impacts across designs is smaller than 0.10 standard deviations – does not account for sampling error. This approach will tend to err on the side judging estimates to be interchangeable even if a formal statistical test might reject the hypothesis that they are equal. However, as Figs. 10–13 show, there was a consistent pattern of close empirical correspondence between RCT and CRD causal estimates and standard errors across three outcome domains (math, literacy, and social behavior) and two estimands (both at and away from the eligibility cutoff). These internal replications should help increase confidence in the results we have presented. They suggest that CRD not only represents a powerful improvement over basic RD, but also can yield valid TOT impact estimates for broad segments of a population when an RCT is infeasible and treatment eligibility can be determined using cutoff scores of an assignment variable.

NOTES

1. In the CRD-Pre design, we could at this point simply take the average of the actual pretests of treatment eligibles instead of relying on the fitted values, $\hat{G}(\mathrm{AV}_i)$. However, as discussed later, additional data on untreated outcomes are ultimately used to estimate the function, and so the average predicted value might not equal the average pretest value in the sample of treatment eligibles. The fitted function $\hat{G}(\mathrm{AV}_i)$ is also of critical importance in CRD-CG as a means of linking untreated outcomes data from the comparison group to treatment-eligible individuals in the basic RD group (by way of the assignment variable).

2. As mentioned previously, all impacts are effect sizes – that is, they are scaled in terms of standard deviations of the outcome. Thus, throughout our exposition of results, we generally refer simply to the point estimates, leaving it implied that they expressed in standard deviations units.

ACKNOWLEDGMENTS

This research was supported by NSF Grants: DRL-1228866 and DGE-1544301. The authors are grateful to Matias D. Cattaneo and two anonymous referees for valuable comments on a draft of this article.

REFERENCES

Angrist, J. D., & Rokkanen, M. (2015). Wanna get away? Regression discontinuity estimation of exam school effects away from the cutoff. *Journal of the American Statistical Association*, *110*(512), 1331–1344.

Calonico, S., Cattaneo, M. D., & Titiunik, R. (2014). Robust nonparametric confidence intervals for regression-discontinuity designs. *Econometrica*, *82*(6), 2295–2326.

Cattaneo, M. D., Titiunik, R., & Vazquez-Bare, G. (in press). Comparing inference approaches for RD designs: A reexamination of the effect of Head Start on child mortality. *Journal of Policy Analysis and Management*.

Cook, T. D., Shadish, W. R., & Wong, V. C. (2008). Three conditions under which experiments and observational studies produce comparable causal estimates: New findings from within-study comparisons. *Journal of Policy Analysis and Management*, *27*(4), 179–750.

Cook, T. D., & Steiner, P. M. (2010). Case matching and the reduction of selection bias in quasi-experiments: The relative importance of pretest measures of outcome, of unreliable measurement, and of mode of data analysis. *Psychological Methods*, *15*(1), 56–68.

Dong, Y., & Lewbel, A. (2015). Identifying the effect of changing the policy threshold in regression discontinuity models. *The Review of Economics and Statistics*, *97*(5), 1081–1092.

Goldberger, A. S. (1972). *Selection bias in evaluating treatment effects: Some formal illustrations*. Madison, WI: University of Wisconsin-Madison.

Hahn, J., Todd, P., & Van der Klaauw, W. (2001). Identification and estimation of treatment effects with a regression-discontinuity design. *Econometrica, 69*(1), 201–209.

Hansen, B. (2008). The prognostic analogue of the propensity score. *Biometrika, 95*(2), 481–488.

Imbens, G., & Kalyanaraman, K. (2012). Optimal bandwidth choice for the regression discontinuity estimator. *The Review of Economic Studies, 79*(3), 933–959.

Kisbu-Sakarya, Y., Cook, T. D., & Tang, Y. (2016). *Comparative regression discontinuity: A stress test of what happens with modest sample sizes*. Manuscript submitted for publication.

LaLonde, R. J. (1986). Evaluating the econometric evaluations of training programs with experimental data. *The American Economic Review, 76*(4), 604–620.

Lee, D. S., & Lemieux, T. (2010). Regression discontinuity designs in economics. *Journal of Economic Literature, 48*(2), 281–355.

Lohr, W. (1972). *An historical view of the research done on the behavioral and organizational factors related to utilization of health services*. Unpublished Paper (January). Rockville, MD: National Center for Health Services Research and Development.

Puma, M., Bell, S., Cook, R., & Heid, C. (2010). *Head Start impact study: Final report*. Washington, DC: United States Department of Health and Human Services, Administration for Children and Families, Office of Planning, Research and Evaluation.

Rubin, D. B. (1974). Estimating causal effects of treatments in randomized and nonrandomized studies. *Journal of Educational Psychology, 66*(5), 688–701.

Rubin, D. B. (2008). Comment: The design and analysis of gold standard randomized experiments. *Journal of the American Statistical Association, 103*(484), 1350–1356.

Schochet, P. Z. (2009). Statistical power for regression discontinuity designs in education evaluations. *Journal of Educational and Behavioral Statistics, 34*(2), 238–266.

Shadish, W. R., Galindo, R., Wong, V. C., Steiner, P. M., & Cook, T. D. (2011). A randomized experiment comparing random and cutoff-based assignment. *Psychological Methods, 16*(2), 179.

St.Clair, T., Cook, T. D., & Hallberg, K. (2014). Examining the internal validity and statistical precision of the comparative interrupted times series design by comparison with a randomized experiment. *American Journal of Evaluation, 35*(3), 311–327.

Tang, Y., Cook, T. D., & Kisbu-Sakarya, Y. (2016). *Statistical power for the comparative regression discontinuity design with a pretest function: Theory and empirical evidence from the National Head Start Impact Study*. Manuscript submitted for publication.

Tang, Y., Cook, T. D., & Kisbu-Sakarya, Y. (2017). Statistical power for the comparative regression discontinuity design with a non-equivalent comparison group. *Psychological Methods*. Advance online publication. doi: 10.1037/met0000118

Wilde, E. T., & Hollister, R. (2007). How close is close enough? Evaluating propensity score matching using data from a class size reduction experiment. *Journal of Policy Analysis and Management, 26*(3), 455–477.

Wing, C., & Cook, T. D. (2013). Strengthening the regression discontinuity design using additional design elements: A within-study comparison. *Journal of Policy Analysis and Management, 32*(4), 853–877.

Wong, V. C., Valentine, J., & Miller-Bains, K. (2017). Empirical performance of covariates in education observational studies. *Journal of Research on Educational Effectiveness, 10*(1), 207–236.

PARTY BIAS IN UNION REPRESENTATION ELECTIONS: TESTING FOR MANIPULATION IN THE REGRESSION DISCONTINUITY DESIGN WHEN THE RUNNING VARIABLE IS DISCRETE

Brigham R. Frandsen

Department of Economics, Brigham Young University, Provo, UT, USA

ABSTRACT

Conventional tests of the regression discontinuity design's identifying restrictions can perform poorly when the running variable is discrete. This paper proposes a test for manipulation of the running variable that is consistent when the running variable is discrete. The test exploits the fact that if the discrete running variable's probability mass function satisfies a certain smoothness condition, then the observed frequency at the threshold has a known conditional distribution. The proposed test is

Regression Discontinuity Designs: Theory and Applications
Advances in Econometrics, Volume 38, 281–315

ISSN: 0731-9053/doi:10.1108/S0731-905320170000038012

applied to vote tally distributions in union representation elections and reveals evidence of manipulation in close elections that is in favor of employers when Republicans control the NLRB and in favor of unions otherwise.

Keywords: Density test; labor unions; political parties

JEL classifications: J50; C21; C14

1. INTRODUCTION

The regression discontinuity (RD) design, a research strategy that exploits plausibly exogenous variation in a treatment assigned via a threshold or cutoff rule, has become one of the most frequently used tools in the empirical economist's kit. Originally developed for, and still commonly applied to evaluation of education interventions, where threshold-based rules are the norm, RD designs have found wide application in labor economics, public finance, environmental economics, political economy, and other diverse settings (see van der Klaauw, 2008, for a survey).

There is good reason for the RD design's popularity. First, in settings where it can be verified, the RD design appears to make good on its promise of delivering unbiased estimates of causal effects. Estimates from RD designs agree with their close cousins, randomized experiments, in numerous within-study comparisons where both methods are available (Black, Galdo, & Smith, 2007; Buddelmeyer & Skoufias, 2003; Cook & Wong, 2008). Another attraction of the RD design lies in the ability to transparently test the plausibility of its identifying assumptions. The RD design relies on the assumption that individuals immediately on either side of a threshold – for example, with test scores just above and below a cutoff – are comparable. This is plausible if individuals cannot precisely manipulate their score. One test of the identifying assumption looks for red flags that individuals are, in fact, manipulating their score by examining the distribution near the RD threshold for discontinuities in the density that would point to manipulation or confounding selection. McCrary's (2008) test exploits this idea using local linear regression of histogram frequencies at the threshold. This test has become part of the recommended battery of analyses for RD practitioners (Imbens & Lemieux, 2008; Lee & Lemieux, 2010). Cattaneo, Jansson, and Ma (2010) propose a similar test based on local linear regression estimates of the density, but avoiding the initial histogram step.

Like the standard RD identifying assumptions in Hahn, Todd, and van der Klaauw (2001), this widely used test assumes the running variable is continuously distributed. In practice, however, many RD designs employ a discrete running variable[1]. When the running variable has a finite number of fixed support points, however, the McCrary test can break down; the local linear regressions that form the basis of the test rely on the number of observed support points near the threshold growing large as the sample size increases, which is the case for a continuously distributed running variable, but not a discrete one with fixed support points. As a result, the test can falsely reject the null of no manipulation at too high a rate (incorrectly sized) or can fail to detect actual anomalies in the running variable's distribution (underpowered).

This paper proposes a test for manipulation of the running variable at the threshold that is consistent when the running variable is discrete with a fixed, finite support. Like McCrary's test, it is based on smooth approximations to the running variable's distribution in the neighborhood of the threshold. Unlike that test, the one proposed here relies only on support points at and immediately adjacent to the RD threshold when the running variable is discrete. It exploits the fact that if the discrete running variable's probability mass function (pmf) satisfies a certain smoothness condition, then the observed frequency at the threshold has a known conditional distribution. This permits tests using only support points immediately adjacent to the threshold, as opposed to local linear regressions that of necessity rely on extrapolation away from the threshold.

Simulation results show the test has correct size and good power even when the smoothness approximation is not exactly correct, while tests based on local linear regression falsely reject the null of no manipulation at much too high a rate. The over-rejection becomes worse as the sample size gets large, and the difference in performance between the two tests grows the coarser the running variable is.

Applying the test to the distribution of National Labor Relations Board (NLRB) union certification election outcomes reveals very strong evidence for manipulation in close elections also described in Frandsen (2015). Elections held during periods when Republicans controlled the NLRB show evidence of selection that favors the employer, with close union victories occurring much less frequently than expected. This advantage for the employer disappears, and in fact manipulation favoring the union appears for elections that could have been reversed in favor of the union by a single ballot challenge when non-Republicans (i.e., Democrats and independents) control the NLRB.

Methodologically, this paper is related to previous work on the problems posed by RD designs when the running variable is discrete. Lee and

Card (2008) show that discrete running variables induce specification errors that can be accounted for in the inference procedure they propose. Kolesár and Rothe (2016) develop an alternative inference procedure for discrete running variables with superior theoretical properties. Dong (2012) develops a bias-corrected estimation procedure to account for rounding error when the discrete running variable is a rounded version of an underlying continuous variable. The current paper complements this previous work on RD estimation and inference in the case of discrete running variables by developing a test of the identifying conditions that can then justify using those tools. This paper also complements work by Gerard, Rokkanen, and Rothe (2015), who consider partial identification in RD designs when manipulation is present.

The testing approach is also related to the finite-sample nonparametric literature on inference for approximately linear functions. The test in this paper is based on a smoothness condition that puts a bound on the finite analog to the second derivative of the running variable's pmf, and derives sharp bounds on the distribution of the test statistic. This approach of performing inference within a functional class defined by bounds on derivatives is also followed by Armstrong and Kolesár (2015) but dates back at least to Sacks and Ylvisaker (1978).

The test proposed in this paper may also be of interest outside of an RD setting. It can be applied to settings where detecting sorting or heaping along a discretely measured dimension is important. For example, the methodology could be used to test behavioral theories that imply bunching or sorting relative to pre-determined benchmarks or norms, such as round numbers in SAT scores or batting averages (Pope & Simonsohn, 2011). The test could also be used to quantify distortions in firm behavior in response to policies based on firm-size thresholds, such as the Family and Medical Leave Act and the Affordable Care Act in the United States or employment protection laws in Europe (Garibaldi, Pacelli, & Borgarello, 2004; Schivardi & Torrini, 2008; Waldfogel, 1999). In applications such as these, the results of the test are of interest in their own right, and not just as specification checks.

2. ECONOMETRIC FRAMEWORK

Consider a standard RD design in which an individual's treatment assignment, D, depends on whether a scalar-valued continuously distributed variable R^*, referred to as the running variable or forcing variable, exceeds some known threshold normalized to zero. Treatment assignment can therefore be written as

$$D = 1(R^* \geq 0).$$

For example, R^* might be the difference between an individual's age and the Medicare eligibility threshold, 65. Let the potential outcome if the individual were not to receive the treatment be $Y(0)$ and if he or she were to receive the treatment, $Y(1)$. The observed outcome is therefore

$$Y = Y(0)(1 - D) + Y(1)\,D.$$

The estimands of interest in the RD design are typically the distributions of $Y(0)$ and $Y(1)$ conditional on $R^* = 0$ or functionals of those distributions such as the conditional average treatment effect at the threshold, $E[Y(1) - Y(0)|R^* = 0]$. The following standard continuity assumption is sufficient to identify the conditional average treatment effect (Frandsen, Frölich, & Melly, 2012; Hahn et al., 2001):

> **Assumption 1.** Regarded as functions of r, $E[Y(0)|R^* = r]$ and $E[Y(1)|R^* = r]$ are continuous at zero.

This assumption rules out discrete jumps in unobservables at the threshold, so that any observed jump in outcomes there can be attributed to the effects of treatment. It implies that individuals immediately on either side of the threshold are comparable. It is analogous to the conditional independence assumptions underpinning standard regression analysis or the exclusion restriction invoked in instrumental variables analysis. The assumption is more plausible when the running variable cannot be chosen or manipulated precisely by the individual. For example, the assumption would be satisfied if R^* were randomly assigned and it had no effect on outcomes other than through its determination of treatment status.

2.1. Standard Tests for Running Variable Manipulation

The RD identifying assumption cannot be tested directly since $Y(0)$ and $Y(1)$ are observable on only one side of the threshold or the other. However, plausible rationales for this assumption imply that the density of R^* should be continuous at zero. For example, suppose R^* can be partially influenced by the individual; one would then expect selection differences to arise on average between treated and untreated individuals. But if R^* has a random component that cannot be precisely manipulated, one would expect individuals immediately on either side of the threshold to be comparable – that is, Assumption 1 should be satisfied (Lee, 2008). A discontinuity in the density of the running variable at the threshold raises a red flag that perhaps individuals can precisely manipulate the running variable after all, and therefore individuals on either side of the threshold may not be comparable.

The idea that anomalies in the running variable at the threshold signal violations of Assumption 1 forms the basis of McCrary's (2008) test, which examines the estimated density of the running variable on either side of the threshold, and has become part of the standard battery for RD practitioners (Imbens & Lemieux, 2008; Lee & Lemieux, 2010). The McCrary test consists of two steps. The first step constructs the running variable's histogram using bins of width b_n, which become narrower as the sample size increases. The second step performs kernel-weighted linear regressions of the log of the histogram frequencies on either side of the threshold using a bandwidth h_n – which also shrinks with the sample size – and tests for a difference in intercepts. A significant difference points to a discontinuity in the running variable density at the threshold, suggesting possible manipulation and violations of Assumption 1.

Cattaneo et al. (2016) propose a test similar in spirit that also tests for differences in local polynomial regression estimates of the density on either side of boundary. This test avoids the initial binning step and instead performs kernel-weighted polynomial regressions of jack-knifed estimates of the empirical cdf.

Both of these tests perform well in terms of size and power in standard RD settings where the running variable's distribution is well approximated by a continuous distribution.

2.2. *Discrete Test for Running Variable Manipulation*

Suppose now that instead of R^*, we observe a discrete running variable, R satisfying the following property:

> **Assumption 2.** R has finite support on points at equally spaced intervals of length $\Delta > 0$.

The discrete running variable R may be taken to be a rounded version of an underlying continuous variable – for example, $R = \lfloor R^*/\Delta r \rfloor \Delta r$, as in age measured in years (Card et al., 2009) – or a naturally discrete variable such as primary school enrollment (Angrist & Lavy, 1999). As above, the treatment threshold's location is normalized so the smallest treated support point is defined to be zero. The intuitive implication of standard RD sufficient identifying conditions that the pmf should be smooth at the threshold still provides the basis for the proposed manipulation test when the observed running variable is discrete. In our smoothness notion, however, finite differences replace derivatives, since our random variable has finite support. Define the second-order finite difference of a function f at a point r as

$$\Delta^{(2)}f(r) := \frac{f(r+\Delta) - 2f(r) + f(r-\Delta)}{\Delta^2},$$

which corresponds to the finite analog of a second derivative. The smoothness condition corresponding to the hypothesis of no manipulation in the discrete case is the following:

Assumption 3. *R* has a probability mass function $f(r)$ with a bounded second-order finite difference at zero that satisfies

$$|\Delta^{(2)}f(0)| \leq k\frac{f(-\Delta) + f(\Delta)}{\Delta^2},$$

for known $k \geq 0$.

Assumption 3 is a finite analog to the standard smoothness condition invoked in, for example, Armstrong and Kolesár (2015), but with the smoothness constant on the right-hand side scaled by $(f(-\Delta) + f(\Delta))/\Delta^2$ to ensure the test statistic defined below is pivotal. The scaling makes no restriction on the class of admissible functions for f beyond a bounded second-order finite difference; for example, suppose the absolute value of the second finite difference were bounded by C. Then define k in Assumption 3 to be $C\Delta^2/(f(-\Delta) + f(\Delta))$. Since f is by definition positive at $-\Delta$ and Δ, this makes no further restriction. Like the inference method proposed in Armstrong and Kolesár (2015), the test proposed here requires the researcher to choose a value for k. One possible difference is in Armstrong and Kolesár's (2015) setting, a conservative choice for the bounding constant does not substantially harm efficiency, which may not be the case here.

Assumption 3 is akin to the local linear approximation in traditional RD settings with a continuously distributed running variable. The smoothness requirement captures the notion of no manipulation as in McCrary (2008). The requirement that it have bounded finite difference is analogous to the differentiability required for RD estimation and inference (Hahn, Todd, & van der Klaauw, 1999). Assumption 3 differs from the conventional assumptions in two important ways, however. First, the condition makes restrictions not only at the threshold, but over a fixed neighborhood. This is a fact of life for discrete running variables; the fixed support points preclude the nonparametric approach of a vanishing bandwidth around the threshold. Second, the condition requires that the bound coefficient, k, be specified. The choice of k determines the degree of departure from linearity around the threshold beyond which manipulation is implied. For example, choosing $k = 0$ specifies that a distribution compatible with no manipulation must be precisely linear around the threshold. Choosing a larger k allows a degree of nonlinearity at the threshold without concluding manipulation. As discussed below, a smaller k leads to a more powerful test, but may also detect manipulation when none is present.

Fig. 1 illustrates the role of Assumption 3. Fig. 1(a), meant to depict a case with no manipulation, shows that $f(r)$ satisfies the bound on the second-order finite difference whose envelope is indicated by the curves. Fig. 1(b) shows an example with manipulation, in which the pmf's nonlinearity exceeds the bounds on the second-order finite difference.

The basis of the test is the observation that when Assumption 3 is satisfied the observed sample frequency at $R = 0$ has a known distribution conditional on $R \in \{-\Delta, 0, \Delta\}$, as the following theorem shows.

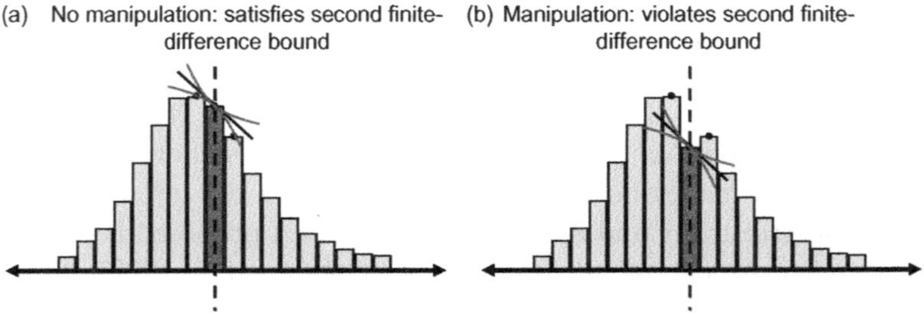

Fig. 1. Example Depicting an Example where the Second Finite Difference Bound in Assumption 3 Is Satisfied (a) and where It Is Violated (b). *Note*: The curves show the envelope for the probability mass function imposed by the bound on the second-order finite difference.

Theorem 1. Suppose Assumption 3 is satisfied. Then

$$\Pr(R = 0|R \in \{-\Delta, 0, \Delta\}) \in \left[\frac{1-k}{3-k}, \frac{1+k}{3+k}\right].$$

Proof. All proofs are in the appendix. ∎

The result says that if $f(r)$ is sufficiently smooth at the threshold, then conditional on R taking on a value at the threshold or an immediately adjacent support point, the probability of being exactly at the threshold is between $(1 - k)/(3 - k)$ and $(1 + k)/(3 + k)$. For $k = 0$, this implies the probability is exactly 1/3. The theorem is exact; the result holds no matter how small the sample.

The intuition for the result is the same as for the McCrary test. If the distribution is smooth then it admits a linear approximation in the neighborhood of the threshold. It turns out that in the linear case with $k = 0$, then $\Pr(R = 0|R \in \{-\Delta, 0, \Delta\}) = 1/3$ is not only necessary but also sufficient for the "no-manipulation" Assumption 3 to hold. The appendix shows this intuition formally.

The theorem's result and its antithesis constitute the null and alternative hypotheses for the proposed test:

$$H_0 : \Pr(R = 0|R \in \{-\Delta, 0, \Delta\}) \in \mathcal{P}(k) := \left[\frac{1-k}{3-k}, \frac{1+k}{3+k}\right];$$

$$H_1 : \Pr(R = 0|R \in \{-\Delta, 0, \Delta\}) \in (0, 1) \backslash \mathcal{P}(k),$$

for known k.

2.2.1. Test Statistic, Null Distribution, Critical Values, and Power

Given an i.i.d. sample of size n from the distribution of R, let $N_r := \sum_{i=1}^{n} 1(R_i = r)$ be the sample frequency at $R = r$, where $r \in \{\ldots, -\Delta, 0, \Delta, \ldots\}$. The test statistic is N_0, the observed sample frequency at $R_i = 0$. The following corollary to Theorem 1 gives the test statistic's distribution under the null hypothesis:

Theorem 2. Suppose $\{R_i\}_{i=1}^{n}$ are independent and identically distributed samples of R satisfying Assumption 2. Then under H_0 and conditional on $m = N_{-\Delta} + N_0 + N_\Delta$,

$$N_0 := \sum_{i=1}^{n} 1(R_i = 0) \sim B(m,p),$$

where $B(m)$, p denotes a Binomial random variable with number of trials m and probability of success $p \in \mathcal{P}(k)$.

Since the null hypothesis specifies a set for the parameter p, simply inverting the Binomial cdf at a given p does not produce uniformly valid critical values. Instead, the following theorem constructs critical values that uniformly control size over $\mathcal{P}(k)$:

Theorem 3. Suppose $\{R_i\}_{i=1}^{n}$ are independent and identically distributed samples of R satisfying Assumption 2. Then under H_0,

$$\Pr\left(C_\alpha^L(k) < N_0 < C_a^U(k) | N_{-\Delta} + N_0 + N_\Delta = m\right) \geq 1 - \alpha,$$

where $C_\alpha^L(k)$ and $C_\alpha^U(k)$ are values in $\{0, \ldots, m\}$ that solve the following discrete minimization problem:

$$\begin{aligned} &\min_{C^L, C^U} |C^U - C^L| \\ &\text{s.t.} \inf_{p \in \mathcal{P}(k)} F_B\left(C^U - 1; m, p\right) - F_B\left(C^L; m, p\right) \geq 1 - \alpha, \end{aligned}$$

and F_B is the Binomial cdf:

$$F_B(x; m, p) := \sum_{j=1}^{\lfloor x \rfloor} \binom{m}{j} p^j (1-p)^{m-j}.$$

The theorem establishes that under H_0 a test that fails to reject when N_0 is strictly between the upper and lower critical values, and rejects otherwise will have size uniformly controlled by α. By minimizing the length of the non-rejection region, the test maintains good power against alternatives that deviate from H_0 in either direction. The following result establishes that the test is consistent against any fixed alternative H_1.

Theorem 4. Suppose $\{R_i\}_{i=1}^{n}$ are independent and identically distributed samples of R satisfying Assumption 2. Suppose H_1 holds with either Pr $(R = 0|R\{\in -\Delta, 0, \Delta\}) = ((1-k)/(3-k))-\delta$ or Pr $(R = 0| R \in \{-\Delta, 0, \Delta\}) = ((1 + k)/(3 + k)) + \delta$ for some $\delta > 0$. Then the probability of rejection approaches unity as the sample size increases.

2.2.2. *Including Additional Support Points*

The proposed test relies on three support points of the discrete running variable. The choice of three is not arbitrary. In principle, a similar test could be constructed using observed frequencies from any odd-numbered (but greater than one) set of support points arranged symmetrically about the threshold. Specifically, consider a test based on an odd number $2d + 1$ support points ($d = 1, 2, 3, \ldots$), the middle one of which is located at the threshold. Assumption 3 implies that the conditional probability of $R = 0$, conditional on R taking on one of the $2d + 1$ values is

$$\Pr(R = 0|R \in \{-d\Delta, \ldots, d\Delta\}) \in \left[\frac{1 - k\sum_{s=1}^{d} s^2}{2d + 1 - k\sum_{s=1}^{d} s^2}, \frac{1 + k\sum_{s=1}^{d} s^2}{2d + 1 + k\sum_{s=1}^{d} s^2}\right].$$

The choice of $d = 1$ (three support points) is preferred for two reasons. First, it requires invoking the smoothness condition over the smallest possible interval, where it is most likely to be reasonable. Second, for a given choice of k, increasing the number of points quickly leads to a test with no power: the lower bound of the null set approaches zero, and the upper bound approaches one.

2.2.3. *Choosing k*

Researchers must choose the parameter $k \geq 0$ in order to implement the test. The choice of k determines the maximal degree of nonlinearity in the pmf that is still considered to be compatible with no manipulation. A large k means the mass at the threshold can deviate substantially from linearity before the test will reject with high probability, while a small k means even small deviations from linearity will lead the test to reject with high probability. Choosing k to be conservatively high will therefore reduce the test's power to detect manipulation. How should researchers choose k? The choice clearly cannot be driven by the data used in the test – that is, observations adjacent to the threshold. The degree of curvature in benchmark distributions may be used to generate rules of thumb. For example, if the running variable was a discretized version of a normally distributed random variable, discretized to coarseness Δ in units of the standard deviation, then the maximal degree of curvature, as a function of Δ, corresponds to a k given by the following expression:

$$k_N^{\max}(\Delta) = \frac{\Delta^3 \phi(\Delta/2)}{2(\Phi(3\Delta/2) - \Phi(\Delta/2))},$$

where ϕ and Φ are the standard normal density and cdf. The maximal k is increasing in the coarseness. For $\Delta = .1$, which corresponds to about 20 support points within one standard deviation of the threshold, the maximal k is .005. For $\Delta = .3$, a very coarse discretization with only about six support points within one standard deviation, the maximal k is .047. A researcher could alternatively take a rule of thumb from the observed distribution of the running variable away from the threshold, but keeping the following in mind: (1) no manipulation can be present elsewhere in the distribution; (2) the degree of curvature elsewhere in the distribution must be informative about the curvature at the threshold; (3) the test then must be taken to be conditional on the observed running variable distribution away from the threshold.

2.3. The McCrary Test with Discrete Data

The McCrary (2008) test, outlined in Section 2.1, generally performs well – that is, it has approximately correct size and good power – when the running variable is continuous, but it can deteriorate when the running variable is discrete.

Nevertheless, the test can be applied to discrete data by taking the support points of the running variable to be the histogram bins, and running a kernel-weighted local linear regression of the log of the observed frequencies on either side of the threshold.

Why might the McCrary test perform poorly with discrete data, when its first step is, in fact, to discretize the running variable? The answer is that the McCrary procedure consistently tests whether two estimands – linear extrapolations of the (log) pmf to the threshold from the left and right – are equal, but equality of those estimands is neither necessary nor sufficient for the no-manipulation smoothness condition in a setting with fixed support points. When the running variable is continuously distributed and the discretization can home in on the threshold as the sample size increases, then equality of the two estimands implies the density is continuous, and the test is consistent, but with a discretely distributed running variable the McCrary procedure can lead to over- or under-rejection of the smoothness condition.

There is an exception, however: if the underlying pmf truly is exactly linear away from the threshold, the McCrary test will be consistent even in a discrete setting, and may have more power, but exact linearity is a strong assumption not commonly made in RD settings, and embodies more than the hypothesis of no manipulation would seem to imply.

3. SIMULATIONS

This section numerically illustrates the theoretical results above concerning the test's size and power.

The first set of simulations illustrates the performance of the test over a range of sample sizes when k is chosen correctly. The running variable is a binned version of an underlying continuous random variable. Let the underlying random variable be $R^* \sim \text{log-}N(\mu, \sigma^2)$, and let the discrete running variable based on this be $R = \lfloor (R^* - r_c)/\Delta \rfloor \times \Delta + r_c$, where r_c is the threshold, and Δ is the support point spacing. The threshold will be placed so that the mode of the distribution, $m = \exp(\mu - \sigma^2)$ is exactly between the treated and untreated support points, that is, $r_c = m + \Delta/2$. For the baseline specification of $\mu = 0$, $\sigma^2 = 1$, and $\Delta = .1$, the value of k is .02. All simulations are based on 1,000 iterations. Fig. 2 plots simulated rejection rates for sample sizes from $n = 200$ to $n = 10,000$, with running variable distribution parameters $\mu = 0$, $\sigma^2 = 1$, and $\Delta = .1$. In this scenario there is no manipulation, so a correctly sized test should reject at a rate equal to the nominal size, $\alpha = .05$. The solid line corresponds to the discrete test with a correctly specified k, and shows that the rejection rate hovers right around the nominal 5 percent. The figure also shows that the McCrary test rejects at only slightly more than the nominal rate at the lowest sample size, but the rejection rate quickly approaches 100 percent for sample sizes above 5,000.

The theoretical results and this numerical illustration show the test has correct size when the researcher knows the k corresponding to the degree of curvature in the running variables distribution in the absence of manipulation. In applications, however, if the choice of k is based on an approximation or rule of thumb, it may be too low – possibly compromising the test's size – or too high – in which case the test's power may suffer.

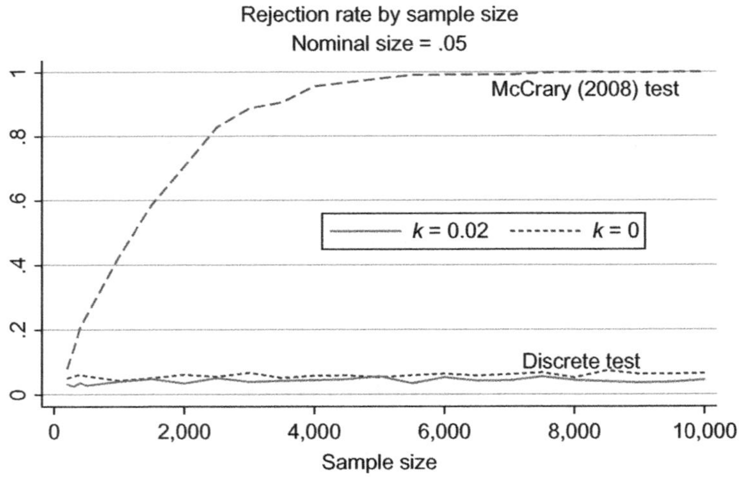

Fig. 2. Monte Carlo Simulation Rejection Rates from a McCrary (2008) Test (Dashed Line) and the Discrete Test Proposed in This Paper (Solid Line) as a Function of the Sample Size (x-Axis). *Notes*: The underlying distribution of the running variable is a discretized log-normal (no manipulation) with a bin width of .1, corresponding to a k parameter of .02. The nominal size of the tests is .05. Based on 1,000 iterations.

The remainder of this section shows how the proposed test performs in terms of size and power when the choice of k is misspecified.

The next set of simulations examines the test's size in a realistic setting when k is chosen too small. The testing procedure will assume $k = 0$ (exactly linear density), but of course as above the true dgp (using the same parameters as the previous simulation) is not linear. Fig. 2 plots the rejection rate for the discrete test with a misspecified k (short-dashed line), and shows that the performance in terms of size across the entire range of sample sizes considered is negligibly affected by the misspecification. Since the null hypothesis in this case is misspecified, for a large enough sample the rejection rate would eventually exceed the nominal size of the test, but the simulations illustrate that for sample sizes common in applied research the distortion is minimal.

The next simulation shows that when the running variable's discreteness is relatively fine the McCrary test has nearly the correct size, but as the variable becomes coarser and coarser the size gets worse and worse, while the discrete test proposed in this paper maintains correct size throughout.

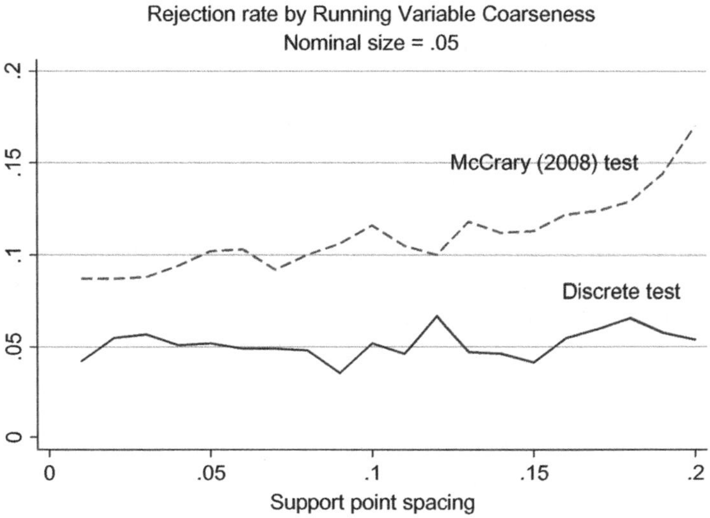

Fig. 3. Monte Carlo Simulation Rejection Rates from a McCrary (2008) Test (Dashed Line) and the Discrete Test Proposed in This Paper (Solid Line) as a Function of the Bin Width of the Discrete Running Variable. *Notes*: The underlying distribution of the running variable is a discretized log-normal (no manipulation) with a bin width indicated on the *x*-axis. The nominal size of the tests is .05. Based on 1,000 iterations with a sample size of 5,000 for each iteration.

Fig. 3 plots the rejection rates for the two tests for bin widths from $\Delta = .01$ to $\Delta = .2$, with a sample size of $n = 5{,}000$, $\mu = 1$, and $\sigma = .5$. The figure shows that when the running variable is the finest, the McCrary test's rejection rate is a little higher than the nominal size. This is as expected, since the McCrary test is designed for a continuously distributed random variable. However, as the variable becomes coarser and coarser, the rejection rate increases, which rejection rates between 15 and 20 percent for bin widths around .2. The discrete test proposed in this paper rejects at near the nominal rate of .05 for the whole range of bin sizes, illustrating that the test has correct size whether the running variable is continuous or discrete.

The next set of simulations explores the impact of asymmetry on the relative performance of the McCrary test and the discrete test, and shows that as the distribution of the running variable becomes more skewed, the McCrary test performs more poorly, while the discrete test maintains good properties. The simulation varies skewness, but holding the variance $V = (\exp \sigma^2 - 1) \exp (2\mu + \sigma^2)$ constant at the level in the first set of simulations, with $\mu = 0$ and $\sigma = 1$ in order to isolate the impact of skewness. This was achieved by varying σ from .05 to 1 (corresponding to skewness from about .15 to over 6), and setting $\mu = -(1/2)(\ln((1/V)(\exp \sigma^2 - 1)) + \sigma^2)$ to hold the variance constant. Skewness depends on the value of σ^2 via the following formula: $\gamma_1 = (\exp \sigma^2 + 2)\sqrt{\exp \sigma^2 - 1}$. The support point spacing is set at $\Delta = .1$ as in the first set of simulations. Fig. 4 plots the rejection rates for the two tests for skewness parameters from .15 to over 6, with a sample size of $n = 3{,}000$. The figure shows that for skewness values below one, both tests perform similarly with rejection rates very close to the nominal .05. As skewness increases, however, the McCrary test rejects more and more frequently, exceeding 80 percent when skewness reaches six.

The previous sets of simulations showed that when the running variable is discrete the McCrary test can falsely reject the null hypothesis at much too high a rate, while the discrete test has approximately correct size even when k is chosen too small. As noted above, the test's good size under

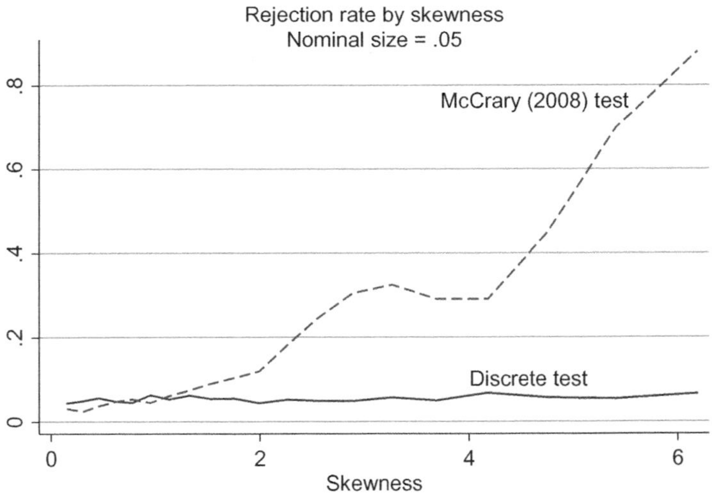

Fig. 4. Monte Carlo Simulation Rejection Rates from a McCrary (2008) Test (Dashed Line) and the Discrete Test Proposed in This Paper (Solid Line) as a Function of the Skewness of the Discrete Running Variable's Distribution. *Notes*: The underlying distribution of the running variable is a discretized log-normal (no manipulation) with a bin width of .1. The nominal size of the tests is .05. Based on 1,000 iterations with a sample size of 3,000 for each iteration.

misspecification is a finite-sample phenomenon: for a large enough sample even small misspecifications will lead to size distortions. A relevant question, then, is whether for the range of sample sizes in which the misspecified test maintains correct size it still has power to detect manipulation. The next set of simulations answers this question. The underlying continuous variable will be altered by introducing a probability of switching from below the threshold to above, where the probability of switching increases near the threshold. To be precise, the altered underlying continuous variable will be $\tilde{R}$ defined as follows:

$$\begin{aligned} \tilde{R} &= R^* + 2(r_c - R^*)S, \\ S &\sim \text{Bernoulli } (p(r))|R^* = r, \\ p(r) &= \beta \exp(-\gamma|r - r_c|) \times 1(R^* < r_c), \\ R^* &\sim \text{log-}N\left(\mu, \sigma^2\right). \end{aligned}$$

The discretized running variable will then simply be $R = \lfloor(\tilde{R} - r_c)/\Delta\rfloor \times \Delta + r_c$. Fig. 5 shows an example of what the resulting distribution looks like.

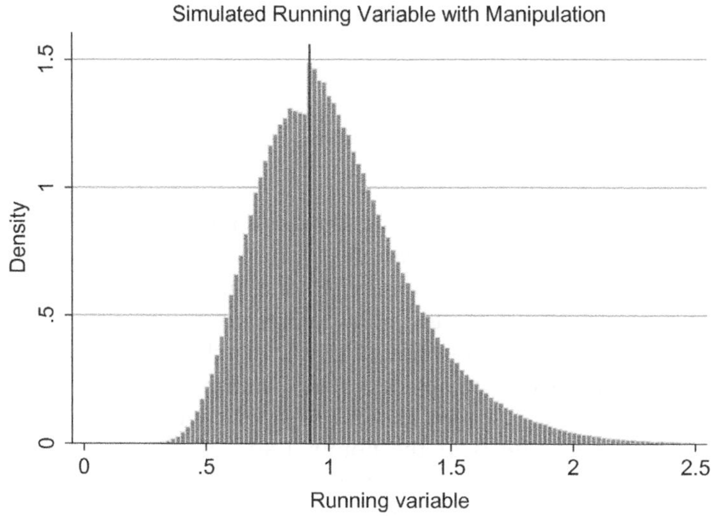

Fig. 5. Simulated Example of a Discrete Running Variable with Manipulation at the Threshold.

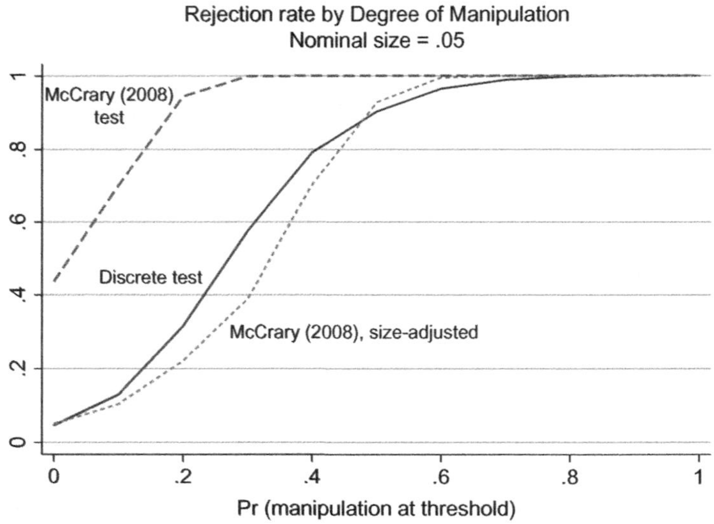

Fig. 6. Monte Carlo Simulation Rejection Rates from a McCrary (2008) Test (Unadjusted: Long Dashed Line; Adjusted: Short Dashed Line) and the Discrete Test Proposed in This Paper (Solid Line) as a Function of the Degree of Manipulation. *Notes*: The underlying distribution of the running variable is a discretized (bin width = .1) log-normal but with a probability of switching to a point above the threshold for draws below the threshold, as described in the text. The maximum probability of switching (x-axis) occurs just below the threshold and fades exponentially at rate one moving away from the threshold. The nominal size of the tests is .05. Based on 1,000 iterations with a sample size of 1,000 for each iteration.

The first power simulation shows the test's rejection rate is minimized where the probability of manipulation is zero (so the test is unbiased) and the power is greater when the probability of manipulation is higher. Fig. 6 plots the rejection rate of the test as a function of β, the maximum probability of manipulation (or switching) at the threshold, and, for reference, the rejection rate of the McCrary test. The running variable's true distribution has parameters $\mu = 0$, $\sigma = 1$, $\Delta = .1$, $\gamma = 1$, and $n = 1,000$. The discrete test's rejection rate is at the nominal size of .05 when $\beta = 0$, as it should be since there the null hypothesis of no manipulation is true. The rejection rate monotonically increases as the probability of manipulation gets higher, reaching 80 percent when $\beta \approx .4$ and 90–100 percent when $\beta = .6$ and higher. The figure also shows for reference that the unadjusted

McCrary test rejects at a higher-than-nominal rate even when there is no manipulation, as previous simulations showed. Adjusting the McCrary test by calculating critical values that give correct rejection rates when $\beta = 0$ (see Lloyd, 2005) makes the power remarkably close to the discrete test's.

The next simulation shows the discrete test is more powerful when the running variable is "more discrete," that is, when the running variable's distribution is coarser. Fig. 7 plots the test's rejection rate as a function of the bin width, Δ, fixing the maximum probability of manipulation at $\beta = .5$, with the other simulation parameters as in the simulation exploring the test's size as a function of Δ: $n = 5{,}000$, $\mu = 1$, and $\sigma = .5$. The figure shows that for the narrowest bins the test has power of about 40 percent, and the power increases rapidly as the bin width gets larger, reaching 80 percent when the bin width is about .04 and higher, and about 100 percent when the bin width is about .1 and higher. This is as expected, since for a fixed

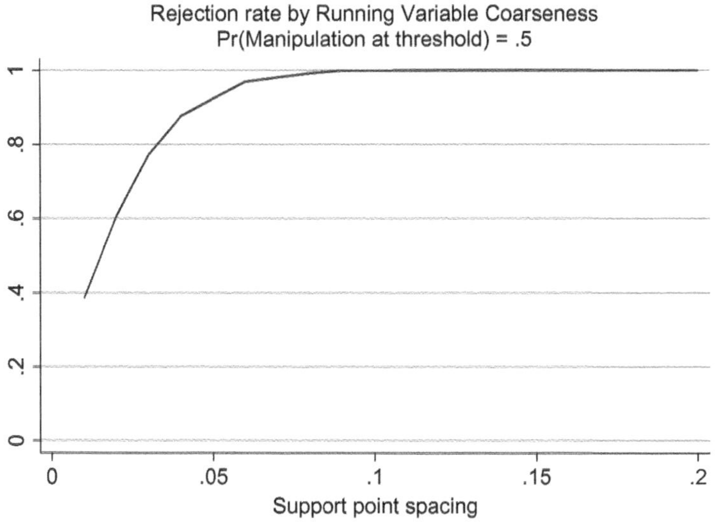

Fig. 7. Monte Carlo Simulation Rejection Rates from the Discrete Test Proposed in This Paper as a Function of the Bin Width of the Discrete Running Variable. *Notes*: The underlying distribution of the running variable is a discretized log-normal but with a probability of switching to a point above the threshold for draws below the threshold, as described in the text. The maximum probability of switching is .5 and occurs just below the threshold and fades exponentially at rate one moving away from the threshold. The nominal size of the tests is .05. Based on 1,000 iterations with a sample size of 5,000 for each iteration.

sample size, when the bins are larger there are more observations at the threshold and adjacent support points. The trade off, of course, is that when the running variable is very coarse, the smoothness approximation may be less exact, even when there is no manipulation, leading to incorrect size. The earlier simulations (Fig. 3) showed that over this range, however, the test maintains correct size.

The next simulation pushes the limits of the approximation ($k = 0$) to show when the test's performance breaks down. The simulation computes the size of the test as a function of the coarseness of the running variable, but over a wider range than in Fig. 3, to show at what point the test begins to significantly over-reject. Fig. 8 plots the rejection rate for bin widths from $\Delta = .01$ to $\Delta = .5$, with the other parameters set as in Fig. 3. The corresponding true values for k range from 4.4×10^{-5} (for $\Delta = .01$) to .044 (for $\Delta = .5$). The upper bound on coarseness of .5 corresponds to a discrete running variable with only 4–5 support points below the threshold and 20–23 above, much coarser than the discrete running variables used in

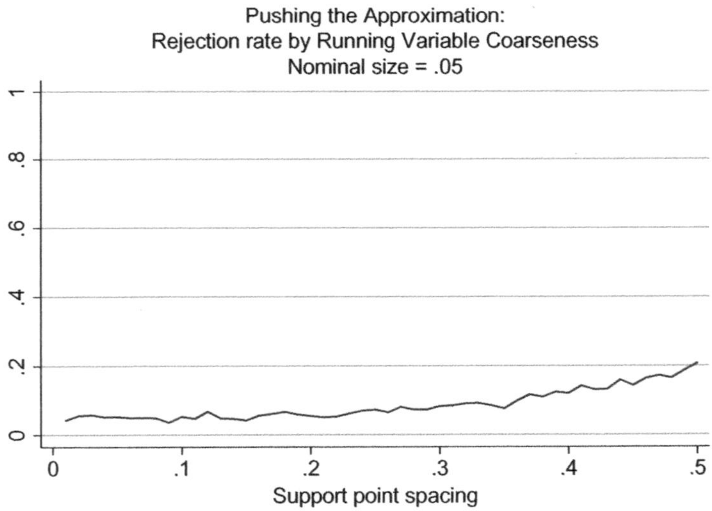

Fig. 8. Monte Carlo Simulation Rejection Rates from the Discrete Test Proposed in This Paper as a Function of the Bin Width of the Discrete Running Variable. *Notes*: The underlying distribution of the running variable is a discretized log-normal (no manipulation) with a bin width indicated on the x-axis. The nominal size of the tests is .05. Based on 1,000 iterations with a sample size of 5,000 for each iteration.

the studies cited in the introduction. The plot shows that for running variables with a support finer than around .25 the test rejects at or very near the nominal rate of 5 percent. For coarseness beyond this point, however, the test begins to over-reject as the approximation becomes less accurate, reaching about 20 percent for support spacing of .5. Thus, while the test breaks down for extremely coarse running variables, it performs well even beyond the range of discreteness seen in practice.

The previous sets of simulations showed that specifying k to be too low – while theoretically compromising the size of the test for a large enough sample size – introduces little size distortion for most typical parameter ranges, but can lead to problems with extremely coarse running variables. What are the consequences of choosing larger values for k? The next simulation answers this question. This simulation sets the running variable distribution's parameters to $\mu = 0$, $\sigma = 1$, $\Delta = .1$, $\gamma = 1$, and $n = 1,000$, and

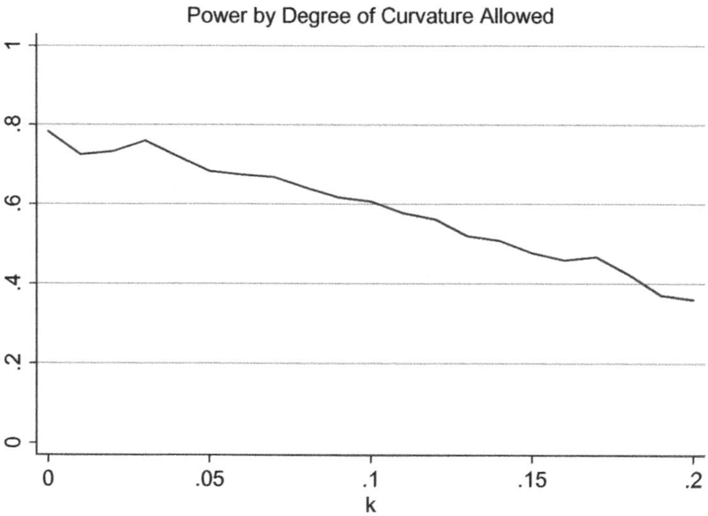

Fig. 9. Monte Carlo Simulation Rejection Rates from the Discrete Test Proposed in This Paper as a Function of the Choice of *k*. *Notes*: The underlying distribution of the running variable is a discretized log-normal but with a probability of switching to a point above the threshold for draws below the threshold, as described in the text. The maximum probability of switching is .4 and occurs just below the threshold and fades exponentially at rate one moving away from the threshold. The nominal size of the tests is .05. Based on 1,000 iterations with a sample size of 1,000 for each iteration.

introduces manipulation as before with $\beta = .4$, where the test with $k = 0$ has power of about 80 percent in the earlier simulations. This simulation considers a range for k from zero to .2. Fig. 9 plots the test's power as a function of k. The test's power decreases as k increases, since larger and larger departures from linearity are still considered consistent with no manipulation as k grows. The power at $k = 0$ is about 80 percent, but decreases to 36 percent at $k = .2$.

To summarize, the simulation results showed that in a setting based on the log-normal distribution the discrete test maintains correct size even when k is chosen to be too small, and appears to have good power properties. Choosing larger and larger k decreases the test's power.

4. EMPIRICAL EXAMPLE

The RD design has been an important tool for studying the impacts of unions on business establishments (DiNardo & Lee, 2004; Lee & Mas, 2012). This strategy exploits the fact that most private sector unions in the United States form via secret ballot representation elections among workers at the establishment who will be part of the potential bargaining unit. Since 1935, these elections have been overseen by the NLRB, which certifies the union as the sole authorized bargaining representative of the workers in the unit if the union obtains a strict majority of the votes. Thus, in close elections, very small differences in vote tallies determine whether an establishment will become unionized or not. If establishments and workers where the union barely wins and barely loses are comparable, then comparisons of post-election outcomes will reflect the causal impact of unionization.

The critical assumption is that the final vote tally in close elections is not manipulable by the union or the employer. One threat to this condition could occur in elections involving a small number of voters, when unions or employers might have more precise knowledge about the likely voting outcomes, and perhaps more influence over the individual voting decisions. In part to overcome this threat, it is common practice in this setting to restrict analysis to elections involving at least 20 voters (DiNardo & Lee, 2004; Frandsen, 2015).

However, even elections involving a large number of voters may be subject to manipulation. The institutional rules governing union representation elections allow for ex post challenges to individual ballots. In elections that come down to a single vote, the losing side would have a great incentive to challenge an opposing vote and use whatever influence it could to have the NLRB throw the ballot out. This kind of manipulation could possibly introduce confounding selection even in close elections involving a large number of voters. Frandsen (2015) shows evidence for just this sort of selection.

This empirical example illustrates the proposed test for manipulation by examining the possibility that the degree of influence unions and employers can have on the outcome of close elections could depend in part on which party controls the NLRB. Evidence for this kind of partisan bias using other methods was also described in Frandsen (2015). The NLRB consists of five members who are appointed by the President to five-year terms, subject to Senate consent. Thus each year one member is replaced. Presidents typically appoint members of their own political party, resulting in a changing party composition of the Board over time. The Board acts as a decision body of last resort for challenges and disputes arising in union certification cases, and thus the Board members could directly or indirectly influence the outcome of contentious certification cases. Close elections are, of course, the most likely to be contentious.

The dataset for the empirical example is drawn from Frandsen (2015) and includes the universe of NLRB union representation election results from 1980 to 2009, merged with the political composition of the NLRB as of the election date. Each election was classified as being held during a time of Republican control if Republicans held a majority of the seats on the Board at the time of the election, and non-Republican (i.e., Democrat or independent) otherwise. Consistent with the literature, the analysis is restricted to elections where at least 20 votes were cast. There are 45,176 elections in the dataset.

Fig. 10 shows the party in control of the NLRB from 1980 to 2009. The graph shows that NLRB party control roughly tracks the President's party, although not perfectly.

The running variable in the analysis will be the union's margin of victory defined as the number of union votes minus the number of votes the union needed for victory. The threshold of this running variable is at $r_c = 0$, corresponding to the union obtaining *exactly* the number of votes it needed to win. As an integer-valued variable, it is suitable for the discrete manipulation test proposed in this paper, while existing tests based on local linear regression would be inconsistent. An alternative choice of running variable frequently used in this setting would be the union's share of the vote. The raw vote share may be closer to continuously valued, but its support depends on firm size in potentially problematic ways: for example, only elections with at least 100 voters can produce 49 percent vote shares. To accommodate

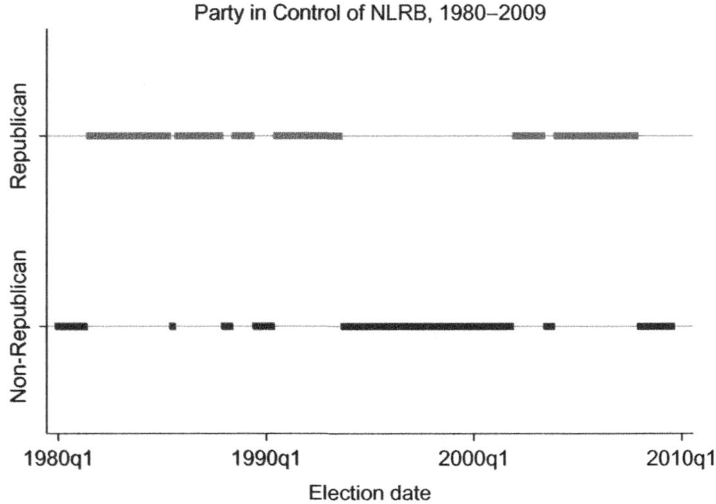

Fig. 10. The Top Series Indicates the Periods when Republicans Held a Majority of the NLRB Seats and the Bottom Series Indicates the Periods when Non-Republicans (i.e., Democrats or Independents) Held a Majority. *Source*: Reproduced from Frandsen (2015). Data are from the NLRB.

the fact that this variable has a different support for elections of different sizes, this variable is commonly binned in 5 percent increments (DiNardo & Lee, 2004), making this alternative running variable discrete, as well. As we will see, the critical elections are those that came down to a single vote, so the focus is on the union's margin in terms of number of votes. A running variable based on number of votes may implicitly give higher weight to small elections, which are more likely to come down to, say, one or two votes, so researchers may want to consider using different transformations of the running variable for manipulation testing and for estimation.

The test will be performed for several choices of k. The rule of thumb value based on the normal distribution (Section 2.2.3) suggests a maximum value for k of .002. The analysis below will show results for a range around this benchmark value: 0, .01, and .02.

The proposed test for manipulation in pooled vote tallies yields suggestive evidence of manipulation at the threshold of union victory that appears to slightly favor the employer. Fig. 11 plots a histogram of the union votes margin. Elections where the union's victory margin was zero (i.e., it obtained exactly the number of votes it needed for victory) appear slightly

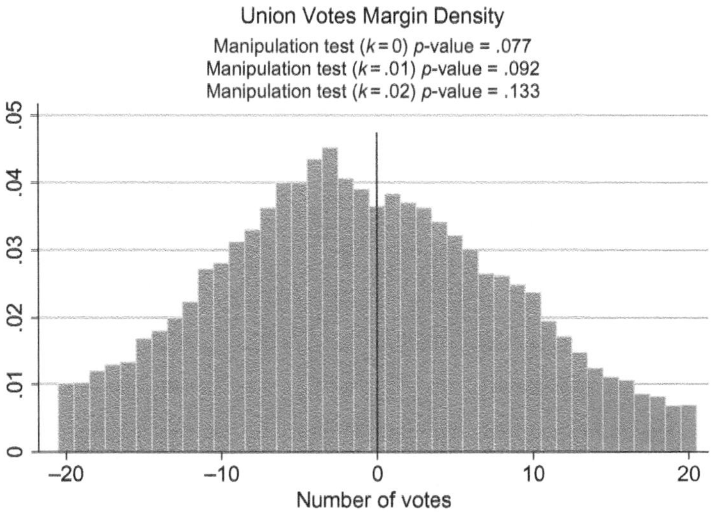

Fig. 11. Density of the Union Margin of Victory in Terms of Number of Votes for Elections with at least 20 Voters. *Source*: Data are from the NLRB. Histogram reproduced from Frandsen (2015).

less frequently than expected. The difference is small enough that even for $k = 0$ the manipulation test's p-value is .077, offering suggestive evidence of manipulation.

If, as seems likely, the mechanism underlying the manipulation is appeals to the NLRB to throw out an opposing ballot, the relative success of this strategy, and whether it tends to favor the employer or the union, may depend in part on which party controls the NLRB. Republican-controlled Boards have been accused of anti-union bias and Democrat-controlled Boards have likewise been accused of anti-employer bias (Cooke & Gautschi, 1982; Issa, 2012). Manipulation tests restricting to elections that were held when one party or another controlled the NLRB show evidence that the manipulation in favor of employers is concentrated during periods when Republicans controlled the NLRB. Fig. 12 plots histograms of the union votes margin for periods when Republicans (Panel a) or non-Republicans – that is, Democrats or independents – (Panel b) controlled the NLRB. Fig. 12(a) shows a clear dip in the density corresponding to the union barely winning when Republicans controlled the NLRB. The test for manipulation strongly rejects the null of no difference, with a p-values ranging from .01 ($k = 0$) to .017 ($k = .02$). Fig. 12(b) shows there is no such dip

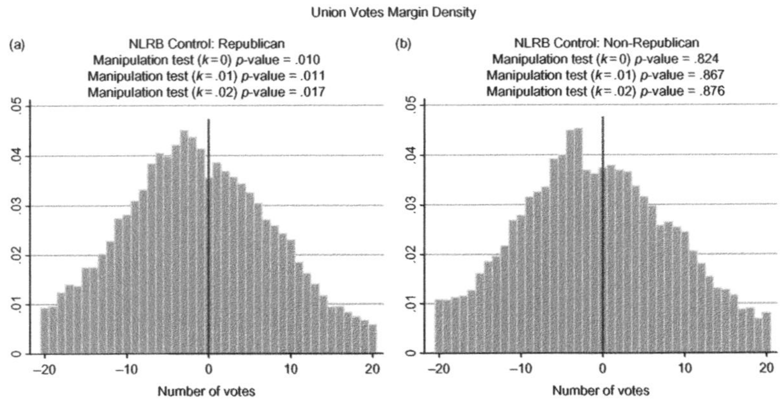

Fig. 12. Density of the Union Margin of Victory in Terms of Number of Votes for Elections with at least 20 Voters. *Source*: Data are from the NLRB. Histogram reproduced from Frandsen (2015). *Notes*: Panel (a) is for elections held during periods when Republicans controlled the NLRB. Panel (b) is for elections held during periods when non-republicans (i.e., Democrats and Independents) controlled the NLRB.

during times of non-Republican control, and the manipulation test gives a p-value of .824 even for $k = 0$, although the histogram is noisy away from the threshold. Thus, behind the modest evidence of manipulation in elections overall is very strong evidence of manipulation in favor of employers when Republicans control the NLRB.

An employer would have the greatest incentive to appeal for a union ballot to be thrown out in close elections with an odd number of voters, since only then could throwing out a single ballot reverse the outcome in the employer's favor. Likewise, the union's incentive to challenge an employer ballot is greater when there is an even number of voters. As has been noted elsewhere (Frandsen, 2015), this leads to manipulation favoring the employers in elections with an odd number of votes cast, and favoring the union in elections with an even number. Fig. 13 plots histograms of the union's votes margin for elections with an odd number of voters (Panel a) and even number of voters (Panel b), reproducing the plots in Frandsen (2015). Fig. 13(a) shows a pronounced dip (p-value <.0005 for all k considered) corresponding to close union victories, implying manipulation that favors the employer, and Fig. 13(b) a less pronounced, but still significant dip (p-value $=$ .039 for $k = 0$, .048 for $k =$ .01 and .063 for $k =$ .02)

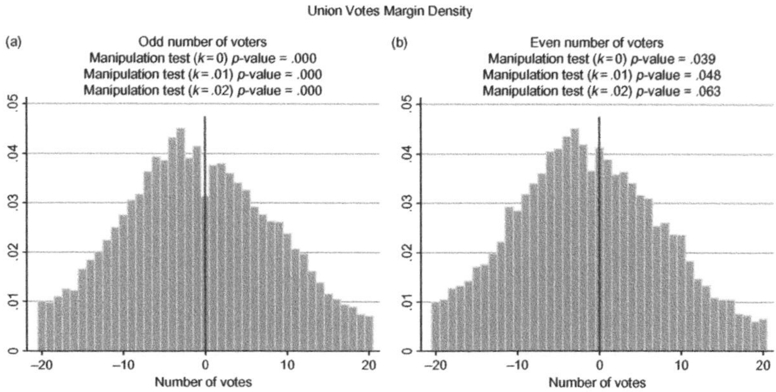

Fig. 13. Density of the Union Margin of Victory in Terms of Number of Votes for Elections with at least 20 Voters. *Source*: Data are from the NLRB. *Notes*: Panel (a) is for elections where an odd number of votes were cast. Panel (b) is for elections where an even number of votes were cast.

corresponding to close union losses, implying manipulation that favors the union.

The advantage in close elections that accrues from whether the number of voters is even or odd interacts with the advantage stemming from the party in control of the NLRB. When Republicans control the NLRB, the advantage to employers in elections with an odd number of voters intensifies, but the advantage to unions in elections with an even number of voters is nullified. Fig. 14 plots histograms of the union votes margin when Republicans control the NLRB. Fig. 14(a) shows strong evidence of manipulation favoring the employer (p-value <.0005 for all k considered) for odd elections during times of Republican control, with a large dip in the frequency of elections where the union barely won. Fig. 14(b) shows no evidence of manipulation in the favor the union for even elections (p-value > .5 for all k). The even-election advantage for the union seen in the right panel of Fig. 13 has been eliminated in times of Republican control.

On the other hand, when non-Republicans control the NLRB, the employer's advantage in odd elections is weakened, but not eliminated, while the union enjoys an even stronger advantage in even elections. Fig. 15 plots histograms of the union votes margin when non-Republicans control the NLRB. The left panel shows evidence of manipulation in odd elections favoring the employer (p-value $=$ .03 to .04) but to a lesser extent than when Republicans control the NLRB. The right panel shows strongest

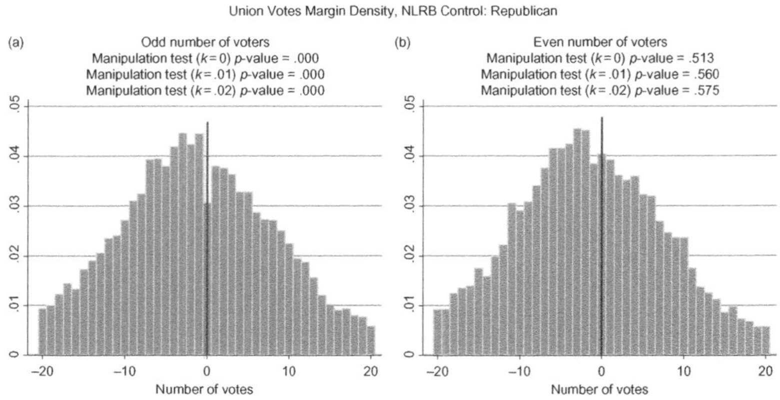

Fig. 14. Density of the Union Margin of Victory in Terms of Number of Votes for Elections with at least 20 Voters and during Periods when Republicans Controlled the NLRB. *Source*: Data are from the NLRB. *Notes*: Panel (a) is for elections where an odd number of votes were cast. Panel (b) is for elections where an even number of votes were cast.

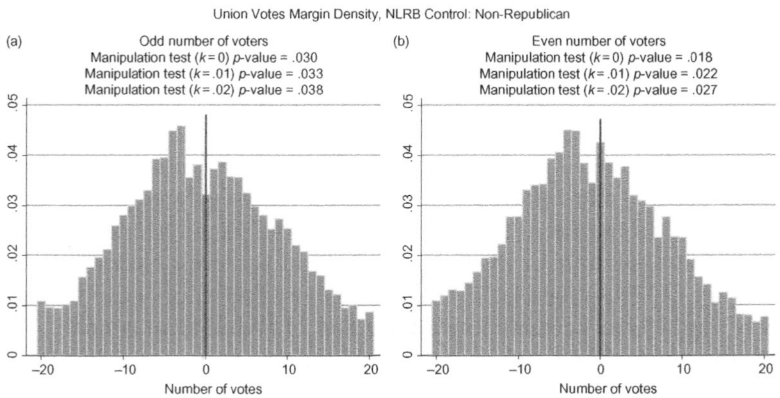

Fig. 15. Density of the Union Margin of Victory in Terms of Number of Votes for Elections with at least 20 Voters and during Periods when Non-Republicans (i.e., Democrats and Independents) Controlled the NLRB. *Source*: Data are from the NLRB. *Notes*: Panel (a) is for elections where an odd number of votes were cast. Panel (b) is for elections where an even number of votes were cast.

evidence yet of manipulation in even elections favoring the union (p-value = .018 to .027).

With additional assumptions, the magnitude as well as the presence of manipulation can be estimated. Under the assumption that $k = 0$ and that manipulation takes the form of discarding unfavorable close elections, one can estimate the fraction of "missing" elections at the threshold that would have to be added back to result in a linear density. For example, the data suggest that 21 percent of elections with an odd number of voters are "missing" from the mass at the threshold (left panel, Fig. 13).[2]

5. CONCLUSION

Many applications of the widely used RD design involve running variables that are discrete. Discrete running variables pose special problems in RD analysis that do not arise in the classical setup where the running variable is continuously distributed, including specification and inference (Lee & Card, 2008). One such challenge is that frequently used density tests for manipulation in the running variable, such as McCrary (2008), are inconsistent when the running variable is discretely distributed.

This paper proposed a test for manipulation that is consistent when the running variable is discrete, and can also be used when the running variable is continuously distributed. Monte Carlo simulations illustrated that the test has correct size and good power even in settings where the smoothness approximation near the threshold is not exact and through the range of parameters relevant in practice. The same simulations also showed that tests designed for a continuous variable will tend to falsely detect manipulation even when there is none.

Applying the test to NLRB union representation election outcomes revealed strong evidence for manipulation in union elections. Overall, the manipulation appears to slightly favor the employer, but this overall slight advantage masks large advantages to the employer when Republicans control the NLRB and advantages to the union otherwise. This evidence should sound a cautionary note to researchers on interpreting comparisons between establishments where the union barely won or barely lost as reflecting the causal impact of unionization, and to policy makers, employers, and workers on the fairness and transparency of the unionizing process.

NOTES

1. Some examples of RD designs based on discrete running variables in the recent literature include the effects of class size based on primary school enrollment (Angrist & Lavy, 1999), the effects of unionization using representation election vote share bins (DiNardo & Lee, 2004; Frandsen, 2015; Lee & Mas, 2012), the effects of Medicaid eligibility based on age in months (Card & Shore-Sheppard, 2004), the effect of Pre-K programs using age (Gormley, Gayer, Phillips, & Dawson, 2005), the effects of summer school on student achievement using discrete test scores (Matsudaira, 2008), the effects of Medicare using age in quarters or years (Card, Dobkin, & Maestas, 2009; Chay, Kim, & Swaminathanchay, 2010), and the effects of college scholarships on student outcomes using categorical test scores (DesJardins, McCall, Ott, & Kim, 2010).

2. There were 744 elections where the union was one vote shy of victory, and 675 where the union had one vote to spare. Assuming $k = 0$, one would expect $(744 + 675)/2 = 709.5$ elections exactly at the threshold of union victory. Only 561 are observed, corresponding to a fraction missing of $1 - 561/709.4 = .21$.

ACKNOWLEDGMENTS

The McCrary tests in the simulations were implemented using the Stata software program DCdensity.ado available at http://emlab.berkeley.edu/jmccrary/DCdensity/. All default options were maintained except the bin width of the discrete random variable was specified. The manipulation test for discrete running variables proposed in this paper is available as a Stata command .ado file from the author upon request.

REFERENCES

Angrist, J. D., & Lavy, V. (1999). Using Maimonides' rule to estimate the effect of class size on scholastic achievement. *The Quarterly Journal of Economics*, *114*(2), 533–575.

Armstrong, T. B., & Kolesár, M. (2015). *Optimal inference in a class of regression models*. Unpublished manuscript.

Black, D., Galdo, J., & Smith, J. (2007, January). *Evaluating the bias of the regression discontinuity design using experimental data*. Mimeo, University of Chicago.

Buddelmeyer, H., & Skoufias, E. (2003). *An evaluation of the performance of regression discontinuity design on PROGRESA*. IZA, Bonn, Germany.

Card, D., Dobkin, C., & Maestas, N. (2009). Does Medicare save lives?*The Quarterly Journal of Economics*, *124*(2), 597–636.

Card, D., & Shore-Sheppard, L. D. (2004). Using discontinuous eligibility rules to identify the effects of the federal Medicaid expansions on low-income children. *The Review of Economics and Statistics*, *86*(3), 752–766.

Cattaneo, M. D., Jansson, M., & Ma, X. (2016). *Simple local regression distribution estimators with an application to manipulation testing*. Working paper. Ann Arbor, Michigan: University of Michigan.

Chay, K. Y., Kim, D., & Swaminathan, S. (2010, February). Medicare, hospital utilization and mortality: Evidence from the program's origins. Unpublished manuscript.

Cook, T. D., & Wong, V. C. (2008). Empirical tests of the validity of the regression-discontinuity design. *Annales d'Economie et de Statistique*, *91/92*, 127–150.

Cooke, W. N., & Gautschi, F. H., III, (1982). Political bias in NLRB unfair labor practice decisions. *Industrial and Labor Relations Review*, *35*(4), 539–549.

DesJardins, S. L., McCall, B. P., Ott, M., & Kim, J. (2010). A quasi-experimental investigation of how the Gates Millennium Scholars program is related to college students' time use and activities. *Educational Evaluation and Policy Analysis*, *32*(4), 456–475.

DiNardo, J., & Lee, D. S. (2004). Economic impacts of new unionization on private sector employers: 1984-2001. *Quarterly Journal of Economics*, *119*(4), 1383–1441.

Dong, Y. (2012). Regression discontinuity applications with rounding errors in the running variable. Unpublished manuscript.

Frandsen, B. R. (2015). The surprising impacts of unionization on establishments: Accounting for selection in close union representation elections. Unpublished manuscript.

Frandsen, B. R., Frölich, M., & Melly, B. (2012). Quantile treatment effects in the regression discontinuity design. *Journal of Econometrics*, *168*(2), 382–395.

Garibaldi, P., Pacelli, L., & Borgarello, A. (2004). Employment protection legislation and the size of firms. *Giornale degli economisti e annali di economia*, pp. 33–68.

Gerard, F., Rokkanen, M., & Rothe, C. (2015, August). Partial identification in regression discontinuity designs with manipulated running variables. Unpublished manuscript.

Gormley, W. T., Jr., Gayer, T., Phillips, D., & Dawson, B. (2005). The effects of universal pre-K on cognitive development. *Developmental Psychology*, *41*(6), 872–884.

Hahn, J., Todd, P., & van der Klaauw, W. (1999, May). *Evaluating the effect of an antidiscrimination law using a regression-discontinuity design*. Working Paper No. 7131, National Bureau of Economic Research.

Hahn, J., Todd, P., & van der Klaauw, W. (2001). Identification and estimation of treatment effects with a regression-discontinuity design. *Econometrica*, *69*(1), 201–209.

Imbens, G. W., & Lemieux, T. (2008). Regression discontinuity designs: A guide to practice. *Journal of Econometrics*, *142*(2), 615–635.

Issa, D. (2012, December). President Obama's pro-union board: The NLRB's metamorphosis from independent regulator to dysfunctional union advocate. *Staff Report, U.S. House of Representatives 112th Congress Committee on Oversight and Government Reform.*

Kolesár, M., & Rothe, C. (2016). *Inference in regression discontinuity designs with a discrete running variable*. Unpublished manuscript.

Lee, D. S. (2008). Randomized experiments from non-random selection in U.S. House elections. *Journal of Econometrics*, *142*(2), 675–697. The regression discontinuity design: Theory and applications.

Lee, D. S., & Card, D. (2008). Regression discontinuity inference with specification error. *Journal of Econometrics*, *142*(2), 655–674.

Lee, D. S., & Lemieux, T. (2010). Regression discontinuity designs in economics. *Journal of Economic Literature*, *48*(2), 281–355.

Lee, D. S., & Mas, A. (2012). Long-run impacts of unions on firms: New evidence from financial markets, 1961–1999. *The Quarterly Journal of Economics*, *127*(1), 333–378.

Lloyd, C. J. (2005). Estimating test power adjusted for size. *Journal of Statistical Computation and Simulation*, *75*(11), 921–933.

Matsudaira, J. (2008). Mandatory summer school and student achievement. *Journal of Econometrics*, *142*(2), 829–850.

McCrary, J. (2008). Manipulation of the running variable in the regression discontinuity design: A density test. *Journal of Econometrics*, *142*, 698–714.

Pope, D., & Simonsohn, U. (2011). Round numbers as goals: Evidence from baseball, SAT takers, and the lab. *Psychological Science*, *22*(1), 71–79.

Sacks, J., & Ylvisaker, D. (1978). Linear estimation for approximately linear models. *Annals of Statistics*, *6*(5), 1122–1137.

Schivardi, F., & Torrini, R. Identifying the effects of firing restrictions through size-contingent differences in regulation. *Labour Economics*, *15*(3), 482–511.

van der Klaauw, W. (2008). Regression discontinuity analysis: A survey of recent developments in economics. *Labour*, *22*(2), 219–245.

Waldfogel, J. (1999). The impact of the Family and Medical Leave Act. *Journal of Policy Analysis and Management*, *18*(2), 281–302.

APPENDIX

A1. Proofs

Proof of Theorem 1. By Bayes' Rule,

$$\Pr(R = 0|R \in \{-\Delta, 0, \Delta\}) = \frac{f(0)}{f(-\Delta) + f(0) + f(\Delta)}. \tag{A.1}$$

Rearranging the definition of a second-order finite difference allows us to write

$$f(0) = \frac{1}{2}(f(-\Delta) + f(\Delta)) - \Delta^{(2)}f(0)\Delta^2/2. \tag{A.2}$$

Substituting for $f(0)$ in Eq. (A.1) using Eq. (A.2) yields

$$\Pr(R = 0|R \in \{-\Delta, 0, \Delta\}) = \frac{f(-\Delta) + f(\Delta) - \Delta^{(2)}f(0)\Delta^2}{3(f(-\Delta) + f(\Delta)) - \Delta^{(2)}f(0)\Delta^2}.$$

Since this expression is monotonic in $\Delta^2 f(0)$, bounds can be obtained by substituting in the upper and lower bounds from Assumption 3, which yields the theorem's result. ■

Proof of Theorem 2. Under H_0, the indicator $1(R_i = 0)$ is by definition conditionally Bernoulli with probability of success $p \in \mathcal{P}(k)$, conditional on $R_i \in \{-\Delta, 0, \Delta\}$. Under the i.i.d. sampling assumed in the corollary's premise, then N_0 consists of m conditionally independent draws of a Bernoulli random variable, each with conditional probability of success p, and therefore is by definition conditionally distributed as $B(m, p)$. ■

Proof of Theorem 3. The minimization problem defining the critical values always has at least one solution, since it involves minimization over a nonempty finite set. The set is nonempty because the constraint is always satisfied over some nontrivial subset (e.g., $C^L = 0$ and $C^U = m$). The constraint then implies the result, since by H_0

$$\begin{aligned}
&\Pr\left(C_\alpha^L(k)<N_0<C_\alpha^U(k)|N_{-\Delta}+N_0+N_\Delta=m\right)\\
&\quad=\Pr\left(N_0<C_\alpha^U(k)|N_{-\Delta}+N_0+N_\Delta=m\right)-\Pr\left(N_0\leq C_\alpha^L(k)|N_{-\Delta}+N_0+N_\Delta=m\right)\\
&\quad\geq \inf_{p\in\mathcal{P}(k)} F_B\left(C^U-1,m,p\right)-F_B\left(C^L;m,p\right)\\
&\quad\geq 1-\alpha,
\end{aligned}$$

where the first equality is by definition, the first inequality follows from Theorem 2 and the definition of a cdf, and the final inequality follows from the constraint. ■

Proof of Theorem 4. Note first that the probability of rejection can be written as

$$\Pr\left(N_0/m\leq C_\alpha^L(k)/m \text{ or } N_0/m\geq C_\alpha^U(k)/m|N_{-\Delta}+N_0+N_\Delta=m\right).$$

Note that $\lim_{n\to\infty}C_\alpha^L(k)/m=(1-k)/(3-k)$ and $\lim_{n\to\infty}C_\alpha^U(k)/m=(1+k)/(3+k)$. Note also that by the weak law of large numbers N_0/m converges in probability to $(1-k)/(3-k)-\delta$ or $(1+k)/(3+k)+\delta$. Then by the definition of convergence in probability

$$\lim_{n\to\infty}\Pr\left(N_0/m\leq C_\alpha^L(k)/m \text{ or } N_0/m|C_\alpha^U(k)/m|N_{-\Delta}+N_0+N_\Delta=m\right)=1.$$

■

A2. Linear Approximation Equivalency

This section shows formally that the result of Theorem 1 for the special case of $k=0$ is equivalent to a linear approximation in a fixed neighborhood of the threshold. To be precise, consider the following definition of a linear pmf in a neighborhood of the threshold:

Definition 1 (Linear probability mass function in a neighborhood of the threshold). The probability mass function of R is linear in a neighborhood of r_c if $\Pr(R=r_c+r)=\Pr(R=r_c)+(\delta r-r_c)$ for $r\in\{r_c-\Delta, r_c, r_c+\Delta\}$.

The key insight is that if R's pmf is linear in a neighborhood of r_c, then the sample frequency at $R=r_c$ has the known conditional Bernoulli distribution given in Theorem 1, and vice versa, as the following proposition shows.

Proposition 5. The probability mass function of R is linear in a neighborhood of r_c if and only if

$$\Pr(R = r_c | R \in \{r_c - \Delta, r_c, r_c + \Delta\}) = 1/3.$$

Proof. First the "only if" direction. By Bayes' rule, $\Pr(R = r_c | R \in \{r_c - \Delta, r_c, r_c + \Delta\})$ is

$$\frac{\Pr(R = r_c)}{\Pr(R = r_c - \Delta) + \Pr(R = r_c) + \Pr(R = r_c + \Delta)}.$$

By the definition of a linear pmf in a neighborhood of r_c, the denominator is $\Pr(R = r_c) - \Delta\delta + \Pr(R = r_c) + \Pr(R = r_c) + \Delta\delta = 3 \times \Pr(R = r_c)$. Now the "if" direction. The result follows if $\Pr(R = \{r_c | R \in r_c - \Delta, r_c, r_c + \Delta\}) = 1/3$ implies that the difference in the conditional probability at $R = r_c + \Delta$ and $R = r_c$ is equal to the difference in the conditional probability at $R = r_c$ and $R = r_c - \Delta$ (the difference would then be δ from the definition). Start with the difference between the conditional probability at $R = r_c + \Delta$ and $R = r_c$. By Bayes' rule and the premise,

$$\begin{aligned}
&\Pr(R=r_c+\Delta | R\in\{r_c-\Delta,r_c,r_c+\Delta\}) - \Pr(R=r_c | R\in\{r_c-\Delta,r_c,r_c+\Delta\}) \\
&= \frac{\Pr(R=r_c+\Delta)}{\Pr(R=r_c-\Delta)+\Pr(R=r_c)+\Pr(R=r_c+\Delta)} - \frac{1}{3} \\
&= 1 - \frac{\Pr(R=r_c-\Delta)+\Pr(R=r_c)}{\Pr(R=r_c-\Delta)+\Pr(R=r_c)+\Pr(R=r_c+\Delta)} - \frac{1}{3} \\
&= 1 - \left(\frac{\Pr(R=r_c-\Delta)}{\Pr(R=r_c-\Delta)+\Pr(R=r_c)+\Pr(R=r_c+\Delta)} + \frac{1}{3}\right) - \frac{1}{3} \\
&= \frac{1}{3} - \frac{\Pr(R=r_c-\Delta)}{\Pr(R=r_c-\Delta)+\Pr(R=r_c)+\Pr(R=r_c+\Delta)} \\
&= \Pr(R=r_c | R\in\{r_c-\Delta,r_c,r_c+\Delta\}) - \Pr(R=r_c-\Delta | R\in\{r_c-\Delta,r_c,r_c+\Delta\}).
\end{aligned}$$

■

Thus, the result of Theorem 1 for the case of $k = 0$ is equivalent to the discrete running variable R having a linear pmf in a neighborhood of the threshold.

TESTING STABILITY OF REGRESSION DISCONTINUITY MODELS

Giovanni Cerulli[a], Yingying Dong[b], Arthur Lewbel[c] and Alexander Poulsen[c]

[a]*IRCrES-CNR, National Research Council of Italy, Rome, Italy*
[b]*Department of Economics, UC Irvine, CA, USA*
[c]*Department of Economics, Boston College, Chestnut Hill, MA, USA*

ABSTRACT

Regression discontinuity (RD) models are commonly used to nonparametrically identify and estimate a local average treatment effect. Dong and Lewbel (2015) show how a derivative of this effect, called treatment effect derivative (TED) can be estimated. We argue here that TED should be employed in most RD applications, as a way to assess the stability and hence external validity of RD estimates. Closely related to TED, we define the complier probability derivative (CPD). Just as TED measures stability of the treatment effect, the CPD measures stability of the complier population in fuzzy designs. TED and CPD are numerically trivial to estimate. We provide relevant Stata code, and apply it to some real datasets.

Regression Discontinuity Designs: Theory and Applications
Advances in Econometrics, Volume 38, 317–339

ISSN: 0731-9053/doi:10.1108/S0731-905320170000038013

Keywords: Regression discontinuity; external validity; threshold changes; marginal effects; treatment effect heterogeneity; program evaluation

JEL classifications: C21; C25

1. INTRODUCTION

Consider a standard regression discontinuity (RD) model, where T is a binary treatment indicator, X is a so-called running or forcing variable, c is the threshold for X at which the probability of treatment changes discontinuously, and Y is some observed outcome. The outcome may be affected both by treatment and by X, though the conditional expectation of potential outcomes given X are assumed to be smooth functions of X. The goal in these models is to estimate the effect of treatment T on the outcome Y, and the main result in this literature is that under weak conditions a local average treatment effect (LATE) can be nonparametrically identified and estimated at the point where $X = c$ (see, e.g., Hahn, Todd, & Van der Klaauw, 2001).

The treatment effect identified by RD models only applies to a small subpopulation, namely, people having $X = c$. In fuzzy RD, the relevant group is even more limited, being just people who both have $X = c$ and are compliers. Compliers are defined to be people who have $T = 1$ if $X \geq c$ and have $T = 0$ if $X < c$. Note that since a person can only have a single value of X and of T, one of these defining conditions is a counterfactual statement, and so we can never know exactly who are the compliers.

Given that the estimated RD treatment effect only applies to people having $X = c$, it is important to investigate the stability of RD estimates, that is, to examine whether people with other values of X near c would have expected treatment effects of similar sign and magnitude. If not, that is, if ceteris paribus a small change in X away from c would greatly change the average effect of treatment, then one would have serious doubts about the general usefulness and external validity of the estimates, since other contexts are likely to differ from the given one in even more substantial ways than a marginal change in X.

In this chapter, we argue that an estimator proposed by Dong and Lewbel (2015) called the TED, for treatment effect derivative, can be used to assess the stability of RD LATE estimates. The TED therefore provides a valuable tool for judging potential external validity of the RD LATE estimator. Dong and Lewbel emphasize coupling TED with a local policy invariance assumption, to evaluate how the RD LATE would change if the threshold changed. In contrast, in this chapter we argue that, regardless of whether the local policy invariance assumption holds or not, the TED provides valuable information regarding stability of RD estimates.

TED is basically the derivative of the RD treatment effect with respect to the running variable. A more precise definition is provided below. We argue that a value of TED that is statistically significant and large in magnitude (see Section 5 for guidance on "how large is large") is evidence of instability and hence a potential lack of external validity. In contrast, having TED near-zero provides some evidence supporting stability of RD estimates.

We therefore suggest that one should estimate the TED in virtually all RD applications, and see how far it is from zero as a way to assess the stability and hence external validity of RD estimates. In addition to TED, we define a very closely related concept called the complier probability derivative, or CPD. Just as TED measures stability of the treatment effect, the CPD measures stability of the population of compliers in fuzzy designs.

Both TED and CPD are numerically trivial to estimate. They can be used to investigate external validity of the RD estimates, without requiring any additional covariates (other than the running variable). We provide easy to use Stata code to implement TED and CPD estimation, and apply it to a couple of real datasets.

It is important to note that TED differs substantially from the regression kink design (RKD) estimand of Card, Lee, Pei, and Weber (2015). The two appear superficially similar, because TED equals the difference in derivatives of a function around the threshold, and the estimate of a kink also corresponds to a difference of derivatives of a function around the threshold. However, RKD is the estimate of treatment effect given a continuous treatment with a kink. TED does not involve a continuous treatment in any way. TED applies when the treatment is binary, not continuous, and TED is not the estimate of a treatment effect, rather, TED is a treatment effect derivative.

TED is also not the same as the kink-based treatment effect estimator of Dong (2016a). Dong (2016a) provides an estimate of a binary treatment effect when the probability of a binary treatment (as a function of the running variable) contains a kink instead of a jump at the threshold. In contrast, TED assumes the standard RD jump in the probability of treatment at the threshold, and equals an estimate not of the treatment effect, but a derivative of the treatment effect.

The next section provides a short literature review. This is followed by sections describing TED for sharp designs, and both TED and CPD in fuzzy designs. We then provide some empirical examples, reexamining two published RD studies to see whether their RD estimates are likely to be stable or not.

2. LITERATURE REVIEW

A number of assumptions are required for causal validity of RD treatment effect estimates. Hahn et al. (2001) provide one formal list of assumptions, though some of their assumptions can be relaxed as noted by Lee (2008) and especially Dong (2016b). One such condition is Rubin's (1978, 1980, 1990) "SUTVA," which assumes that treatment of one set of individuals does not affect the potential outcomes of others. Another restriction is that potential outcomes, if they depend directly on X, are continuous functions of X. This relaxes the usual Rubin (1990) unconfoundedness condition, and so is one of the attractions of the RD method. In RD, one instead depends on the "no manipulation" assumption, which is generally investigated using the McCrary (2008) density and covariate smoothness tests.

One of the assumptions in Hahn et al. (2001) is a local independence assumption. This assumption says that treatment effects are independent of X in a neighborhood of the cutoff. Dong (2016b) shows that validity of RD does not actually require this condition, and that it can be replaced by some smoothness assumptions. Dong also shows that the local independence assumption implies that TED equals zero. So one use for TED is to test whether the local independence assumption holds.

Most tests of internal or external validity of treatment effect estimates require covariates with certain properties. For example, one check of validity is the falsification test, which checks whether estimated treatment effects equal zero when the RD estimator is applied after replacing the outcome Y with predetermined covariates. Angrist and Fernandez-Val (2013) assess external validity by investigating how LATE estimates vary across different conditioning sets of covariates. Angrist and Rokkanen (2015) provide conditions that allow RD treatment effects to be applied to individuals away from the cutoff, to expand the population to which RD estimates can be applied, and thereby increase external validity. Angrist and Rokkanen require local independence after conditioning on covariates. Wing and Cook (2013) bring in an additional indicator of being an untreated group, while Bertanha and Imbens (2014) look at conditioning on types. In contrast to all of these, a nice feature of TED is that it does not require any covariates other than X.

More generally, identification and estimation of TED require no additional data or information beyond what is needed for standard RD models. The only additional assumptions required to identify and estimate TED are slightly stronger smoothness conditions than those needed for standard RD, and these required differentiability assumptions are already imposed in practice when one uses standard RD estimators such as local quadratic regression.

TED focuses on changes in slope of the function $E(Y|X)$ around the cutoff $X = c$. Other papers that also examine or exploit slope changes in RD models include Dong (2016a) and Calonico, Cattaneo, and Titiunik (or CCT, 2014).

3. SHARP DESIGN TED

The intuition behind TED is simple. Let

$$Y = g_0(X) + \pi(X)T + e,$$

where $g_0(X)$ is the average untreated outcome, given X, $\pi(x)$ is the treatment effect for compliers who have $X = x$, and e is an error term that embodies all heterogeneity across individuals. Let $\pi'(x) = \partial\pi(x)/\partial x$. The treatment effect estimated by RD designs is $\pi(c)$, and TED is just $\pi'(c)$.

Let $Z = I(X \geq c)$, so Z equals one if the running variable is at or exceeds the cutoff, and is zero otherwise. Sharp RD design has $T = Z$, so $Y = g_0(X) + \pi(X)Z + e$. By just looking at individuals in a small neighborhood of c, we can approximate $g_0(X)$ and $\pi(X)$ with linear functions making

$$Y \approx \beta_1 + Z\beta_2 + (X - c)\beta_3 + (X - c)Z\beta_4 + e. \tag{1}$$

Local linear estimation with a uniform kernel consists precisely of selecting only individuals who have X observations close to (within one bandwidth of) c, and using just those people to obtain estimates $\widehat{\beta}_1$, $\widehat{\beta}_2$, $\widehat{\beta}_3$, and $\widehat{\beta}_4$ in this regression by ordinary least squares (for local quadratic estimation, see below). Under the standard RD and local linear estimation assumptions, shrinking the bandwidth at an appropriate rate as the sample size grows, we get $\widehat{\beta}_2 \rightarrow^p \pi(c)$ and $\widehat{\beta}_4 \rightarrow^p \pi'(c)$ (see Dong & Lewbel, 2015 for details). As a result, $\widehat{\beta}_2$ is the usual estimate of the RD treatment effect, and $\widehat{\beta}_4$ is the estimate of TED.

For any function h and small $\varepsilon > 0$, define the left and right limits of the function h as

$$h_+(x) = \lim_{\varepsilon \rightarrow 0} h(x+\varepsilon) \quad \text{and} \quad h_-(x) = \lim_{\varepsilon \rightarrow 0} h(x-\varepsilon).$$

Similarly, define the left and right derivatives of the function h as

$$h'_+(x) = \lim_{\varepsilon \rightarrow 0} \frac{h(x+\varepsilon) - h(x)}{\varepsilon} \quad \text{and} \quad h'_-(x) = \lim_{\varepsilon \rightarrow 0} \frac{h(x) - h(x-\varepsilon)}{\varepsilon}.$$

Let $g(x) = E(Y|X = x)$. Formally, the sharp RD design treatment effect is defined by $\pi(c) = g_+(c) - g_-(c)$, and Dong and Lewbel (2015) show that the sharp RD design TED, defined by $\pi'(c)$, satisfies the equation $\pi'(c) = g'_+(c) - g'_-(c)$. The above-described local linear estimator is nothing more than a nonparametric regression estimator of $\pi(c)$ and its derivative $\pi'(c)$.

Local quadratic estimation just adds squared terms to Eq. (1). That is, local quadratic regression adds $(X - c)^2\beta_5 + (X - c)^2 Z\beta_6$ to the right side of Eq. (1). But $\widehat{\beta}_2$ will still be a consistent estimate of the treatment effect, and $\widehat{\beta}_4$ will still be a consistent estimate of the TED. Empirical applications also often make use of nonuniform kernels. These correspond exactly to estimating the above regression using weighted least squares instead of ordinary least squares, where the weight of any observation i (given by the choice of kernel) is a function of the distance $|x_i - c|$, with observations closest to c getting the largest weight.

4. FUZZY DESIGN TED AND CPD

For fuzzy RD, where T is the treatment indicator and $Z = I(X \geq c)$ is the instrument, in addition to the outcome Eq. (1) we have the additional linear approximating equation

$$T \approx \alpha_1 + Z\alpha_2 + (X - c)\alpha_3 + (X - c)Z\alpha_4 + u, \tag{2}$$

where u is an error term. Once again, this equation can be estimated by ordinary least squares using only individuals who have X close to c, corresponding to a uniform kernel, or by weighted least squares given a different kernel. Equation (2) is a local linear approximation to $r(x) = E(T|X = x)$. Since T is binary, this equals the probability of treatment for an individual that has $X = x$.

Define a complier as an individual for whom T and Z are the same random variable, so a complier has $T = 1$ if and only if he has $Z = 1$. Equivalently, a complier is treated if and only if his value of X is greater than or equal to c. Let $p(x) = \Pr(T = Z|X = x)$. Note that this is not the probability that $T = Z = 0$, or that $T = Z = 1$, rather, this is the probability that T and Z are the same random variable, conditional on $X = x$. So $p(x)$ is the conditional probability that someone is a complier, conditioning on that person having $X = x$.

Let $p'(x) = \partial p(x)/\partial x$. By the same logic as in sharp design estimation in the previous section $p(c) = r_+(c) - r_-(c)$ and $p'(c) = r'_+(c) - r'_-(c)$. Under standard assumptions for fuzzy RD design and local linear estimation, we then have $\widehat{\alpha}_2 \to^p p(c)$ and $\widehat{\alpha}_4 \to^p p'(c)$. So $\widehat{\alpha}_4$ is a consistent estimator of what we will call the CPD.

Equation (2) is exactly the same as Eq. (1), replacing the outcome Y with T, replacing $g(x)$ with $r(x)$, and replacing the treatment effect $\pi(c)$ with the complier probability $p(c)$. So the exact same TED machinery as in the sharp design can be applied to Eq. (2), and the CPD is then just the TED when we replace Y with T. Also as before, for local quadratic estimation we just add $(X - c)^2\alpha_5 + (X - c)^2Z\alpha_6$ to Eq. (2), and doing so does not change the consistency of $\widehat{\alpha}_2$ and the CPD $\widehat{\alpha}_4$.

Let $q(x) = E(Y(1)|X = x) - E(Y(0)|X = x)$, so $q(c) = g_+(c) - g_-(c)$. The standard fuzzy design treatment effect is given by $\pi_f(c) = q(c)/p(c)$, and so is consistently estimated by

$$\widehat{\pi}_f(c) = \widehat{\beta}_2/\widehat{\alpha}_2.$$

Applying the formula for the derivative of a ratio,

$$\pi'_f(x) = \frac{\partial \pi_f(x)}{\partial x} = \frac{\partial \big(q(x)/p(x)\big)}{\partial x} = \frac{q'(x)}{p(x)} - \frac{q(x)p'(x)}{p(x)^2} = \frac{q'(x) - \pi_f(x)p'(x)}{p(x)}. \tag{3}$$

So, as shown by Dong and Lewbel (2015), the fuzzy design TED $\pi_f'(c)$ is consistently estimated by

$$\widehat{\pi}_f'(c) = \frac{\widehat{\beta}_4 - \widehat{\pi}_f(c)\widehat{\alpha}_4}{\widehat{\alpha}_2}.$$

5. STABILITY

We can now see how the TED, $\pi'(c)$, measures stability of the RD treatment effect, since $\pi(c+\varepsilon) \approx \pi(c) + \varepsilon\pi'(c)$ for small ε (A related expansion appears in Dinardo and Lee (2011) in a different context, that of extrapolating average treatment effects on the treated). If the TED is zero, then the average treatment effect of an individual with x close to but not equal to c will be $\pi(c+\varepsilon) \approx \pi(c)$, indicating stability of the estimated effect. However, if $\pi'(c)$ is large in magnitude, then people who are almost the same in every way to those at the cutoff, differing only in having a marginally lower or higher value of X, will have dramatically different treatment effects on average. A large value of TED therefore indicates instability.

The exact same logic applies also to fuzzy designs, with $\pi_f(c+\varepsilon) \approx \pi_f(c) + \varepsilon\pi_f'(c)$. However, in fuzzy designs there are two potential sources of instability. As Eq. (3) shows when evaluated at $x = c$, the fuzzy treatment effect could be unstable because $g'(c)$ is far from zero, indicating a true change in the effect on the average complier. Alternatively, the fuzzy treatment effect could be unstable because $p'(c)$ is far from zero. This latter condition is what the CPD tests. Like TED, the CPD is a measure of stability, since having the CPD near zero suggests potential stability of the complier population, whereas a large positive or negative value of the CPD says that the population of compliers changes dramatically with small changes in X. Note that in sharp design $p(c) = 1$ and therefore $p'(c) = 0$. So the CPD is always zero in sharp designs, and therefore only needs to be estimated in fuzzy designs.

Additional support for TED as a stability measure comes indirectly from Gelman and Imbens (2014). They argue that high-order local polynomials should not be used for estimating RD models, because the resulting estimates can be unstable. Unstable estimates from polynomial orders that are too high will typically result in very different slope estimates above and below the threshold, and hence a large estimated TED value.

Either $|\pi(c)/\pi'(c)|$ or $|\pi_f(c)/\pi_f'(c)|$ is approximately how large ε would need to be to change the sign of the estimated treatment effect. The smaller this value is, the more unstable is the RD LATE. Define the relative TED as

$$\text{sharp relative TED} = \left|\frac{\pi(c)}{\pi'(c)b}\right| = \left|\frac{\beta_2}{\beta_4 b}\right|,$$
$$\text{fuzzy relative TED} = \left|\frac{\pi_f(c)}{\pi_f'(c)b}\right| = \left|\frac{\alpha_2\beta_2}{(\alpha_2\beta_4 - \alpha_4\beta_2)b}\right|, \tag{4}$$

where b is the bandwidth used for estimation (meaning that the data used for estimation have values of X in the range from $c-b$ to $c+b$). A relative TED smaller than one implies that the treatment effect would change sign for some subset of people in this range. A simple rule of thumb might be that the RD LATE is unstable if TED is statistically significant and if the relative TED is smaller than about one or two. The CPD cannot change sign, but one might similarly be concerned about stability of the complier population if CPD is significant and the relative CPD, given by $|p(c)/(p'(c)b)|$ and estimated by $|\alpha_2/(\alpha_4 b)|$, were smaller than one.

It is important to note that instability does not mean that the RD estimates are invalid, but rather that they need to be interpreted cautiously. In contrast, a finding of stability (i.e., a small TED) suggests some external validity, since it implies some other people, those away from but near the cutoff, likely have treatment effects of similar magnitudes to those right at the cutoff.

6. EMPIRICAL EXAMPLES

In this section, we provide empirical RD examples that illustrate: (i) estimating LATE, TED and CPD; (ii) testing their significance, and (iii) graphically visualizing the results. We show that the same representation typically used to graph the discontinuity of the outcome (and/or of the probability) at the threshold can be readily extended to include the TED, so one can simultaneously visualize both the magnitude of the treatment effect (LATE) and its stability (TED and CPD). We present two examples drawn from existing RD empirical literature, one using a sharp design, and the other using a fuzzy design.

6.1. Sharp RD Example

The first example we consider is from Haggag and Paci (2014). These authors use a sharp RD design to examine the effect of suggested tip levels offered by credit card machines on consumers' actual tipping behavior, based on data from around 13 million New York City taxi rides. The RD design exploits different tip suggestions offered by the credit card machine depending on whether the fare was above or below $15. For rides under $15, tip suggestions are $2, $3, and $4, while for rides above $15 consumers are presented with 20%, 25%, and 30% tip suggestions. At the $15 threshold, the shift represents an increase in the suggested tip levels of approximately $1, $0.75, and $0.50, respectively. Haggag and Paci find that the suggested tips have a large local treatment effect. They find that this discontinuous increase in suggested tip amounts yields an increase of $0.27–0.30 in actual tips, which is more than a 10% increase in the average tip at the $15 threshold.

Here, we use TED to investigate stability of this $0.27–0.30 treatment effect. TED provides information on whether and how much this estimated local treatment effect is likely to change for fares that are slightly higher or lower than the $15 cut off. Treatment effects are estimated in two different ways; either by measuring outcomes (tips) in terms of dollar amounts, or as fractions of the total fare. The estimated RD LATE and TED are presented in Table 1. Columns 1 and 4 (RD1) correspond to the original specifications used in Haggag and Paci (2014, Column 2 of Table 2, Column 1 of Table 3) which is a third-order local polynomial, with a bandwidth that limits fares to be between $5 and $25, and controlling for driver fixed effects, pickup day of the week, pickup hour, pickup location, and drop-off location. We then also provide, and focus on, local quadratic regressions (RD2 and RD3), using correspondingly smaller bandwidths (limiting fares to be between $10 and $20 or between $12 and $18). We use local quadratic regressions because, as noted by Porter (2003), lower than quadratic order local polynomials suffer from boundary bias in RD estimation, while Gelman and Imbens (2014) report that higher than quadratic order local polynomials tend to be less accurate. In this application, the differences across these different polynomials and different bandwidth choices are rather small, probably because the sample size is large.

In Table 1, the estimated RD LATE for the tip amount (tips measured in dollars) ranges from $0.274–0.287, and the TED estimates are $0.052–0.061. These TED estimates are relatively large, suggesting that the average impact of the treatment could be five or six cents higher or lower

Table 1. TED and Sharp RD Treatment Effects of Defaults on Tipping.

	Tip Amount			Tip Percent		
	RD1	RD2	RD3	RD1	RD2	RD3
RD LATE	0.276	0.274	0.287	2.025	1.861	1.816
	(0.006)***	(0.008)***	(0.006)***	(0.038)***	(0.038)***	(0.050)***
TED	0.061	0.052	0.056	0.589	0.204	0.231
	(0.006)***	(0.006)***	(0.013)***	(0.038)***	(0.038)***	(0.081)***
Bandwidth	10	5	3	10	5	3
N	6,218,196	2,246,689	1,184,411	6,218,196	2,246,689	1,184,411
Polynomial order	3	2	2	3	2	2

Notes: This table uses the data from Haggag and Paci (2014); The sample is limited to Vendor-equipped cab rides without tolls, taxes, or surcharges; As in Haggag and Paci (2014), all specifications include fixed effects for driver, pickup day of the week, pickup hour, pickup location borough, and drop-off location borough; Columns 1 and 4 (RD1) are the original specifications used in Haggag and Paci (2014). Bandwidth equal to 10 corresponds to $5 ≤ fare ≤ $25 (the original bandwidth used in Haggag & Paci, 2014); Bandwidth equal to 5 corresponds to $10 ≤ fare ≤ $20; Bandwidth equal to 3 corresponds to $12 ≤ fare ≤ $18; Robust standard errors clustered at each fare value ($0.40 intervals).
***Significant at the 1% level.

with just a dollar change in fares. Defining the tip outcome and associated treatment effect in terms of percentage of the fare tells a similar story. The estimated RD treatment effect for tip percentage is 1.816–2.025 percentage points, meaning that the discontinuous jump in suggested tips (around the $15 fare level) increased actual tips by about 2%. The associated TED estimates are 0.204–0.589. This suggests that if the fare were 1% higher, the RD LATE might increase from around 2% to anywhere from 2.204% to 2.589%. Whether the change in the treatment effect would actually be this large if the threshold were actually increased depends on whether the local policy invariance assumption holds in this context (see the next section for details).

These TED estimates are all statistically significant. At the middle bandwidth of 5, the estimated relative TED is 1.05 or 1.82, which is near the borderline suggesting instability of the RD LATE estimates. The magnitude of the TED here means there is a good chance that the magnitude of the treatment effect could be quite different at somewhat lower or higher values of the threshold.

This relatively large TED value (and the associated instability it implies) can be seen in Fig. 1, which uses the tip percentage as the outcome variable. The circles in this figure show cell means, and the curves are the fitted local polynomials. Since this is a sharp design the left dark curve is an estimate of $E(Y(1) \mid X)$ and the right dark curve is an estimate of $E(Y(0) \mid X)$, where $Y(t)$ is the potential outcome given treatment t. As usual, the RD LATE equals the gap between the left and right curves at the cutoff point. The two straight lines in the figure show the estimated slopes of these local polynomials evaluated at the cutoff. Since this is a sharp design, TED just equals the difference in the slopes of these tangent lines. One can see from the figure that the slope of the tangent lines decreases quite a bit from the left to the right side of the threshold. As a result, if one extrapolated the curve on the left a little to the right of the threshold, and evaluated the gap between the curves at this new point (say, at 16 instead of 15), then this new RD LATE would be around 2.3% instead of around 2%. This is the instability that the TED measures.

To derive the RD LATE, Hahn et al. (2001) invoke a local independence assumption. This assumption says that treatment effects are independent of X in a neighborhood of the cutoff. As noted earlier, Dong (2016b) shows that validity of RD does not actually require this condition, that it can be replaced by some smoothness assumptions, and that the local independence

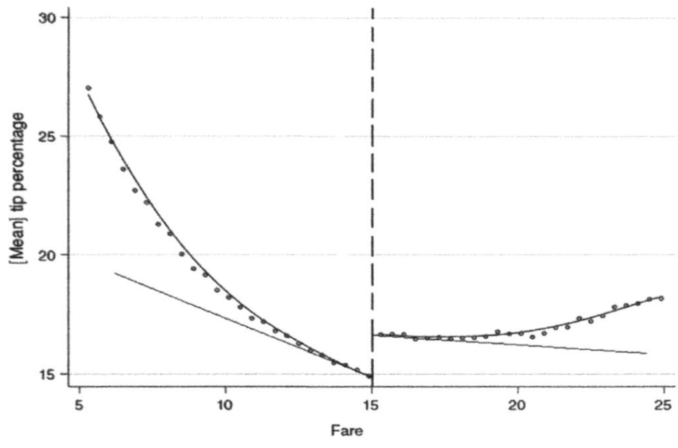

Fig. 1. Sharp RD Discontinuity in the Outcome Variable (Percentage Tips) and Tangents Lines at Threshold. *Source*: Dataset from Haggag and Paci (2014). *Note*: Each dot is the average within a discrete fare amount.

assumption implies that TED equals zero. So in the present application, our finding of a significantly nonzero TED means that this assumption is violated.

All the TED estimates in this application are significantly different from zero at the 1% significance level. Note that the sample size is very large in this application. With very large samples, even a behaviorally tiny estimate of TED could be statistically significant. This shows why it is important to consider the magnitude of the estimated TED, and not just its statistical significance, in judging stability.

6.2. Fuzzy RD Example

We now consider the fuzzy design RD model in Clark and Martorell (2010, 2014), which evaluates the signaling value of a high school diploma. In about half of US states, high school students are required to pass an exit exam to obtain a diploma. Clark and Martorell assume the random chance that leads to students falling on either side of the passing score threshold generates a credible RD design. They use this exit exam rule to evaluate the impact on earnings of having a high school diploma, since the difference in average actual learning between students with or without the diploma should be negligible, when only considering student who had grades very close to the passing grade cutoff. In this application a fuzzy RD design is appropriate, because students need to fulfill other requirements in addition to passing the exit exam in order to obtain a diploma, and some eligible students can be exempted from taking the exit exam. These other requirements include, for example, maintaining a 2.0 GPA and earning a required number of course credits. See Clark and Martorell (2010) for more details on these requirements and exemptions.

Using Texas and Florida school administrative data combined with the earnings information from the Unemployment Insurance (UI) records, Clark and Martorell find that having a high school diploma per se has little impact on earnings. This is an important finding for comparing human capital theory to signaling theory as possible explanations of the returns to education. We therefore want to investigate whether their estimates appear stable near the RD cutoff. Here, the outcome Y is the present discounted value of earnings 7 years after one takes the last round of exit exams. The treatment T is whether a student receives a high school diploma or not. The running variable X is the exit exam score (centered at the threshold passing score). Following Clark and Martorell (2010, 2014), we focus on the last chance sample, that is, those who take the last round of exit exams in high school. In our sample, X ranges from -100 to 50. About 46.7% of these students receive a high school diploma, and their average earnings are \$25,721. Detailed information on the construction of the sample can be found in Clark and Martorell (2010).

Figs. 2 and 3 show, respectively, the probability of receiving a high school diploma and earnings as a function of the exit exam score. As we can see from Fig. 2, the probability of receiving a high school diploma changes from about 40% to about 90% at the threshold passing score. This figure shows a modest change in slope at the threshold, from slightly increasing to slightly decreasing, indicating a small negative CPD. Fig. 3 shows very little if any discontinuity in outcomes around the threshold, which is the basis of Clark and Martorell's finding that having a high school diploma *per se* has little impact on earnings. The tangent lines shown in Fig. 3 are close to parallel, indicating that TED (which depends on the difference in the slopes of these lines) is also close to zero. This suggests that Clark and Martorell's results are stable, and not just a quirk of

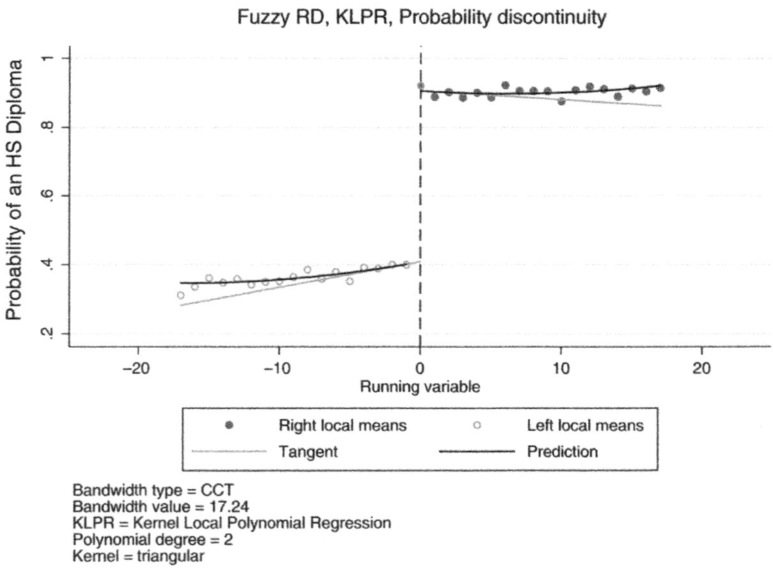

Fig. 2. Fuzzy RD Discontinuity in the Probability and Tangents Lines at Threshold. *Source*: Dataset from Clark and Martorell (2010).

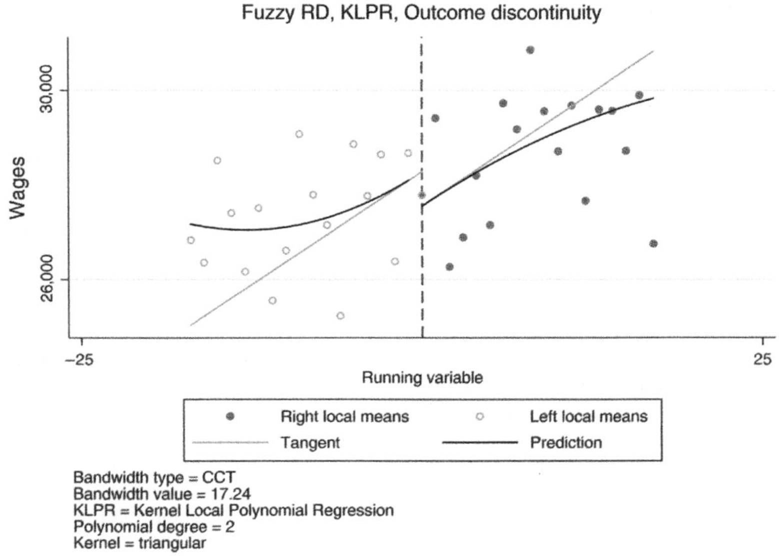

Fig. 3. Fuzzy RD Discontinuity in the Outcome and Tangents Lines at Threshold. *Source*: Dataset from Clark and Martorell (2010).

where the threshold is located. This small TED suggests that if the threshold had been somewhat lower or higher, the estimated LATE would likely still have been close to zero.

Our numerical estimates confirm what is seen in these figures. Table 2 presents both the estimated first-stage discontinuity and the RD LATE, while Table 3 provides the CPD and the TED. We compare estimates based on three popular bandwidth selectors: the CCT (Calonico et al., 2014), the IK (Imbens & Kalyanaraman, 2012), and the CV (Ludwig & Miller, 2007) bandwidths. The table also considers two different kernel functions, the triangular kernel, which is shown to be optimal for estimating the conditional mean at a boundary point (Fan & Gijbels, 1996) and the uniform kernel, which is commonly used for its convenience. As noted earlier, we use local quadratic regressions based on Porter (2003) and Gelman and Imbens (2014). As discussed in Dong and Lewbel (2015), for fuzzy designs one can estimate TED either by a local two-stage least squares (2SLS), using Z, ZX, and ZX^2 as excluded IVs for T and TX in the outcome model to get point estimates, or one can estimate local quadratic regressions separately for the reduced-form outcome and treatment equations, and then construct TED

Table 2. Fuzzy RD Estimates of the Impacts of HS Diploma on Wages.

	CCT1	IK1	CV1	CCT2	IK2	CV2
First-stage discontinuity	0.497	0.499	0.516	0.497	0.502	0.523
	(0.016)***	(0.012)***	(0.008)***	(0.017)***	(0.012)***	(0.007)***
RD LATE	−1556.1	−1659.0	−113.8	−2742.3	−1067.5	−45.4
	(2640.7)	(2418.0)	(1602.7)	(2728.9)	(2384.9)	(1592.3)
Bandwidth	11.63	19.58	50.00	8.29	15.39	42.50
N	13,364	21,694	40,795	10,051	17,744	37,715

Notes: This table uses the Florida data from Clark and Martorell (2014); All RD LATE estimates are based on the bias-corrected robust inference proposed by Calonico et al. (2014) using local linear regressions; CCT refers to the optimal bandwidth by CCT; IK refers to the optimal bandwidth proposed by Imbens and Kalyanaraman (2012); CV refers to the cross validation optimal bandwidth proposed by Ludwig and Miller (2007); 1 uses a triangular kernel, and 2 uses a uniform kernel; Standard errors are in parentheses.
*Significant at the 10% level, **significant at the 5% level, ***significant at the 1% level.

Table 3. TED and CPD of Fuzzy RD Treatment Effects of HS Diploma on Wages.

	CCT1	IK1	CV1	CCT2	IK2	CV2
CPD	−0.006	−0.006	−0.005	−0.010	−0.006	−0.004
	(0.003)*	(0.003)*	(0.001)***	(0.006)*	(0.003)*	(0.001)***
TED	287.9	296.0	44.3	509.3	360.3	−23.0
	(529.1)	(499.2)	(243.7)	(464.2)	(702.3)	(194.2)
Bandwidth	24.41	25.16	50.00	23.46	21.04	50.00
N	21,694	26,846	40,795	17,744	23,460	41,220

Notes: This table uses the Florida data from Clark and Martorell (2014); All estimates are based on local quadratic regressions; CCT refers to the optimal bandwidth proposed by Calonico et al. (2014); IK refers to the optimal bandwidth proposed by Imbens and Kalyanaraman (2012); CV refers to the cross validation optimal bandwidth proposed by Ludwig and Miller (2007); 1 uses a triangular kernel, and 2 uses a uniform kernel; Bandwidth and sample size *N* refer to those of the outcome equation; Bootstrapped Standard errors based on 500 simulations are in parentheses.
*Significant at the 10% level, **significant at the 5% level, ***significant at the 1% level.

from the estimated intercepts and slopes in the two equations as described in Section 4. For convenience, we chose the latter method, using the bootstrap to calculate standard errors.

Consistent with findings in Clark and Martorell (2010), Table 2 shows that the probability of receiving a high school diploma increases by about 50% at the threshold, which is statistically significant at the 1% level and is largely insensitive to different bandwidth and kernel choices. In contrast, all of the RD LATE estimates in Table 2 are numerically small and statistically not significant.

The estimates of CPD in Table 3 range from −0.004 to −0.010, which are all statistically significant. The normalized exam score ranges from −100 to 50. These estimates suggest that, given a 10-point decrease in the exit exam score, the percent of students who are compliers would increase from about 50% to somewhere between 54% and 60%. The relative CPD implied by the estimates in Tables 2 and 3 is around five. So the set of compliers looks unstable. However, as one would guess from Fig. 3, the estimates of TED in Table 3 are rather small and not statistically

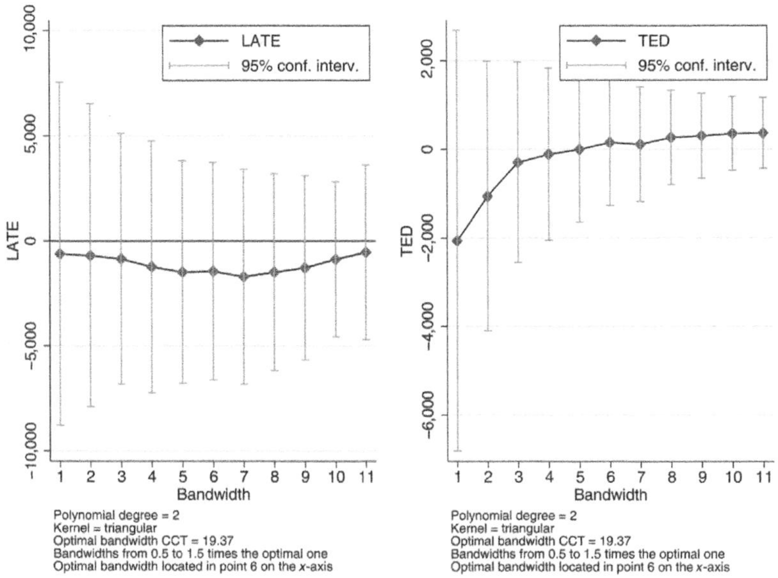

Fig. 4. Fuzzy RD LATE and TED Point Estimations and Confidence Intervals over a Range of Bandwidths. Bootstrap Confidence Intervals Based on 200 Replications. *Source*: Dataset from Clark and Martorell (2010).

significant. The implied relative TED estimates are all well below one. Together these results indicate that although the set of compliers is not stable, Clark and Martorell's conclusion (that among students with test scores near the cutoff, there is little effect of having diploma or not) does appear stable.

Finally, in Fig. 4, we check how sensitive these estimates are to bandwidth choice (a comparable exercise was not needed in our previous empirical application because the extremely large sample size there resulted in very little dependence on bandwidth). Consistent with Tables 2 and 3, Fig. 4 shows that the point estimates of both RD LATE and TED at varying bandwidths are almost all near zero. The only exception is that the TED estimate moves away from zero at the lowest bandwidth, however, the confidence band around the estimate also widens considerably at that bandwidth, so in all cases both the RD LATE and the TED are statistically insignificant.

7. COVARIATES

As noted earlier, one advantage of TED over other tools for evaluating RD estimates is that TED does not require covariates. Nevertheless, one often does have covariates, and one may readily estimate TED (and CPD) conditioning on covariate values. For example, given a binary covariate W, one may replace each β_j in Eq. (1) with $\beta_{j0}(1-W)+\beta_{j1}W$. Then in sharp designs β_{20} and β_{21} are the average treatment effects conditioning on $W=0$ and $W=1$ respectively, and similarly β_{40} and β_{41} are conditional TED values, conditioning on $W=0$ and $W=1$ respectively. For fuzzy design, one would also replace each α_j in Eq. (2) with $\alpha_{j0}(1-W)+\alpha_{j1}W$. Then $\pi_{fw}(c)=\beta_{2w}/\alpha_{2w}$ is fuzzy conditional average treatment effect, conditioning on $W=w$ (for w equal to zero or one), and it follows immediately that the corresponding fuzzy design conditional TED is $\left(\beta_{4w}-\pi_{fw}(c)\alpha_{4w}\right)/\alpha_{2w}$. The extension to more covariate values is straightforward.

One possible use for conditional TED values is that RD estimates could turn out to be relatively stable for some sets of covariate values but not others. One might then have more confidence in making policy recommendations in other contexts based on the more stable RD estimates. Alternatively, one would have still more confidence in RD estimates if they appear stable not just unconditionally, but also conditionally on each covariate value.

8. TED AND MTTE

We have stressed the use of TED as a stability measure, but as derived in Dong and Lewbel (2015) (see their paper for more details) under a local policy invariance assumption TED equals the MTTE, that is the marginal threshold treatment effect. Here, we provide a little more insight into what TED means, by comparing the difference between TED and MTTE.

For sharp design RD, define $S(x,c) = E[Y(1) - Y(0)|X = x, C = c]$, that is, $S(x,c)$ is the conditional expectation of the treatment effect $Y(1) - Y(0)$, conditioning on having the running variable X equal the value x, and conditioning on having the threshold equal the value c. If the design is fuzzy, then assume $S(x,c)$ is defined as conditioning on $X = x$, $C = c$, and on being a complier. It follows from this definition that for sharp designs $\pi(x) = S(x,c)$, and for fuzzy designs $\pi_f(x) = S(x,c)$. So $S(c,c)$ is the LATE that is identified by standard RD estimation.

The level of the cutoff c is a policy parameter. This notation makes explicit that the RD LATE depends on this policy. When $x \neq c$, the function $S(x,c)$ is a counterfactual. It defines what the expected treatment effect would be for a complier who is not actually at the cutoff c, but instead has his running variable equal to x, despite having the policy for everyone be that the cutoff equals c.

Define the function $s(x)$ by $s(x) = S(x,x)$. What TED equals is given by

$$\text{TED} = \pi'(c) = \frac{\partial S(x,c)}{\partial x}\bigg|_{x=c}.$$

In contrast, the MTTE is defined by

$$\text{MTTE} = \frac{ds(c)}{dc} = \frac{\text{d}S(c,c)}{dc} = \frac{\partial S(x,c)}{\partial x}\bigg|_{x=c} + \frac{\partial S(x,c)}{\partial c}\bigg|_{x=c} = \pi'(c) + \frac{\partial S(x,c)}{\partial c}\bigg|_{x=c}.$$

As the above equations show, TED equals MTTE if and only if $\left[\partial S(x,c)/\partial c\right]\big|_{x=c} = 0$, that is, if one's expected average treatment effect at a given value x of the running variable would not change if c marginally changed. This is the local policy invariance assumption defined by Dong and Lewbel (2015). This assumption is similar to, but weaker than, the general policy invariance condition defined by Abbring and Heckman (2007).

If we knew the MTTE, then we could evaluate how the treatment effect would change if the cutoff marginally changed from c to $c + \varepsilon$, using

$$s(c+\varepsilon) \approx s(c) + \frac{ds(c)}{dc}\varepsilon.$$

To illustrate these concepts, consider the empirical applications analyzed in the previous section. In the Haggag and Paci (2014) application, local policy invariance implies that, holding my taxi fare fixed at $c = \$15$, the difference in the amount I would tip depending on which set of tip suggestions I saw would not change if, for everybody else, the threshold for switching between the two sets of tip suggestions changed from c to $c+\varepsilon$ for a small ε. In this application, local policy invariance is quite plausible, since it is unlikely that people even notice the discontinuity at all. If local policy invariance does hold here, then we can use our estimate of TED to estimate how the tip suggestion LATE would change if the cutoff at which the change in tip suggestions occurred were raised or lowered a little.

In contrast, consider the Clark and Martorell (2010, 2014) application. There, local policy invariance roughly means that, holding my grade level fixed at $x = c$, my expected earnings difference between having a diploma or not would not change if, for all other compliers, the grade cutoff changed marginally from c to $c+\varepsilon$ for a tiny ε. In this application, local policy invariance might not hold due to general equilibrium effects, for example, a change in c might change employer's perception of the value of a diploma. More generally, local policy invariance might not hold for the same reasons that Rubin's (1978) stable unit treatment value assumption (SUTVA) could be violated, that is, if treatment of one individual might affect the outcomes of others. However, suppose local policy invariance does hold for this application (or that $\left[\partial S(x,c)/\partial c\right]\big|_{x=c}$ is close to zero, as it would be if these general equilibrium effects are small). Then the TED we reported would not just be a stability measure, but it would also tell us how much the RD treatment effect of getting a diploma would change if the cutoff grade were raised or lowered by a small amount.

These cutoff change calculations are relevant because many policy debates center on whether to change thresholds. Examples of such policy thresholds include minimum wage levels, the legal age for drinking, smoking, voting, medicare, or pension eligibility, grade levels for promotions, graduation, or scholarships, and permitted levels of food additives or of environmental pollutants. In addition to measuring stability, TED at minimum provides information that is relevant for these debates, since even if local policy invariance does not hold, TED at least comprises one component of the MTTE.

9. CONCLUSIONS

We reconsider the TED estimator defined in Dong and Lewbel (2015), and we define a related CPD estimator for fuzzy RD designs. We note how both of these are nonparametrically identified and can be easily estimated using only the same information that is needed to estimate standard RD models. No covariates or other outside information is needed to calculate the TED or CPD.

Dong and Lewbel (2015) focus on using the TED to evaluate the impact of a hypothetical change in threshold given a local policy invariance assumption. In contrast, we show here that both the TED and CPD provide valuable information, regardless of whether local policy invariance holds or not. In particular, we claim that the TED should be examined in nearly all RD applications, as a way of assessing the stability and hence the potential external validity of RD estimates. Additionally, in fuzzy RD applications, one should examine the CPD (in addition to the TED), since the CPD can be used to evaluate the stability of the complier population. We provide a simple rule of thumb for determining if TED or CPD are large enough to indicate instability.

We illustrate these claims using two different empirical applications, one sharp design and one fuzzy design. We find that the sharp RD treatment effects of taxi tip suggestions reported in Haggag and Paci (2014) are not stable, indicating that the magnitudes of their estimated treatment effects might change significantly at slightly lower or higher tip levels. The second application we consider is the fuzzy RD model of Clark and Martorell (2014). The CPD of their model suggests that the set of compliers is not stable, but the TED is numerically small and statistically insignificant. This near-zero TED suggests that their finding of almost no effect of receiving a diploma on wages (conditional on holding the level of education, as indicated by test scores, fixed) is stable, and so would likely remain valid at somewhat lower or higher test score levels.

ACKNOWLEDGEMENT

We would like to thank Damon Clark for sharing the Florida data.

REFERENCES

Abbring, J. H., & Heckman, J. J. (2007). Econometric evaluation of social programs, part III: Distributional treatment effects, dynamic treatment effects, dynamic discrete choice, and general equilibrium policy evaluation. In J. J. Heckman, & E. E. Leamer (Eds.), *Handbook of Econometrics* (Vol. 6). Berlin; Heidelberg, Germany: Elsevier.

Angrist, J. D., & Fernandez-Val, I. (2013). ExtrapoLATE-ing: External validity and overidentification in the LATE framework. In *Advances in economics and econometrics: Theory and applications, tenth world congress, volume III: Econometrics*. Econometric Society Monographs.

Angrist, J. D., & Rokkanen, M. (2015). Wanna get away? Regression discontinuity estimation of exam school effects away from the cutoff. *Journal of the American Statistical Association*, *110*(512), 1331–1344.

Bertanha, M., & Imbens, G. W. (2014). *External validity in fuzzy regression discontinuity designs*. No. w20773. National Bureau of Economic Research.

Calonico, S., Cattaneo, M. D., & Titiunik, R. (2014). Robust nonparametric confidence intervals for regression discontinuity designs. *Econometrica*, *82*, 2295–2326.

Card, D., Lee, D. S., Pei, Z., & Weber, A. (2015). Inference on causal effects in a generalized regression kink design. *Econometrica*, *83*(6), 2453–2483.

Clark, D., & Martorell, P. (2010). *The signaling value of a high school diploma*. Princeton IRS Working Paper No. #557.

Clark, D., & Martorell, P. (2014). The signaling value of a high school diploma. *Journal of Political Economy*, *122*(2), 282–318.

Dinardo, J., & Lee, D. S. (2011). Program evaluation and research designs. In O. Ashenfelter, & A. Card (Eds.), *Handbook of labor economics* (Vol. 4a, pp. 463–536). Amsterdam: Elsevier.

Dong, Y. (2016a). *Jump or kink? Regression probability jump and kink design for treatment effect evaluation*. Unpublished Manuscript.

Dong, Y. (2016b). *An alternative assumption to identify LATE in regression discontinuity designs*. Unpublished Manuscript.

Dong, Y., & Arthur Lewbel. (2015). Identifying the effect of changing the policy threshold in regression discontinuity models. *Review of Economics and Statistics*, *97*(5), 1081–1092.

Fan, J., & Gijbels, I. (1996). *Local polynomial modelling and its applications*. London: Chapman and Hall.

Gelman, A., & Imbens, G. (2014). *Why high-order polynomials should not be used in regression discontinuity designs*. NBER Working Paper No. 20405.

Haggag, K., & Paci, G. (2014). Default tips. *American Economic Journal: Applied Economics*, *6*(3), 1–19.

Hahn, J., Todd, P. E., & Van der Klaauw, W. (2001). Identification and estimation of treatment effects with a regression-discontinuity design. *Econometrica*, *69*, 201–209.

Imbens, G., & Kalyanaraman, K. (2012). Optimal bandwidth choice for the regression discontinuity estimator. *The Review of Economic Studies*, *79*(3), 933–959.

Lee, D. S. (2008). Randomized experiments from non-random selection in US House elections. *Journal of Econometrics*, *142*(2), 675–697.

Ludwig, J., & Miller, D. L. (2007). Does head start improve children's life chances? Evidence from a regression discontinuity design. *The Quarterly Journal of Economics*, *122*, 159–208.

McCrary, J. (2008). Manipulation of the running variable in the regression discontinuity design: A density test. *Journal of Econometrics*, *142*(2), 698–714.

Porter, J. R. (2003). *Estimation in the regression discontinuity model*. Unpublished Manuscript.

Rubin, D. B. (1978). Bayesian inference for causal effects: The role of randomization. *The Annals of Statistics 6*, 34–58.

Rubin, D. B. (1980). Using empirical Bayes techniques in the law school validity studies. *Journal of the American Statistical Association, 75*(372), 801–816.

Rubin, D. B. (1990). Formal mode of statistical inference for causal effects. *Journal of Statistical Planning and Inference, 25*(3), 279–292.

Wing, C., & Cook, T. D. (2013). Strengthening the regression discontinuity design using additional design elements: A within-study comparison, *Journal of Policy Analysis and Management, 32*, 853–877.

REGRESSION KINK DESIGN: THEORY AND PRACTICE

David Card[a,b,c], David S. Lee[b,d], Zhuan Pei[c,e] and Andrea Weber[c,f]

[a]*UC Berkeley, CA, USA*
[b]*NBER, Cambridge, MA, USA*
[c]*IZA, Bonn, Germany*
[d]*Princeton University, Princeton, NJ, USA*
[e]*Cornell University, Ithaca, NY, USA*
[f]*Central European University, Budapest, Hungary*

ABSTRACT

A regression kink design (RKD or RK design) can be used to identify casual effects in settings where the regressor of interest is a kinked function of an assignment variable. In this chapter, we apply an RKD approach to study the effect of unemployment benefits on the duration of joblessness in Austria, and discuss implementation issues that may arise in similar settings, including the use of bandwidth selection algorithms and bias-correction procedures. Although recent developments in nonparametric estimation (Calonico, Cattaneo, & Farrell, 2014; Imbens & Kalyanaraman, 2012) are sometimes interpreted by practitioners as pointing to a default estimation procedure, we show that in any given

Regression Discontinuity Designs: Theory and Applications
Advances in Econometrics, Volume 38, 341–382

ISSN: 0731-9053/doi:10.1108/S0731-905320170000038016

application different procedures may perform better or worse. In particular, Monte Carlo simulations based on data-generating processes that closely resemble the data from our application show that some asymptotically dominant procedures may actually perform worse than "sub-optimal" alternatives in a given empirical application.

1. INTRODUCTION

Regression discontinuity designs (RDDs) have become part of the toolkit used by applied microeconomists to identify causal effects in observations settings. Among the recent literatures that have successfully adopted RDD techniques are studies of educational outcomes (Angrist & Lavy, 1999; Chay, McEwan, & Urquiola, 2005), health (Card, Dobkin, & Maestas, 2009; McCrary & Royer, 2011), election outcomes (Lee, 2008; Lee, Moretti, & Butler, 2004), unions (DiNardo & Lee, 2004; Lee & Mas, 2012), and unemployment (Card, Chetty, & Weber, 2007; Schmieder, von Wachter, & Bender, 2012). Concurrent with this growth in empirical applications is the widespread adoption of local polynomial methods for estimation and inference, following the recommendations of Porter (2003), Imbens and Lemieux (2008), and Lee and Lemieux (2010), and a series of recent papers offering guidance on how to select appropriate local polynomial specifications (Calonico, Cattaneo, & Farrell, 2014; 2016; Imbens & Kalyanaraman, 2012).

In this chapter, we discuss how the methods that are now widely used in RDD estimation can be extended to implement a *regression kink design* (RKD), a term first coined by Nielsen, Sørensen, and Taber, 2010. Like an RDD, an RKD arises when a policy variable of interest is determined (at least in part) by a known assignment rule. While RDD posits a discontinuity in the assignment rule, in an RKD the policy rule is assumed to have a *kink* in the relationship between the policy variable and the underlying assignment variable. Such kinks arise, for example, when a benefit formula is subject to minimum or maximum values, or when marginal tax rates exhibit discrete jumps. The idea of the RKD is to examine the slope of the relationship between the outcome of interest and the assignment variable at the exact location of the kink in the policy formula. Provided that individuals on either side of the kink threshold are "similar," any kink in the outcome can be attributed to the treatment effect of the policy variable. As in an RDD, implementing an RKD involves the estimation of quantities "close" to a threshold using (local) polynomial regressions. Instead of estimating a shift in the intercept, however, we are interested in estimating a slope change.

We illustrate the approach by applying a "fuzzy" RKD to study the effects of unemployment insurance (UI) benefits on the duration of joblessness in Austria. The UI benefit schedule in Austria, as in other countries, is an increasing function of past earnings, subject to a minimum and a maximum benefit. The kinks in the benefit schedule generated by the minimum and maximum benefit levels allow us to estimate the elasticity of jobless duration with respect to benefit levels for a subpopulation of claimants with earnings near the kink point.

The application is an extension of the example presented in Card, Lee, Pei, and Weber (2015a) – hereafter, CLPW. This chapter presents a more extensive discussion of the potential issues that arise in applying local polynomial methods in the Austrian UI setting. We also use a different measure of joblessness as our key outcome variable, and compare results at the kinks created by the maximum and minimum benefit rules. As in CLPW, we present a range of alternative estimates of the elasticity of joblessness with respect to UI benefits using local linear and quadratic polynomial models, several bandwidth selection algorithms – including a rule of thumb procedure based on Fan and Gijbels (1996) (hereafter, FG) and Calonico, Cattaneo, Farrell, and Titiunik (Forthcoming) (hereafter, CCFT) – and bias-corrected robust confidence intervals proposed by Calonico et al. (2014) (hereafter, CCT). These procedures have made a large impact on recent empirical work and are often interpreted by practitioners as providing a default estimation approach that yields "preferred" results. In our application, however, we find that alternative procedures give rise to widely disparate findings. Moreover, the default method suggested by CCT (and CCFT) – local quadratic estimator with regularized bandwidth selector and bias-correction – is typically uninformative, despite visual evidence of kinks in both the UI benefit assignment rule and in the outcomes of interest.

We then offer some suggestions for assessing the performance of alternative estimators by conducting Monte Carlo simulations of data-generating processes (DGPs) that are closely based on the actual data in the specific application. In our setting, this approach suggests that local quadratic estimators have substantially larger (asymptotic) mean-squared errors (MSE) than local linear estimators. Moreover, the bias correction procedure suggested by CCT leads to a large loss in precision with only modest offsetting reductions in bias. Of course, as we emphasize throughout this chapter these conclusions are specific to our empirical context: The simulations and empirical RD examples of Calonico et al. (2014), for example, show limited precision loss associated with bias correction.

We close our discussion with a few brief comments on a related study by Card, Johnston, Leung, Mas, and Pei (2015). Card, Johnston, et al. (2015) apply an RKD approach to study the effect of UI benefit levels on the duration of unemployment, using administrative data from the state of Missouri. In contrast to our Austrian example, the Missouri application has a much sharper "first-stage" relationship between benefit levels and baseline earnings, leading to more stable estimates across alternative estimators.

The rest of the chapter is structured as follows. We review the main identification results from CLPW in Section 2, and the theory of estimation and inference in Section 3. Our empirical analysis on the impact of Austrian UI benefits follows in Section 4, where we discuss institutional background (Section 4.1), our data (Section 4.2), graphical and estimation results (Sections 4.3 and 4.4), and related findings from Card, Johnston, et al. (2015) (Section 4.5). Section 5 concludes.

2. REVIEW OF IDENTIFICATION

In this section, we review identification results in RKDs. Following the notation in CLPW, we use B to denote the treatment variable of interest (e.g., the unemployment benefit level), V the assignment variable (e.g., earnings in the period prior to the start of the benefit spell), U the error term and $Y \equiv y(B, V, U)$ the outcome of interest (e.g., the duration of time until the next job). We are interested in the causal effect of increasing B on Y, that is, the partial derivative of y with respect to its first argument, which we denote by $y_1(B, V, U)$.

Identification of the treatment effect of B is formally investigated in Nielsen et al. (2010) and CLPW under the "sharp" RKD, where B is a *deterministic* function of V, that is, $B = b(V)$ with a kink (slope change) at $V = 0$.[1] Nielsen et al. (2010) prove identification in the case where the function $y(B, V, U)$ is separably additive, while CLPW extend the result to the nonseparable case. In particular,

$$\frac{\lim_{v_0 \to 0^+} \frac{dE[Y|V=v]}{dv}\big|_{v=v_0} - \lim_{v_0 \to 0^-} \frac{dE[Y|V=v]}{dv}\big|_{v=v_0}}{\lim_{v_0 \to 0^+} \frac{db(v)}{dv}\big|_{v=v_0} - \lim_{v_0 \to 0^-} \frac{db(v)}{dv}\big|_{v=v_0}} = E[y_1(b_0, 0, U)|V = 0], \quad \text{where } b_0 = b(0). \tag{1}$$

Equation (1) states that the RKD estimand – the slope change in the outcome variable (numerator) scaled by the change in the first stage (denominator) – identifies an average treatment effect. This treatment effect parameter is what Florens et al. (2008) call the treatment-on-the-treated (TOT) and what Altonji and Matzkin (2005) call the local average response (LAR). Analogous to Lee (2008), CLPW interpret the identified TOT/LAR parameter as a weighted average of the treatment effects across the population, where individuals receive higher weights for having a higher likelihood of being at the threshold ($V = 0$). Sufficient regularity conditions are proposed for identification. Intuitively speaking, these conditions require that everything evolves smoothly across the threshold *except* the derivative of the first-stage relationship, $b(\cdot)$. Similar to an RDD, these conditions can be tested by examining the smoothness (in the first derivative) of the observed density of V, f_V, and the conditional expectation of any predetermined covariate, $E[X|V = v]$.

CLPW also prove identification in a "fuzzy" generalization of the sharp design, where the observed relationship between B and V deviates from the statutory rule, $b(\cdot)$. The deviation can arise due to measurement errors, noncompliance with the policy formula, or dependence of the formula for B on other unobserved variables. In this more general setup, the RK estimand still identifies a weighted average of the treatment effect subject to an expanded set of regularity conditions. As with the sharp case, CLPW show that the validity of these conditions can be tested by examining the smoothness of f_V and $E[X|V = v]$.[2]

In our experience (Card, Lee, & Pei 2009; Card, Lee, Pei, & Weber 2012; Card et al., 2015), potential applications of the regression kink idea using administrative data usually require a fuzzy design approach. In these datasets, it is almost inevitable that there are unexplained deviations from the prescribed benefit schedule. Nevertheless, the majority of the data points typically follow the statutory schedule exactly. As we will see below, the strength of this first-stage relationship is important for obtaining precise and robust empirical estimates.

3. REVIEW OF ESTIMATION AND INFERENCE

In this section, we review the theory of estimation and inference in an RKD, as developed in CCT and CLPW. The kinks in the outcome and treatment variable are measured by estimating local polynomial regressions of order p to the left and right of the kink point ($V = 0$) with bandwidth h and kernel K. Denoting the resulting first-stage slope estimators above and below the threshold by $\widehat{\kappa}_1^-$ and $\widehat{\kappa}_1^+$ and outcome slope estimators by $\widehat{\beta}_1^+$ and $\widehat{\beta}_1^-$, the fuzzy RKD estimator is defined as

$$\widehat{\tau} = \frac{\widehat{\beta}_1^+ - \widehat{\beta}_1^-}{\widehat{\kappa}_1^+ - \widehat{\kappa}_1^-}. \tag{2}$$

Estimation in a sharp design can be considered as a special case in which the estimators $\widehat{\kappa}_1^+$ and $\widehat{\kappa}_1^-$ are equated to the *known* slopes in the first stage: $\widehat{\kappa}_1^+ = \lim_{v \to 0^+} b'(v)$ and $\widehat{\kappa}_1^- = \lim_{v \to 0^-} b'(v)$.

When implementing the RKD estimator in practice, one must make choices for the kernel K, bandwidth h, and polynomial order p. Regarding K, Cheng, Fan, and Marron (1997) show that the triangular kernel is boundary optimal. In our experience based on actual applications and Monte Carlo examples, however, the efficiency losses from using a uniform kernel are small. Moreover, the ease of implementation of a uniform kernel – which converts the RKD problem to a simple OLS/2SLS problem using data within a window of width h from the threshold – makes it an attractive choice, and we use it in our applications below.[3]

Following IK, CCT propose an MSE-optimal bandwidth selector for the sharp RKD. A feature of the IK procedure, also adopted by CCT, is the inclusion of a "regularization" term in the bandwidth formula that prevents the selection of large bandwidths. CLPW extend these selectors for the fuzzy design and experiment with removing the regularization term in the default CCT bandwidth. The presence or absence of the regularization term does not affect the optimality features of the bandwidth selector, since the term shrinks toward 0 in large samples. In the samples used in our application, however, the regularization term is often relatively large, leading to 30–70% reductions in the size of the bandwidth.

Very recent papers by Calonico et al. (2016) and Cattaneo, Farrell, and Titiunik (2016) and CCFT propose additional bandwidth choices. Building on Calonico, Cattaneo, and Farrell (Forthcoming); Calonico et al. (2016) derive a bandwidth selector in the RD/RK context that minimizes the asymptotic coverage error (CE) of the confidence interval; this CE-optimal bandwidth is obtained by shrinking the MSE-optimal bandwidth by a factor that depends on the sample size n and polynomial order p.[4] Calonico et al. (2016) study the identification and local estimation of RDD and RKD after incorporating predetermined covariates in an additively separable way, and propose corresponding MSE-optimal and CE-optimal bandwidth selectors. CCFT implement Calonico et al. (2016,b) as well as additional bandwidth selectors in Stata and R.

Data-dependent estimation approaches that build on MSE-optimal bandwidth selectors will in general introduce an asymptotic bias in the resulting estimator. CCT propose a procedure to estimate this bias term and construct robust confidence intervals that are re-centered *and* account for the estimation error in the bias correction term. Alternatively we can construct a bias-corrected estimator $\hat{\tau}^{bc}$ by subtracting the estimated bias term from $\hat{\tau}$, and an associated standard error that incorporates the uncertainty in the bias term.[5]

A final issue is the choice of the polynomial order, p. Standard arguments suggest that a local quadratic ($p = 2$) is preferred to local linear ($p = 1$) in estimating the boundary derivatives in the RK design, because the former leads to a lower order of bias than the latter, allowing a sequence of bandwidths that shrinks at a slower rate and delivering a smaller-order asymptotic MSE. As noted by Ruppert and Wand (1994) and Fan and Gijbels (1996), however, arguments based on asymptotic rates do not imply a universally preferred choice. Instead, as we argue in Card, Lee, Pei, and Weber (2014) the best choice for p depends on the sample size and the derivatives of the conditional expectation functions, $E[Y|V = v]$ and $E[B|V = v]$, in the particular data set of interest.[6]

In the following section, we illustrate the implementation of a fuzzy RKD approach with applications using UI data from Austria over the 2001–2012 period. We exploit the kinks in the unemployment benefit schedule at the minimum and maximum benefit levels to estimate the effect of UI benefits on the duration of joblessness. We present a variety of estimates of this effect, which are obtained by using local linear and local quadratic regressions with several bandwidth selectors, with and without bias correction: the MSE-optimal bandwidth, calculated including the regularization term (which is the "default" option in the software package distributed by CCFT), the MSE-optimal bandwidth without regularization, the MSE-optimal bandwidth for a fuzzy design (without regularization) and the FG bandwidth as defined in Card et al. (2012). Using simulated data with DGP's closely resembling our actual data, we assess the performance of these alternative estimators.

4. EMPIRICAL APPLICATION: THE EFFECT OF UI BENEFITS ON UNEMPLOYMENT DURATIONS

4.1. The UI Benefit Schedule in Austria

A UI claimant's benefit entitlement in Austria depends on his or her earnings in the "base period" prior to layoff. This period is either the previous calendar year or the second most recent year (depending on whether the claim starts early or late in the year). The UI replacement rate is 55%, subject to a maximum benefit level. Claimants with dependent family members are also eligible for a family allowance that is added to their basic UI benefit. For low-wage claimants, there is also a minimum benefit level that applies as long as benefits do not exceed 60% (for a single individual) or 80% (for a claimant with dependents) of base period earnings. The end result of these rules is a schedule with multiple kink points.

To illustrate, Fig. 1 plots actual daily UI benefits against annual base year earnings for a sample of UI claimants in 2004. The large fraction of claimants whose observed UI benefits are exactly equal to the amount predicted by the formula leads to a series of clearly discernible lines in the figure, though there are also many observations scattered above and below these lines.[7] Specifically, in the middle of the figure, there are five distinct upward-sloping linear segments, corresponding to claimants with 0, 1, 2, 3, or 4 dependents.

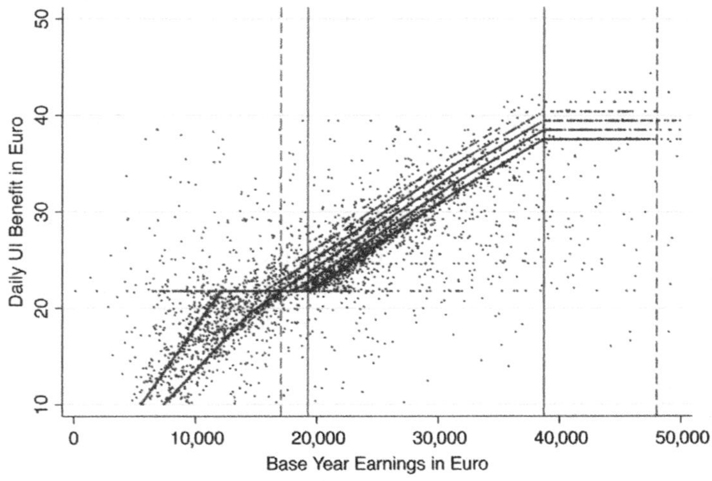

Fig. 1. UI Benefits in 2004 Austria.

These schedules all reach an upper kink point at the maximum benefit threshold which we call $T^{\max}$, shown in the graph by a solid vertical line on the right side of the figure. At the lower end, the situation is more complicated: each of the upward-sloping segments reaches the minimum daily benefit (€22 per day in 2004) at a different level of earnings, reflecting the fact that the basic benefit includes family allowances, but the minimum does not. We define the minimum benefit $T^{\min}$ – shown by the solid vertical line on the left side of the figure – as the level of earnings at which a single claimant receives the basic minimum benefit amount.[8] Finally, very low-paid claimants receive a subminimum benefit, which is again an upward-sloping function of base period earnings. These subminimum claimants fall largely in two groups: single claimants (whose benefit is 60% of their base earnings, net of taxes) and those with dependents (whose benefits are 80% of their net base year earnings).[9]

4.2. Data and Analysis Sample

Our data are drawn from the Austrian Social Security Database and contain information on base period earnings, the level of benefits actually received, benefit duration, and the starting and ending dates of employment spells, as well as demographic information including age, gender, education, marital status, job tenure, and industry.[10] Our analysis sample includes qualifying claims filed between 2001 and 2012. Among other sample restrictions, we focus on claimants whose earnings are high enough to place them above the subminimum portion of the benefit schedule for a single claimant (i.e., above the level indicated by the vertical dashed line on the left side of Fig. 1) and below the Social Security earnings cap (denoted by the vertical dashed line on the right side of Fig. 1).

We study the effects of the kinks induced by the minimum and maximum benefit levels separately by defining two subsamples. Specifically, we divide the claimants in each year into two (roughly) equal groups based on their gross base year earnings: those below the 50th percentile are assigned to the "bottom kink" sample, while those above this threshold are assigned to the "top kink" sample. Since earnings have a right-skewed distribution, there is narrower support for our observed assignment variable (annual base year earnings) around the bottom kink relative to the top kink. We re-center base year earnings for observations in the bottom kink subsample around $T^{\min}$ (indicated by the solid vertical line just below €20,000 in Fig. 1), and base year earnings for those in the top kink subsample around

T^{max} (indicated by the solid vertical line just below €40,000 in Fig. 1), so both kinks occur at $V = 0$. Each sample contains about 275,000 observations. As mentioned above, the existence of family allowances in Austria generates several distinct benefit schedules that all share the same upper kink location, but have different bottom kink thresholds. Therefore, in the top kink sample, the change in the *slope* of the benefit schedule is the same for all claimants at T^{max}. At the lower end of the benefit schedule, the kink location and magnitudes differ for families with or without dependents: To the left of T^{min}, benefits for claimants with no dependents are constant, whereas benefits for claimants with dependents continue to fall. Since we cannot distinguish between claimants who do or do not have dependents in our data, the measured kink in the average benefit at T^{min} will be proportional to the number of claimants without dependents.

Table 1 reports basic summary statistics for the bottom and top kink samples. Mean base year earnings for the bottom kink group are about €22,000, with a relatively narrow range of variation (standard deviation = €2,800), while mean earnings in the top kink group are higher (mean = €34,000) and more dispersed (standard deviation = €6,700). Mean daily UI benefits are €25.2 for the bottom kink group (implying an annualized benefit of €9,200, about 44% of T^{min}), while mean benefits for the top kink sample are €33.5 (implying an annualized benefit of €12,300, about 28% of T^{max}). Claimants in the bottom kink sample are more likely to be female, are a little younger, less likely to be married, more likely to have had a blue-collar occupation, and are less likely to have post-secondary education. Despite the differences in demographic characteristics and mean pay, the means of our main outcome variable, the time to next job, are quite similar in the two samples: the average duration of joblessness is around 150 days.[11] Only about 10% of claimants exhaust their regular UI benefits.

4.3. Graphical Overview of the Effect of Kinks in the UI Benefit Schedule

The first-stage relationships between (re-centered) base period earnings and UI benefits are presented in Fig. 2(a) and (b), which divide base period earnings into small bins and plot the average UI benefits in each bin.[12] These figures average away some of the noise seen in the raw scatter plots of Fig. 1 and make it easier to visualize the conditional expectation function $E[B|V]$. Kinks are visible in both figures: we see an increase in slope as earnings go across T^{min} in Fig. 2(a) and a decrease as they go across T^{max} in Fig. 2(b). As mentioned above, the UI schedule is only kinked for single claimants at T^{min},

Table 1. Summary Statistics for Bottom and Top Kink Samples of UI Claimants.

	Bottom Kink Sample		Top Kink Sample	
	Mean (1)	Std. Dev. (2)	Mean (3)	Std. Dev. (4)
Baseline earnings (€)	22,142	2,805	33,847	6,730
Daily UI benefit (€)	25.2	3.0	33.5	5.7
Time to next job (days)[a]	148.2	129.1	147.4	131.7
Duration of initial UI spell (days)[b]	77.6	67.0	79.8	70.3
Total days of UI received	1944.4	1707.0	2679.7	2480.1
Fraction exhausted benefits[c]	0.09	0.29	0.09	0.29
Fraction with time to next job censored	0.19	0.39	0.20	0.40
Fraction eligible for extended benefits	0.78	0.41	0.90	0.29
Fraction female	0.38	0.48	0.21	0.41
Mean age	33.06	8.48	36.04	7.45
Fraction Austrian nationals	0.80	0.40	0.86	0.35
Fraction married	0.34	0.47	0.40	0.49
Fraction blue-collar occupation	0.65	0.48	0.56	0.50
Fraction with higher education	0.11	0.31	0.18	0.38
Fraction in Vienna	0.19	0.39	0.21	0.41
Tenure in most recent job (Years)	3.39	3.27	4.23	4.20
Recalled to last job	0.20	0.40	0.25	0.44
Industry				
Construction	0.13	0.34	0.23	0.42
Manufacturing	0.20	0.40	0.23	0.42
Trade	0.20	0.40	0.15	0.35
Services	0.26	0.44	0.23	0.42
Number of observations	275,293		275,665	

[a]Time to next job is censored at 365 days.
[b]Claim duration is censored at 39 weeks (maximum entitlement).
[c]Indicator equals 1 if claim duration = maximum entitlement.
Notes: Sample contains UI claimants under the age of 50 with claims in 2001–2012, who had at least 1 year of tenure on their previous job, began their claim within 4 weeks of losing their past job, and had a valid UI claim record and non-missing earnings in the base period prior to the claim. Observations in the bottom kink sample have base period earnings in a range around the bottom kink in the UI benefit schedule; observations in the top kink sample have base period earnings in a range around top kink. See text.

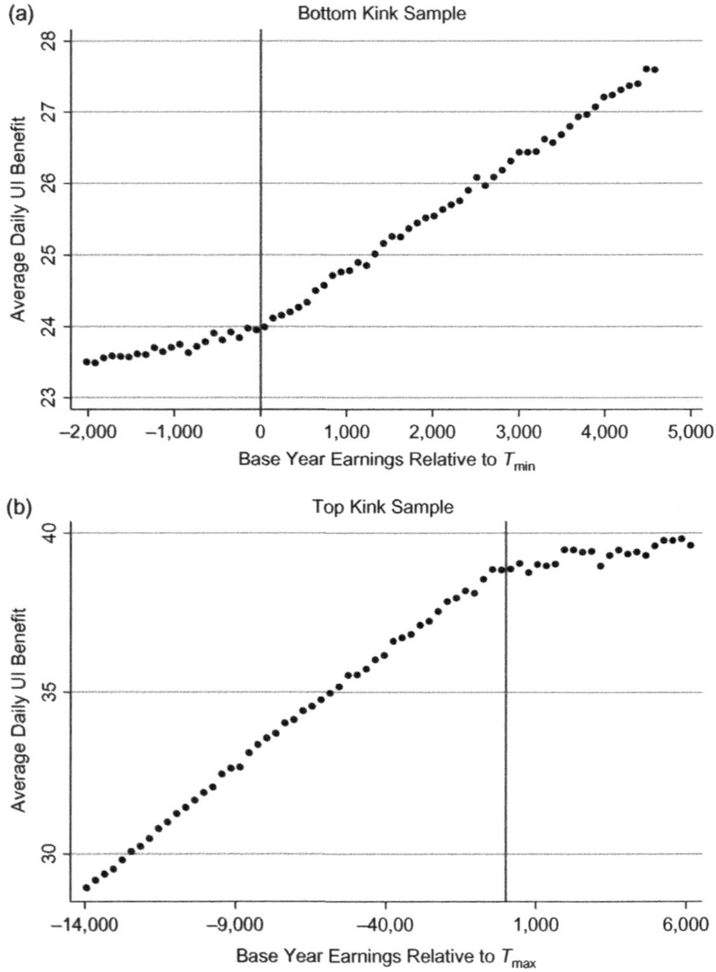

Fig. 2. Daily UI Benefits. (a) Bottom Kink Sample and (b) Top Kink Sample.

whereas it is kinked for all claimants at $T^{\max}$. As a result, the slope change at $T^{\max}$ appears sharper than that at $T^{\min}$.[13] Fig. 3(a) and (b) present analogous plots for our main outcome variable, the logarithm of time to the next job. These figures also show discernible kinks, though there is clearly more variability than in Fig. 2(a) and (b).

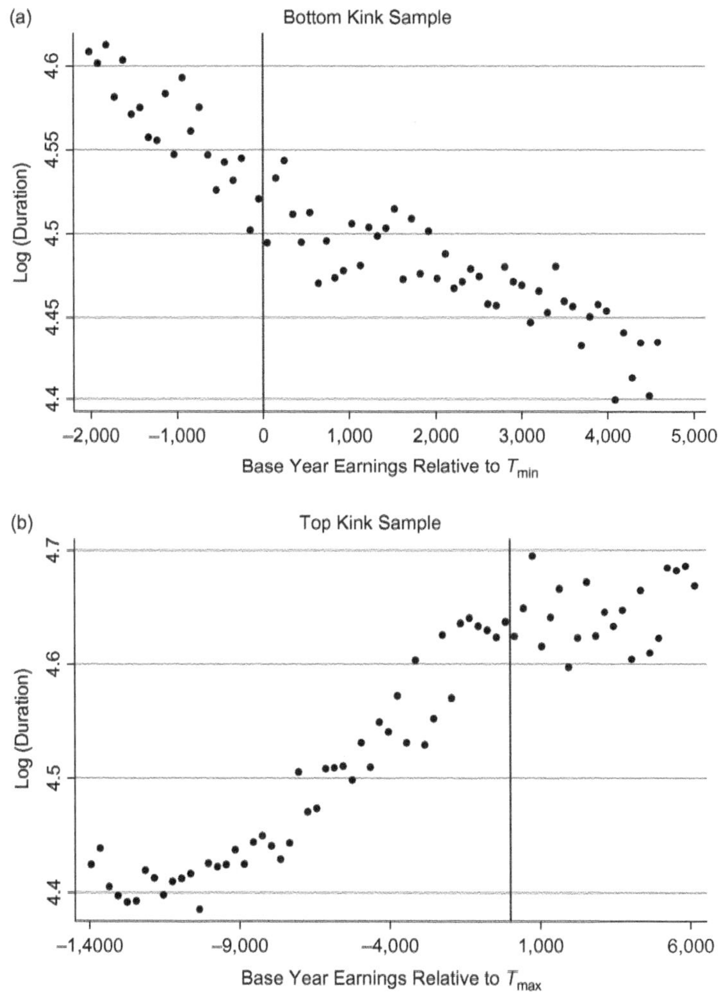

Fig. 3. Log Time to Next Job. (a) Bottom Kink Sample and (b) Top Kink Sample.

To check the validity of the kink design, we examine the patterns of the predetermined covariates around T^{min} and T^{max}. We do this by constructing a "covariate index" equal to the predicted value from a simple linear regression of the log of the time to next job on a total of 59 predetermined covariates, including gender, occupation, age, previous job tenure, quintile of the previous daily wage, industry, region, year of the claim, previous firm size, and the recall rates of the previous employer.[14] This estimated covariate index function can be interpreted as the best linear prediction of mean log time to next job given the vector of predetermined variables. Fig. 4(a) and (b) plot the mean values of the estimated covariate indices

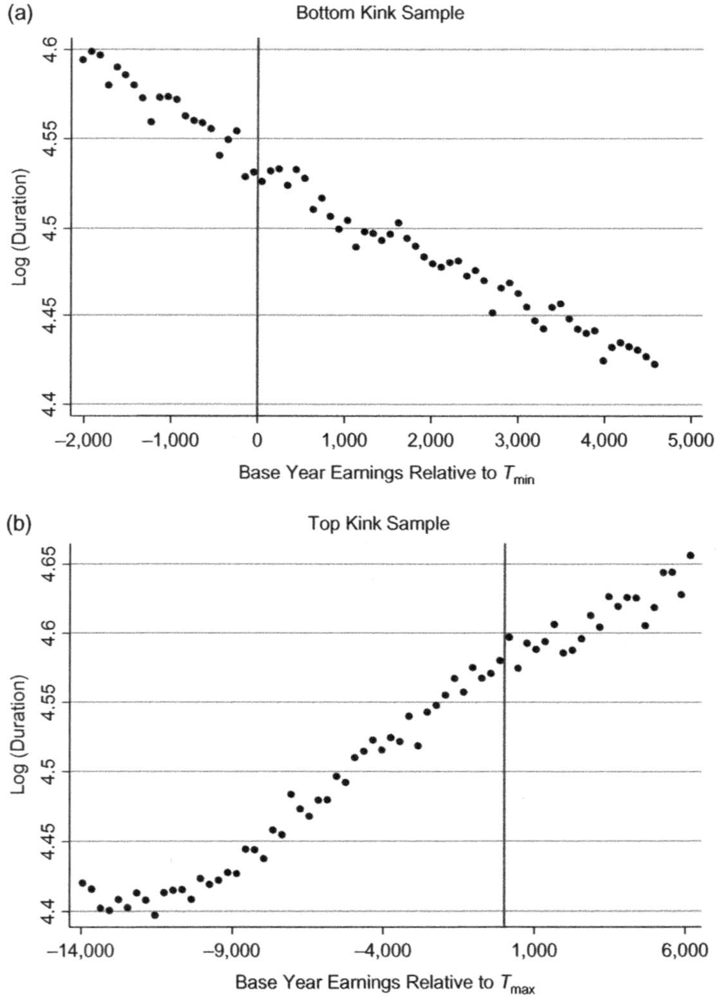

Fig. 4. Predicted Time to Next Job. (a) Bottom Kink Sample and (b) Top Kink Sample.

around the top and bottom kinks, and they evolve relatively smoothly through both the top and bottom kinks.

4.4. RKD Estimation Results

4.4.1. Reduced Form Kinks in Treatment and Outcome Variables

Table 2a presents the first stage and the reduced form estimates of the kinks in our endogenous policy variable (log daily benefits) and our main outcome variable (log of time to next job) around T^{min} and T^{max}. For each variable we show results using three different bandwidth selection procedures: the MSE-optimal choice with the regularization term included (i.e., the default procedure suggested by CCT procedure, updated using the procedures distributed by CCFT); the MSE-optimal bandwidth selection procedure without regularization; and the FG bandwidth. We show the estimated kink arising from each selected bandwidth, as well as the corresponding bias-corrected estimate and the associated robust 95% confidence interval. We present estimates from local linear models in columns 1 and 3, and from local quadratic models in columns 2 and 4.

Despite the visual evidence of kinks in the benefit formula in Fig. 2, especially in the top kink sample, an examination of the estimated "first stage" kinks in Panel A of Table 2a suggests that not all the procedures yield statistically significant kink estimates. In particular, the default MSE-optimal bandwidth selector with regularization chooses relatively small bandwidths for the local linear model and yields an insignificant estimate of the bottom kink ($t = 1.6$) and a marginally significant estimate of the top kink ($t = -2.4$). The corresponding bias-corrected kink estimates are substantially less precise, with confidence intervals that are about 3 times wider. Although the default MSE-optimal procedure chooses somewhat larger bandwidths for the local quadratic models, this potential gain in efficiency is offset by the difficulty of precisely estimating the slopes on either side of the kink point once the quadratic terms are included, and neither the estimated bottom kink nor the estimated top kink in the quadratic models is close to being significant. As with the local linear models, the corresponding bias-corrected quadratic kink estimates are even less precise, with very wide confidence intervals.

Relative to the default MSE-optimal bandwidth selector, the MSE-optimal selector *without regularization* yields much larger bandwidths – 3–5 times larger for the local linear models, and 1.3–2.7 times larger for

Table 2a. First Stage and Reduced Form Estimated Kinks.

	Bottom Kink		Top Kink	
	Local linear (1)	Local quad. (2)	Local linear (3)	Local quad. (4)
Panel A: First-stage model (dependent variable = log daily benefit)				
MSE optimal (with regularization)				
Main (pilot) bandwidth	461 (940)	818 (1,346)	1,400 (2,808)	1,571 (2,619)
Estimated kink	1.5	1.9	−1.6	−1.1
(conventional std. error)	(0.9)	(1.5)	(0.7)	(2.2)
Estimated kink – bias corrected	1.1	1.1	−1.8	−2.2
[robust confidence interval]	[−1.6,3.9]	[−2.8,5.0]	[−3.8,0.2]	[−7.6,3.1]
MSE optimal with no regularization				
Main (pilot) bandwidth	2,388 (2,022)	2,243 (1,770)	5,189 (6,746)	2,034 (3,037)
Estimated kink	2.2	2.4	−1.5	−3.1
(conventional std. error)	(0.1)	(0.4)	(0.1)	(1.5)
Estimated kink – bias corrected	2.2	2.0	−1.4	−4.4
[robust confidence interval]	[1.4,3.0]	[−1.1,5.2]	[−1.9,−0.9]	[−8.5,−0.3]
FG				
Main (pilot) bandwidth	2,466 (2,380)	2,628 (4,146)	11,603 (6,343)	7,003 (13,081)
Estimated kink	2.2	2.2	−1.9	−1.4
(conventional std. error)	(0.1)	(0.3)	(0.4)	(0.2)
Estimated kink – bias corrected	2.4	1.9	−1.4	−0.8
[robust confidence interval]	[1.7,3.1]	[0.4,3.4]	[−2.1,−0.7]	[−1.9,0.3]

Panel B: Reduced form model (dependent variable = log time to next job)

MSE optimal (with regularization)				
Main (pilot) bandwidth	497 (983)	599 (987)	1,725 (3,574)	2,571 (4,041)
Estimated kink	6.4	30.7	1.4	6.9
(conventional std. error)	(7.5)	(22.8)	(3.0)	(6.6)
Estimated kink – bias corrected	6.1	42.5	3.2	8.7
[robust confidence interval]	[−17.1,29.3]	[−15.9,101]	[−5.6,12.0]	[−9.0,26.5]
MSE optimal with no regularization				
Main (pilot) bandwidth	1,866 (1,097)	1,418 (1,647)	3,208 (9,972)	3,717 (5,191)
Estimated kink	3.3	6.1	−3.0	−1.8
(conventional std. error)	(1.0)	(6.3)	(1.2)	(3.8)
Estimated kink – bias corrected	4.5	11.6	−4.5	1.6
[robust confidence interval]	[−25.2,34.2]	[−13.3,36.5]	[−7.9,-1.1]	[−10.1,13.2]
FG				
Main (pilot) bandwidth	2,501 (3,858)	4,259 (4,918)	4,465 (7,255)	8,001 (12,383)
Estimated kink	3.0	4.8	−3.0	−4.1
(conventional std. error)	(0.8)	(2.4)	(0.7)	(1.5)
Estimated kink – bias corrected	4.2	9.9	−5.2	0.0
[robust confidence interval]	[−1.4,9.8]	[−3.6,23.4]	[−8.4,−1.9]	[−7.0,6.9]

Notes: Estimated kinks, conventional standard errors, and robust confidence intervals are all multiplied by 10. Point estimates and standard errors are obtained from regressions described in Card et al. (2012). The MSE optimal bandwidths and the Robust C.I.'s are obtained by the Stata package "rdrobust" by Calonico et al. (2016).

Table 2b. Bias and Variance Tradeoff – Conventional versus Bias-Corrected Estimators.

	Bottom Kink		Top Kink	
	Local linear (1)	Local quad. (2)	Local linear (3)	Local quad. (4)
Panel A: First-stage model (dependent variable = log daily benefit)				
Default CCT (with regularization)				
Estimated bias	0.4	0.8	0.2	1.1
Estimated bias squared	0.2	0.7	0.0	1.3
Change in estimated variance	1.1	1.6	0.6	2.8
CCT with no regularization				
Estimated bias	0.0	0.4	−0.1	1.2
Estimated bias squared	0.0	0.1	0.0	1.5
Change in estimated variance	0.2	2.5	0.1	2.2
FG				
Estimated bias	−0.1	0.3	−0.2	−0.5
Estimated bias squared	0.0	0.1	0.0	0.3
Change in estimated variance	0.1	0.5	0.1	0.2
Panel B: Reduced Form Model (dependent variable = log time to next job)				
Default CCT (with regularization)				
Estimated bias	0.2	−11.8	−1.8	−1.8
Estimated bias squared	0.0	139.5	3.2	3.4
Change in estimated variance	83.5	367.5	11.1	38.0
CCT with no regularization				
Estimated bias	−1.2	−5.4	1.5	−3.4
Estimated bias squared	1.4	29.7	2.1	11.6
Change in estimated variance	228.4	122.1	1.5	21.0
FG				
Estimated bias	−1.3	−5.5	2.2	−4.4
Estimated bias squared	1.7	30.3	4.6	18.9
Change in estimated variance	7.5	39.6	2.3	10.0

Notes: Estimated bias is the difference between the value of the conventional linear estimator and its bias-corrected counterpart, which is the center of the robust confidence interval. The change in estimated variance is the difference between the variance of the bias corrected estimator and its conventional counterpart. Estimated bias is multiplied by 10^5, and estimated bias squared and change in estimated variance are both multiplied by 10^{10}.

the local quadratic models. These larger bandwidths lead to first-stage kink estimates from the local linear models that are relatively precise (with t-ratios in excess of 15 in absolute value). The estimated kinks from the local quadratic estimates, however, are still relatively noisy ($t = 6.5$ for the bottom kink and $t = -2.1$ for the top kink). As is the case when the regularization term is included, the bias correction adds substantial imprecision, leading to confidence intervals that are 8 times wider for the bottom kink estimates, and 3–5 times wider for the top kink estimates.

The third bandwidth selection procedure – the simple rule of thumb plug-in approach of FG – yields bandwidths for the lower kink that are very close to those selected by the MSE-optimal procedure without regularization, but 2–3 times bigger for the upper kink. Without accounting for bias correction, the implied kink estimates are relatively precise, even using local quadratic polynomials. However, the bias term is imprecisely estimated and the bias-corrected FG estimates have confidence intervals that are relatively wide.

Turning to the reduced form outcome results in Panel B, the estimated kinks in the duration of joblessness are less precise than the first-stage kinks in UI benefits. Again, the default MSE-optimal selector with regularization chooses relatively small bandwidths and yields imprecise estimates of the kink. The bandwidths under the MSE-optimal procedure without regularization are substantially larger (3–4 times larger for the bottom kink, and 1.5–1.8 times larger for the top kink) and yield marginally significant estimated kinks in the outcome variable from the local linear models (with t-ratios in excess of 2.5 in absolute value). The FG bandwidths are even larger, and yield estimated kinks that are significant or marginally significant in both the linear and quadratic models. Nevertheless the bias-corrected confidence intervals for all three procedures are relatively wide and include 0 in all but two cases, reflecting the additional uncertainty associated with the bias correction term.[15] In finite samples, the use of higher-order polynomial models and bias correction may come at the cost of an increase in variance relative to lower-order uncorrected models. In the remainder of this subsection we examine the tradeoff between bias and variance associated with the CCT bias correction procedure. We defer a discussion of the polynomial order choice to Section 4.4.3.

The intent of the bias correction is to eliminate the bias in the p-th-order polynomial estimator $\hat{\tau}_p$ by subtracting the estimated asymptotic bias, $h^p\hat{\rho}_p$, where the constant ρ_p in part depends on the $(p+1)$th derivative of $E[Y|V=v]$ and $E[B|V=v]$. The cost of bias correction, however, is that the bias term is imprecisely estimated, leading to an increase in the overall variance of the corrected estimator $\hat{\tau}_p^{bc}$ relative to the uncorrected estimator $\hat{\tau}_p$.[16] The usual metric for trading off bias and variance is the (asymptotic) MSE, or AMSE, of the estimator, which is the sum of its squared bias and its variance. By Lemma A1 and Theorem A1 of CCT, the AMSE of $\hat{\tau}_p$ and $\hat{\tau}_{p+1}^{bc}$ are $\text{AMSE}(\hat{\tau}_p) = (h^p \rho_p)^2 + o_p(h^{2p}) + \text{var}(\hat{\tau}_p)$ and $\text{AMSE}(\hat{\tau}_p^{bc}) = o_p(h^{2p}) + \text{var}_p^{bc}$. It follows that the *change* in the AMSE associated with bias correction is asymptotically $-(h^p \rho_p)^2 + \text{var}_p^{bc} - \text{var}(\hat{\tau}_p)$.

In Table 2b, we report the estimated bias $h^p \hat{\rho}_p$, its square, and the change in estimated variance, $\widehat{\text{var}}_p^{bc} - \widehat{\text{var}}(\hat{\tau}_p)$, for the first stage and reduced form estimators presented in Table 2a. The increase in variance is larger than the estimated squared bias for both the local linear and local quadratic estimates using either the default CCT bandwidth selector or the alternative version that ignores the regularization term. This is also the case for the FG bandwidth in the bottom kink sample. In the top kink sample, however, the estimated bias for the FG bandwidth is relatively large (−0.5), and bias correction appears to decrease the AMSE. This suggests that bias correction could be important for estimators based on the FG bandwidth in the top kink sample. At the same time, since the bias term is estimated (and the estimate is relatively imprecise), it may deviate from the actual bias. In Section 4.4.3, we evaluate the performance of the various estimators in Monte Carlo simulations using DGP's approximating our data, where we can directly obtain the MSEs without having to estimate the bias.

4.4.2. Fuzzy RKD Estimates

In this subsection, we present fuzzy RKD estimates of the elasticity of the duration of joblessness with respect to the level of UI benefits. Building on the discussion in Sections 2 and 4.1, we interpret the fuzzy RK estimate in the bottom kink sample as an average behavioral response for claimants whose base period earnings are close to T^{min} and who follow the benefit schedule intended for single claimants. We interpret the fuzzy RK estimate in the top kink sample as the average elasticity for claimants whose base period earnings are close to T^{max} and who follow any of the benefit schedules seen in Fig. 1. Therefore, applying a regression kink approach to the two samples allows us to estimate the elasticity of joblessness with respect to UI benefit generosity for two very different subpopulations.

Table 3 presents estimated elasticities for the local linear and quadratic estimators and the three bandwidth selection procedures discussed above.

For reference, we also report the associated first-stage estimates for each of the structural estimates. As in Tables 2a and 2b, we show both the conventional estimates and the bias-corrected estimates with robust confidence intervals that take account of the sampling variability in bias correction.

A first observation is that the MSE-optimal procedure with regularization yields highly variable and imprecise estimates of the elasticity of the duration of joblessness with respect to UI benefits. The MSE-optimal procedure without regularization yields somewhat more precise estimates, particularly from the local linear specifications, though accounting for bias correction the confidence intervals include 0 for all specifications except the local linear estimate for the top kink. Again, as we saw in Tables 2a and 2b, the FG bandwidth selection procedure gives rise to estimates that are not different from those obtained using the MSE-optimal selector without the regularization term, but the bias-corrected FG procedure yields relatively imprecise estimates.

Given the wide range of estimates in Table 3, which specification should we pick? The most conservative choice is the default CCFT procedure, which uses a richer model (local quadratic), bias correction, and small bandwidths in our application. For the bottom kink sample, this choice suggests that the elasticity of joblessness with respect to UI benefits is 18.6, whereas for the top kink sample the estimate is wrong-signed and equal to −3.9. However, given the wide confidence intervals both estimates are essentially uninformative. At the opposite extreme, the uncorrected estimates from local linear specifications using the FG bandwidth selector are 1.4 for the bottom kink sample (conventional standard error = 0.4) and 2.4 for the top kink sample (conventional standard error = 1.2). In both cases, the estimated bias correction term suggests that the uncorrected estimates are "too small" in magnitude, though the bias correction terms are imprecise.

4.4.3. Comparison of Alternative Estimators

To gain additional insight into the performance of the alternative estimators, we conduct a series of Monte Carlo simulations based on DGP's that closely resemble our actual samples, in the same spirit as the IK and CCT Monte Carlo simulations. Distinct from IK and CCT, however, we impose the first-stage kink parameter and the elasticity parameter in constructing the DGP's because we are interested in the power of the candidate estimators. For the bottom kink sample, we impose the first-stage kink as $\tau_B = 2.3\times10^{-5}$ and the elasticity $\tau_{\text{FRKD}} = 1.3$, which implies a reduced-form kink of $\tau_Y = \tau_B{\cdot}\tau_{\text{FRKD}} = 3.0\times10^{-5}$. For the top kink sample, we set $\tau_B = -1.4\times10^{-5}$ and $\tau_{\text{FRKD}} = 2.0$ with an implied $\tau_Y = -2.8\times10^{-5}$.

Table 3. Estimates of Benefit Elasticity (Fuzzy RK), Alternative Estimators and Bandwidths.

	Bottom Kink				Top Kink			
	Local linear		Local quadratic		Local linear		Local quadratic	
	First stage (Coeff $\times 10^5$) (1)	Struct. model (2)	First stage (Coeff $\times 10^5$) (3)	Struct. model (4)	First stage (Coeff $\times 10^5$) (5)	Struct. model (6)	First stage (Coeff $\times 10^5$) (7)	Struct. model (8)
MSE optimal (with regularization)								
Main (pilot) bandwidth	548 (1,094)		619 (992)		1,678 (992)		2,591 (4060)	
Estimated kink	1.7	3.5	0.9	31.0	−1.9	−0.7	−1.6	−3.1
(conventional std. error)	(0.7)	(4.1)	(2.3)	(81.7)	(0.5)	(1.7)	(1.0)	(4.6)
Estimated kink – bias corrected	1.5	3.9	1.7	18.6	−1.5	−1.8	−1.9	−3.9
[robust conf. interval]	[−0.6,3.6]	[−8.7,16.43]	[−4.4,7.8]	[−196.3,233.4]	[−2.9,−0.1]	[−6.6,3.0]	[−4.6,0.8]	[−16.2,8.4]
MSE optimal with no regularization								
Main bandwidth (pilot)	1,414 (1,614)		1,294 (1,062)		4,893 (6,921)		7,174 (6,233)	
Estimated kink	2.2	0.7	2.3	−0.1	−1.5	2.1	−1.4	4.0
(conventional std. error)	(0.2)	(0.7)	(0.8)	(3.2)	(0.1)	(0.4)	(0.2)	(1.4)
Bias-corrected estimate	1.9	0.0	0.7	17.4	−1.5	3.6	−0.8	−0.6
[robust conf. interval]	[0.8,3.0]	[−4.6,4.6]	[−6.4,7.8]	[−11.6,46.4]	[−2.0,−1.0]	[1.2,5.9]	[−2.3,0.6]	[−9.1,7.9]
FG								
Main bandwidth (pilot)	2,501 (3,858)		4,259 (4,918)		4,465 (7,255)		8,011 (12,383)	
Estimated kink	2.2	1.4	2.0	2.0	−1.4	2.4	−1.4	2.7
(conventional std. error)	(0.1)	(0.4)	(0.3)	(0.5)	(0.1)	(1.2)	(0.2)	(1.1)
Bias-corrected estimate	2.1	2.0	1.9	5.0	−1.0	3.7	−1.1	1.2
[robust conf. interval]	[1.5, 2.7]	[−0.6,4.5]	[0.4, 3.4]	[−1.9,11.9]	[−1.9,−0.1]	[1.2,6.1]	[−2.0, 0.2]	[−4.5,6.9]

Notes: Point estimates and standard errors are obtained from 2SLS regressions described in Card et al. (2012). The MSE optimal bandwidths and the Robust C.I.'s are obtained by the Stata package "rdrobust" by Calonico et al. (2016). The fuzzy FG bandwidths are authors' calculations.

As with the Monte Carlo simulations in IK and CCT, we specify the DGP's for $E[B|V]$ and $E[Y|V]$ as separate quintics on each side of the threshold. The parameters of the quintics are estimated by regressing, respectively, $B - \tau_B \cdot D \cdot V$ and $Y - \tau_Y \cdot D \cdot V$ on the polynomial terms V^j and $D \cdot V^k$ where $D = 1_{[V \geq 0]}$, $j = 0, 1, \ldots, 5$ and $k = 2, \ldots, 5$.[17] For our simulation, we sample V from its empirical distribution and the errors (ε_B, ε_Y) jointly from the residuals from the quintic regressions, and construct $B = E[B|V] + \varepsilon_B$ and $Y = E[Y|V] + \varepsilon_Y$. We draw 1,000 repeated samples in our Monte Carlo exercise; each simulated sample contains the same number of observations as in the original dataset.

Panels A and B of Table 4 summarize the performance of the alternative estimators for the simulated bottom kink and top kink samples, respectively. The two polynomial orders (linear and quadratic), two bias-correction choices (uncorrected and bias-corrected) and four bandwidth selection procedures (MSE-optimal, MSE-optimal with no regularization, Fuzzy MSE-optimal with no regularization, and FG) give rise to 16 candidate estimators in total. For each estimator, we report the average value of the associated bandwidth(s) and its performance in estimating the first-stage kink (columns 3–5) and the structural elasticity parameter (columns 6–11).

With respect to the first-stage kink estimates, we show 3 statistics: the fraction of replications in which the first stage would be declared "insignificant" (i.e., the confidence interval for the first-stage kink estimate includes 0); the root-mean-squared error (RMSE) of the estimator (expressed as a fraction of the true value of the first-stage kink); and the coverage rate of the confidence interval. With respect to the structural elasticity estimate we show the RMSE (with and without trimming the 5% of the simulation sample with the greatest deviation between $\hat{\tau}_{\text{FRKD}}$ and τ_{FRKD}), the confidence interval coverage rate, the bias and squared bias of the estimator, and the variance of the parameter estimate.

A comparison across the rows of Table 4 reveals several interesting patterns about the performance of the various estimators in simulated datasets designed to mimic our actual data. First, for a given bandwidth procedure and bias correction approach, local quadratic models have higher average RMSE's in both the first stage kink estimates and the structural elasticity estimates. Typically, the quadratic model has two to 3 times larger RMSE than the corresponding linear model. Moreover, the mean-squared bias associated with the quadratic estimator is typically larger than the mean-squared bias for the linear estimator with the same bandwidth selection and bias correction.[18] A second consistent pattern is that the introduction

Table 4. Summary of Monte Carlo Studies: DGP with Kink.

	First-Stage Model Estimation Summary					Elasticity Estimation Summary					
	Median main b.w. (1)	Median pilot b.w. (2)	Fraction of Replications: C.I. includes 0 (3)	RMSE/ true value (trimmed) (4)	C.I. Coverage Rate (5)	RMSE/true value (6)	RMSE/ true value (trimmed) (7)	C.I. Coverage Rate (8)	Bias (trimmed) (9)	Bias^2 (trimmed) (10)	Variance (trimmed) (11)
Panel A: DGP design based on bottom kink sample											
A.1. Local linear, no bias correction											
MSE-optimal	456	–	0.34	0.42	0.94	35.9	4.13	1.00	1.90	3.61	25.3
MSE-opt, no reg	878	–	0.05	0.20	0.90	5.8	1.82	0.95	1.01	1.03	4.55
Fuzzy MSE-opt, no reg	943	–	0.04	0.21	0.90	7.45	1.47	0.95	0.74	0.55	3.10
FG	2,437	–	0.00	0.07	0.66	0.28	0.24	0.96	0.11	0.01	0.08
A.2. Local linear, bias-corrected											
MSE-optimal	456	947	0.63	0.63	0.94	3778	7.28	1.00	3.31	11.0	78.6
MSE-opt, no reg	878	1,223	0.38	0.47	0.94	56.7	3.71	0.96	2.01	4.04	19.3
Fuzzy MSE-opt, no reg	943	1,317	0.34	0.48	0.92	496.0	3.53	0.95	2.03	4.11	17.0
FG	2,437	3,795	0.00	0.16	0.89	1.22	1.07	0.90	0.39	0.16	1.77
A.3. Local quadratic, no bias correction											
MSE-optimal	676	–	0.79	0.94	0.95	99	10.4	1.00	1.24	1.54	181
MSE-opt, no reg	1,186	–	0.38	0.53	0.93	183.9	6.15	0.99	2.53	6.40	57.5
Fuzzy MSE-opt, no reg	1,261	–	0.32	0.50	0.91	23.7	4.58	0.98	2.46	6.04	29.5
FG	4,190	–	0.00	0.15	0.84	1.48	1.32	0.79	0.99	0.98	1.98
A.4. Local quadratic, bias-corrected											
MSE-optimal	676	1,076	0.85	1.26	0.94	2.05E+08	17.7	1.00	2.26	5.11	527
MSE-opt, no reg	1,186	1,373	0.77	1.14	0.93	9188	11.7	0.99	3.21	10.28	222
Fuzzy MSE-opt, no reg	1,261	1,450	0.75	1.05	0.93	2.07E+06	10.9	0.97	3.65	13.3	188
FG	4,190	4,914	0.25	0.36	0.91	4.46	3.93	0.77	3.88	15.0	11.1

Panel B: DGP design based on top kink sample

B.1. Local linear, no bias correction											
MSE-optimal	1,402	–	0.60	0.59	0.79	57.6	3.88	0.99	0.36	0.13	60.1
MSE-opt, no reg	2,757	–	0.11	0.32	0.72	22.84	1.08	0.92	0.23	0.05	4.66
Fuzzy MSE-opt, no reg	2,708	–	0.15	0.37	0.72	7.9	1.11	0.93	0.31	0.10	4.9
FG	4,464	–	0.00	0.15	0.79	0.49	0.36	0.84	0.11	0.01	0.51
B.2. Local linear, bias-corrected											
MSE-optimal	1,402	2,871	0.87	0.90	0.77	2777	10.6	0.99	2.16	4.68	446
MSE-opt, no reg	2,757	3,779	0.65	0.77	0.65	44280	2.73	0.93	0.62	0.39	29.3
Fuzzy MSE-opt, no reg	2,708	3,856	0.62	0.76	0.69	1092	3.01	0.92	0.91	0.82	36
FG	4,464	7,256	0.01	0.19	0.89	1.21	0.96	0.81	1.52	2.32	1.40
B.3. Local quadratic, no bias correction											
MSE-optimal	2,024	–	0.89	1.14	0.87	96.3	8.05	0.99	−1.37	1.89	258
MSE-opt, no reg	3,481	–	0.65	0.76	0.74	209.9	7.50	0.97	−0.09	0.01	225
Fuzzy MSE-opt, no reg	3,551	–	0.63	0.74	0.71	66.3	5.28	0.96	0.09	0.01	112
FG	8,012	–	0.00	0.15	0.92	1.09	0.94	0.82	1.61	2.60	0.93
B.4. Local quadratic, bias-corrected											
MSE-optimal	2,024	3,188	0.91	1.43	0.90	3776	13.6	1.00	−0.27	0.07	743
MSE-opt, no reg	3,481	4,064	0.93	1.41	0.79	1.12E+05	30.4	0.98	4.74	22.44	3686
Fuzzy MSE-opt, no reg	3,551	4,302	0.92	1.99	0.76	1.40E+05	23.6	0.97	0.26	0.1	2228
FG	8,012	12,409	0.70	0.64	0.61	2.05	1.55	0.93	0.54	0.30	9.3

Notes: Based on 1,000 simulations. In Panel A, DGP is based on fifth-order polynomial approximation of bottom kink sample. True kink in first stage is: 2.3 $\times 10^{-5}$; true elasticity is: 1.3. In Panel B, DGP is based on fifth-order polynomial approximation of top kink sample. True kink in first stage is: -1.4×10^{-5}; true elasticity is: 2.0. The trimmed statistic are obtained by first trimming the 5% sample in which the estimates deviate the most from the true parameter value.

of bias correction leads to higher RMSE's for both linear and quadratic models and all four bandwidth selection procedures, and larger mean-squared bias in every case but two (the exceptions are the quadratic model for the top kink sample using either the MSE-optimal bandwidth with regularization or the FG bandwidth). Typically bias correction doubles the RMSE of the kink estimates and leads to a 200–300% increase in the RMSE of the structural elasticity estimates. A third finding is that the inclusion of the regularization term in the calculation of the MSE-optimal bandwidth results in higher RMSE's for both the first-stage kink estimates and (in all but one case) also leads to higher RMSE's for the structural elasticity estimate.[19] Finally, comparing across bandwidth procedures for a given polynomial choice and a given choice over bias correction or not, the FG bandwidth has uniformly the lowest RMSE for both the first stage and structural estimates. Indeed, only the FG procedure delivers structural estimates with an RMSE relative to the true value of the parameter that is less than 0.5. The other bandwidth procedures have proportional RMSE's in excess of 1 for the structural elasticity – a degree of variability which suggests that none of these procedures can provide useful information from our sample.[20]

Although simple comparisons of RMSE suggest that in our application a local linear specification using the FG procedure and no bias correction may be a plausible approach, in the top kink sample the confidence intervals for this estimator are too small, leading to an 11 percentage point under-coverage rate. Thus, we conclude that in these DGP's none of the procedures achieves both lowest RMSE and correct coverage rates.

In a parallel simulation study, we assess the performance of the estimators in DGP's where we impose the true elasticity τ_{FRKD} to be zero: the results are summarized in Table 5. In terms of RMSE, the results show that local linear specifications still dominate quadratic models even when there is no kink in the outcome variable; estimators without bias correction perform better than the corresponding bias-corrected estimators; the MSE-optimal bandwidth selector without regularization performs better than the selector with the regularization term included (with the exception of the quadratic model with bias correction for the top kink sample); and the FG bandwidth selection procedure yields the lowest RMSE among the alternative bandwidth selectors. In terms of coverage rates, however, local linear models using the FG selection procedure without bias correction perform quite poorly: the coverage rate is 2% for the bottom kink simulation and 19% for the top. The bias corrected local linear model with FG bandwidth selection has better coverage rates, but still below 90%.

Table 5. Summary of Monte Carlo Studies: DGP without Kink.

	First-Stage Model Estimation Summary					Elasticity Estimation Summary					
	Median main b.w. (1)	Median pilot b.w. (2)	Fraction of replications: C. I. includes 0 (3)	RMSE/true value (trimmed) (4)	C.I. coverage rate (5)	RMSE (6)	RMSE (trimmed) (7)	C.I. coverage rate (8)	Bias (trimmed) (9)	$Bias^2$ (trimmed) (10)	Variance (trimmed) (11)
Panel A: DGP design based on bottom kink sample											
A.1. Local linear, no bias correction											
MSE-optimal	464	–	0.33	0.41	0.94	24.3	4.46	1.00	0.35	0.12	19.8
MSE-opt, no reg	1,015	–	0.03	0.18	0.89	5.3	1.83	0.75	0.67	0.45	2.88
Fuzzy MSE-opt, no reg	976	–	0.03	0.20	0.88	135.9	1.71	0.78	0.68	0.46	2.48
FG	2,602	–	0.00	0.08	0.59	1.40	1.36	0.02	1.33	1.77	0.09
A.2. Local linear, bias-corrected											
MSE-optimal	464	959	0.63	0.62	0.94	1050.03	7.5	1.00	0.35	0.12	56.6
MSE-opt, no reg	1,015	1,323	0.34	0.46	0.93	15.11	3.60	0.97	0.29	0.09	12.9
Fuzzy MSE-opt, no reg	976	1,371	0.32	0.47	0.93	16242.67	3.72	0.94	0.25	0.06	13.8
FG	2,602	3,364	0.00	0.16	0.89	1.91	1.70	0.83	0.80	0.64	2.24
A.3. Local quadratic, no bias correction											
MSE-optimal	683	–	0.79	0.94	0.94	168	14.6	1.00	−0.02	0.00	213
MSE-opt, no reg	1,223	–	0.36	0.49	0.92	305	6.4	0.98	0.27	0.07	41.4
Fuzzy MSE-opt, no reg	1,261	–	0.33	0.50	0.92	71	5.0	0.98	0.26	0.07	25.2
FG	3,714	–	0.00	0.14	0.88	2.13	1.91	0.71	1.14	1.29	2.34
A.4. Local quadratic, bias-corrected											
MSE-optimal	683	1,087	0.85	1.26	0.94	51580.95	25.0	1.00	0.61	0.37	625
MSE-opt, no reg	1,223	1,401	0.76	1.17	0.94	4.33E+07	13.0	1.00	0.22	0.05	168
Fuzzy MSE-opt, no reg	1,261	1,444	0.76	1.05	0.94	26966.60	12.0	0.99	−0.16	0.03	145
FG	3,714	4,947	0.24	0.36	0.90	3.60	3.07	0.95	0.06	0.00	9.45

Table 5. (*Continued*)

	First-Stage Model Estimation Summary					Elasticity Estimation Summary					
	Median main b.w. (1)	Median pilot b.w. (2)	Fraction of replications: C. I. includes 0 (3)	RMSE/true value (trimmed) (4)	C.I. coverage rate (5)	RMSE (6)	RMSE (trimmed) (7)	C.I. coverage rate (8)	Bias (trimmed) (9)	Bias^2 (trimmed) (10)	Variance (trimmed) (11)
Panel B. DGP design based on top kink sample											
B.1. Local linear, no bias correction											
MSE-optimal	1,405	–	0.60	0.60	0.77	38.0	7.34	1.00	0.79	0.62	53.3
MSE-opt, no reg	2,998	–	0.10	0.31	0.71	25.3	2.18	0.66	0.98	0.97	3.79
Fuzzy MSE-opt, no reg	2,798	–	0.13	0.35	0.74	14.8	2.21	0.71	1.03	1.07	3.8
FG	5,621	–	0.00	0.18	0.66	1.31	1.23	0.19	1.12	1.25	0.25
B.2. Local linear, bias-corrected											
MSE-optimal	1,405	2,877	0.86	0.91	0.76	5.17E+06	20.9	0.99	1.23	1.52	435
MSE-opt, no reg	2,998	3,951	0.61	0.73	0.70	2883.03	4.87	0.93	0.86	0.73	23.0
Fuzzy MSE-opt, no reg	2,798	3,931	0.62	0.75	0.68	243.42	5.34	0.92	0.98	0.96	27.6
FG	5,621	9,306	0.00	0.17	0.90	2.02	1.84	0.65	1.54	2.37	1.02
B.3. Local quadratic, no bias correction											
MSE-optimal	2,037	–	0.88	1.10	0.87	102	17.6	1.00	0.06	0.00	309
MSE-opt, no reg	3,605	–	0.60	0.74	0.70	342	13.9	0.95	0.90	0.80	192
Fuzzy MSE-opt, no reg	3,634	–	0.61	0.74	0.70	257	9.1	0.94	0.67	0.45	82
FG	10,274	–	0.00	0.16	0.81	3.18	3.00	0.26	2.77	7.69	1.32
B.4. Local quadratic, bias-corrected											
MSE-optimal	2,037	3,219	0.91	1.40	0.89	2.8E+07	32.9	1.00	0.02	0.00	1085
MSE-opt, no reg	3,605	4,134	0.93	1.53	0.77	3.68E+05	44.2	0.99	1.15	1.33	1951
Fuzzy MSE-opt, no reg	3,634	4,335	0.92	1.66	0.76	20482.07	32.8	0.98	−0.07	0.01	1078
FG	10,274	13,142	0.70	0.66	0.57	3.65	3.02	0.92	0.82	0.67	8.44

Notes: Based on 1,000 simulations. In Panel A, DGP is based on fifth-order polynomial approximation of bottom kink sample. True kink in first stage is: 2.3 $\times 10^{-5}$; true elasticity is: 0. In Panel B, DGP is based on fifth-order polynomial approximation of top kink sample. True kink in first stage is: -1.4×10^{-5}; true elasticity is: 0. The trimmed statistic are obtained by first trimming the 5% sample in which the estimates deviate the most from the true parameter value.

In another exercise, we provide further evidence on the choice between linear and quadratic specifications in RKD applications by directly estimating the AMSE's for the local linear and quadratic estimators, per Card et al. (2014). In results available upon request, we find that using the MSE-optimal bandwidth selection procedure with regularization, the AMSE for the local quadratic model is at least an order of magnitude larger than the AMSE for the local linear model. For the other bandwidth selectors, the linear AMSE is also much smaller (at least 68% smaller). These results lead us to conclude that in our dataset, local linear models are likely to be preferred to local quadratic models.

As a final robustness check, we investigate how sensitive the elasticity estimates are with respect to the choice of bandwidth. Fig. 5(a) and (b) plot the local linear elasticity estimates for the bottom and top kink samples associated with a range of potential bandwidths. Ruppert (1997) argues that one can use the relationship between the point estimates and the bandwidth choice as an indicator of potential bias, with stability in the estimate indicating the absence of significant bias. Fig. 5(a) shows that the estimated elasticity of time to next job with respect to UI benefits around the bottom kink is relatively stable at close to 1.4 for a very wide range of bandwidths. Fig. 5(b) shows that the estimated elasticity around the top kink is a little more sensitive to bandwidth choice, with a larger estimate (between 2 and 3) for lower bandwidths, but an elasticity of 2 or less for bandwidths above €5,000.

4.4.4. Estimates with Covariates

As pointed out by Lee and Lemieux (2010) and Calonico et al. (2016), incorporating covariates may improve the precision of the estimates. In Table A.1 of the online Supplemental Material (Card et al., 2016), we present results with 79 covariates included by implementing the estimators in Calonico et al. (2016) and Calonico et al. (2016a). Compared to their unadjusted counterparts in Table 3, all but one of the covariate-adjusted MSE-optimal main bandwidths in Table A.1 are narrower, by a factor ranging from 6% to 39%. The exception is the local quadratic estimator for the bottom kink sample, for which the covariate-adjusted main bandwidth is 32% wider.

In terms of the precision of the RK estimates, the results are mixed. Covariate adjustment increases the conventional first stage t-statistic in exactly half of the cases. In two of the eight cases, the first stage robust confidence interval no longer covers 0 after covariate adjustment, but the adjustment also has the opposite effect in another case. For the structural elasticity estimator, covariate adjustment drastically reduces the standard

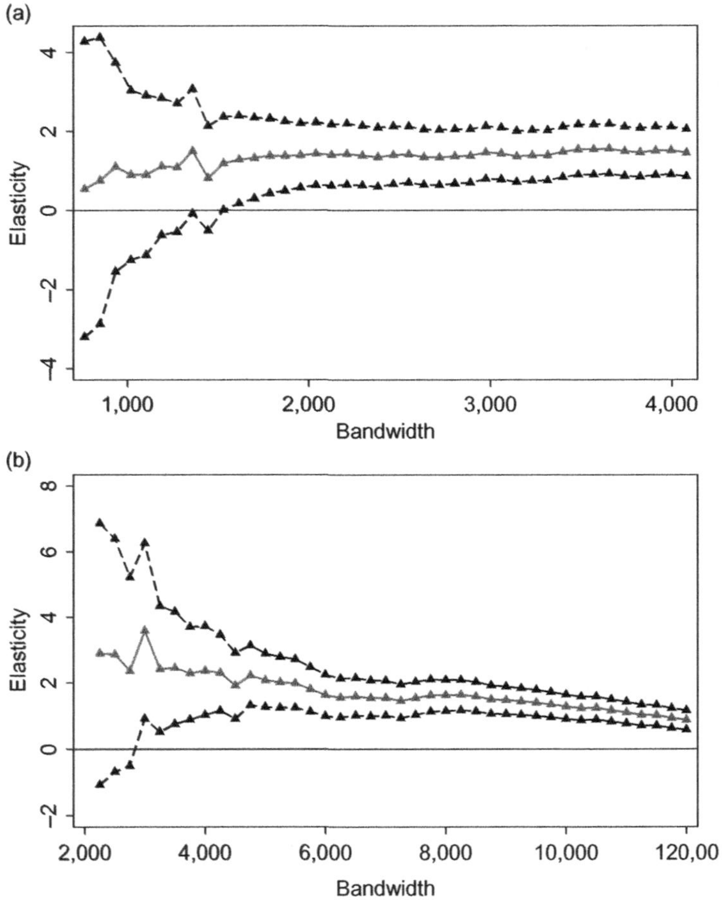

Fig. 5. Fuzzy RKD Estimation with Varying Bandwidth. Log Time to Next Job (a) Bottom Kink Sample and (b) Top Kink Sample. *Notes*: Local linear estimation, estimated coefficients (solid) with confidence bounds (dash).

error in the local-quadratic-bottom-kink case to 7% of its original level, as a result of the larger bandwidth. However, in all seven other cases, the conventional standard error actually *increases* after covariate adjustment by as much as 150%. Similarly, the width of the robust confidence interval increases in six out of eight cases after covariate adjustment.

Despite its inability to improve precision in the Austrian application, accounting for covariates may be important for other inference procedures, for example, the permutation test suggested by Ganong and Jäger (2014). As a starting point for our discussion, note that if we plot the relationship between baseline earnings and the log time to next job not just locally around the kink points as in Fig. 3(a) and (b), but globally for the whole range of baseline earnings available in the data, the relationship is highly nonlinear. The upper- and lower-left panels in Fig. 6 show the raw data plots, with observations centered at the annual cutoff values for the bottom and top-kink points in the unemployment benefit formula. Over the full range of baseline earnings, the relationship appears to follow a u-shape, and this general pattern renders the kinks at the policy thresholds less salient.

Ganong and Jäger (2014) raise concerns about the sensitivity of the RKD estimates when the relationship between the running variable and the outcome is highly nonlinear. To assess this sensitivity, they suggest a permutation-type test which shifts the threshold value for the kink across the support of the running variable and estimates slope changes in these placebo samples. One example that they use to demonstrate their method is based on the raw data plot for the bottom kink sample in Fig. 6. Ganong and Jäger (2014) argue that because of the curvature in the global relationships many of the placebo estimates result in large and significant slope changes.

We replicate this exercise based on the raw individual level data. Following Ganong and Jäger (2014), we randomly draw 200 cutoff values for both the bottom and top kink samples over the range of baseline earnings where there is no kink in the benefit formula. The empirical c.d.f.'s of the coefficient estimate and the corresponding *t*-statistic are shown by the solid lines in Fig. 7, where the graphs in the top row correspond to the bottom kink sample and those in the bottom row correspond to the top kink sample.[21] The c.d.f. of the placebo coefficient estimate is similar to Fig. 3 in Ganong and Jäger (2014), although the median value is somewhat lower in our sample. But the placebo estimates find a high share of coefficient estimates that are significantly different from zero. The same pattern appears for the top kink sample, although in this case the sign of the placebo estimates is the opposite of that of the reduced form RKD estimate in Table 2.

It is clear from visual inspection of Fig. 6 that the global curvature of the outcome variable is driving the high number of statistically significant placebo estimates. In the right panels of Fig. 6, we show that much of the curvature can be explained by the changing composition of workers over the distribution of baseline earnings. To do this, we regress log time to the next job on all observable characteristics on a 5% subsample over the full range of baseline earnings.[22] We then plot the residuals from predictions based on this regression for the remaining 95% of observations. The right panels of Fig. 6

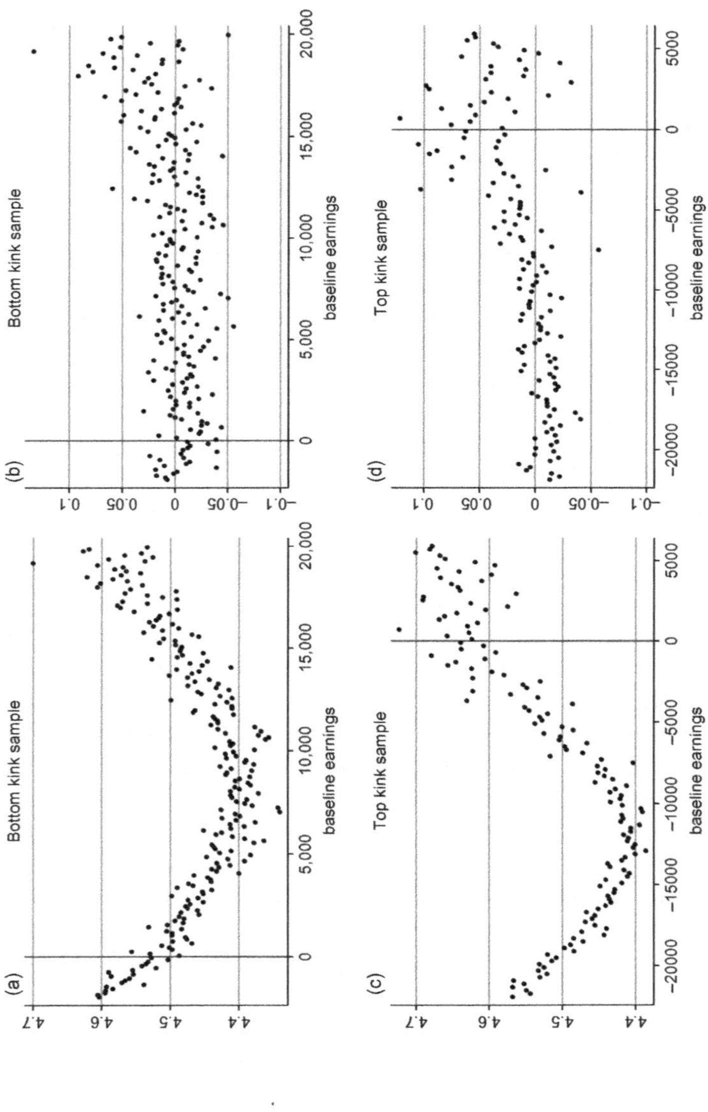

Fig. 6. Global Relationship between Log Time to Next Job and Baseline Earnings.

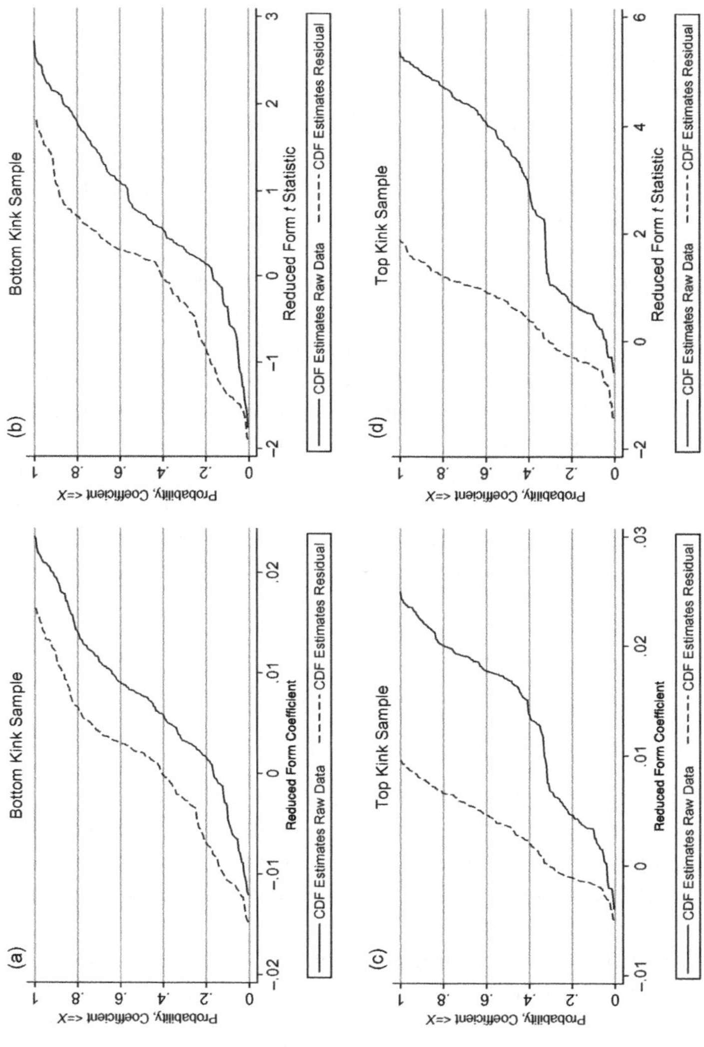

Fig. 7. Distribution of the Coefficient Estimate and *t*-Statistic in the Permutation Test.

show that most of the curvature in the global relationships can be explained by changes in observable characteristics. The residual plots are mostly flat along the range of baseline earnings, and the nonlinearity at the kink points becomes even more salient. We then repeat the permutation test on the residuals. The corresponding c.d.f.'s of the placebo coefficient estimate and t-statistic are plotted using the dashed lines in the graphs in Fig. 7. The placebo point estimates now have a high density around zero, and the absolute values of the t-statistics are below 2.

Our exercise highlights the importance of the role of curvature heterogeneity in the permutation test of Ganong and Jäger (2014). Specifically, the test may not be informative when the curvature of the conditional expectation function $\mu(v) \equiv E[Y|V = v]$ changes. Suppose $\mu(v)$ is locally piecewise linear around the threshold $v = 0$ with a non-zero kink but has substantial curvature far away from the threshold, as appears to be true in our data. In this case, the permutation test may fail to detect the kink at the threshold, as the curvature away from the threshold can lead to many significant local linear estimates. Based on a similar argument, the permutation test may also have poor size control when the curvature of $\mu(v)$ changes. Hence, if researchers wish to conduct the permutation test, it will be important to control for confounding nonlinearities by taking the distribution of observables into account.[23] We conclude that after controlling for covariates our design passes the Ganong and Jäger (2014) permutation test.

4.5. Comparison with the Missouri RKD Application in Card, Johnston, et al. (2015)

In this subsection, we compare our Austrian analysis with a related study by Card, Johnston, et al. (2015) – hereafter CJLMP. CJLPM are also interested in the effects of the UI benefit level on unemployment duration, and implement an RKD using variation from a kinked benefit schedule in the state of Missouri. As in Austria, the Missouri UI benefit schedule is a piecewise linear function of past earnings, except past earnings in this case are defined as the highest quarterly earnings in the first four of the preceding five calendar quarters. It is simpler than Austria's, however, as it exhibits only one kink and does not change with dependents.

CJLMP examine how responsiveness to UI benefit generosity varies before and after the Great Recession and conduct their RKD analysis separately using data from 2003 to 2007 and from 2008 to 2013. The data come from the Missouri Department of Labor and Industrial Relations, and the two analysis samples – "pre-" and "post-recession" contain 295,639 and 409,753 observations respectively. Fig. 8(a) and (b) reproduces the two first-stage plots from CJLPM over the 2 time periods, which show the average weekly UI benefit in each small bin of the normalized base period earnings. The sharpness of these figures, as compared to their Austrian counterparts in Fig. 2(a) and (b), are due to a much lower proportion of off-schedule observations in Missouri. In fact, Fig. 8(b) appears deterministic to the naked eye even though a small fraction of observations still deviate from the prescribed UI rule. Fig. 8(a) looks noisier to the right of the kink point, but this is due to the changing maximum UI benefit over the pre-recession period. As we will see below, the stronger first stage in Missouri yields both improved precision and stability in the empirical estimates.

Reproducing Fig. 2 of CJLMP, Fig. 9(a) and (b) depicts the relationship between the outcome, log UI claim duration, and base period earnings in Missouri. Both plots show pronounced slope changes at the threshold where the benefit cap starts to bind, and the kink is especially salient in the 2008–2013 sample. Even more compelling, Fig. A1 in the Supplemental Material (reproducing Fig. 5 of CJLMP) shows that when the benefit kink threshold increased over time, the UI duration kink had moved along with it.

Estimated benefit elasticities are shown in Table A2 in the Supplemental Material (based on Appendix Table 1 of CJLMP) using a variety of estimators.[24] Overall, the results point to a larger benefit elasticity in the post-recession period relative to the pre- period. Looking at the local linear estimates, which yielded the smallest MSE's in the simulations, the estimated elasticities for the pre-period center around 0.35, whereas the post-period estimates indicate that the elasticity is at least 0.68. Compared to Austria, both conventional and robust confidence intervals from Missouri are much narrower, thanks in part to the sharp first stage. Here again, however, we see that the local quadratic estimates tend to be less precise than local linear, yielding robust confidence intervals that cover or nearly cover zero for all bandwidth choices.

5. CONCLUSION

In this chapter, we discuss and highlight potential issues in the implementation of a fuzzy RKD approach by studying the effect of unemployment

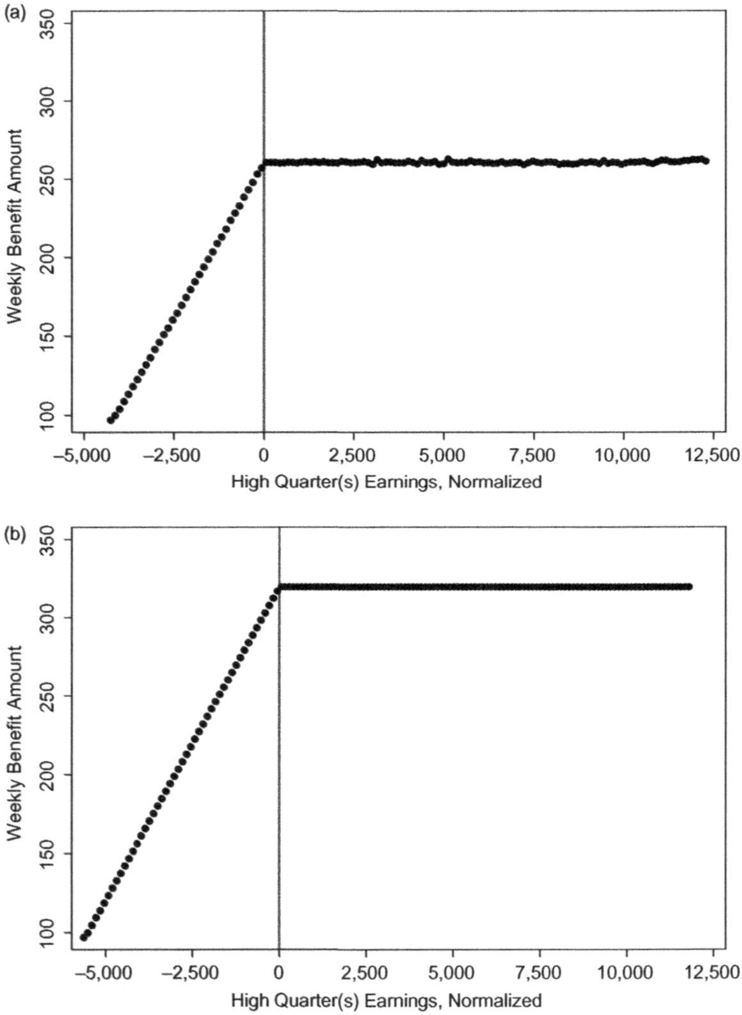

Fig. 8. Missouri First Stage (a) 2003–2007 and (b) 2008–2013. *Source*: Reproduced from Card, Johnston, et al. (2015).

benefits on the duration of joblessness in Austria, where the benefit schedule has kinks at the minimum and maximum benefit levels. We present graphical evidence that these benefit kinks are associated with kinks in the mean duration of total joblessness between the end of the previous job and the start of

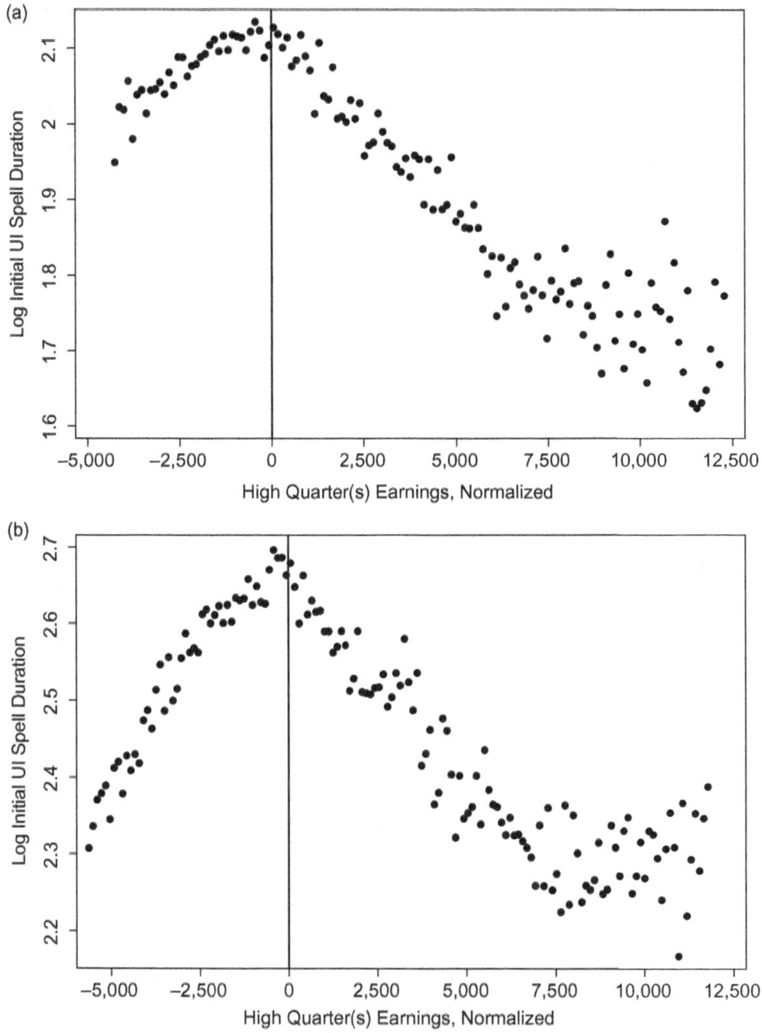

Fig. 9. Log Initial UI Spell Duration, Missouri (a) 2003–2007 and (b) 2008–2013. *Source*: Reproduced from Card, Johnston, et al. (2015).

the next job, as well as a number of tests of the RKD identifying assumptions. We show that in our particular application, estimates are sensitive to alternative estimators, and discuss ways to evaluate the empirical performance of these estimators using Monte Carlo simulations. We also discuss another application in Card, Johnston, et al. (2015) using data from Missouri, where the RKD produces more robust results across alternative specifications.

Through the exercises in this chapter, we emphasize the following points. Although the burgeoning literature on nonparametric estimation in RDDs (e.g., IK, CCT, and CCFT) may be interpreted by practitioners as pointing to a "default" estimation procedure, there is no single "optimal" approach. In practice, different estimators may have very different point estimates and sampling errors, requiring the researcher to try to "sift" the evidence from different estimators. We argue that some useful information is provided by the magnitude and sampling error of the asymptotic bias correction terms for different estimators. Calculation of these terms helps to interpret the differences between uncorrected and corrected estimators.

One way of choosing among alternative estimators is to conduct Monte Carlo simulations based on DGPs that closely resemble the data. In our own simulations, we find that some of the asymptotically dominant procedures may actually perform worse than "suboptimal" alternatives in a given empirical application, even though the sample size is very large by conventional standard for parametric estimators. However, we also caution against generalizing conclusions based on specific Monte Carlo exercises to overarching guide for practitioners. While the recently proposed data-driven nonparametric procedures may provide a good starting point, we are skeptical of a one-size-fits-all solution and advocate for judiciously combining the art of sensitivity analysis and the science of econometric theory.

NOTES

1. Unless otherwise mentioned, the kink threshold is normalized to zero throughout this chapter.

2. In this chapter, we focus on tests of the smoothness of the predetermined covariates at the kink. CLPW discuss the smoothness of the density of UI claimants around the kink point in the benefit formula in Austria.

3. As noted in Card et al. (2012) and CLPW, imposing continuity in $E[Y|V = v]$ and $E[B|V = v]$, which follows from the RKD identifying assumptions, does not affect the asymptotic properties of $\widehat{\tau}$ when a uniform kernel and symmetric bandwidths are used in estimation. However, imposing continuity may change the asymptotic properties of $\widehat{\tau}$ when *asymmetric* bandwidths are used. For our conventional estimates below, we use symmetric bandwidths and a uniform kernel and report results obtained by imposing continuity, and we find that removing this restriction has little impact on the estimates. In certain applications, there may be both a level and slope change in the first-stage relationship. CLPW show that in

general the RK estimand does not identify a treatment effect, though it is possible to recover a causal parameter with additional model restrictions (Turner, 2013). In these applications, it is clearly sensible to not impose continuity.

4. It is worth noting that the motivation of (Calonico, Cattaneo, & Farrell, Forthcoming) and Calonico et al. (2016) echos the message of this chapter that there is not a single procedure "optimal" for all empirical applications.

5. For the bias-corrected estimates, we follow the CCFT procedure which does not impose continuity in $E[Y|V = v]$ or $E[B|V = v]$.

6. In a related study, Ganong and Jäger (2014) raise concerns about the sensitivity of the RKD estimates when the relationship between the running variable and the outcome is highly nonlinear. They propose a permutation test to account for the estimation bias. We perform their test on our data and discuss the details in Section 4.4.4.

7. The existence of these off-the-schedule observations is attributable to some combination of errors in the calculation of base year earnings (due to errors in the calculation of the claim start date, for example), errors in the Social Security earnings records that are overridden by benefit administrators, and mis-reported UI benefits. Similar errors have been found in many other settings – for example, Kapteyn and Ypma (2007).

8. Note that T^{min} is only a kink point for single claimants: the schedules for claimants with one or more dependents have kinks at earnings thresholds below T^{min}. Since we do not know the number of dependents a claimant reports, however, we use the kink point for the largest group of claimants, that is, those with no dependents.

9. The line for low-earning single claimants actually bends, reflecting the earnings threshold at which a single claimant begins paying income taxes.

10. See Card, Lee, Pei, and Weber (2015b) for more details on the data and analysis samples.

11. In CLPW, the main outcome is registered unemployment duration, an alternative measure of unemployment.

12. The bin sizes are €100 and €300 for the bottom (Fig. 2(a)) and top kink (Fig. 2(b)) samples, respectively. Fig. 2(b) is reproduced from Fig. 2 of CLPW. See Calonico, Cattaneo, and Titiunik (Forthcoming) for nonparametric procedures for picking the bin size in RD-type plots.

13. As noted in CLPW, the slopes in the empirical benefit functions to the left of T^{min}, and to the right of T^{max} are mainly due to dependent allowances. Moving left from T^{min}, for example, the average number of dependent allowances is falling, as claimants with successively higher numbers of dependents hit the minimum benefit level (Fig. 1).

14. We fit a single prediction model using the pooled bottom kink and top kink samples. The inclusion of the recall variable is motivated by the observation that many workers cycle between seasonal jobs (often with the same employer each year) and UI – see Del Bono and Weber (2008).

15. The exceptions are the bias-corrected local linear estimates for the top kink sample with the two larger bandwidths. In these cases, the robust confidence intervals are relatively wide but the bias-corrected point estimates are also large in magnitude.

16. Remark 5 of CCT states that the variance of $\widehat{\tau}_p^{\text{bc}}$ is smaller than that of $\widehat{\tau}_p$ for a large n, but this asymptotic advantage may not materialize in a given finite sample and does not appear to hold in our data.

17. There is obviously room for discretion in specifying the approximating DGP, and we believe this is unavoidable. That said, we will also specify an alternative DGP below that imposes zero elasticity, in order to examine the sensitivity of our conclusions with respect to the choice of DGP.

18. An exception is the bias-corrected estimators applied to the top kink sample – though both the linear and quadratic bias corrected estimators perform poorly in the top kink sample.

19. The exception is the quadratic model with bias correction for the top kink sample.

20. In results available upon request, we have also examined the performance of the coverage-error optimal bandwidth selector of Calonico et al. (2016) and the asymmetric fuzzy MSE-optimal bandwidth selectors in Calonico et al. (2016a), and their corresponding RMSE's are also consistently dominated by that of FG in our simulations.

21. We estimate local linear models with a fixed bandwidth of 2,500 for the bottom kink sample and a fixed bandwidth of 4,000 for the top kink sample.

22. The list of covariates includes: indicator variables for gender, marital status, Austrian citizenship, education (six categories), deciles for age, tenure, size, and recall rate at the firm of the last job, monthly wage last job (quintiles), days worked over the last five years (quartiles), eight industry groups interacted with blue collar status, region, year, and month dummies.

23. Ando (2016) makes a similar point regarding the role of covariates in RK estimation.

24. The Fuzzy IK bandwidth in Table A.2 refers to the fuzzy extension of the Imbens and Kalyanaraman (2012) bandwidth selector developed by Card, Lee, et al. (2015a).

ACKNOWLEDGMENTS

We thank Sebastian Calonico, Matias Cattaneo, Pauline Leung, Tim Moore, two anonymous referees, and seminar participants at CUFE, Econometric Society China Meeting, Hanyang, Tsinghua and Sichuan University for helpful comments. Suejin Lee provided outstanding research assistance.

REFERENCES

Altonji, J. G., & Rosa L. M. (2005). Cross section and panel data estimators for nonseparable models with endogenous regressors. *Econometrica*, *73*(4), 1053–1102.

Ando, M. (2016). How much should we trust regression-kink-design estimates? *Empirical Economics*, 1–36.

Angrist, J. D., & Lavy, V. (1999). Using maimonides' rule to estimate the effect of class size on scholastic achievement. *Quarterly Journal of Economics, 114*(2), 533–575.

Bono, E., & Weber, A. (2008). Do wages compensate for anticipated working time restrictions? Evidence from seasonal employment in Austria. *Journal of Labor Economics, 26*(1), 181–221.

Calonico, S., Cattaneo, M. D., & Farrell, M. H. (Forthcoming). On the effect of bias estimation on coverage accuracy in nonparametric inference. *Journal of the American Statistical Association.*

Calonico, S., Cattaneo, M. D., & Farrell, M. H. (2016). *Coverage error optimal confidence intervals for regression discontinuity designs.* Working paper.

Calonico, S., Cattaneo, M. D., & Titiunik, R. (2014). Robust nonparametric confidence intervals for regression-discontinuity designs. *Econometrica, 82*(6), 2295–2326.

Calonico, S., Cattaneo, M. D., & Titiunik, R. (2015). Optimal data-driven regression discontinuity plots. *Journal of the American Statistical Association, 110*(512), 1753–1769.

Calonico, S., Cattaneo, M. D., Farrell, M. H., & Titiunik, R. (Forthcoming). rdrobust: Software for regression discontinuity designs. *Stata Journal.*

Calonico, S., Cattaneo, M. D., Farrell, M. H., & Titiunik, R. (2016). *Regression discontinuity designs using covariates.* Working paper.

Card, D., Johnston, A., Leung, P., Mas, A., & Pei, Z. (2015). The effect of unemployment benefits on the duration of unemployment insurance receipt: New evidence from a regression kink design in Missouri, 2003-2013. *American Economic Review: Papers and Proceedings, 105*(5), 126–130.

Card, D., Dobkin, C., & Maestas, N. (2009). Does medicare save lives? *Quarterly Journal of Economics, 124*(2), 597–636.

Card, D., Lee, D. S., & Pei, Z. (2009, November). Quasi-experimental identification and estimation in the regression kink design. Princeton University Industrial Relations Section Working Paper No. 553.

Card, D., Lee, D. S., Pei, Z., & Weber, A. (2012, November). *Nonlinear policy rules and the identification and estimation of causal effects in a generalized regression kink design.* NBER Working Paper No. 18564.

Card, D., Lee, D. S., Pei, Z., & Weber, A. (2014, October). *Local polynomial order in regression discontinuity designs.* Working paper.

Card, D., Lee, D. S., Pei, Z., & Weber, A. (2015a). Inference on causal effects in a generalized regression kink design. *Econometrica, 83*(6), 2453–2483.

Card, D., Lee, D. S., Pei, Z., & Weber, A. (2015b). Inference on causal effects in a generalized regression kink design. Technical Report 15-218, Upjohn Institute Working Paper.

Card, D., Lee, D. S., Pei, Z., & Weber, A. (2016). *Supplemental material to "Regression Kink Design: Theory and Practice".* Retrieved from https://sites.google.com/site/peizhuan/files/rkd-aie-supplemental-material.pdf

Card, D., Chetty, R., & Weber, A. (2007). Cash-on-hand and competing models of intertemporal behavior: New evidence from the labor market. *Quarterly Journal of Economics, 122*(4), 1511–1560.

Chay, K. Y., McEwan, P. J., & Urquiola, M. (2005). The central role of noise in evaluating interventions that use test scores to rank schools. *American Economic Review, 95*(4), 1237–1258.

Cheng, M.-Y., Fan, J., & Marron, J. S. (1997). On automatic boundary corrections. *The Annals of Statistics, 25*(4), 1691–1708.

DiNardo, J., & Lee, D. S. (2004). Economic impacts of new unionization on private sector employers: 1984-2001. *Quarterly Journal of Economics*, *119*(4), 1383–1441.
Fan, J., & Gijbels, I. (1996). Local polynomial modelling and its applications. London: Chapman and Hall.
Florens, J.-P., Heckman, J. J., Meghir, C., & Vytlacil, E. J. (2008). Identification of treatment effects using control functions in models with continuous, endogenous treatment and heterogeneous effects. *Econometrica*, *76*(5), 1191–1206.
Ganong, P., & Jäger, S. (2014, June). A permutation test and estimation alternatives for the regression kink design.
Imbens, G. W., & Kalyanaraman, K. (2012). Optimal bandwidth choice for the regression discontinuity estimator. *Review of Economic Studies*, *79*(3), 933–959.
Imbens, G. W., & Lemieux, T. (2008). Regression discontinuity designs: A guide to practice. *Journal of Econometrics*, *142*(2), 615–635.
Kapteyn, A., & Ypma, J. Y. (2007). Measurement error and misclassification: A comparison of survey and administrative data. *Journal of Labor Economics*, *25*(3), 513–551.
Lee, D. S. (2008). Randomized experiments from non-random selection in U.S. house elections. *Journal of Econometrics*, *142* (2), 675–697.
Lee, S. D., & Mas, A. (2012). Long-run impacts of unions on firms: New evidence from financial markets, 1961-1999. *Quaterly Journal of Economics*, *127* (1), 333–378.
Lee, S. D., & Lemieux, T. (2010). Regression discontinuity designs in economics. *Journal of Economic Literature*, *48*(2), 281–355.
Lee, S. D., Moretti, E., & Butler, M. (2004). Do voters affect or elect policies? Evidence from the U.S. House. *Quarterly Journal of Economics*, *119*(3), 807–859.
McCrary, J., & Royer, H. (2011). The effect of female education on fertility and infant health: Evidence from school entry policies using exact date of birth. *American Economic Review*, *101*(1), 158–195.
Nielsen, H. S., Sørensen, T., & Taber, C. R. (2010). Estimating the effect of student aid on college enrollment: Evidence from a government grant policy reform. *American Economic Journal: Economic Policy*, *2*(2), 185–215.
Porter, J. *Estimation in the regression discontinuity model*. Department of Economics, University of Wisconsin, 2003.
Ruppert, D. (1997). Empirical-bias bandwidths for local polynomial nonparametric regression and density estimation. *Journal of the American Statistical Association*, *92*(439), 1049–1062.
Ruppert, D., & Wand, M. P. (1994). Multivariate locally weighted least squares regression. *The Annals of Statistics*, *22*(3), 1346–1370.
Schmieder, J. F., von Wachter, T., & Bender, S. (2012). The effects of extended unemployment insurance over the business cycle: Evidence from regression discontinuity estimates over 20 years. *The Quarterly Journal of Economics*, *127*(2), 701–752.
Turner, L. J. (2013). *The road to pell is paved with good intentions: The economic incidence of federal student grant aid*. Technical Report, University of Maryland.

REGRESSION DISCONTINUITY DESIGNS WITH CLUSTERED DATA

Otávio Bartalotti[a] and Quentin Brummet[b]

[a]*Department of Economics, Iowa State University, Ames, IA, USA*
[b]*Center for Administrative Records Research and Applications, United States Census Bureau, Washington, DC, USA*

ABSTRACT

Regression discontinuity designs have become popular in empirical studies due to their attractive properties for estimating causal effects under transparent assumptions. Nonetheless, most popular procedures assume i.i.d. data, which is unreasonable in many common applications. To fill this gap, we derive the properties of traditional local polynomial estimators in a fixed-G setting that allows for cluster dependence in the error term. Simulation results demonstrate that accounting for clustering in the data while selecting bandwidths may lead to lower MSE while maintaining proper coverage. We then apply our cluster-robust procedure to an application examining the impact of Low-Income Housing Tax Credits on neighborhood characteristics and low-income housing supply.

Keywords: Regression discontinuity designs; local polynomials; clustering; optimal bandwidth selection

JEL classifications: C13; C14; C21

Regression Discontinuity Designs: Theory and Applications
Advances in Econometrics, Volume 38, 383–420

ISSN: 0731-9053/doi:10.1108/S0731-905320170000038017

1. INTRODUCTION

Regression discontinuity (RD) designs have become one of the leading empirical strategies in economics, public policy evaluation, and other social sciences due to their ability to provide consistent estimation of causal effects under transparent assumptions. Nonetheless, the current literature on bandwidth selection and inference in RD designs typically assumes that the observations around the cutoff are independent. While many researchers explicitly consider cluster dependence when conducting inference, commonly used bandwidth selection procedures such as those in Imbens and Kalyanaraman (2012) and Calonico, Cattaneo, and Titiunik (2014) explicitly assume i.i.d. data. Therefore, researchers performing RD designs with clustered data are left with the choice of either using an ad hoc bandwidth or relying on a bandwidth selection procedure whose assumptions are clearly violated.

We fill this gap in the RD literature by analyzing the properties of local polynomial estimators of treatment effects in RD designs with clustering and deriving cluster-robust optimal bandwidth selectors. We derive these formulas under two setups. In our first general analysis, we use fixed-G calculations to derive the properties of estimators in a setup that allows for unrestricted dependence among observations within clusters. These results demonstrate that the widely used "cluster-robust" standard error formulas are appropriate in this setting. This finding relates to the results in Imbens and Lemieux (2008) and Calonico, Cattaneo, and Farrell (2017) (henceforth, "CCF") for the usual i.i.d. case. It is also connected to Lee and Card (2008), who suggest the use of cluster-robust standard errors to account for specification errors in RD designs with discrete running variables. Our analysis demonstrates that in the context of our framework, the intuitive idea of using cluster-robust standard errors is valid even when using non-parametric local polynomial estimators. In addition, we extend the mean-squared error (MSE)-optimal bandwidth choice procedures proposed by Imbens and Kalyanaraman (2012) and Calonico et al. (2014) (henceforth, "IK" and "CCT," respectively) to allow for unrestricted dependence structure within cluster, where the resulting optimal bandwidth collapses to traditional optimal bandwidth selectors when observations are independent.

Additionally, we consider the special case in which clustering is defined at the running variable level, for which asymptotic approximations can be obtained. The implementation of this optimal bandwidth selector is identical to the analysis above, and this analysis provides further insight into the effects of clustering on inference and bandwidth selection in RD designs. This is particularly relevant for two scenarios that are common in the applied literature. First, researchers may wish to use microdata to implement an RD design based on a higher-level running variable. For example, a researcher might be interested in using student-level microdata to examine a policy implemented based on a school-level running variable. A framework that considers clustering allows the researcher to select bandwidths, estimate parameters, and perform tests in a way that is compatible with the use of microdata in RD designs. Another salient example is that of RD designs with a discrete running variable. A cluster-robust bandwidth choice procedure allows researchers using discrete running variables to select bandwidths in a manner that is compatible with the cluster-robust inference procedure suggested by Lee and Card (2008).

We then present a simulation study which demonstrates that a cluster-robust bandwidth choice procedure outperforms traditional bandwidth choices in terms of MSE in many practical settings without any noticeable decline in coverage. We then illustrate the empirical importance and usefulness of the procedure in an application analyzing the impact of Low-Income Housing Tax Credits (LIHTC) on neighborhood characteristics. The outcomes in this application are observed at the person level, but the running variable is defined at the census tract level, generating clustering issues. The results show that accounting for clustering in the data when choosing bandwidths can lead to practically significant changes in the interpretation of empirical results.

The remainder of this chapter is structured as follows. Section 2 presents the setup and notation. Section 3 then presents our main results related to MSE and optimal bandwidth selection and discusses the special case in which the clusters are defined at the running variable level. Section 4 shows the distributional approximations of the statistics used for performing inference, and Section 6 then provides a small simulation study. Finally, Section 7 presents the application to the impacts of LIHTC on neighborhood characteristics, and Section 8 concludes.

2. SETUP AND NOTATION

In the typical sharp RD setting, a researcher wishes to estimate the local causal effect of treatment at a given threshold. The running variable, X, has density given by $f(X)$ and determines treatment assignment. Given a known threshold, $\bar{x}$, set to zero without loss of generality, a unit receives treatment if $X_i \geq 0$ or does not receive treatment if $X_i < 0$. Let $Y_i(1)$ and $Y_i(0)$ denote the potential outcomes for unit i given it receives treatment and in the absence of treatment, respectively. Hence, the observed sample is comprised of the running variable, X_i, and

$$Y_i = Y_i(0)\mathbb{1}\{X_i < 0\} + Y_i(1)\mathbb{1}\{X_i \geq 0\}, \tag{1}$$

where $\mathbb{1}\{\cdot\}$ denotes the indicator function. For convenience, define

$$m(x) = \mathbb{E}[Y_i | X_i = x]. \tag{2}$$

In most cases, the population parameter of interest is the average treatment effect at the threshold, $\tau = \mathbb{E}[Y(1) - Y(0) | X = \bar{x}]$. Under continuity and smoothness conditions on both the conditional distribution of X_i and the first moments of $Y(0)$ and $Y(1)$ at the cutoff,[1] τ is nonparametrically identified (Hahn, Todd, & Van der Klaauw, 2001) by:

$$\tau = m_+ - m_-$$
$$\text{where } m_+ = \lim_{x \to 0^+} m(x) \text{ and } m_- = \lim_{x \to 0^-} m(x).$$

In general, one might also be interested in the discontinuity of a higher-order derivative of the conditional expectation at the threshold such as in the "regression kink" literature (Card, Lee, & Pei, 2009). Let $m^{(\eta)}(x) = \frac{d^\eta m(x)}{dx^\eta}$ be the ηth derivative of the unknown regression function and define $m_+^{(\eta)} = \lim_{x \to 0^+} m^{(\eta)}(x)$ and $m_-^{(\eta)} = \lim_{x \to 0^-} m^{(\eta)}(x)$. The parameter of interest in these cases is given by $\tau^{(\eta)} = m_+^{(\eta)} - m_-^{(\eta)}$.

The estimation of $\tau^{(\eta)}$ in RD designs focuses on the problem of approximating $\mathbb{E}[Y(1)|X = x]$ and $\mathbb{E}[Y(0)|X = x]$ near the cutoff. Due to its desirable properties when estimating regression functions at the boundary, the most common approach fits separate kernel-weighted local polynomial regressions in neighborhoods on both sides of the threshold (Calonico et al., 2016; Fan & Gijbels, 1992; Hahn et al., 2001; Porter, 2003). For a local polynomial of order p, we use the following estimator:

$$\widehat{\tau}^{(\eta)} = \widehat{m}_{+}^{(\eta)} - \widehat{m}_{-}^{(\eta)},$$

$$\left(\widehat{\beta}_{+}, \widehat{\beta}_{+}^{(1)}, \ldots, \widehat{\beta}_{+}^{(p)}\right) = \underset{b_0, b_1, \ldots, b_p}{\operatorname{argmin}} \sum_{i=1}^{N} \mathbb{1}\{X_i \geq 0\}\left(Y_i - b_0 - b_1 X_i - \cdots - b_p X_i^p\right)^2 \cdot K_h(X_i),$$

$$\left(\widehat{\beta}_{-}, \widehat{\beta}_{-}^{(1)}, \ldots, \widehat{\beta}_{-}^{(p)}\right) = \underset{b_0, b_1, \ldots, b_p}{\operatorname{argmin}} \sum_{i=1}^{N} \mathbb{1}\{X_i < 0\}\left(Y_i - b_0 - b_1 X_i - \cdots - b_p X_i^p\right)^2 \cdot K_h(X_i),$$

where the kernel function is given by $K_h(x_{ig}) = K\left(\frac{x_{ig}}{h}\right)\frac{1}{h}$, and $\widehat{m}^{(\eta)} = \eta! \widehat{\beta}^{(\eta)}$.

Building on this traditional RD setup, we now turn to the setting where there is cluster dependence in the data. Consider sampling from a large number of clusters and, for each group g, we observe data on the outcome, running variable and potential covariates for N_g observations (Wooldridge, 2010, p. 864). This sampling scheme is assumed to generate observations that are independent across clusters. Then, for a random sample of G groups of size N_g, we observe

$$Y_{ig} = m(x_{ig}) + \epsilon_{ig}, \tag{3}$$

where the subscript ig refers to unit i in cluster g.

The notation used in this chapter follows closely the definitions established in CCF. Let $x_{h,ig} = \frac{x_{ig} - \bar{x}}{h}$ where $\bar{x}$ is the treatment cutoff which is set to zero without loss of generality. Also, let $r_p(u) = (1, u, u^2, \ldots, u^p)'$ and $R_p = [r_p(x_{h,11}), r_p(x_{h,21}), \ldots, r_p(x_{h,N_G G})]'$ and e_η' be a vector of zeros except for the $(\eta + 1)^{\text{th}}$ entry equal to one, for example, $e_0' = (1 \quad 0 \quad \cdots \quad 0)$. Define the weighting matrix $W_{+,p,g} = diag(h^{-1}\mathbb{1}[x_{1g} > 0]K(x_{h,1g}), h^{-1}\mathbb{1}[x_{2g} > 0] K(x_{h,2g}), \ldots, h^{-1}\mathbb{1}[x_{N_g g} > 0]K(x_{h,N_g g}))$ and $W_{-,p,g} = diag(h^{-1}\mathbb{1}[x_{1g} \leq 0] K(x_{h,1g}), h^{-1}\mathbb{1}[x_{2g} \leq 0] \; K(x_{h,2g}), \ldots, h^{-1}\mathbb{1}[x_{N_g g} \leq 0]K(x_{h,N_g g}))$. Then, let $W_{+,p}$ and $W_{-,p}$ be the block diagonal matrices with g-th block $W_{+,p,g}$ and $W_{-,p,g}$ for $g = 1, \ldots, g$, respectively. Also, define $H_p = diag(1, h^{-1}, \ldots, h^{-p})$. Finally, let $\Gamma_{+,p} = R_p' W_{+,p} R_p$, $\Gamma_{-,p} = R_p' W_{-,p} R_p$, $\Lambda_{+,p} = R_p' W_{+,p}\left[x_{h,11}^{p+1}, \ldots, x_{h,N_G G}^{p+1}\right]'$ and $\Lambda_{-,p} = R_p' W_{-,p}\left[x_{h,11}^{p+1}, \ldots, x_{h,N_G G}^{p+1}\right]'$. Also define $\Psi_{+,p} = R_p' W_{+,p} \sum W_{+,p} R_p$ and $\Psi_{-,p} = R_p' W_{-,p} \sum W_{-,p} R_p$, where $\sum$ is the block diagonal matrix with typical block given by $\Omega_g(x) = \text{Var}(Y_g | X)$. It is useful to establish notation for the (i, s) elements of the variance matrix Ω_g as σ_{gis} and its limits around the threshold $\sigma_{+,is} = \lim_{x \to 0^+} \sigma_{gis}$ and $\sigma_{-,is} = \lim_{x \to 0^-} \sigma_{gis}$ for describing the cluster at the running variable case. Note that this allows Ω_g to vary over X, and we do not restrict cluster size to be identical across clusters.

More details are provided in the appendix. Some useful calculations can be obtained by using the fixed-G expectation over the support of X to some of the objects defined below. Hence, adopting the notation on CCF, let $\tilde{\Lambda}_{+,p} = \mathbb{E}[\Lambda_{+,p}]$, and similarly for $\Lambda_{-,p}$, $\Gamma_{+,p}$, $\Gamma_{-,p}$, $\Psi_{+,p}$, and $\Psi_{-,p}$ (see Lemmas A.1 and A.2).

In Section 3, we present two sets of results. First, to address the issue of variance estimation and bandwidth selection in the general clustering case, we use fixed-G expectations calculated conditionally on the observed values of the running variable, following Calonico et al. (2016). This provides a valid MSE function and accompanying optimal bandwidth rule that take account of cluster dependence in the data.

Second, we focus on the special case in which the clusters are defined at the running variable level, implying that all units in a cluster have the same value, X_g, for the running variable. In this case, cluster dependence does not vanish asymptotically even as the bandwidth shrinks. Our asymptotic approximations based on $G \to \infty$, $h \to 0$, and $Gh \to \infty$ provide insight on the sources of improvements achieved by recognizing the presence of cluster dependence.

3. MAIN RESULTS

To capture the impact of cluster dependence on the behavior of the estimator, $\hat{\tau}^{(\eta)}$, we first utilize fixed-G calculations for variance and MSE obtained conditional on the observed values of the running variable following the approach in CCF.

Theorem 3.1 presents the MSE approximation for $\hat{\tau}^{(\eta)}$ and its corresponding MSE-optimal bandwidth for the general clustering case, as well as the conditional bias and variance formulas and their limiting expressions. Assumptions and proofs are provided in the appendix.

Theorem 3.1. *Under Assumptions 1–6, the conditional MSE for $\hat{\tau}^{(\eta)}$ has fixed-G approximation given by,*

$$E[(\hat{\tau}^{(\eta)} - \tau^{(\eta)})^2 | x_{11}, \ldots, x_{N_G G}] = \frac{1}{Gh^{2\eta+1}}[\mathbf{V_n}] + h^{2(p+1-\eta)}[\mathbf{B}_\mathbf{n}^2 + o_p(1)] \quad (4)$$

$$= \frac{1}{Gh^{2\eta+1}} \left[\tilde{\mathbf{V}} + o_p(1) \right] + h^{2(p+1-\eta)} \left[\tilde{\mathbf{B}}^2 + o_p(1) \right], \tag{5}$$

where

$$\mathbf{V_n} = \eta!^2 e'_\eta \left[\Gamma^{-1}_{+,p} \Psi_{+,p} \Gamma^{-1}_{+,p} + \Gamma^{-1}_{-,p} \Psi_{-,p} \Gamma^{-1}_{-,p} \right] e_\eta, \tag{6}$$

$$\mathbf{B_n} = \frac{\eta!}{(p+1)!} e'_\eta \left[m_+^{(p+1)} \Gamma^{-1}_{+,p} \Lambda_{+,p} - m_-^{(p+1)} \Gamma^{-1}_{-,p} \Lambda_{-,p} \right], \tag{7}$$

$$\tilde{\mathbf{V}} = \eta!^2 e'_\eta \left[\tilde{\Gamma}^{-1}_{+,p} \tilde{\Psi}_{+,p} \tilde{\Gamma}^{-1}_{+,p} + \tilde{\Gamma}^{-1}_{-,p} \tilde{\Psi}_{-,p} \tilde{\Gamma}^{-1}_{-,p} \right] e_\eta, \tag{8}$$

$$\tilde{\mathbf{B}} = \frac{\eta!}{(p+1)!} e'_\eta \left[m_+^{(p+1)} \tilde{\Gamma}^{-1}_{+,p} \tilde{\Lambda}_{+,p} - (-1)^{p+1+\eta} m_-^{(p+1)} \tilde{\Gamma}^{-1}_{-,p} \tilde{\Lambda}_{-,p} \right]. \tag{9}$$

Additionally, if $\mathbf{B_n} \neq 0$*, the approximated AMSE-optimal bandwidth is*

$$h^{\text{opt}} = \left[\frac{2\eta + 1}{2(p+1-\eta)} \frac{\tilde{\mathbf{V}}}{\tilde{\mathbf{B}}} \right]^{\frac{1}{2p+3}} G^{-\frac{1}{2p+3}}. \tag{10}$$

The optimal bandwidth formula presented above follows very closely the format of similar results for the local linear estimator by IK and the more general case developed by CCT (Lemma 1, p. 11) for the i.i.d. data. Similarly, the rate at which the optimal bandwidth shrinks is similar to the one presented in those papers, with the number of clusters replacing the full sample size ($h^{\text{opt}} \propto G^{-\frac{1}{2p+3}}$). As pointed out by CCT, this rate has important implications for the control of the leading term of the bias, which we discuss in detail in Section 4.

Note that both the "pre-asymptotic" conditional variance leading component, $\mathbf{V}_n$, and its limit, $\tilde{\mathbf{V}}$, through Ψ_n and $\tilde{\Psi}$, correctly capture the information about the cluster structure and dependence. In Eq. (10), the within-cluster data dependence is captured by $\tilde{\mathbf{V}}$ and would be ignored if conventional bandwidth choice procedures for i.i.d. data were implemented. With these traditional procedures, the researcher minimizes an incomplete MSE, and the resulting bandwidth does not correctly assess the trade-off between bias and variance.

In order to implement the optimal bandwidth, we need to replace $\tilde{\mathbf{V}}$ and $\tilde{\mathbf{B}}$ with feasible estimates. This can be done by replacing $\mathbf{V}_n$ and $\mathbf{B}_n$, using a preliminary bandwidth and obtaining estimates of the unknown parameters $m_+^{(p+1)}$, $m_-^{(p+1)}$, and $\sum$. CCF point out that using standard errors based on fixed-G calculations, as opposed to first-order asymptotic approximations, can achieve higher-order refinements in terms of coverage error. These authors also discuss the related issue of obtaining bandwidths that provide optimal coverage rates. Finally, refer to Calonico, Cattaneo, Farrell, and Titiunik (2017) for details on the implementation of the feasible optimal bandwidth choice using the "pre-asymptotics" formulas described above.[2]

3.1. Clustering at the Running Variable Level

In this section, we analyze the special case in which the clusters are defined at the running variable level. Note first that the usual asymptotic approximations based on shrinking the bandwidth as the number of clusters increase often fail to adequately capture the dependence structure present in the data. Typically, as the number of clusters grows and bandwidth shrinks to zero the covariance terms in the asymptotic variance vanish, and the clustering issue disappears. This result is similar to the situation described by Bhattacharya (2005) in the context of multistage sampling. Intuitively, the proportion of units from a given cluster within the bandwidth goes to zero. However, when the cluster is defined at the running variable level and all units in a cluster share the same value of X_g, the clustering issue remains asymptotically.

This setup encompasses multiple interesting applied cases. In particular, as discussed in the introduction, researchers are often faced with situations where either the running variable is discrete or where the running variable is defined at a higher level of aggregation than the unit of observation. In addition, this setup is important as it provides additional intuition on the theoretical and practical challenges of incorporating data clustering into local identification strategies such as RD designs.

Given this setup, we derive the asymptotic properties of $\hat{\tau}^{(\eta)}$.[3] The results emphasize the similarities with the usual asymptotic results for i.i.d. cases described in Porter (2003), IK, and CCT; making clear the intuition about the pitfalls of ignoring data clustering when performing inference or choosing benchmark bandwidths.[4] The asymptotic variance of the estimator is given by the following:[5]

$$\mathrm{Var}[\hat{\tau}^{(\eta)} - \tau^{(\eta)}|X] = C_{2,\eta}\left[\frac{\sum_{i=1}^{N_g}\sum_{s=1}^{N_g}\sigma_{is}^{+}}{GN_g^2h^{2\eta+1}f(0)} + \frac{\sum_{i=1}^{N_g}\sum_{s=1}^{N_g}\sigma_{is}^{-}}{GN_g^2h^{2\eta+1}f(0)}\right]\{1 + o_p(1)\}$$

This follows directly from Theorem 3.1 and is analogue to the results for i.i.d. data in CCT (Lemma A.1, p. 16). This formula makes clear that ignoring dependence in the data misrepresents the variance of the estimators by ignoring the terms in $\sum_{i=1}^{N_g}\sum_{i\neq s}^{N_g}\sigma_{is}^{+}$ for clusters on both sides of the threshold.

As a secondary point, note that the form of the leading term of the asymptotic bias,

$$E[\hat{\tau}^{(\eta)}|X] = \tau^{(\eta)} + h^{p+1-\eta}C_{1,\eta}\left(m_{+}^{(p+1)} - (-1)^{(p+1+\eta)}m_{-}^{(p+1)}\right) + o_p\left(h^{p+1-\eta}\right),$$

is not directly affected by the presence of clustering as described here. As pointed out by IK and CCT, the leading bias term depends on the higher-order derivatives of the conditional outcome above and below the cutoff and not on the form of the asymptotic variance.

The infeasible MSE-optimal bandwidth choice in this case is then given by the following:

$$h_{\text{opt}} = \left[C_{\kappa\eta}\frac{\frac{\sum_{i=1}^{N_g}\sum_{s=1}^{N_g}\sigma_{+,is}}{GN_g^2f(0)} + \frac{\sum_{i=1}^{N_g}\sum_{s=1}^{N_g}\sigma_{-,is}}{GN_g^2f(0)}}{\left[m_{+}^{(p+1)} - (-1)^{(p+1+\eta)}m_{-}^{(p+1)}\right]^2}\right]^{\frac{1}{2p+3}}. \tag{11}$$

These results follow very closely those presented for the local linear estimator by IK and CCT (Lemma 1, p.11) for the i.i.d. data case.[6] If the errors are indeed i.i.d., this bandwidth collapses to the traditional MSE-optimal bandwidth choice.

For further insight, consider the case of a linear local estimator ($p = 1$) in the standard RD design ($\eta = 0$) with a constant group-level shock, c_g, and Ω taking the familiar "random effects" structure:

$$\Omega_g = \begin{pmatrix} \sigma_c^2 + \sigma_u^2 & \sigma_c^2 & \cdots & \sigma_c^2 \\ \sigma_c^2 & \sigma_c^2 + \sigma_u^2 & \cdots & \sigma_c^2 \\ \vdots & \vdots & \ddots & \vdots \\ \sigma_c^2 & \sigma_c^2 & \cdots & \sigma_c^2 + \sigma_u^2 \end{pmatrix}.$$

Under this setup, Eq. (11) can be written as follows:

$$h_{\text{opt}} = \left(\frac{C_{2,0}}{4C_{1,0}}\right)^{\frac{1}{5}} \left[\frac{\left(\sigma_{u,+}^2 + N_g\sigma_{c,+}^2\right) + \left(\sigma_{u,-}^2 + N_g\sigma_{c,-}^2\right)}{f(0)\left[\mu_+^{(2)} - \mu_-^{(2)}\right]^2}\right]^{\frac{1}{5}} N^{-1/5}. \tag{12}$$

This rewrite makes clear that the key components driving differences in the cluster-robust and conventional optimal bandwidth formulas are cluster size and within-cluster dependence. As cluster size or within-cluster dependence increase, the current approach produces bandwidths that differ from the usual formulas based on i.i.d. data. Intuitively, if there is strong within-cluster dependence each observation provides relatively less information to the researcher than if the observations were independent.

In this case where all units in a cluster share the same value for the running variable, an alternative approach to address within-cluster data dependence is to aggregate the data to the cluster level. This approach, introduced by Moulton (1987) and discussed in Angrist and Pischke (2009), relies on the basic intuition that averaging the data to the cluster-level treats each cluster as an independent observation. By collapsing the data, one can absorb the within-cluster dependence into each of the group-level observations. Hence, one could propose to collapse the data and obtain the benchmark bandwidth using a traditional bandwidth choice procedure based on i.i.d. data. In practice, this approach has some pitfalls. First, it forfeits the use of variation in outcomes (or covariates) at a lower level than the cluster. Also, it requires that the cluster-level observations be weighted by their relative cluster sizes. In addition, this aggregation method would not address any across-cluster heteroskedasticity, which could potentially induced by the averaging of different sized groups. Last, and most importantly, this approach would obviously not be valid if the running variable was allowed to change within clusters. With these caveats, careful implementation of a strategy based on aggregating the data to the cluster level could be a valid alternative. However, given the developments in this chapter and the ease of implementation of these procedures, researchers might prefer to use microdata while explicitly accounting for the clustering.

4. DISTRIBUTIONAL APPROXIMATION

In order to perform inference about the parameter of interest, we require an approximation to its distribution. While one could obtain an approximation based on the "usual" studentized statistic $\frac{\widehat{\tau}^{(\eta)} - \tau^{(\eta)}}{\sqrt{\frac{1}{Gh^{2\eta+1}}\mathbf{V}_n}}$, this statistic converges to a standard normal distribution only if the bandwidth is allowed to shrink fast enough to assure the leading bias term is negligible ($Gh^{2p+3} \to 0$). However, as seen in Section 3, the MSE-optimal bandwidth is proportional to $G^{\frac{1}{2p+3}}$, which leaves a first-order bias in the approximated distribution, as described by CCT for i.i.d. data.

While a common practice in this situation is to "undersmooth" by choosing an arbitrarily smaller bandwidth than the value suggested by h_{opt}, an alternative proposed by CCT and CCF is to directly remove the leading term of the bias, $h^{p+1-\eta}\mathbf{B}_n$. This bias-corrected studentized statistic behavior is adequately approximated by a standard normal distribution if $Gh^{2p+5} \to 0$, which is compatible with the proposed optimal bandwidth choice discussed above, and is described in Theorem 4.1.

Theorem 4.1. *Under Assumptions 1–6, if* $Gh \to \infty$ *and* $Gh^{2p+5} \to 0$. *Then,*

$$\frac{\widehat{\tau}^{(\eta)} - \tau^{(\eta)} - h^{p+1-\eta}\mathbf{B}_n}{\sqrt{\frac{1}{Gh^{2\eta+1}}\mathbf{V}_n}} \to_d N(0, 1). \tag{13}$$

Notice that Theorem 4.1 is based on the infeasible pre-asymptotic formulas for bias and standard errors, $\mathbf{B}_n$ and $\sqrt{\mathbf{V}_n}$. As pointed out by CCT and CCF for the i.i.d. case, these "fixed-G" standard errors are valid and produce statistics that achieve higher-order refinements relative to approximations based on a first-order asymptotic approximation. We expect that similar refinements would benefit the statistics described above when employing "pre-asymptotic" cluster-robust standard errors. In addition, note that this result can also be cast as the cluster analogue of the point made by Imbens and Lemieux (2008) that the usual parametric heteroskedasticity-robust standard errors may be used in traditional RD designs with i.i.d. data.[7]

Additionally, we propose the use of natural, direct plug-in, feasible estimators for $\mathbf{B}_n$ and $\sqrt{\mathbf{V}_n}$ that replace the unknown quantities $m_+^{(p+1)}$, $m_-^{(p+1)}$, and $\sum$ with consistent estimates. For $\sum$, it is useful to note that, as it is usual in clustering, we are not interested in estimating each σ_{gis} term in the relevant variance–covariance matrix and focus on estimating Ψ_p based on the pre-asymptotic formulas $\widehat{\Psi}_{+,p} = R_p' W_{+,p} \widehat{\epsilon\epsilon'} W_{+,p} R_p$ and $\widehat{\Psi}_{-,p} = R_p' W_{-,p} \widehat{\epsilon\epsilon'} W_{-,p} R_p$.

As described in detail by CCT, consistent estimates of the leading bias term can be obtained by plugging in $\widehat{m}_+^{(p+1)}(b)$ and $\widehat{m}_-^{(p+1)}(b)$ from a $(p+1)$-order local polynomial around the threshold using a potentially different bandwidth, b.[8] This bias correction strategy introduces non-negligible variability to the bias-corrected statistic and requires that the denominator account for the contribution of both $\widehat{\tau}^{(\eta)}$ and the bias estimate.

Recently, CCF argue that explicit bias-correction coupled with an analytical adjustment to the standard errors "yields confidence intervals with coverage that is accurate, or better, than the best possible undersmoothing approach" (Calonico et al., 2016, p. 2). It is straightforward to allow for cluster-robust bandwidth selection in the context of bias correction, as the lead bias term is not meaningfully affected by the cluster dependence, as shown in Theorem 3.1. In addition, the standard error correction proposed in CCT follows similarly with the addition of off-diagonal variance terms that account for cluster dependence. The details of both discussions are beyond the scope of this chapter, and the reader is referred to CCT and CCF.

Note that the cluster-robust variance estimator and several bandwidth choice algorithms have recently been incorporated into the rdrobust package in STATA and R, making it easily available to practitioners. For a detailed discussion on different implementations of both the standard errors, analytical bias correction, and optimal bandwidth choice in RD settings including clustered data, see Calonico et al. (2016a).

5. EXTENSIONS

A main conclusion of the analysis above is that cluster dependence should be taken into account not only when performing inference, but also when selecting bandwidths. This insight can be transferred directly to a variety of other results in the RD literature.

First, the cluster-dependent variance results presented above extend naturally to fuzzy RD designs, as the usual estimates can be rewritten as a linear combination of the discontinuities observed at the outcome and treatment status at the cutoff. As discussed above, the leading term of the bias is not affected by clustering in the setting we analyze. The asymptotic variance formulas for fuzzy RD designs follow the descriptions in IK Section 5.1 and CCT's Lemma 2 with the simple variance terms being replaced by the cluster-robust formulas above. One would naturally want to allow the outcome and treatment status to be potentially correlated within cluster by allowing $\mathrm{Cov}(Y, T|X)$ to have non-zero off diagonal elements within clusters.[9]

Second, instead of relying on optimal-MSE bandwidths, CCF propose bandwidth choice methods for RD designs that focus on delivering confidence intervals with optimal coverage error rates. While the authors note that the bandwidth formulas for optimal coverage rate results are "prohibitively cumbersome" (Calonico et al., 2016, p. 29), they suggest a rule-of-thumb bandwidth choice based on an adjusted MSE-optimal bandwidth, which provides the correct coverage error rate. A straightforward modification of these procedures to account for cluster dependence can be incorporated for researchers wishing to use these procedures.

6. SIMULATIONS

To illustrate the practical importance of adequately accounting for clustering when performing RD designs, we present a simulation study based on two data-generating processes (DGPs). Throughout this section, estimation is performed using a local linear estimator, the preferred method in most applications. All simulation results use the implementation software developed by Calonico et al. (2016a) for STATA.[10] For clarity, the setup follows a random effects structure:

$$Y_{ig} = m(x_{ig}) + c_g + u_{ig}.$$

The simulations we present are based on a conditional mean function in CCT that modifies the DGP based on Lee (2008)'s data to induce bias near the cutoff. This choice highlights the potential trade-off between bias and variance when examining MSE in a clustering context. This setting demonstrates the properties of our cluster-robust bandwidth selection even in a situation where accounting for bias with small bandwidths is particularly important. For both DGPs presented here, $m(x)$ takes the following form:

$$m_1(x) = \begin{cases} 0.48 + 1.27x - 0.5\times 7.18x^2 + 0.7\times 20.21x^3 \\ \quad + 1.1\times 21.54x^4 + 1.5\times 7.33x^5 \quad \text{if} \quad x<0 \\ 0.52 + 0.84x - 0.1\times 3.00x^2 - 0.3\times 7.99x^3 \\ \quad - 0.1\times 9.01x^4 + 3.56x^5 \qquad \text{if} \quad x\geq 0. \end{cases} \tag{14}$$

For the first set of simulations, the running variable is constructed so that x is a linear combination of x_1 and x_2. x_1 is defined at the group level, while x_2 is defined at the individual level. Hence, while clusters span multiple values of the running variable, there is within-cluster correlation in x. Each x_j is drawn from a half normal distribution, $N_f(0,4)$ and clusters are constrained so that they do not contain units from both sides of the treatment threshold.[11] Both u and c are normally distributed, the variance of u is set to 0.1295^2 and the variance of c is adjusted to obtain the desired value of $\rho \equiv \frac{\sigma_c^2}{\sigma_c^2+\sigma_u^2}$. Simulations are run for various values of within cluster dependence, ρ.

Fig. 1 presents MSE in this simulation for three procedures: the cluster-robust and traditional bandwidth selection procedures described in Section 3, a traditional bandwidth selection procedure that does not account for clustering, and a cluster-robust bandwidth choice that optimizes coverage error (CER) as discussed in Section 5. Each panel in Fig. 1 plots the empirical MSE of each procedure for different values of ρ, where panels are separated by cluster size and number of clusters. Based on the discussion in Section 3, we expect accounting for clustering to become more important as cluster size (N_g) or within-cluster dependence (ρ) increase. The simulations support this intuition. As N_g or ρ increase, the clustering problem becomes more important, emphasizing the need for approximations and procedures capable of capturing the cluster dependence. In addition, note that the cluster-robust CER bandwidth procedure produces higher MSE than the other procedures, which is to be expected given that it is the only procedure not explicitly designed to minimize MSE. Fig. 2 plots coverage for each of the procedures for different levels of ρ. Across all panels, coverage is similar between

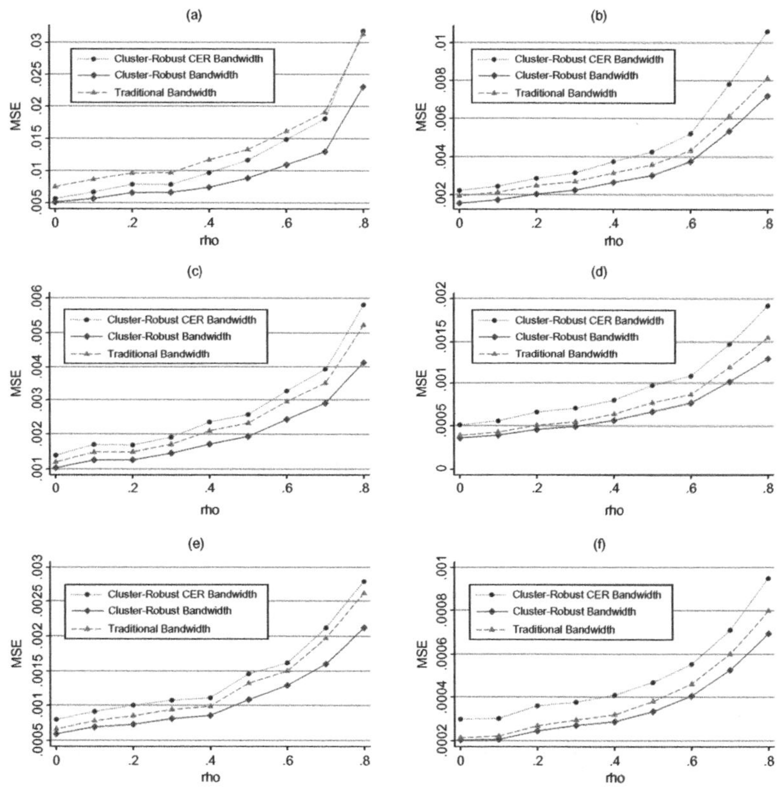

Fig. 1. MSE Performance in Simulated Data – Data-Generating Process 1. (a) Size = 5, Number of Clusters = 250. (b) Size = 5, Number of Clusters = 1,000. (c) Size = 25, Number of Clusters = 250. (d) Size = 25, Number of Clusters = 1,000. (e) Size = 50, Number of Clusters = 250. (f) Size = 50, Number of Clusters = 1,000. *Note*: Results are not plotted if the MSE in the traditional bandwidth procedure is more than 25 times the cluster-robust procedure.

the three procedures. Note that while each procedure picks a difference bandwidth, all procedures are based on robust bias-corrected confidence intervals with inference that accounts for clustering. Therefore, it is potentially unsurprising that the bandwidth choices lead to similar coverage.

The second simulation showcases the properties of our procedure in a setting where there exists clustering at the level of the running variable. Here, $m(x_g)$ is the mean function shared by all individuals in cluster g, c_g is

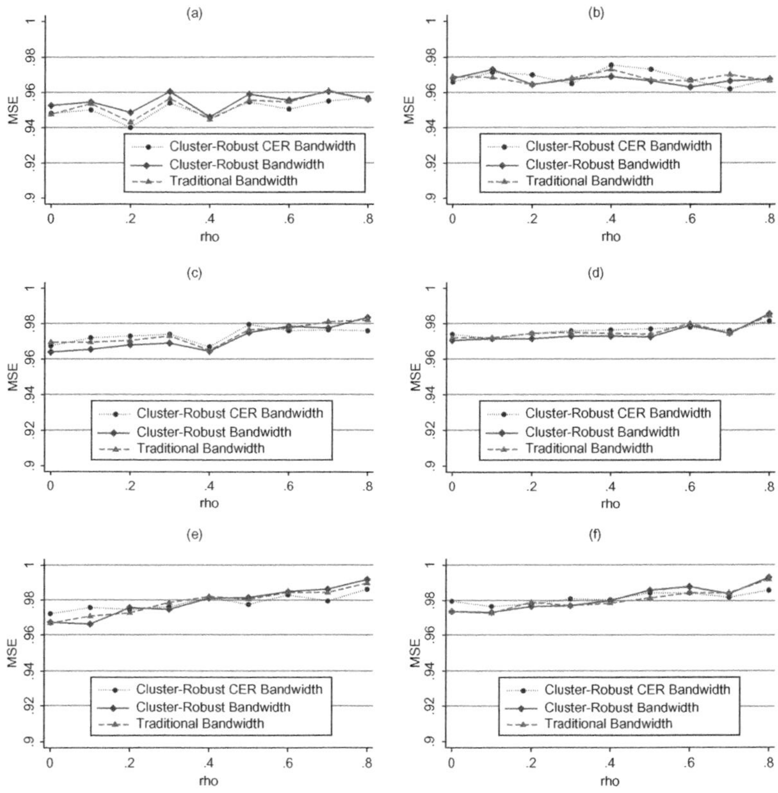

Fig. 2. Coverage Performance in Simulated Data – Data-Generating Process 1. (a) Size = 5, Number of Clusters = 250. (b) Size = 5, Number of Clusters = 1,000. (c) Size = 25, Number of Clusters = 250. (d) Size = 25, Number of Clusters = 1,000. (e) Size = 50, Number of Clusters = 250. (f) Size = 50, Number of Clusters = 1,000.

group-level shock with variance σ_c^2, and u_{ig} is idiosyncratic error term with variance σ_u^2. Note that the functional form of $m(.)$ is the same defined above in Eq. (14). Both u and c are normally distributed, and as before the variance of u is set to 0.1295^2 with the variance of c being adjusted to obtain the desired value of ρ.

Additionally, in this set of simulations we report results with data aggregated to the running variable level using the traditional bandwidth choice. This procedure averages all observations in a given cluster and performs estimation with the aggregated data, thereby treating each cluster as a single observation. This approach is sometimes used by researchers such as Ahn and Vigdor (2014) when facing clustering issues in RD designs. By aggregating the data to the running variable level, the researcher collapses the dependence structure in the data, exploiting the fact that clusters are independent from each other, as described in Section 3.1.

These simulation results are presented in Fig. 3. As expected, higher levels of within-cluster dependence, ρ, lead to situations where the cluster-robust approach dominates procedures using traditional bandwidth selection algorithms in terms of empirical MSE. Moreover, as the size of clusters or rho increase, the cluster-robust procedure far outperforms traditional bandwidth choices using the microdata. In particular, both columns show that the cluster-robust and traditional procedures perform similarly well for small values of ρ, and the former is significantly better as ρ increases, when accounting for clustering becomes more important with larger dependence. As before, the CER bandwidth choice leads to larger empirical MSE across all panels. Last, the ad hoc procedure using aggregate data performs quite similar to the cluster-robust procedure in terms of MSE, indicating that this procedure may also be able to account for clustering in RD designs in the special case where clusters are defined at the level of the running variable. Fig. 4 presents coverage across these four procedures. While there are some slight differences in coverage between procedures when $G = 250$, the four procedures all produce very similar coverage when $G = 1,000$, indicating that accounting for clustering when selecting a bandwidth does little to change coverage.

7. APPLICATION: LIHTC AND NEIGHBORHOOD CHARACTERISTICS

We now demonstrate the usefulness of these new methods using an empirical application that examines the effect of low-income housing subsidies on housing development and neighborhood characteristics. In particular, we focus the effects of the LIHTC, a program that has provided funding for roughly one third of all new units in multifamily housing built in the United States over the past 30 years (Khadduri, Climaco, & Burnett, 2012). We exploit a discontinuity in program eligibility rules designating whether a particular census tract becomes a Qualified Census Tract (QCT). As discussed in Hollar and Usowski (2007), Baum-Snow and Marion (2009) and

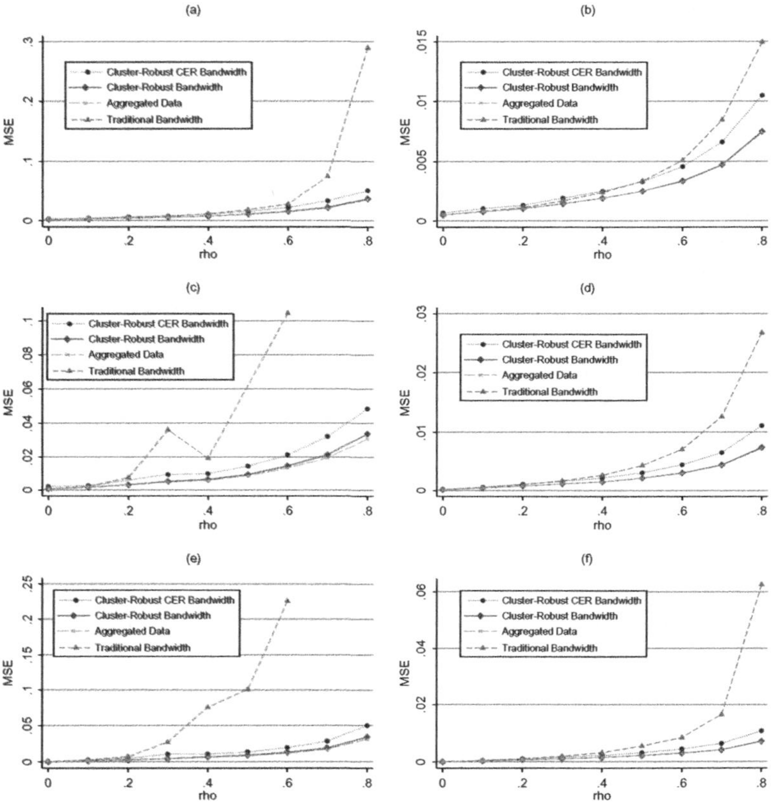

Fig. 3. MSE Performance in Simulated Data – Data-Generating Process 2. (a) Size = 5, Number of Clusters = 250. (b) Size = 5, Number of Clusters = 1,000. (c) Size = 25, Number of Clusters = 250. (d) Size = 25, Number of Clusters = 1,000. (e) Size = 50, Number of Clusters = 250. (f) Size = 50, Number of Clusters = 1,000. *Note*: Results are not plotted if the MSE in the traditional bandwidth procedure is more than 25 times the cluster-robust procedure.

Freedman and McGavock (2015), projects located in QCTs are eligible for up to 30 percent larger tax credits than projects in tracts not labeled as QCTs. Importantly, during the time period studied this designation is based on the fraction of households whose income falls below 60 percent of Area Median Gross Income (AMGI). If the majority of households in a census tract have household income less than 60 percent of AMGI, the tract becomes eligible to receive QCT status. Therefore, the percent of

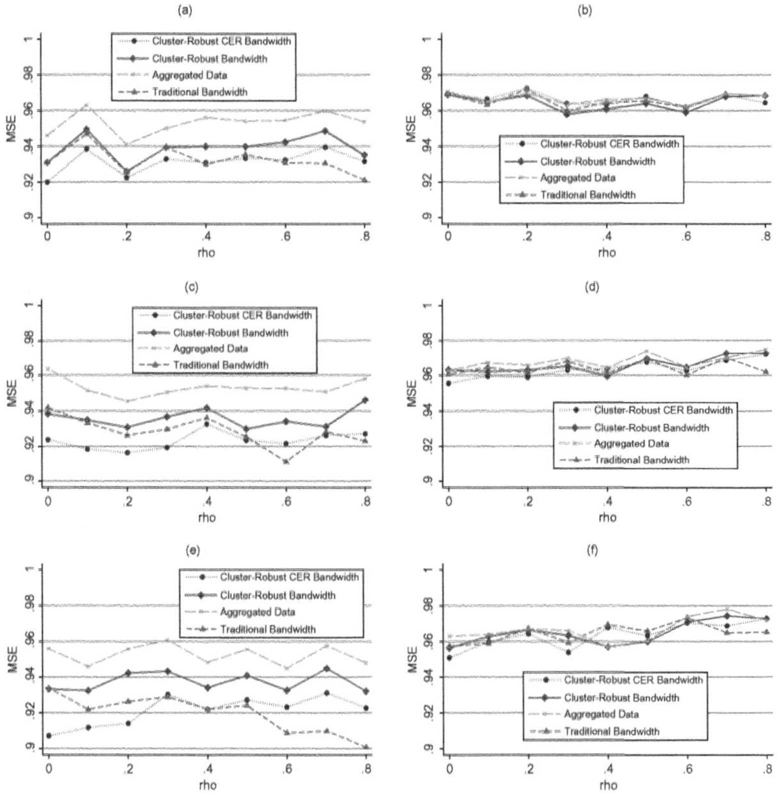

Fig. 4. Coverage Performance in Simulated Data – Data-Generating Process 2. (a) Size = 5, Number of Clusters = 250. (b) Size = 5, Number of Clusters = 1,000. (c) Size = 25, Number of Clusters = 250. (d) Size = 25, Number of Clusters = 1,000. (e) Size = 50, Number of Clusters = 250. (f) Size = 50, Number of Clusters = 1,000.

households below 60 percent AMGI forms our running variable and the cutoff is 50 percent. By comparing only individuals that lived in tracts with a similar percentage of households below 60 percent of AMGI, we exploit random variation in QCT designation near the cutoff to identify the impact of the tax credits on housing development and neighborhood outcomes.

We perform this application using restricted access individual-level data from Census 2000 long form microdata.[12] As in Baum-Snow and Marion (2009), we restrict to census tracts in metropolitan areas, and exclude Alaska and Hawaii. Table 1 displays descriptive statistics for this data set.

The number of LIHTC units and projects variables refers to the number of these units in the census tract. Clearly, QCT tracts contain much more disadvantaged populations than non-QCT tracts, a fact that is obvious due to the construction of the QCT status. In addition, note that QCT tracts have much larger numbers of LIHTC units and projects than non-QCT tracts. However, these descriptive differences between QCT and non-QCT tracts are not necessarily caused by LIHTC development or QCT designation, motivating the use of an RD design.

Table 2 displays results of four estimation procedures applied to the data. All estimates represent the results of local linear regressions using a triangular kernel, with standard errors that are robust to clustering at the tract level. In other words, both procedures using the microdata perform inference with the same "cluster-robust" standard error formulas, while tract-level regressions utilize heteroskedasticity-robust standard errors and

Table 1. Descriptive Statistics.

	QCT	Non-QCT
Homeownership	0.3316	0.6984
	(0.4708)	(0.4590)
Fraction non-white	0.7565	0.2778
	(0.4292)	(0.4479)
High school diploma or higher	0.5744	0.8367
	(0.4944)	(0.3696)
Bachelors degree or higher	0.1110	0.2819
	(0.3142)	(0.4499)
Employment population ratio	0.4808	0.6363
	(0.4996)	(0.4811)
Number of LIHTC projects	0.2714	0.1096
	(0.7147)	(0.5675)
Number of LIHTC units	16.8094	8.8745
	(55.7813)	(43.7748)
Running variable	0.1155	−0.2514
	(0.0913)	(0.1097)
N	3,063,042	27,879,680
N Clusters	6,778	37,938

Source: Microdata from the long form of the 2,000 decennial census.
Notes: Cells contain sample means. Standard deviations are in parentheses.

Table 2. Local Linear Estimates of the Effect of QCT Status.

Dependent Variable	Cluster-Robust Bandwidth	Traditional Bandwidth	Tract-Level	Cluster-Robust Bandwidth Region FE
Homeownership	−0.0089	−0.0335	−0.0061	−0.0115
	(0.0130)	(0.0217)	(0.0094)	(0.0117)
	[−0.0402, 0.0186]	[−0.0789, 0.0099]	[−0.0281, 0.0162]	[−0.0406, .0121]
	4,537,956	1,179,593	8,761	4,656,057
	$w = 0.098$	$w = 0.027$	$w = 0.107$	$w = 0.100$
Fraction non-white	0.0024	−0.0048	0.0042	0.0024
	(0.0173)	(0.0386)	(0.0153)	(0.0167)
	[−0.0429, 0.0311]	[−0.0825, 0.0707]	[−0.0361, 0.0320]	[−0.0407, 0.0303]
	5,022,420	945,413	8,600	4,936,642
	$w = 0.107$	$w = 0.022$	$w = 0.105$	$w = 0.106$
High school diploma	−0.0073	−0.0006	−0.0105*	−0.0076
or higher	(0.0078)	(0.0123)	(0.0061)	(0.0074)
	[−0.0217, 0.0129]	[−0.0248, 0.0256]	[−0.0243, 0.0043]	[−0.0211, 0.0115]
	3,828,193	1,304,465	10,109	3,941,003
	$w = 0.129$	$w = 0.052$	$w = 0.121$	$w = 0.131$
Bachelors degree	0.0039	0.0041	0.0024	0.0039
or higher	(0.0057)	(0.0077)	(0.0064)	(0.0058)
	[−0.0098, 0.0177]	[−0.0133, 0.0222]	[−0.0123, 0.0178]	[−0.0096, 0.0181]
	3,285,861	1,747,085	8,336	3,207,979
	$w = 0.115$	$w = 0.067$	$w = 0.103$	$w = 0.112$
Employment rate	0.0081	0.0218**	−0.0028	0.0051
	(0.0054)	(0.0093)	(0.0045)	(0.0051)
	[−0.0028, 0.0221]	[0.0040, 0.0421]	[−0.014, 0.0075]	[−0.0057, 0.0181]
	2,602,616	811,030	10,104	2,839,725
	$w = 0.082$	$w = 0.027$	$w = 0.121$	$w = 0.089$
	7.959***	12.729***	4.628**	7.808***

Table 2. *(Continued)*

Dependent Variable	Cluster-Robust Bandwidth	Traditional Bandwidth	Tract-Level	Cluster-Robust Bandwidth Region FE
Number of LIHTC				
units	(2.956)	(5.809)	(2.180)	(2.975)
	[1.543, 14.884]	[1.177, 24.512]	[−0.464, 9.856]	[1.357, 14.770]
	4,618,876	778,252	8,338	4,523,254
	$w = 0.099$	$w = 0.018$	$w = 0.103$	$w = 0.098$
Number of LIHTC	0.0341	−0.0473	0.0662*	0.0351
projects	(0.0581)	(0.1020)	(0.0258)	(0.0580)
	[−0.0995, 0.1475]	[−0.2484, 0.1579]	[0.0020, 0.1204]	[−0.0985, 0.1482]
	2,961,625	234,210	10,222	2,974,891
	$w = 0.068$	$w = 0.006$	$w = 0.122$	$w = 0.068$
N	30,330,540	30,330,540	45,294	30,330,540
N clusters	44,716	44,716	45,294	44,716

Source: Microdata and tract-level data from the long form of the 2,000 decennial census.
Notes: Standard errors in parentheses are adjusted for clustering at the tract level. Robust bias-corrected confidence intervals following CCT are shown in brackets just above effective sample size used in the local polynomial regression. "w" refers to bandwidth, where tract-level regressions use CCT bandwidths unadjusted for clustering. All point estimates are not bias corrected, and are derived from local linear regressions using a triangular kernel.
**Indicates significance at the 0.05 level, ***indicates significance at the 0.1 level.

are weighted to account for differential cluster sizes. The first column presents the results applying cluster-robust bandwidth selection procedures, as described above, applied to the microdata. Next, the second column presents results using the traditional bandwidth selection algorithm that does not account for clustering at the tract level. The third column presents results from applying this same procedure to data that has been aggregated to the tract level. These estimates are intended to replicate what a researcher would do when only aggregate data is available and the clustering issue is sidestepped. Last, the fourth column investigates whether including covariates has an effect on the estimates. To do this, we include census region fixed effects as covariates using a cluster-robust procedure based on Calonico, Cattaneo, Farrell, & Titiunik (2016). For each regression, we present coefficient estimates, standard errors in parentheses, robust bias-corrected confidence intervals in brackets, effective sample size, and local polynomial bandwidth choice.

The results show that accounting for potential dependence in outcomes within a census tract can substantially change the benchmark optimal-MSE bandwidth. As argued in Sections 3.1 and 6, the cluster-robust optimal bandwidth should be similar to the usual bandwidth choices in the absence of data dependence. The sizable differences between the bandwidth values suggest that the usual algorithms potentially misrepresent the MSE bias/variance trade-off by failing to capture the dependence in the data.

In terms of the point estimates, the results show little evidence of a discontinuity in neighborhood characteristics at the QCT designation threshold. Note that this analysis differs from Baum-Snow and Marion (2009) in that it considers levels of neighborhood characteristics in 2000 instead of changes in characteristics from 1990 to 2000. Therefore, the two analyses are not directly comparable. However, there is clear evidence of jumps in the implementation of new LIHTC units at the boundary, indicating that the QCT policy is indeed producing increases in LIHTC construction. This is one area where the cluster-robust procedure leads to different empirical results than the traditional IK bandwidth selection. In particular, the procedure that imposes i.i.d. data on the microdata produces a negative and statistically insignificant estimate of the effect of QCT status on the number of LIHTC projects in the tract, whereas both the aggregated data and the current procedure produce estimates that suggest that there is a positive effect of QCT status on the number of LIHTC projects in a tract, as intended by policymakers. Also, when focusing on the number of LIHTC units available in the tract, the four approaches provide very different optimal bandwidths and estimates, even though they all indicate a positive effect of the tax credits as the policy was designed to achieve. Last, note that the census region fixed effects do little to change the estimates, confidence intervals, or bandwidth choice. This result suggests that sorting of census tracts around the cutoff on the basis of census tract is not an issue for the empirical analysis.

Turning to standard error estimates, we see that applying the cluster-robust bandwidth choice procedure to the microdata produces estimates that are more precise than those obtained using a traditional bandwidth selection algorithm. This result is unsurprising, as accounting for the clustering will typically lead to larger bandwidth choices when there is positive dependence within clusters. When comparing the cluster-robust and aggregated data procedures, there is no clear relationship between the magnitude of the standard error estimates. Again, this reinforces the idea that both the cluster-robust and the aggregated data procedure account for clustering. On the whole, the cluster-robust, aggregated data procedures, and cluster-robust procedure with region fixed effects all provide similar results, and give a different empirical perspective than simply applying the IK bandwidth selection algorithm to the microdata.

8. CONCLUSION

Even though many researchers utilizing RD designs perform inference using cluster-robust standard error estimates, the justification for these methods is typically ad hoc. Moreover, current bandwidth selection procedures do not account for potential dependence among observations, creating a conflict in the assumptions between the bandwidth selection algorithm and inference procedure.

To fill this gap in the literature, we derive the distributional properties of local polynomial estimators in RD designs under a fixed-G approximation and provide accompanying optimal bandwidth selection rules. In addition, we analyze the special case of data clustered at the running variable level, providing further insight into our results. Both sets of results lead to the same implementation of cluster-robust optimal bandwidth selectors, which extend the prior results in IK and CCT to account for clustering. The insights provided by this analysis can also be applied directly to fuzzy and kink RD designs, as well as the robust bias-corrected methods in CCT.

Simulation results indicate that in some practically important settings failing to account for dependence among observations leads to non-trivial increases in MSE due to bandwidth choices that fail to capture adequately the bias-variance trade-off. We also present a simple application that demonstrates the practical importance of the cluster-robust optimal bandwidth choice algorithm by analyzing the impact of LIHTCs on neighborhood characteristics.

NOTES

1. The assumptions used in the derivations and results presented here closely follow IK and are discussed in Appendix A.1.

2. The cluster-robust variance estimator and several bandwidth choice algorithms have recently been incorporated into the rdrobust package in STATA, making it easily available to practitioners (Calonico et al., 2017).

3. The asymptotic derivations in this section assume that N_g identical across clusters, the results can be easily extended to the case where cluster size is allowed to vary.

4. For simplicity all asymptotic results are presented for the case in which clusters have equal sizes, N_g, general case involves cumbersome notation and formulas and can be similarly derived.

5. See Corollary A.1 for the exact formulas for the terms $C_{1,\eta}$, $C_{2,\eta}$, and $C_{\kappa,\eta}$, which are constants that depend on the kernel function and polynomial order selected by the researcher.

6. See Corollary A.1.

7. See Section 3.1 of CCF for a careful treatment of this issue.

8. For details on the implementation of the bias estimation and correction, see CCT and Calonico et al. (2017).

9. Note that the fuzzy RD design implementation is available in STATA through the rdrobust package (Calonico et al., 2017).

10. We use the software's version updated in April 2016, available for download at https://sites.google.com/site/rdpackages/rdrobust

11. This simulation is based in part on Hagemann (2016), and defines $x = (\sqrt{\rho_x})x_1 + (\sqrt{1-\rho_x})x_2$, with $\rho_x = 0.8$. For clusters below the threshold the signs of x_1 and x_2 are negative, and conversely above the cutoff, guaranteeing there is no crossing within clusters.

12. Since QCT classification and eligibility to extra tax credits was based on 1990 census tracts, location in 2000 is converted to tract location in 1990 using U.S. Census Bureau tract relationship files available at https://www.census.gov/geo/maps-data/data/relationship.html

ACKNOWLEDGMENTS

We are especially grateful to Matias Cattaneo for suggestions that significantly improved the chapter. We also thank Gary Solon, Helle Bunzel, Valentin Verdier, Thomas Fujiwara, Maggie Jones, and participants at the 2014 North American Summer Meetings of the Econometric Society, 2014 Midwest Econometrics Group, 2015 Econometric Society World Congress, Advances in Econometrics RD Conference, and U.S. Census Bureau for helpful comments. We also are grateful to Matthew Freedman and Emily Owens for providing us with data and code to compute QCT eligibility for all census tracts. The views expressed within are those of the authors and not necessarily those of the U.S. Census Bureau. Any errors or omissions are our own.

REFERENCES

Ahn, T., & Vigdor, J. (2014). *The impact of no child left behind's accountability sanctions on school performance: Regression discontinuity evidence from North Carolina.* NBER Working Paper No. 20511.

Angrist, J., & Pischke, J. (2009). *Mostly harmless econometrics: An empiricists' companion.* Princeton, NJ: Princeton University Press.

Baum-Snow, N., & Marion, J. (2009). The effects of low income housing tax credit developments on neighborhoods. *Journal of Public Economics*, *93*(5), 654–666.

Bhattacharya, D. (2005). Asymptotic inference from multi-stage samples. *Journal of Econometrics*, *126*(1), 145–171.

Card, D. E., Lee, D., & Pei, Z. (2009). *Quasi-experimental identification and estimation in the regression kink design.* Working Paper, Princeton University.

Calonico, S., Cattaneo, M. D., & Farrell, M. H. (2017). On the effect of bias estimation on coverage accuracy in nonparametric inference. *Journal of the American Statistical Association*, Forthcoming.

Calonico, S., Cattaneo, M. D., Farrell, M. H., & Titiunik, R. (2017). rdrobust: Software for regression discontinuity designs. *Stata Journal*, Forthcoming.

Calonico, S., Cattaneo, M. D., Farrell, M. H., & Titiunik, R. (2016). *Regression discontinuity designs using covariates.* Working Paper, University of Michigan.

Calonico, S., Cattaneo, M. D., & Titiunik, R. (2014). Robust nonparametric confidence intervals for regression-discontinuity designs. *Econometrica*, *82*(6), 2295–2326.

Fan, J., & Gijbels, I. (1992). Variable bandwidth and local linear regression smoothers. *Annals of Statistics*, *20*(4), 1669–2195.

Freedman, M., & McGavock, T. (2015). Low-income housing development, poverty concentration, and neighborhood inequality. *Journal of Policy Analysis and Management*, *34*(4), 805–834.

Hagemann, A. (2016). Cluster-robust bootstrap inference in quantile regression models. *Journal of the American Statistical Association*, Forthcoming.

Hahn, J., Petra T., & Van der Klaauw, W. (2001). Identification and estimation of treatment effects with a regression-discontinuity design. *Econometrica*, *69*(1), 201–209.

Hollar, M., & Usowski, K. (2007). Low-income housing tax credit qualified census tracts. *Cityscape*, *9*(3), 153–159.

Imbens, G. W., & Kalyanaraman, K. (2012). Optimal bandwidth choice for the regression discontinuity estimator. *Review of Economic Studies*, *79*(3), 933–959.

Imbens, G. W., & Lemieux, T. (2008). Regression discontinuity designs: A guide to practice. *Journal of Econometrics*, *142*(2), 615–635.

Khadduri, J., Climaco, C., & Burnett, K. (2012). *What happens to low-income housing tax credit properties at year 15 and beyond?* U.S. Department of Housing and Urban Development, Office of Policy Development and Research.

Lee, D. S. (2008) Randomized experiments from non-random selection in U.S. house elections. *Journal of Econometrics*, *142*(2), 675–697.

Lee, D. S., & Card, D. (2008). Regression discontinuity inference with specification error. *Journal of Econometrics*, *142*(2), 655–674.

Moulton, B. R. (1987). Diagnostics for group effects in regression analysis. *Journal of Business & Economic Statistics*, *5*(2), 275–282.

Porter, J. (2003). *Estimation in the regression discontinuity model.* Unpublished Manuscript, Department of Economics, University of Wisconsin – Madison, pp. 5–19.

Wooldridge, J. M. (2010). *Econometric analysis of cross section and panel data* (2nd ed.). Cambridge, MA: MIT Press.

APPENDIX A: ASSUMPTIONS AND PROOFS

A.1. Assumptions

We use the following standard assumptions in the RD literature. For some $\kappa_0 > 0$, the following holds in the neighborhood $(-\kappa_0, \kappa_0)$ around the threshold $\bar{x} = 0$.

1. We have G independent and identically distributed clusters, with data $(Y_g, X_g)'$, where Y_g and X_g are $1 \times N_g$ vectors for $g = 1, \ldots, g$.
2. $m(x) = E[Y|X]$ is at least $p + 2$ times continuously differentiable.
3. The density of the forcing variable X, denoted $f(X)$, is continuous and bounded away from zero.
4. The conditional variance–covariance matrix, $\sum$, is block diagonal with typical block given by $\Omega_g(x) = \mathrm{Var}(Y_g|X)$, is bounded, and right and left continuous at $\bar{x}$. The right and left limit at the threshold exist and are positive definite.
5. The kernel $K(\cdot)$ is non-negative, bounded, differs from zero on a compact interval $[0, \kappa]$, and is continuous on $(0, \kappa)$ for some $\kappa > 0$.
6. Individuals on the same cluster are assumed to be on the same side of the cutoff.

A.2. Results

For analyzing the properties of the estimator $\hat{\tau}^{(\eta)} = \hat{m}_+^{(\eta)} - \hat{m}_-^{(\eta)}$, note that $y_{ig} = m(x_{ig}) + \epsilon_{ig}$. Let $x_{h,ig} = \frac{x_{ig} - \bar{x}}{h}$ where $\bar{x}$ is the treatment cutoff which is set to zero without loss of generality. Also, let $r_p(u) = (1, u, u^2, \ldots, u^p)'$ and $R_p = [r_p(x_{h,11}), r_p(x_{h,21}), \ldots, r_p(x_{h,N_G G})]'$ and e'_η be a vector of zeros except for the $(\eta + 1)$-th entry equal to one, for example, $e'_0 = (1 \quad 0 \quad \cdots \quad 0)$. Define the weighting matrix $W_{+,p,g} = \mathrm{diag}(h^{-1}\mathbf{1}[x_{1g} > 0]K(x_{h,1g}), h^{-1}\mathbf{1}[x_{2g} > 0] \quad K(x_{h,2g}), \ldots, h^{-1}\mathbf{1}[x_{N_g g} > 0]K(x_{h,N_g g}))$ and $W_{-,p,g} = \mathrm{diag}(h^{-1}\mathbf{1}[x_{1g} \leq 0]K(x_{h,1g}), h^{-1}\mathbf{1}[x_{2g} \leq 0] \quad K(x_{h,2g}), \ldots, h^{-1}\mathbf{1}[x_{N_g g} \leq 0]K(x_{h,N_g g}))$. Then, let $W_{+,p}$ and $W_{-,p}$ be the block diagonal matrices with g-th block $W_{+,p,g}$ and $W_{-,p,g}$ for $g = 1, \ldots, g$, respectively. Also, define $H_p = \mathrm{diag}(1, h^{-1}, \ldots, h^{-p})$. Finally, let $\Gamma_{+,p} = R'_p W_{+,p} R_p$, $\Gamma_{-,p} = R'_p W_{-,p} R_p$, $\Lambda_{+,p} = R'_p W_{+,p} \left[x_{h,11}^{p+1}, \ldots, x_{h,N_G G}^{p+1}\right]'$ and $\Lambda_{-,p} = R'_p W_{-,p} \left[x_{h,11}^{p+1}, \ldots, x_{h,N_G G}^{p+1}\right]'$. Then,

$$\widehat{m}_{+}^{(\eta)} = \eta!\widehat{\beta}_p = \eta! e'_\eta H_p \Gamma_{+,p}^{-1} R'_p W_{+,p} Y. \tag{A.1}$$

And analogously for $\widehat{m}_{-}^{(\eta)}$. Also define $\Psi_{+,p} = R'_p W_{+,p} \sum W_{+,p} R_p$ and $\Psi_{-,p} = R'_p W_{-,p} \sum W_{-,p} R_p$. It is useful to to establish notation for the (i,s) elements of the variance matrix Ω_g as σ_{gis} and its limits around the threshold $\sigma_{+,is} = \lim_{x \to 0^+} \sigma_{gis}$ and $\sigma_{-,is} = \lim_{x \to 0^-} \sigma_{gis}$ for describing the cluster at the running variable case.

Moreover, some useful approximations can be obtained by using the fixed-G expectation over the support of X to some of the objects defined above. Following the notation on CCF, let $\tilde{\Lambda}_{+,p} = \mathbb{E}[\Lambda_{+,p}]$, and similarly for $\Lambda_{-,p}$, $\Gamma_{+,p}$, $\Gamma_{-,p}$, $\Psi_{+,p}$, and $\Psi_{-,p}$ (See Lemma A.1 and Lemma A.2).

Theorem 3.1. *Under Assumptions 1–6, the conditional MSE for $\widehat{\tau}^{(\eta)}$ has fixed-G approximation given by,*

$$E[(\widehat{\tau}^{(\eta)} - \tau^{(\eta)})^2 | x_{11}, \ldots, x_{N_G G}] = \frac{1}{Gh^{2\eta+1}}[\mathbf{V_n}] + h^{2(p+1-\eta)}\left[\mathbf{B_n^2} + o_p(1)\right] \tag{A.2}$$

$$= \frac{1}{Gh^{2\eta+1}}\left[\tilde{\mathbf{V}} + o_p(1)\right] + h^{2(p+1-\eta)}\left[\tilde{\mathbf{B}}^2 + o_p(1)\right], \tag{A.3}$$

$$\mathbf{V_n} = \eta!^2 e'_\eta \left[\Gamma_{+,p}^{-1}\Psi_{+,p}\Gamma_{+,p}^{-1} + \Gamma_{-,p}^{-1}\Psi_{-,p}\Gamma_{-,p}^{-1}\right] e_\eta, \tag{A.4}$$

$$\mathbf{B_n} = \frac{\eta!}{(p+1)!} e'_\eta \left[m_{+}^{(p+1)}\Gamma_{+,p}^{-1}\Lambda_{+,p} - m_{-}^{(p+1)}\Gamma_{-,p}^{-1}\Lambda_{-,p}\right], \tag{A.5}$$

$$\tilde{\mathbf{V}} = \eta!^2 e'_\eta \left[\tilde{\Gamma}^{-1}_{+,p}\tilde{\Psi}_{+,p}\tilde{\Gamma}^{-1}_{+,p} + \tilde{\Gamma}^{-1}_{-,p}\tilde{\Psi}_{-,p}\tilde{\Gamma}^{-1}_{-,p}\right] e_\eta, \tag{A.6}$$

$$\tilde{\mathbf{B}} = \frac{\eta!}{(p+1)!} e'_\eta \left[m_+^{(p+1)}\tilde{\Gamma}^{-1}_{+,p}\tilde{\Lambda}_{+,p} - (-1)^{p+1+\eta} m_-^{(p+1)}\tilde{\Gamma}^{-1}_{-,p}\tilde{\Lambda}_{-,p}\right]. \tag{A.7}$$

Additionally, if $\mathbf{B_n} \neq 0$, the approximated MSE-optimal bandwidth is

$$h^{\text{opt}} = \left[\frac{2\eta + 1}{2(p + 1 - \eta)} \frac{\tilde{\mathbf{V}}}{\tilde{\mathbf{B}}}\right]^{\frac{1}{2p+3}} G^{-\frac{1}{2p+3}}. \tag{A.8}$$

Proof. Start by rewriting the estimator as,

$$\widehat{m}_+^{(\eta)} = \eta!\widehat{\beta}_p = \eta! e'_\eta H_p \Gamma_p^{-1} R'_p W_p Y = \eta! e'_\eta H_p \Gamma_p^{-1} R'_p W_p [m(x) + \epsilon] \tag{A.9}$$

$$= \eta! e'_\eta H_p \Gamma_p^{-1} R'_p W_p m(x) + \eta! e'_\eta H_p \Gamma_p^{-1} R'_p W_p \epsilon. \tag{A.10}$$

For the conditional, fixed-G representation of the bias we proceed as follows. Define $M = \left(m(x_{11}), \ldots, m\left(x_{N_G G}\right)\right)'$.

$$\mathbf{E}[\widehat{m}_+^{(\eta)} | x_{11}, \ldots, x_{N_G G}] = \eta! e'_\eta H_p \Gamma_p^{-1} R'_p W_p M. \tag{A.11}$$

Taking a Taylor expansion of $m(\cdot)$ around $\bar{x} = 0$, we can rewrite:

$$m(x_{ig}) = m(0) + m^{(1)}(0)x_{h,ig} + \frac{1}{2}\cdot m^{(2)}(0)x_{h,ig}^2 + \cdots + \frac{1}{(p+1)!}\cdot m^{(p+1)}(0)x_{h,ig}^{p+1} + T_{ig},$$

where $|T_{ig}| \leq \sup_x |m^{(p+2)}(x)x_{h,ig}^{p+2}|$. Then,

$$M = R\begin{pmatrix} m(0) \\ \vdots \\ \dfrac{m^{(p)}(0)}{p!} \end{pmatrix} + S + T,$$

where $S_{ig} = \frac{1}{(p+1)!} m^{(p+1)}(0)x_{h,ig}^{p+1}$.
From Lemma A.1, we know that $\eta! e_\eta' H_p \Gamma_p^{-1} R_p' W_p T = o_p(h^{p+1-\eta})$. Then, the conditional bias is given by

$$\mathbb{E}\left[\widehat{m}_+^{(\eta)} | x_{11}, \ldots, x_{N_G G}\right] - m_+^{(\eta)} = h^{p+1-\eta} m_+^{(p+1)} \frac{\eta!}{(p+1)!} e_\eta' \Gamma_{+,p}^{-1} \Lambda_{+,p} + o_p(h^{p+1-\eta}).$$

Note that this form of the bias is similar to the one derived by CCT and CCF, and the clustered nature of the data does not affect the basic structure of the bias term as one would expect. Similarly for the conditional variance term, still under a fixed-G approximation:

$$\mathbb{V}\left[\widehat{m}_+^{(\eta)} | x_{11}, \ldots, x_{N_G G}\right] = \frac{1}{G} \eta!^2 e_\eta' H_p \Gamma_{+,p}^{-1} R_p' W_{+,p} S W_{+,p} R_p \Gamma_{+,p}^{-1} H_p e_\eta \quad \text{(A.12)}$$

$$= \frac{1}{G h^{2\eta+1}} \eta!^2 e_\eta' \Gamma_{+,p}^{-1} R_p' W_{+,p} S W_{+,p} R_p \Gamma_{+,p}^{-1} e_\eta. \quad \text{(A.13)}$$

And similarly for $\mathbb{E}[\widehat{m}_-^{(\eta)}|x_{11},\ldots,x_{N_GG}] - m_-^{(\eta)}$ and $\mathbb{V}[\widehat{m}_-^{(\eta)}|x_{11},\ldots,x_{N_GG}]$. Then, the conditional MSE can be described as:

$$
\begin{aligned}
E[(\widehat{\tau}^{(\eta)} - \tau^{(\eta)})^2|x_{11},\ldots,x_{N_GG}] &= \mathbb{V}\left[\widehat{\tau}^{(\eta)} - \tau^{(\eta)}|x_{11},\ldots,x_{N_GG}\right] \\
&\quad + \left\{E\left[\widehat{\tau}^{(\eta)} - \tau^{(\eta)}|x_{11},\ldots,x_{N_GG}\right]\right\}^2 \\
&= \mathbb{V}\left[\widehat{m}_+^{(\eta)} - m_+^{(\eta)}|x_{11},\ldots,x_{N_GG}\right] \\
&\quad + \mathbb{V}\left[\widehat{m}_-^{(\eta)} - m_-^{(\eta)}|x_{11},\ldots,x_{N_GG}\right] \\
&\quad + \left\{\mathbb{E}\left[\widehat{m}_+^{(\eta)} - m_+^{(\eta)}|x_{11},\ldots,x_{N_GG}\right]\right. \\
&\quad \left. + \mathbb{E}\left[\widehat{m}_-^{(\eta)} - m_-^{(\eta)}|x_{11},\ldots,x_{N_GG}\right]^2\right\} \\
&= \frac{1}{Gh^{2\eta+1}}[\mathbf{V_n}] + h^{2(p+1-\eta)}\left[\mathbf{B_n^2} + o_p(1)\right],
\end{aligned}
$$

where

$$
\mathbf{V_n} = \eta!^2 e_\eta' \left[\Gamma_{+,p}^{-1} R_p' W_{+,p} S W_{+,p} R_p \Gamma_{+,p}^{-1} + \Gamma_{-,p}^{-1} R_p' W_{-,p} S W_{-,p} R_p \Gamma_{-,p}^{-1}\right] e_\eta, \tag{A.14}
$$

$$
\mathbf{B_n} = \frac{\eta!}{(p+1)!} e_\eta' \left[m_+^{(p+1)} \Gamma_{+,p}^{-1} \Lambda_{+,p} + m_-^{(p+1)} \Gamma_{-,p}^{-1} \Lambda_{-,p}\right]. \tag{A.15}
$$

If we replace the conditional expectations by their nonrandom fixed-G expectations, we obtain:

$$
E[(\widehat{\tau}^{(\eta)} - \tau^{(\eta)})^2|x_{11},\ldots,x_{N_GG}] = \frac{1}{Gh^{2\eta+1}}\left[\tilde{\mathbf{V}} + o_p(1)\right] + h^{2(p+1-\eta)}\left[\tilde{\mathbf{B}}^2 + o_p(1)\right] \tag{A.16}
$$

$$\tilde{\mathbf{V}} = \eta!^2 e_\eta' \left[\tilde{\Gamma}_{+,p}^{-1} \tilde{\Psi}_{+,p} \tilde{\Gamma}_{+,p}^{-1} + \tilde{\Gamma}_{-,p}^{-1} \tilde{\Psi}_{-,p} \tilde{\Gamma}_{-,p}^{-1} \right] e_\eta \tag{A.17}$$

$$\tilde{\mathbf{B}} = \frac{\eta!}{(p+1)!} e_\eta' \left[m_+^{(p+1)} \tilde{\Gamma}_{+,p}^{-1} \tilde{\Lambda}_{+,p} - (-1)^{p+1+\eta} m_-^{(p+1)} \tilde{\Gamma}_{-,p}^{-1} \tilde{\Lambda}_{-,p} \right]. \tag{A.18}$$

To obtain the formula for the conditional optimal bandwidth, minimize the conditional MSE expression with respect to h conditional on $\mathbf{B}_n$ and $\mathbf{V}_n$. The calculations are exactly the same as presented by CCT Lemma 3.2. ∎

Theorem 4.1. *Under Assumptions 1–6, if $Gh \to \infty$ and $Gh^{2p+5} \to 0$. Then,*

$$\frac{\widehat{\tau}^{(\eta)} - \tau^{(\eta)} - h^{p+1-\eta} \mathbf{B}_n}{\sqrt{\frac{1}{Gh^{2\eta+1}} \mathbf{V}_n}} \to_d N(0,1). \tag{A.19}$$

Proof.

$$\frac{\widehat{m}_+^{(\eta)} - m_+^{(\eta)} - h^{p+1-\eta} B_{+,n}}{\sqrt{\frac{1}{Gh^{2\eta+1}} V_{+,n}}} = \frac{\widehat{m}_+^{(\eta)} - E\left[\widehat{m}_+^{(\eta)} | X\right] + E\left[\widehat{m}_+^{(\eta)} | X\right] - m_+^{(\eta)} - h^{p+1-\eta} B_{+,n}}{\sqrt{\frac{1}{Gh^{2\eta+1}} V_n}}$$

$$= \varepsilon_{1,n} + \varepsilon_{2,n} = \varepsilon_{1,n} + o_p(1),$$

where $V_{+,n} = \eta!^2 e_\eta' \Gamma_{+,p}^{-1} R_p' W_{+,p} \sum W_{+,p} R_p \Gamma_{+,p}^{-1}$ and $B_{+,n} = \frac{\eta!}{(p+1)!} m_+ (p+1) e_\eta' \Gamma_{+,p}^{-1} \Lambda_{+,p}$. Since,

$$\varepsilon_{2,n} = \frac{E\left[\widehat{m}^{(\eta)} | X\right] - m^{(\eta)} - h^{p+1-\eta} B_{+,n}}{\sqrt{\frac{1}{Gh^{2\eta+1}} V_{+,n}}} = O_p\left(\sqrt{Gh^{5+2p}}\right) = o_p(1).$$

Then,

$$\varepsilon_{1,n} = \left(\mathrm{Var}\left[\widehat{m}_+^{(\eta)} - m_+^{(\eta)}|X\right]\right)^{-\frac{1}{2}}\left(\widehat{m}_+^{(\eta)} - E\left[\widehat{m}_+^{(\eta)}|X\right]\right)$$
$$= \left(\frac{1}{Gh^{2\eta+1}}V_{+,n}\right)^{-\frac{1}{2}}\left(\frac{\eta! e_\eta' H_p \Gamma_{+,p}^{-1} R_p' W_{+,p}\epsilon}{G}\right)$$

and, following the same arguments as CCT, $\varepsilon_1 = \tilde{\varepsilon}_{1,n} + o_p(1)$, where $\tilde{\varepsilon}_{1,n} = \sum_{g=1}^{G} \omega_g' \epsilon_g$ where $\epsilon_g = \left[\epsilon_{g1}, \ldots, \epsilon_{gN_g}\right]$ for all $g = 1, \ldots, g$. Let $R_{p,g} = \left[r_p(x_{h,1g}), r_p(x_{h,2g}), \ldots, r_p(x_{h,N_g g})\right]'$ and

$$\omega_g = \left(\frac{1}{Gh^{2\eta+1}}V_n\right)^{-1/2} \frac{\eta! e_\eta' H_p \Gamma_{+,p}^{-1} R_{p,g}' W_{+,p,g}}{G}.$$

Since the vector of disturbances is independent across clusters and the clusters are randomly sampled we have that $E\left[\tilde{\varepsilon}_{1,n}\right] = 0$ and $V\left[\tilde{\varepsilon}_{1,n}\right] \to I$. Hence, it will follow a central limit theorem converging to an $N(0, 1)$. And similar results holds for $\widehat{\mu}_-^{(\eta)}$. Combining both results concludes the proof. ■

The following Corollary presents the asymptotic results for the special case in which the clusters are defined at the running variable level.

Corollary A.1. (Clustering at the Running Variable) *Suppose Assumptions 1–6 hold and the clusters are defined at the running variable level, formally, $X_{ig} = X_g \forall i = 1, \ldots, g$. Also, let $Gh \to \infty$.*

1. **(B)** *If $h \to 0$, then*

$$E\left[\widehat{\tau}^{(\eta)}|X\right] = \tau^{(\eta)} + h^{p+1-\eta}C_{1,\eta}\left(m_+^{(p+1)} - (-1)^{(p+1+\eta)}m_-^{(p+1)}\right) + o_p\left(h^{p+1-\eta}\right)$$

with $C_{1,\eta}\frac{\eta!}{(p+1)!}e_\eta'\Gamma_p^{-1}\Lambda_p$.

2. **(V)** *If $h \to 0$, then*

$$\mathrm{Var}\left[\widehat{\tau}^{(\eta)} - \tau^{(\eta)}|X\right] = C_{2,\eta}\left[\frac{\sum_{i=1}^{N_g}\sum_{s=1}^{N_g}\sigma_{is}^+}{GN_g^2h^{2\eta+1}f(0)} + \frac{\sum_{i=1}^{N_g}\sum_{s=1}^{N_g}\sigma_{is}^-}{GN_g^2h^{2\eta+1}f(0)}\right]\{1 + o_p(1)\}$$

with $C_{2,\eta} = \eta!^2 e_\eta'\Gamma_p^{-1}\Psi_p\Gamma_p^{-1}e_\eta$.

3. **(MSE)**

$$\mathrm{MSE}(h) = \frac{1}{Gh^{2\eta+1}} C_{2,\eta} \left[\frac{\sum_{i=1}^{N_g}\sum_{s=1}^{N_g}\sigma_{is}^{+}}{N_g^2 f(0)} + \frac{\sum_{i=1}^{N_g}\sum_{s=1}^{N_g}\sigma_{is}^{-}}{N_g^2 f(0)} \right]$$

$$+h^{2(p+1-\eta)} C_{1,\eta}^2 \left[m_{+}^{(p+1)} - (-1)^{(p+1+\eta)} m_{-}^{(p+1)} \right]^2 + o_p\left(\frac{1}{Gh^{2\eta+1}} + h^{2(p+1-\eta)} \right).$$

4. **(Optimal Bandwidth)** *If* $m_{+}^{(p+1)} \neq (-1)^{(p+1+\eta)} m_{-}^{(p+1)}$, *then the optimal bandwidth that minimizes the asymptotic approximation to* MSE(h) *is*

$$h_{\mathrm{opt}} = \left[C_{\kappa\eta} \frac{ \frac{\sum_{i=1}^{N_g}\sum_{s=1}^{N_g}\sigma_{+,is}}{GN_g^2 f(0)} + \frac{\sum_{i=1}^{N_g}\sum_{s=1}^{N_g}\sigma_{-,is}}{GN_g^2 f(0)} }{ \left[m_{+}^{(p+1)} - (-1)^{(p+1+\eta)} m_{-}^{(p+1)} \right]^2 } \right]^{\frac{1}{2p+3}}, \tag{A.20}$$

where $C_{\kappa\eta} = \frac{2\eta+1}{2(p+1-\eta)} \frac{C_{2\eta}}{C_{1\eta}^2}$.

Proof. The proof follows directly from Theorem 3.1 by replacing $\Gamma_{+,p}$, $\Gamma_{-,p}$, $\Lambda_{+,p}$, $\Lambda_{-,p}$, $\Psi_{+,p}$ and $\Psi_{-,p}$ by their asymptotic approximations obtained in Lemmas A.1 and A.2. ■

A.3. Auxiliary Lemmas

The proofs and notation presented in this appendix are based and follow as close as possible the proofs in IK and CCT regarding the asymptotic properties of the local polynomial estimators used in RD designs as well as the choice of MSE-optimal bandwidths. Lemma A.1 is equivalent to Lemma SA1 on CCT's Technical Supplement.

Let $\nu_j = \int_0^\infty u^j K(u)\,du$ and $\pi_j = \int_0^\infty u^j K^2(u)\,du$ be deterministic functions of the kernel function chosen by the researcher. Additionally, define Γ_p and Ψ_p as $(p+1)\times(p+1)$ matrices with element (i,j) given by ν_{i+j-2} and π_{i+j-2}, respectively. Finally, Λ_p is the column vector with typical element $(j,1)$ given by ν_{j+p+1}.

Lemma A.1. *Under Assumptions 1–6 and $Gh \to \infty$.*

$$
\begin{aligned}
G^{-1}\Lambda_{+,p} &= \tilde{\Lambda}_{+,p} + o_p(1) \\
&\quad \text{where } \tilde{\Lambda}_{+,p} = N_g \int_0^\infty K(u) r_p(u) u^{p+1} f(uh)\, \mathrm{d}u, \\
G^{-1}\Lambda_{-,p} &= (-1)^{p+1} H_p(-1) \tilde{\Lambda}_{-,p} + o_p(1) \\
&\quad \text{where } \tilde{\Lambda}_{-,p} = N_g \int_0^\infty K(u) r_p(u) u^{p+1} f(-uh)\, \mathrm{d}u, \\
G^{-1}\Gamma_{+,p} &= \tilde{\Gamma}_{+,p} + o_p(1) \\
&\quad \text{where } \tilde{\Gamma}_{+,p} = N_g \int_0^\infty K(u) r_p(u) r_p(u)' f(uh)\, \mathrm{d}u, \\
G^{-1}\Gamma_{-,p} &= H_p(-1) \tilde{\Gamma}_{-,p} H_p(-1) + o_p(1) \\
&\quad \text{where } \tilde{\Gamma}_{-,p} = N_g \int_0^\infty K(u) r_p(u) r_p(u)' f(-uh)\, \mathrm{d}u,
\end{aligned}
$$

where $H_p(-1) = \mathrm{diag}\left(1, (-1)^{-1}, \ldots, (-1)^{-p}\right)$. *Also, if* $h \to 0$.

$$
\begin{aligned}
\tilde{\Lambda}_{+,p} &= N_g f(0) \Lambda_p + o(1), \\
\tilde{\Lambda}_{-,p} &= N_g f(0) \Lambda_p + o(1), \\
\tilde{\Gamma}_{+,p} &= N_g f(0) \Gamma_p + o(1), \\
\tilde{\Gamma}_{-,p} &= N_g f(0) \Gamma_p + o(1).
\end{aligned}
$$

Proof. To obtain the results, note that $\Lambda_{+,p}$, $\Lambda_{-,p}$, $\Gamma_{+,p}$, and $\Gamma_{-,p}$ can be written as functions of $F_{+,j}$ and $F_{-,j}$ with $F_{+,j} = \frac{1}{Gh}\sum_{g=1}^{G} \sum_{i=1}^{N_g} \mathbf{1}\left[X_{ig} > 0\right] \; K\left(X_{h,ig}\right) X_{h,ig}^{j} = \frac{1}{G}\sum_{g=1}^{G} N_g \frac{1}{N_g h}\sum_{i=1}^{N_g} \mathbf{1}\left[X_{ig} > 0\right] \; K\left(X_{h,ig}\right) X_{h,ig}^{j} = \frac{1}{G}\sum_{g=1}^{G} N_g A_{+,jg}$, where $A_{+,jg} = \frac{1}{N_g h}\sum_{i=1}^{N_g} \mathbf{1}\left[X_{ig} > 0\right] K\left(X_{h,ig}\right) X_{h,ig}^{j}$. And similarly for $F_{-,j}$ with $\mathbf{1}\left[X_{ig} \leq 0\right]$ replacing $\mathbf{1}\left[X_{ig} > 0\right]$. If N_g is equal for all G clusters, then $F_{+,j} = \frac{1}{G}\sum_{g=1}^{G} A_{jg}$. Under Assumptions 1–6, (i) for non-negative integer j

$$
\begin{aligned}
E\left[A_{+,jg}\right] &= E\left[\frac{1}{N_g h}\sum_{i=1}^{N_g} \mathbf{1}\left[X_{ig} > 0\right] K\left(X_{h,ig}\right) X_{h,ig}^{j}\right] = h^{-1}\int_0^\infty K\left(\frac{x}{h}\right)\left(\frac{x}{h}\right)^{j} f(x)\, \mathrm{d}x \\
&= \int_0^\infty K(u) u^{j} f(uh)\, \mathrm{d}u = \tilde{A}_{jg}
\end{aligned}
$$

and $\tilde{F}_{+,j} = N_g \int_0^\infty K(u)u^j f(uh)\, du$. When $h \to 0$,

$$\tilde{A}_{+,jg} = f(0)v_j + O(1).$$

Then,

$$\tilde{F}_{+,j} = E\left[F_{+,j}\right] = \frac{1}{G}\sum_{g=1}^{G} N_g \tilde{A}_{+,jg} = N_g f(0)v_j + O(1).$$

For the variance,

$$\begin{aligned} \operatorname{Var}\left[A_{+,jg}\right] &= E\left[A_{+,jg}^2\right] - E\left[A_{+,jg}\right]^2 \\ &\leq \frac{1}{N_g^2 h^2} E\left[\sum_{i=1}^{N_g} \mathbf{1}\left[X_{ig} > 0\right] K^2\left(X_{h,ig}\right) X_{h,ig}^{2j}\right] \\ &= \frac{1}{N_g h}\int_0^\infty K^2(u)u^{2j} f(uh)\, du = O\left(h^{-1}\right). \end{aligned}$$

By noting that A_{jg} are independent across clusters.

$$\begin{aligned} \operatorname{Var}\left[F_{+,j}\right] &= \operatorname{Var}\left[\frac{1}{G}\sum_{g=1}^{G} N_g A_{jg}\right] = \frac{1}{G^2}\sum_{g=1}^{G} N_g^2 \operatorname{Var}\left[A_{jg}\right] = \frac{1}{G^2}\sum_{g=1}^{G} O\left(h^{-1}\right) \\ &= O\left(G^{-1}h^{-1}\right) = o(1). \end{aligned}$$

Then,

$$F_{+,j} = \tilde{F}_{+,j} + o_p(1).$$

Noting that $G^{-1}\Gamma_{+,p}$ and $G^{-1}\Gamma_{-,p}$ have typical element (i, j) given by $F_{+,i+j-2}$ and $F_{-,i+j-2}$, respectively gives the result. Similarly, $G^{-1}\Lambda_{+,p}$ and $G^{-1}\Lambda_{-,p}$ are column vectors with typical element $(j, 1)$ given by $F_{+,p+1+j}$ and $F_{-,p+1+j}$, respectively. ■

The following discussion concerns the variance terms which are altered to allow for clustering.

Lemma A.2. *Under Assumptions 1–6 above and $Gh \to \infty$.*

$$
\begin{aligned}
G^{-1}\Psi_{+,p} = &\ \tilde{\Psi}_{+,p} + o_p(1) \text{ where } \tilde{\Psi}_{+,p} \\
= &\ \int_0^\infty \int_0^\infty K(u)K(w)r_p(u)\sum r_p(w)' f(uh, wh)\, du\, dw, \\
G^{-1}\Psi_{-,p} = &\ H_p(-1)\tilde{\Psi}_{-,p}H_p(-1) + o_p(1) \text{ where } \tilde{\Psi}_{-,p} \\
= &\ \int_0^\infty \int_0^\infty K(u)K(w)r_p(u)\sum r_p(w)' f(-uh, -wh)\, du\, dw.
\end{aligned}
$$

If the cluster is defined at the tunning variable level, that is, $X_{ig} = X_g$ for all $i = 1, \ldots, N_g$, and $h \to 0$. Then,

$$
\begin{aligned}
\tilde{\Psi}_{+,p} &= f(0)\Psi_p \sum_{i=1}^{N_g}\sum_{s=1}^{N_g} \sigma_{+,is} + o(1), \\
\tilde{\Psi}_{-,p} &= f(0)\Psi_p \sum_{i=1}^{N_g}\sum_{s=1}^{N_g} \sigma_{-,is} + o(1).
\end{aligned}
$$

Proof. Start by noticing that the typical element (t, j) in matrix $G^{-1}\Psi_{+,p}$ is given by $Q_{+,tj} = \frac{1}{Gh^2}\sum_{g=1}^{G}\sum_{i=1}^{N_g}\sum_{s=1}^{N_g} \mathbf{1}[x_{ig} > 0]\mathbf{1}[x_{sg} > 0]$ $K(x_{h,ig})K(x_{h,sg})x^t_{h,ig}x^j_{h,sg}\sigma_{gis}$, where σ_{gtj} is the term in the t-th line and j-th column in Ω_g. Then,

$$
\begin{aligned}
E[Q_{+,tj}] &= E\left[\frac{1}{Gh^2}\sum_{g=1}^{G}\sum_{i=1}^{N_g}\sum_{s=1}^{N_g}\mathbf{1}[x_{ig} > 0]\mathbf{1}[x_{sg} > 0]K(x_{h,ig})K(x_{h,sg})x^t_{h,ig}x^j_{h,sg}\sigma_{gis}\right] \\
&= \frac{1}{h^2}\int_0^\infty \int_0^\infty \sum_{i=1}^{N_g}\sum_{s=1}^{N_g} K(x_{h,ig})K(x_{h,sg})x^t_{h,ig}x^j_{h,sg}\sigma_{gis} f(x_i, x_s)\, dx_i\, dx_s \\
&= \int_0^\infty \int_0^\infty \sum_{i=1}^{N_g}\sum_{s=1}^{N_g} K(u)K(w)u^t w^j \sigma_{gis} f(uh, wh)\, du\, dw = \tilde{Q}_{+,tj}.
\end{aligned}
$$

The $\mathrm{Var}(Q_{+,tj})$ can be bounded by arguments similar to the ones used in Lemma A.1. Then,

$$
Q_{+,tj} = \tilde{Q}_{+,tj} + o_p(1).
$$

Furthermore, if the cluster is defined at the tunning variable level, that is, $X_{ig} = X_g$ for all $i = 1, \ldots, N_g$, and $h \to 0$. Then,

$$\tilde{Q}_{+,tj} = E\left[Q_{tj}\right] = f(0)\pi_{t+j} \sum_{i=1}^{N_g} \sum_{s=1}^{N_g} \sigma_{+,is}(0) + O(1), \text{ and}$$
$$= f(0)\pi_{t+j} \sum_{i=1}^{N_g} \sum_{s=1}^{N_g} \sigma_{+,is} + o_p(1).$$

Noting that $G^{-1}\Psi_{+,p}$ and $G^{-1}\Psi_{-,p}$ have typical element (i, j) given by $Q_{+,tj}$ and $Q_{-,tj} = \frac{1}{Gh^2} \sum_{g=1}^{G} \sum_{i=1}^{N_g} \sum_{s=1}^{N_g} \mathbf{1}\left[x_{ig} \leq 0\right] \mathbf{1}\left[x_{sg} \leq 0\right] \; K\left(x_{h,ig}\right) K\left(x_{h,sg}\right) x_{h,ig}^{t} x_{h,sg}^{j} \sigma_{gis}$, respectively gives the result. ■

BOOTSTRAP CONFIDENCE INTERVALS FOR SHARP REGRESSION DISCONTINUITY DESIGNS*

Otávio Bartalotti, Gray Calhoun and Yang He

Department of Economics, Iowa State University, Ames, IA, USA

ABSTRACT

This chapter develops a novel bootstrap procedure to obtain robust bias-corrected confidence intervals in regression discontinuity (RD) designs. The procedure uses a wild bootstrap from a second-order local polynomial to estimate the bias of the local linear RD estimator; the bias is then subtracted from the original estimator. The bias-corrected estimator is then bootstrapped itself to generate valid confidence intervals (CIs). The CIs generated by this procedure are valid under conditions similar to Calonico, Cattaneo, and Titiunik's (2014) analytical correction – that is, when the bias of the naive RD estimator would otherwise prevent valid inference. This chapter also provides simulation evidence that our method

*Source code for the Monte Carlo experiments and empirical component of this chapter are available online at https://github.com/grayclhn/boot-rd

Regression Discontinuity Designs: Theory and Applications
Advances in Econometrics, Volume 38, 421–453

ISSN: 0731-9053/doi:10.1108/S0731-905320170000038018

is as accurate as the analytical corrections and we demonstrate its use through a reanalysis of Ludwig and Miller's (2007) Head Start dataset.

Keywords: Regression discontinuity; bias correction; wild bootstrap; iterated bootstrap

JEL classification: C14; C15; C21

1. INTRODUCTION

Regression discontinuity (RD) designs have emerged in the last decade as an important and popular research design strategy for analyzing the causal impact of policies and interventions in several fields of the social sciences, including economics, political science, public policy, and sociology. This research strategy exploits the fact that many programs use a threshold based on a numeric score to determine whether or not to provide a treatment.[1] In its basic version, *sharp RD*, individuals or groups with score above the threshold are treated while those below the threshold are left untreated. The identification of the treatment effect at the threshold is then based on comparing treated and untreated units at the cutoff. When a subject's position just above or below the cutoff is credibly not related to unobserved characteristics that would affect the outcome of interest, differences between treated and untreated individuals at the cutoff can be plausibly attributed to the treatment alone. As a practical matter this involves comparing units within a bandwidth just above and just below the threshold. Other RD strategies exist that can exploit different forms of discontinuities as well.

The RD design strategy was introduced by Thistlethwaite and Campbell (1960) to study educational outcomes, and many of its recent applications in economics were to estimate the effects of other educational policies: evaluating the impact of investments in school facilities, class sizes, remedial education, early childhood education, and financial aid effects on student achievement and later outcomes, for example.[2] But the underlying identification strategy has proven to apply much more widely and RD has been used in health economics,[3] political science,[4] and labor economics,[5] among other fields. Imbens and Lemieux (2008) and Lee and Lemieux (2010) provide recent overviews of this literature with many more examples.

In these studies, identification occurs exactly at the cutoff, so the treatment effect is typically estimated by fitting separate local linear models above and below the cutoff, then extrapolating the models to the exact point of discontinuity. The difference between the estimated outcomes at that point is taken to be an estimate of the treatment effect. As a practical matter, a key econometric issue is determining the bandwidth for the local linear models. One very popular choice is the bandwidth estimator proposed by Imbens and Kalyanaraman (2012) and extended by Calonico, Cattaneo, and Titiunik (2014), henceforth "CCT," which minimizes the asymptotic mean-squared error (AMSE) of difference in the models' point estimators at the cutoff. But, as observed by CCT, the AMSE-optimal bandwidth has the serious drawback that it produces invalid confidence intervals (CIs) and hypothesis tests. Local polynomial estimators of the treatment effect are generally biased in finite samples because the functional form of the local conditional expectation that they need to approximate is unknown. The unmodeled component of the conditional expectation becomes smaller as the bandwidth itself becomes smaller, so the estimator's bias vanishes asymptotically as long as the bandwidth shrinks as the sample size increases. AMSE-optimal bandwidth shrinks as the sample size increases, so its estimator of the treatment effect is consistent, but the bandwidth shrinks slowly enough that the remaining bias term is large enough to affect the asymptotic distribution of the estimator. Consequently, the usual "naive" CIs for the RD treatment effects are invalid and can have coverage well below their nominal level.

CCT show that the bias resulting from undersmoothing can be estimated and they provide a bias-corrected treatment effect estimator that remains asymptotically unbiased even when the bandwidth converges to zero at the AMSE-optimal rate. They also show that this bias-correction term contributes to the asymptotic variance of the resulting treatment effect estimator and provide a new formula for the asymptotic variance of the bias-corrected estimator. The resulting CIs have accurate coverage even when the naive RD interval does not. Furthermore, as shown by Calonico, Cattaneo, and Farrell (2017), these intervals can achieve faster convergence rates than previously believed, outperforming intervals produced by undersmoothing in some settings while maintaining greater robustness to the choice of bandwidth parameter. Previous research by Hall (1993) and Neumann (1997) has found that undersmoothing outperforms naive bias-corrected CIs, but Calonico et al. (2017) show that the naive variance estimator implicitly requires more restrictions than CCT's expression and those restrictions artificially limit the intervals' performance.

In this chapter, we propose a bootstrap alternative to CCT's analytical corrections. CCT motivate their estimator by showing that the bias and variance components for the local linear estimator can be accounted for by estimating a local second-order polynomial with bandwidth of the same order.[6] They use a Taylor expansion around the cutoff to show that the bias associated with the second-order polynomial converges to zero at a faster rate, fast enough that the bias of the local linear model can be estimated and removed using the second-order polynomial. Additionally, that approximation provides fast enough convergence that it can be used to estimate the correct variance correction as well.

Our approach exploits CCT's theoretical insight through a new wild bootstrap. In particular, we propose estimating the local linear model as usual, but generating bootstrap datasets from an approximation provided by a local second-order polynomial. Since the second-order polynomial is the true data-generating process (DGP) for the bootstrapped dataset, its estimate of the treatment effect is the true value of the treatment effect under the distribution induced by this bootstrap. The bias of the linear model is therefore known under this distribution and can be calculated by averaging the error of the linear model's estimates across many bootstrap replications. This approach is described in detail by our Algorithm 1 and the resulting bias-corrected estimator is shown to be equivalent to CCT's, up to simulation error in the bootstrap, and asymptotically normal with mean zero in our Theorem 1 under AMSE-optimal bandwidth rates.

Just as in CCT, our bias correction step introduces additional variability. However, the second-order polynomial again adequately estimates the features of the true DGP that are necessary for estimating and accommodating that additional variability. So we propose an iterated bootstrap procedure (Hall & Martin, 1988): use the second-order polynomial bootstrap to produce many bootstrap replications of the bias-corrected estimator, and then use the resulting bootstrap distribution to produce CIs. This procedure, which requires bootstrapping the datasets produced by an initial bootstrap, is described in Algorithm 2 and the resulting CIs are shown to be asymptotically valid in Theorem 2.[7] Simulation results presented in Section 4 indicate that this iterated bootstrap provides very similar results in finite samples to CCT's analytical CIs as well.

The methods presented in this chapter are intended as a "proof of concept" that demonstrates that the analytical correction proposed by CCT can also be implemented through a new bootstrap procedure. This procedure is promising because it, like other bootstrap methods, can allow applied researchers to construct accurate CIs in settings where correct formulas for the asymptotic variance have not yet been derived, as can be the case when there is an unusual form of dependence – the dependence can be accommodated in principle by modifying the resampling algorithm to one that matches the dependence in the dataset. The specific version of the bootstrap presented in this chapter, however, does not achieve that level of generality, and one limitation should be noted in particular. The CIs produced by this bootstrap depend on two separate bandwidth parameters that must be generated elsewhere. Since these bandwidths are generated using analytical methods, the CIs could easily be produced using the same methods. Moreover, in this chapter we focus on the baseline case of sharp RD with a local linear model, however, the results and procedures can be easily extended to the general local polynomial and higher order derivatives cases (e.g., Regression Kink Designs, see Card, Lee, & Pei, 2009; Card, Lee, Pei, & Weber, 2015, 2017).

The chapter is organized as follows. Section 2 describes the basic RD approach, its usual implementation, and the explicit analytical bias correction approach in the literature. Section 3 presents our proposed bootstrap bias-corrected RD algorithm and discusses its asymptotic properties. Simulation evidence that the bootstrap procedure provides valid CIs and its relative performance to the analytical bias correction are presented in Sections 4 and 5 demonstrate the estimator's usage by applying it to the Head Start dataset used by Ludwig and Miller (2007).[8] Finally, Section 6 concludes.

2. BACKGROUND

This section provides additional details of RD estimators in general and of CCT's proposed bias correction. It also defines some of the notation and presents the assumptions that will be used for our theoretical analysis in Section 3. We have adopted CCT's notation where possible to aid readers familiar with that paper.

In the typical sharp RD setting, a researcher wishes to estimate the local causal effect of treatment at a given threshold. A running variable, X_i, determines treatment assignment. Given a known threshold, which we will set to zero without loss of generality, the ith subject receives the treatment of interest if $X_i \geq 0$ and does not receive treatment if $X_i < 0$.

Subject i's potential outcomes are denoted by the variable $Y_i(\cdot)$; $Y_i(1)$ is the subject's outcome under treatment and $Y_i(0)$ is the outcome without treatment. Since only one of the two outcomes is observed, the sample is comprised of the running variable, X_i, and the observed outcome Y_i, where

$$Y_i = Y_i(0)\mathbb{1}\{X_i < 0\} + Y_i(1)\mathbb{1}\{X_i \geq 0\}$$

and $\mathbb{1}\{\cdot\}$ denotes the indicator function.

In most cases, the population parameter of interest is the average treatment effect (ATE) at the cutoff, which we will denote τ. This parameter is the difference in expected potential outcomes given $X_i = 0$; formally,

$$\tau = \mathbb{E}(Y(1) - Y(0)|X = 0)$$

with $\mathbb{E}$ representing the expectation. Hahn, Todd, and Van der Klaauw (2001) show that the effect τ is identified under continuity and smoothness conditions on the joint distribution of X_i, $Y_i(0)$, and $Y_i(1)$ around the cutoff $X_i = 0$. Under these conditions, which are made precise in our Assumption 1, τ is equal to

$$\tau = \lim_{x \to 0+} \mu(x) - \lim_{x \to 0-} \mu(x),$$

where

$$\mu(x) = \mathbb{E}(Y_i|X_i = x).$$

For later convenience, also define the derivatives

$$\mu^{(\eta)}(x) = \frac{\mathrm{d}^{\eta}\mu(x)}{\mathrm{d}x^{\eta}}$$

and let

$$\begin{aligned} \mu_+(x) &= \mathbb{E}(Y_i(1)|X_i = x), \quad \mu_-(x) = \mathbb{E}(Y_i(0)|X_i = x), \\ \sigma^2_+(x) &= \mathbb{V}(Y_i(1)|X_i = x), \quad \sigma^2_-(x) = \mathbb{V}(Y_i(0)|X_i = x), \end{aligned}$$

and

$$\mu^{(\eta)}_+ = \lim_{x \to 0^+} \mu^{(\eta)}(x), \;\; \mu^{(\eta)}_- = \lim_{x \to 0^-} \mu^{(\eta)}(x),$$

where the symbol $\mathbb{V}(\cdot)$ represents the variance. Assumption 1 lists standard conditions in the RD literature which ensure that both μ_- and μ_+ can be estimated consistently.

Assumption 1. (Behavior of the DGP near the cutoff). *The random variables $\{Y_i, X_i\}_{i=1}^n$ form a random sample of size n. There exists a positive number κ_0 such that the following conditions hold for all x in the neighborhood $(-\kappa_0, \kappa_0)$ around zero:*

1. *The density of X_i is continuous and bounded away from zero at x.*
2. *$\mathbb{E}(Y_i^4|X_i = x)$ is bounded.*
3. *$\mu_+(x)$ and $\mu_-(x)$ are both 3 times continuously differentiable.*
4. *$\sigma_+^2(x)$ and $\sigma_-^2(x)$ are both continuous and bounded away from zero.*

Part 1 and the continuity component of Part 3 of Assumption 1 are crucial for identifying τ – informally, they ensure that the number of points arbitrarily close to the cutoff increases with the sample size and that there is a small neighborhood around the cutoff in which the change in the conditional expectation can be attributed to the change in treatment status. (See Hahn et al., 2001; Porter, 2003, for more detailed discussion of identification in RD designs.) The additional smoothness conditions in Part 3 allow higher order polynomials to be used for estimation and Parts 2 and 4 ensure that the estimated local polynomials are well behaved.

Since these conditions only need to hold in a neighborhood around the cutoff, μ_+ and μ_- can be estimated by extrapolating from a local polynomial regression. We will focus here on local linear regression.[9] For this model, if h represents a bandwidth parameter and $K(\cdot)$ a density function, the estimator of τ, $\widehat{\tau}(h)$, is defined as

$$\widehat{\tau}(h) = \widehat{\mu}_+(h) - \widehat{\mu}_-(h)$$

with

$$\widehat{\mu}_+(h) = \arg\min_{\beta_0} \min_{\beta_1} \sum_{i=1}^{n} \mathbb{1}\{X_i \geq 0\}\left(Y_i - \beta_0 - X_i\beta_1\right)^2 \frac{1}{h} K\left(X_i/h\right)$$

and

$$\widehat{\mu}_-(h) = \arg\min_{\beta_0} \min_{\beta_1} \sum_{i=1}^{n} \mathbb{1}\{X_i < 0\}\left(Y_i - \beta_0 - X_i\beta_1\right)^2 \frac{1}{h} K\left(X_i/h\right).$$

Conventional (naive) CIs can be calculated by using an asymptotic approximation for $\widehat{\tau}(h)$. In particular, if

$$\frac{\hat{\tau}(h) - \tau}{\sqrt{V(h)}} \to^d N(0,\ 1), \tag{1}$$

with

$$V(h) = \mathbb{V}(\hat{\tau}(h)|X_1, \ldots, X_n),$$

then valid CIs can be constructed through the usual method of inverting the t-test.[10] This procedure gives the widely used interval estimator

$$\hat{\tau}(h)\ \pm\ q_{1-\alpha/2}V(h)^{1/2},$$

where $q_{1-\alpha/2}$ is the $1-\alpha/2$ quantile of the standard normal distribution.

The statistical properties of these estimators, however, clearly depend on the bandwidth parameter h, and bandwidths that have desirable properties for point estimation may not have desirable properties for hypothesis testing or interval estimation. In particular, for Eq. (1) to hold, h must satisfy $nh \to \infty$ and $nh^5 \to 0$ (Hahn et al., 2001; Porter, 2003). Otherwise, the finite-sample bias of $\hat{\tau}(h)$ does not converge in probability to zero quickly enough and it contributes to the asymptotic distribution in Eq. (1). This result holds even though $\hat{\tau}(h)$ can be consistent under those conditions. These bandwidth issues are relevant in practice because many widely used bandwidth selection procedures, most notably the AMSE-optimal bandwidth and cross-validation bandwidth (Imbens & Kalyanaraman, 2012), do not produce $o_p\left(n^{-1/5}\right)$ bandwidths.

CCT solve this problem by deriving the analytical form of the first-order bias and explicitly recentering $\hat{\tau}(h)$. Under weaker assumptions on the asymptotic behavior of the bandwidth, which we will specify in Assumption 2, CCT show that the approximate bias of $\hat{\tau}(h)$ has the form

$$\mathbb{E}\left(\hat{\tau}(h)|X_1, \ldots, X_n\right) - \tau = h^2\left[\frac{\mu_+^{(2)}}{2}\mathfrak{B}_+(h) - \frac{\mu_-^{(2)}}{2}\mathfrak{B}_-(h)\right]\left(1 + o_p(1)\right),$$

where $\mathfrak{B}_+(h)$ and $\mathfrak{B}_-(h)$ are observed quantities that depend on the kernel, bandwidth, and running variables $X_1, \ldots, X_n$; formal definitions of these terms are given in the Mathematical Appendix. The plug-in bias-corrected estimator then requires estimates for the second derivatives of the conditional mean from above and below the cutoff, $\mu_+^{(2)}$ and $\mu_-^{(2)}$, and CCT show that these derivatives can be estimated by fitting a second-order local polynomial, that is, one order higher than the polynomial used to obtain $\hat{\tau}$, using a (potentially) different pilot bandwidth b. Their procedure gives the bias-corrected estimator

$$\hat{\tau}^{bc}(h, b) = \hat{\tau}(h) - h^2 \left[\frac{\hat{\mu}_+^{(2)}(b)}{2} \mathfrak{B}_+(h) - \frac{\hat{\mu}_-^{(2)}(b)}{2} \mathfrak{B}_-(h) \right].$$

The variance introduced by the bias-correction term does not vanish, so the naive CI needs not only to be re-centered to correct the bias, but also rescaled to allow for the additional variability introduced by the bias correction, resulting in the following asymptotic approximation:

$$\frac{\hat{\tau}^{bc}(h, b) - \tau}{V^{bc}(h, b)^{1/2}} \to^d N(0, 1)$$

with $V^{bc}(h, b) = V(h) + C(h, b)$ and $C(h, b)$ an additional variance component generated by the bias-correction term.[11] This new approximation can be used instead of Eq. (1) to construct "bandwidth robust" CIs, and CCT provide simulation evidence that their intervals perform well in finite samples even when the naive interval performs badly.

Assumption 2 specifies the bandwidth and kernel conditions assumed in this chapter, which are standard in this literature.

> **Assumption 2. (Bandwidth and kernel).** *Let h be the bandwidth used to estimate the local linear model and let b be the bandwidth used to estimate a second local quadratic model. Then $nh \to \infty$, $nb \to \infty$, and $n \times \min(h, b)^5 \times \max(h, b)^2 \to 0$ as $n \to \infty$.*[12] *The kernel function K is positive, bounded, and continuous on the interval $[-\kappa, \kappa]$ and zero outside that interval for some $\kappa > 0$.*

To simplify notation, let $m = \min(h, b)$ and define the scaled and truncated kernel functions

$$K_{+,h}(x) = \frac{1}{h} K(x/h) \mathbb{1}\{x \geq 0\} \quad K_{-,h}(x) = \frac{1}{h} K(x/h) \mathbb{1}\{x < 0\}$$

and

$$K_{+,b}(x) = \frac{1}{b} K(x/b) \mathbb{1}\{x \geq 0\} \quad K_{-,b}(x) = \frac{1}{b} K(x/b) \mathbb{1}\{x < 0\}.$$

Assumption 2 allows for the vast majority kernel functions commonly used in practice. Note that different kernel functions could be used for the treatment effect estimator and the higher order approximation used for the bootstrap.[13] For simplicity, however, we use the same function for all steps of estimation in this chapter. Potentially different kernel functions and bandwidths could also be used for estimation below and above the cutoff, but we again focus on the simpler case here.

In the next section, we build upon the insight provided by CCT's bias-corrected interval and propose a simple bootstrap procedure that can directly construct the robust CIs without requiring the derivation of analytical formulas and direct estimators for the bias, variance and covariance terms, while relying on the same first-order bias correction approximation.

3. BOOTSTRAP BIAS CORRECTION

This section presents our theoretical contributions. We propose two algorithms in this section. The first uses a wild bootstrap based on a second-order local polynomial to estimate the bias of the local linear model. That estimate can be subtracted from the biased original estimator to provide an asymptotically unbiased estimator with the same asymptotic distribution as CCT's. As in CCT, this bias correction term introduces a new source of variance, invalidating standard (naive) critical values. Consequently, the second algorithm we propose uses an iterated bootstrap to estimate the correct critical values of the bias-corrected estimator. The intuition behind both procedures is straightforward. CCT show that a local second-order polynomial captures the aspects of the DGP necessary for constructing valid CIs. Our proposed algorithms estimate and embed the second-order behavior in the bootstrap DGP through a wild bootstrap.

Throughout this section and the rest of the chapter, we will let $\mathbb{E}^*$, $\mathbb{P}^*$, etc. denote expectations and probabilities taken with respect to the distribution induced by the bootstrap (which implicitly conditions on $X_1, \dots, X_n$ and $Y_1, \dots, Y_n$) and let parameters with * superscripts be the parameter values under the distribution induced by the bootstrap. Two * superscripts indicate that the parameter or probability measure corresponds to a secondary bootstrap distribution.

Algorithm 1 explains the bias-correction steps in detail.

Algorithm 1. (Bias estimation). *Assume h and b are bandwidths as defined by Assumption 2.*

1. *Estimate local second-order polynomials* $\hat{g}_-$ *and* $\hat{g}_+$ *with least squares using* $K_{-,b}$ *and* $K_{+,b}$ *for weights:*

$$\hat{g}_-(x) = \widehat{\beta}_{-,0} + \widehat{\beta}_{-,1}x + \widehat{\beta}_{-,2}x^2, \qquad \hat{g}_+(x) = \widehat{\beta}_{+,0} + \widehat{\beta}_{+,1}x + \widehat{\beta}_{+,2}x^2 \quad (2)$$

with

$$\left(\widehat{\beta}_{-,0}, \widehat{\beta}_{-,1}, \widehat{\beta}_{-,2}\right)' = \underset{\beta_0,\beta_1,\beta_2}{\arg\min} \sum_{i=1}^{n} \left(Y_i - \beta_0 - \beta_1 X_i - \beta_2 X_i^2\right)^2 K_{-,b}(X_i),$$

$$\left(\widehat{\beta}_{+,0}, \widehat{\beta}_{+,1}, \widehat{\beta}_{+,2}\right)' = \underset{\beta_0,\beta_1,\beta_2}{\arg\min} \sum_{i=1}^{n} \left(Y_i - \beta_0 - \beta_1 X_i - \beta_2 X_i^2\right)^2 K_{+,b}(X_i).$$

Let

$$\hat{g}(x) = \begin{cases} \hat{g}_-(x) & \textit{if} \quad x<0 \\ \hat{g}_+(x) & \textit{otherwise} \end{cases}$$

and calculate the residuals $\widehat{\varepsilon}_i = Y_i - \hat{g}(X_i)$ *for all i.*

2. *Repeat the following steps* B_1 *times to produce the bootstrap estimates* $\widehat{\tau}_1^*(h), \ldots, \widehat{\tau}_{B_1}^*(h)$. *For the kth replication:*
   (a) *Draw i.i.d. random variables* e_i^* *with mean zero, variance one, and bounded fourth moments independent of the original data and construct*

   $$\varepsilon_i^* = \widehat{\varepsilon}_i e_i^*,$$

   and

   $$Y_i^* = \hat{g}(X_i) + \varepsilon_i^*$$

   for all i.
   (b) *Calculate* $\widehat{\mu}_+^*(h)$ *and* $\widehat{\mu}_-^*(h)$ *by estimating the local linear model on the bootstrap data set using* $K_{+,h}$ *and* $K_{-,h}$ *for weights:*

$$\widehat{\mu}^*_-(h) = \arg\min_{\mu}\min_{\beta}\sum_{i=1}^{n}\left(Y_i^* - \mu - \beta X_i\right)^2 K_{-,h}(X_i),$$

$$\widehat{\mu}^*_+(h) = \arg\min_{\mu}\min_{\beta}\sum_{i=1}^{n}\left(Y_i^* - \mu - \beta X_i\right)^2 K_{+,h}(X_i).$$

(c) Save $\widehat{\tau}^*_k(h) = \widehat{\mu}^*_+(h) - \widehat{\mu}^*_-(h)$.

3. *Estimate the bias as*

$$\Delta^*(h, b) = \frac{1}{B_1}\sum_{k=1}^{B_1}\widehat{\tau}^*_k(h) - \left[\hat{g}_+(0) - \hat{g}_-(0)\right]. \tag{3}$$

Note that $\hat{g}_+(0) - \hat{g}_-(0)$ is the true treatment effect under the distribution induced by this bootstrap. The bootstrap estimator works by constructing an approximate DGP with known properties. As the dataset gets larger, the approximate DGP mimics the unknown real DGP more closely, and the population parameter values in the bootstrap DGP can become accurate estimates of the true parameter values in the real DGP.

Under Assumptions 1 and 2, the procedure described by Algorithm 1 provides a consistent estimator of the bias component that converges fast enough in probability that it can be used as a correction. As is standard in the bootstrap literature, we will assume that the number of bootstrap replications, B_1, is large enough that the simulation error can be ignored. Theorem 1 presents the result formally.

Theorem 1. *Under Assumptions 1 and 2,*

$$\Delta^*(h, b) = h^2\left[\frac{\widehat{\mu}^{(2)}_+(b)}{2}\mathfrak{B}_+(h) - \frac{\widehat{\mu}^{(2)}_-(b)}{2}\mathfrak{B}_-(h)\right] \tag{4}$$

almost surely and

$$\frac{\widehat{\tau}(h) - \Delta^*(h, b) - \tau}{V^{\mathrm{bc}}(h, b)^{1/2}} \to^d N(0, 1), \tag{5}$$

where $\Delta^*(h, b)$ *is defined by Eq. (3) and* $V^{\mathrm{bc}}(h, b) = V(h) + C(h, b)$, *both which are defined in Eqs. (A.4) and (A.5).*

For practical purposes, Theorem 1 implies that our bias-corrected estimator has the same asymptotic distribution as CCT's. This equivalence should be unsurprising; both estimators use a second-order polynomial to directly estimate the bias of the local linear model, so they should behave very similarly. Equation (4) implies an even stronger relationship between the two results, that the bias correction is equivalent to CCT's up to simulation error in the bootstrap.

Since the second-order polynomial captures the relevant aspects of the DGP for estimating the variance as well as the bias, the asymptotic distribution of the bias-corrected estimator $\widehat{\tau}(h) - \Delta^*(h, b)$ can also be approximated with a bootstrap. We propose bootstrapping $\widehat{\tau}(h) - \Delta^*(h, b)$ using the same wild bootstrap method used in Algorithm 1. Algorithm 2 provides the details of our procedure and Theorem 2 establishes its theoretical properties.

Algorithm 2 (Distribution). *Assume h and b are bandwidths as defined by Assumption 2 and Algorithm 1.*

1. *Estimate $\hat{g}_+$ and $\hat{g}_-$ and generate $\hat{g}(\cdot)$ and the residuals $\widehat{\varepsilon}_i$ just as in Algorithm 1.*

2. *Repeat the following steps B_2 times to produce bootstrap estimates of the bias-corrected estimate. For the kth replication:*
   (a) Draw i.i.d. random variables e_i^* with mean zero, variance one, and bounded fourth moments independent of the original data and construct

$$\varepsilon_i^* = \widehat{\varepsilon}_i e_i^*,$$

and

$$Y_i^* = \hat{g}(X_i) + \varepsilon_i^*.$$

for all $i = 1, \ldots, n$.
   (b) Calculate $\widehat{\mu}_+^*(h)$ and $\widehat{\mu}_-^*(h)$ by estimating the local linear model on the bootstrap dataset using $K_{+,h}$ and $K_{-,h}$ for weights:

$$\widehat{\mu}_-^*(h) = \arg\min_{\mu} \min_{\beta} \sum_{i=1}^{n} \left(Y_i^* - \mu - \beta X_i\right)^2 K_{-,h}(X_i),$$

$$\widehat{\mu}_+^*(h) = \arg\min_{\mu} \min_{\beta} \sum_{i=1}^{n} \left(Y_i^* - \mu - \beta X_i\right)^2 K_{+,h}(X_i).$$

(c) *Apply Algorithm 1 to the bootstrapped data set* $(Y_1^*, X_1), \ldots, (Y_n^*, X_n)$ *set using the same bandwidths h and b that are used in the rest of this algorithm but reestimating all of the local polynomials on the bootstrap data. Generate* B_1 *new bootstrap samples and let* $\Delta^{**}(h, b)$ *represent the bias estimator returned by Algorithm 1.*

(d) *Save the estimator of* τ *and its bias,* $\widehat{\tau}_k^*(h) = \widehat{\mu}_+^*(h) - \widehat{\mu}_-^*(h)$ *and* $\Delta_k^{**}(h, b)$.

3. *Define* $\tau^* = \hat{g}_+(0) - \hat{g}_-(0)$ *and use the empirical CDF of* $\widehat{\tau}_1^*(h) - \Delta_1^{**}(h, b) - \tau^*, \ldots, \widehat{\tau}_{B_2}^*(h) - \Delta_{B_2}^{**}(h, b) - \tau^*$ *as the sampling distribution of* $\widehat{\tau}(h) - \Delta^*(h, b) - \tau$.

Theorem 2 establishes that this iterated bootstrap approximates the asymptotic distribution of the bias-corrected statistic proposed by Algorithm 1 and justifies this second algorithm. As before, we assume that B_1 and B_2 are large enough that simulation error can be ignored.

Theorem 2. *Under Assumptions 1 and 2,*

$$\mathbb{V}^*\left(\widehat{\tau}^*(h) - \Delta^{**}(h, b)\right)/V^{\text{bc}}(h, b) \to^p 1$$

and

$$\sup_x \left| \mathbb{P}^*\left[\frac{\widehat{\tau}^*(h) - \Delta^{**}(h, b) - \tau^*}{\mathbb{V}^*\left(\widehat{\tau}^*(h) - \Delta^{**}(h, b)\right)^{1/2}} \le x\right] - \mathbb{P}\left[\frac{\widehat{\tau}(h) - \Delta^*(h, b) - \tau}{V^{\text{bc}}(h, b)^{1/2}} \le x\right]\right| \to^p 0.$$

Two brief remarks on implementing these algorithms. First, many of the matrix operations required to estimate the local polynomials only involve the regressors and, therefore, only need to be computed once. For example, the parameter estimates for the linear model on the treated group are equal to

$$\left(\mathbf{X}'\mathbf{K}_+\mathbf{X}\right)^{-1}\mathbf{X}'\mathbf{K}_+\mathbf{Y} \equiv \mathbf{W}_{+,1}\mathbf{Y},$$

where

$$\mathbf{X} = \begin{pmatrix} 1 & X_1 \\ 1 & X_2 \\ \vdots & \vdots \\ 1 & X_n \end{pmatrix} \mathbf{Y} = \begin{pmatrix} Y_1 \\ Y_2 \\ \vdots \\ Y_n \end{pmatrix} \mathbf{K}_+ = \begin{pmatrix} K_{+,h}(X_1) & 0 & \cdots & 0 \\ 0 & K_{+,h}(X_2) & \cdots & 0 \\ 0 & 0 & \ddots & 0 \\ 0 & 0 & \cdots & K_{+,h}(X_n) \end{pmatrix}.$$

The matrix $\mathbf{W}_{+,1}$ is not affected by the bootstrap so it can be computed once initially and reused in the bootstrap calculations. Then each bootstrap replication just requires a single matrix–vector multiplication. The other parameter estimates have this property as well. Consequently, the computational burden of this iterated bootstrap is much lower than might appear.[14]

Second, if the error terms have pronounced heteroskedasticity, it can be advantageous to transform the residuals before applying the bootstrap, as discussed by Davidson and Flachaire (2008) – these transformations typically do not affect the first-order asymptotic properties of the bootstrap distribution but can provide improvements in finite samples analogous to alternative heteroskedasticity-robust variance estimators (MacKinnon & White, 1985; MacKinnon, 2013). Since RD estimators rely on a small subset of the data available in any application, these finite-sample adjustments can be especially effective in improving performance, and we use the "HC3" transformation proposed by Davidson and Flachaire (2008) in this chapter's Monte Carlo simulations and empirical exercise; that is, Algorithms 1 and 2 use the transformed residuals

$$\frac{\widehat{\varepsilon}_i}{1 - H_{ii}} \tag{6}$$

instead of $\widehat{\varepsilon}_i$, with H_{ii} the (i, i) element of the matrix

$$\mathbf{H} = \mathbf{K}_+^{1/2}\mathbf{X}(\mathbf{X}'\mathbf{K}_+\mathbf{X})^{-1}\mathbf{X}'\mathbf{K}_+^{1/2}. \tag{7}$$

The matrix $\mathbf{H}$ is the projection matrix onto the kernel-weighted design matrix.

Evidence of the usefulness of the procedures proposed above and their relative performance to the analytical bias correction proposed in CCT are presented in a series of Monte Carlo simulations in Section 4.

4. SIMULATION EVIDENCE

This section presents evidence from Monte Carlo simulations that the bootstrap procedures proposed in Section 3 produce valid, robust CIs similar to those obtained by the analytical procedures established in CCT. We also observe that all of the bias-corrected methods can offer substantial improvements over estimators that do not bias correct.

The Monte Carlo experiments have a similar structure. For all of them, we generate 500 i.i.d. observations from the DGP

$$\begin{aligned} Y_i &= \mu_j(X_i) + \varepsilon_i, \\ X_i &\sim 2 \times \text{beta}(2, 4) - 1, \\ \varepsilon_i &\sim N\left(0,\ 0.1295^2 \times \sigma_i^2\right), \\ \sigma_i &= z(\gamma)\mu_j(X_i)^\gamma, \end{aligned}$$

where j indexes the specific DGP and $z(\gamma)$ is a scale parameter that ensures that the average variance in a neighborhood around the cutoff equals the unconditional value under homoskedasticity; that is, $z(\gamma)$ satisfies

$$z(\gamma)^2 = \frac{\#\{X_i\ :\ \min\left(-\bar{h}, -\bar{b}\right) \leq X_i \leq \max(\bar{h}, \bar{b})\}}{\sum_{i=1}^{n} \mu_j(X_i)^{2\gamma} \mathbb{1}\{\min(-\bar{h}, -\bar{b}) \leq X_i \leq \max(\bar{h}, \bar{b})\}}$$

For $\gamma = 0$, we have $z(\gamma) = 1$ and the errors are homoskedastic.[15] This is the experimental design used by Imbens and Kalyanaraman (2012) and CCT for homoskedastic simulations, which we adopt here to make our simulation results directly comparable with theirs and with the rest of the literature. The form of heteroskedasticity is motivated by MacKinnon (2013), and we consider γ equal to 0 and 2 for the simulations. We use the same three functional forms for μ_j as CCT, which we describe below.

The first DGP is designed to match features of Lee's (2008) analysis of U.S. congressional elections. Lee estimates the incumbency advantage in electoral races for the House of Representatives – parties that received the largest vote share in the previous election are the incumbents, which creates the discontinuity. The conditional expectation is a fifth-order polynomial fit to that dataset (after excluding a small number of extreme observations; see Imbens & Kalyanaraman, 2012, or CCT for further details) giving

$$\mu_1(x) = \begin{cases} 0.48 + 1.27x + 7.18x^2 + 20.21x^3 + 21.54x^4 + 7.33x^5 & \text{if} \quad x<0 \\ 0.52 + 0.84x - 3.00x^2 + 7.99x^3 - 9.01x^4 + 3.56x^5 & \text{otherwise.} \end{cases}$$

The population ATE for this DGP is $0.04\,(= 0.52 - 0.48)$.

The second DGP is based on Ludwig and Miller's (2007) analysis of the Head Start program. As described in Section 5, eligibility to grant-writing assistance is determined at the county level using the county's historical poverty rate, with a sharp threshold that determines the provision of services. We use the fifth-order polynomial estimated on Ludwig and Miller's dataset as the conditional expectation for the second DGP:

$$\mu_2(x) = \begin{cases} 3.71 + 2.30x + 3.28x^2 + 1.45x^3 + 0.23x^4 + 0.03x^5 & \text{if} \quad x<0 \\ 0.26 + 18.49x - 54.81x^2 + 74.30x^3 - 45.02x^4 + 9.83x^5 & \text{otherwise} \end{cases}$$

and the population ATE is $-3.45(= 0.26 - 3.71)$.

Finally, for the third DGP, we use CCT's modification of μ_1, given by

$$\mu_3(x) = \begin{cases} 0.48 + 1.27x + 3.59x^2 + 14.147x^3 + 23.694x^4 + 10.995x^5 & \text{if} \quad x<0 \\ 0.52 + 0.84x - 0.30x^2 + 2.397x^3 - 0.901x^4 + 3.56x^5 & \text{otherwise} \end{cases}$$

and the population ATE is again 0.04. CCT introduce this DGP because it has high curvature and local linear models are likely to exhibit high bias, making it a natural test case for both their analytical corrections and our bootstrap.

To estimate the finite-sample coverage of our new bootstrap-based CI, we simulate 5,000 samples from each of the three DGPs (with $\gamma = 0$ and $\gamma = 2$) and calculate nominal 95% two-sided CIs. We use 999 bootstrap replications (B_2) to calculate the asymptotic distribution of the bias-corrected estimator, and each of those replications uses an additional 500 replications (B_1) to estimate the bias. The bandwidths, h and b, are chosen using the AMSE-optimal rule proposed by CCT and the wild bootstrap uses the "HC3" transformation of Davidson and Flachaire (2008), defined in Eqs. (6) and (7). In these simulations, we use Mammen's (1993) two-point distribution for e_i^* in the bootstrap:

$$e_i^* = \begin{cases} \dfrac{1+\sqrt{5}}{2} & \text{with probability } \dfrac{\sqrt{5}-1}{2\sqrt{5}} \\ \dfrac{1-\sqrt{5}}{2} & \text{otherwise.} \end{cases} \tag{8}$$

We include two other procedures for comparison: CCT's bias-corrected interval and a "naive" uncorrected estimator. We estimate CCT's bias-corrected procedure with AMSE-optimal bandwidth and estimate the "naive" uncorrected interval with the optimal bandwidth h proposed by Imbens and Kalyanaraman (2012). We use triangular kernel throughout all the three procedures to facilitate direct comparisons.[16]

Table 1 presents the results of these simulations. It consists of two panels, with results for homoskedastic data in Panel A and heteroskedastic data in Panel B. The row labeled "Wild bootstrap" shows results for our proposed bootstrap based interval; "CCT" shows results for the analytically corrected interval and "Naive" shows results for the analytically uncorrected interval. The first three columns give the bias, standard deviation, and Root-MSE of the point estimators corresponding to each approach. As expected, both bias-corrected estimators have bias close to zero, with their standard deviation largely determining the estimators' RMSE. As we remark earlier, the bias correction from the wild bootstrap is equivalent to CCT's bias correction up to simulation error from the bootstrap and this equivalence is apparent in the Monte Carlo. This equivalence also explains why the bootstrap and analytically corrected estimators have essentially the same standard deviation across all of the DGPs.[17]

The column "CI Coverage" lists the experimental coverage of each procedure. In Panel A, all the intervals produced by the two robust estimators are reasonably close to their nominal coverage, although they slightly under-cover the true treatment effect. Our proposed bootstrap procedure consistently performs well and it is about a percentage point closer to nominal coverage than the analytically corrected interval. The uncorrected interval performs substantially worse, with coverage as low as 31% (in the second DGP) and never higher than 86%.

Panel B presents results for the heteroskedastic case. The coverage obtained is similar to that in Panel A and highlights the robustness of the wild bootstrap procedure proposed to heteroskedasticity. The two robust procedures again perform similarly and have coverage close to the nominal level. The bootstrap explicitly accounts for heteroskedasticity in the DGP;

Table 1. Experimental Coverage Probabilities for Each Interval Estimator Based on 5,000 Simulations; Nominal Coverage Probabilities Are 95% for Each Estimator.

DGP	Method	Bias	SD	RMSE	CI Coverage (%)	CI Length
Panel A: Homoskedastic data with $\gamma = 0$						
1.	Wild bootstrap	0.011	0.067	0.068	0.926	0.247
	CCT	0.011	0.067	0.068	0.914	0.239
	Naive	0.037	0.041	0.055	0.823	0.152
2.	Wild bootstrap	0.008	0.086	0.086	0.938	0.334
	CCT	0.008	0.086	0.086	0.932	0.346
	Naive	0.145	0.065	0.159	0.311	0.224
3.	Wild bootstrap	0.005	0.065	0.065	0.945	0.252
	CCT	0.005	0.065	0.065	0.934	0.244
	Naive	−0.034	0.051	0.061	0.851	0.185
Panel B: Heteroskedastic data with $\gamma = 2$						
1.	Wild bootstrap	0.010	0.073	0.074	0.935	0.276
	CCT	0.010	0.073	0.074	0.923	0.265
	Naive	0.035	0.046	0.058	0.838	0.165
2.	Wild bootstrap	0.007	0.084	0.084	0.922	0.312
	CCT	0.007	0.084	0.084	0.929	0.340
	Naive	0.144	0.064	0.158	0.277	0.211
3.	Wild bootstrap	0.002	0.081	0.081	0.940	0.311
	CCT	0.002	0.081	0.081	0.926	0.297
	Naive	−0.033	0.059	0.068	0.835	0.194

Notes: The column "CI Coverage" lists the coverage frequency in these simulations and "CI Length" lists the average length of the CI across simulations.

CCT's procedure implicitly accommodates smooth heteroskedasticity by estimating the variance in a neighborhood around the cutoff.

The final column, "CI Length," indicates that intervals from the two robust procedures have average lengths close to each other in all of the DGPs, while interval from the naive procedure is much shorter on average.

These are DGPs where the naive CIs are known to perform poorly and this set of simulation results indicates that our proposed bootstrap approach is competitive with analytical methods for producing robust intervals. Overall, the bootstrap bias-correction procedure proposed in this chapter provides a simple alternative to obtain valid robust CIs in RD designs and performs similarly well compared to the analytical bias correction procedures proposed by CCT.

5. APPLICATION

In this section, we apply the bootstrap procedure to the data used in Ludwig and Miller (2007).[18] In their chapter, the effects of Head Start application assistance on health and schooling were investigated under a sharp RD design.

In 1965, the Head Start program was established to help poor children aged three to five and their families. The program elements include parent involvement, nutrition, social services, mental health services, and health services. To promote this program in the most needing area, the Office of Economic Opportunity provided grant-writing assistance to the poorest 300 counties in the United States based on the 1960 poverty rate. So the poverty rate of the 300th poorest county serves as a sharp cutoff of treatment. It is shown in Ludwig and Miller (2007) that the 228 "treatment" counties with poverty rates 10 percentage points above this cutoff have average Head Start spending per four-year old as twice of that for 349 "control" counties with poverty rates 10 percentage points below this cutoff.

Ludwig and Miller (2007) utilize this fact to estimate the "intent-to-treat" effect: the effect of proposal developing assistance on health and schooling. They use mortality as their health outcome measure and use data from the National Vital Statistics System of the National Center for Health Statistics, which provide information on cause of death and age at death. Ludwig and Miller (2007) limited the causes of death to those which could be affected by Head Start health services and found a large drop in mortality rates of children five to nine years of age over the period of 1973–1983. They also found some evidence for a positive effect on schooling from decennial census data.

We reestimate the ATE on health and schooling with robust procedures. To be specific, we apply both the robust procedure proposed by CCT and the bootstrap procedure introduced in this chapter. Our bootstrap estimator uses the AMSE-optimal bandwidths proposed by CCT, the uniform kernel, and the HC3 wild bootstrap from Davidson and Flachaire (2008) with Mammen's (1993) two-point distribution on e_i^*; see Eqs. (6) and (7) of this chapter and their surrounding remarks as well as Eq. (8) for formal definitions. In the bootstrap procedure, we use 500 bootstraps for bias correction and 999 to calculate the CIs. The analytical estimator using CCT's bias correction and variance estimator uses their AMSE-optimal bandwidths as well.

For completeness, we also report the original results in Ludwig and Miller (2007).[19] Since our research focuses on CIs, we calculate and report the conventional unadjusted RD CI for each of the bandwidths used by Ludwig and Miller (2007). (Ludwig and Miller use a range of bandwidths for their analysis.) They also use a paired bootstrap algorithm to generate p-values, and we report those p-values for completeness.

Table 2 shows the results of the Head Start program on mortality of children five to nine years of age.[20] Instead of choosing an optimal bandwidth, Ludwig and Miller (2007) adopted three candidate bandwidths 9, 18, and 36. Their estimates indicate that Head Start assistance lowers the targeted mortality rate by −1.895, −1.198, and −1.114 respectively, which are not very sensitive to the choice of bandwidths in this range. These ATEs are also significantly different from zero (with p-value 0.036, 0.081, and 0.027) based on Ludwig and Miller's percentile-t bootstrapped p-value. The statistical inference changes when conventional analytical CI is used, which includes zero for bandwidth 9 and 18.

Results from the two robust procedures are similar to each other but greatly differ from the original estimates. Both ATE estimates are significantly different from zero and indicate a reduction in mortality, although the CIs are also very wide, which is likely to be due to the much smaller bandwidths used relative to the other estimators ($h = 3.888$, $b = 6.807$). The ATE estimates reported are the bias-corrected point estimates, and their value indicates that the naive estimates are subject to substantial bias.

Table 2. The Effect of Head Start Assistance on Mortality.

	ATE	95% CI	h	b	p-Value
LM (2007)	−1.895	(−3.930, 0.139)	9		0.036
LM (2007)	−1.198	(−2.561, 0.165)	18		0.081
LM (2007)	−1.114	(−2.138, −0.090)	36		0.027
CCT	−3.795	(−7.040, −0.551)	3.888	6.807	
Wild bootstrap	−3.800	(−6.781, −0.408)	3.888	6.807	

Notes: The first three rows come from Table 3 in Ludwig and Miller (2007) except "95% CI," which is calculated using the conventional asymptotic interval estimator. The last two rows list results from two robust procedures.

Table 3 presents the effect of the program on schooling for the cohort aged 18–24 in 1990. The measurement of schooling is the fraction of people with high school or more in Panel A and the fraction of people with some college or more in Panel B. A bandwidth of 7 is used in Ludwig and Miller (2007). Their estimates suggest that Head Start assistance increases the fraction of people with high school or more by 3% and the fraction of people with some college or more by 3.7%. Both are significantly different from zero based on both their p-value and the conventional analytical CI.

The two robust procedures are again very similar to each other and give statistically significant, positive, estimates. The CIs tend to shift relative to the original intervals, increasing both the range of effects consistent with the data as well as the lower bound on plausible effects. In contrast to the striking differences between the conventional and bias-corrected procedures for the mortality estimates, the differences in Table 3 are much smaller.

To briefly summarize, the bootstrap procedure performs similar to the robust estimator proposed by CCT in the above applications. Both provide somewhat dissimilar answers from the classical point and interval estimators that do not account for the estimator's bias, but our estimates largely support the direction and the statistical significance of Ludwig and Miller's (2007) empirical findings.

Table 3. The Effect of Head Start Assistance on Education for Cohort 18–24 in 1990.

	ATE	95% CI	h	b	p-Value
Panel A: Fraction "high school or more"					
LM (2007)	0.030	(0.003, 0.057)	7		0.032
CCT	0.055	(0.014, 0.096)	3.671	8.618	
Wild bootstrap	0.055	(0.010, 0.097)	3.671	8.618	
Panel B: Fraction "some college or more"					
LM (2007)	0.037	(0.002, 0.073)	7		0.032
CCT	0.051	(0.004, 0.099)	5.076	10.251	
Wild bootstrap	0.052	(0.006, 0.092)	5.076	10.251	

Notes: Panel A uses the fraction of people with high school or more as dependent variable. Panel B uses the fraction of people with some college or more as dependent variable. The first row in each panel comes from Table 4 in Ludwig and Miller (2007) except "95% CI," which is calculated using the conventional asymptotic interval estimator. The last two rows in each panel list result from two robust procedures.

6. CONCLUSION

This chapter proposes a novel bootstrap procedure to obtain robust bias-corrected CIs in sharp RD designs. The approach proposed builds upon the developments and intuition advanced by CCT and is based on a first-order bias correction. We exploit CCT's theoretical insight through a new wild bootstrap. In particular, we propose estimating the local linear model as usual, then estimating a local second-order polynomial and generating bootstrap datasets based on that polynomial. This bootstrap allows the bias of the linear model to be estimated and removed, and the bootstrap can be repeated to accurately estimate the sampling distribution of the bias-corrected estimator. These results demonstrate that there are bootstrap procedures that can generate asymptotically valid CIs in the same settings as CCT's.

NOTES

1. This score is often referred to as the *running variable* in this literature.
2. See, for example, Van der Klaauw (2002), Jacob and Lefgren (2004), Ludwig and Miller (2007), Urquiola and Verhoogen (2009), and Cellini, Ferreira, and Rothstein (2010).
3. Card, Dobkin, and Maestas (2009), Barreca, Guldi, Lindo, and Waddell (2011), and Keele, Lorch, Passarella, Small, and Titiunik (2017).
4. Lee and Card (2008), Caughey and Sekhon (2011), Keele and Titiunik (2015), Erikson and Titiunik (2015), and Fujiwara (2011, 2015).
5. Schmieder, Von Wachter, and Bender (2012) and Frandsen (2017).
6. More generally, they show that the bias and variance of a local polynomial of order p can be accounted for by estimating the $p+1$ local polynomial. We will restrict our analysis to the case with $p=1$ in this chapter because of its widespread use.
7. Since our bootstrap resamples the residuals but not the running variable, many of the quantities used to produce the estimates do not change under the bootstrap distribution and can be reused from the original estimate. Exploiting this feature makes the iterated bootstrap much less computationally burdensome; see the discussion following Algorithm 2.
8. The simulations and empirical analysis were carried out in the R programming language (R Core Team, 2015) and rely on the *rdrobust* (Calonico, Cattaneo, & Titiunik, 2015; Calonico, Cattaneo, Farrell, & Titiunik, 2017), *doParallel*, *foreach* (Revolution Analytics and Weston, 2015a,b), and *doRNG* (Gaujoux, 2014) packages. Our source code is available online at https://github.com/grayclhn/boot-rd
9. See Hahn et al. (2001), Porter (2003), or Fan and Gijbels (1992) for discussions of the properties of local polynomial regressions for boundary problems. The bootstrap algorithm proposed in this chapter can be extended to accommodate

higher order polynomials discontinuities in the derivatives of the conditional expectation, at the cost of stronger moment and smoothness assumptions.

10. The Mathematical Appendix of this chapter gives a precise definition for $V(h)$.

11. $C(h, b)$ and $V(h)$ are defined precisely in the Mathematical Appendix.

12. Unless otherwise stated, all limits in this chapter are assumed to hold as $n \to \infty$.

13. It would be a slight abuse of notation for $K_{+,h}$, $K_{-,h}$, $K_{+,b}$, and $K_{-,b}$ to all use different underlying kernels but would not affect our proofs.

14. An open-source implementation of this algorithm written in R, which was used for the empirical and Monte Carlo components of this chapter, is available online at https://github.com/grayclhn/boot-rd

15. The values $\bar{h}$ and $\bar{b}$ are taken as the average values over 500 homoskedastic simulations of the bandwidths returned by CCT's optimal bandwidth selector. This ensures that the relevant variance around the cutoff is approximately the same for the heteroskedastic and homoskedastic DGPs.

16. Simulation results for other kernels and variance estimators produced similar outcomes.

17. Note that this table reports the estimators' variance across simulations, not the *estimated* variance used for the CIs.

18. The data is publicly available from http://faculty.econ.ucdavis.edu/faculty/dlmiller/statafiles

19. One minor issue arose in reproducing Ludwig and Miller's results: their chapter presents results for the *triangular* kernel, but we were only able to recover their ATE estimate using the *uniform* kernel. Results for the triangular and uniform kernel were essentially the same, and this discrepancy does not affect Ludwig and Miller's conclusions.

20. We, like Ludwig and Miller, focus on the 1973–1983 period.

21. As we mention in Section 2, we have adopted CCT's notation where possible and these terms originate in that paper.

ACKNOWLEDGMENTS

The authors would like to thank Matias Cattaneo, Juan Carlos Escanciano, Sebastian Calonico, Max Farrell, Helle Bunzel, Brent Kreider, Quentin Brummet, seminar participants at the Advances in Econometrics conference on Regression Discontinuity Designs, Third Annual Conference of the International Association for Applied Econometrics, 2016 North American Summer Meeting of the Econometric Society and the 69th European Meeting of the Econometric Society, and especially the two anonymous referees for substantial help and advice on earlier versions of this chapter.

REFERENCES

Barreca, A. I., Guldi, M., Lindo, J. M., & Waddell, G. R. (2011). Saving babies? Revisiting the effect of very low birth weight classification. *The Quarterly Journal of Economics, 126*(4), 2117–2123.

Calonico, S., Cattaneo, M. D., & Farrell, M. H. (2017). On the effect of bias estimation on coverage accuracy in nonparametric inference. *Journal of the American Statistical Association*, Forthcoming.

Calonico, S., Cattaneo, M. D., Farrell, M. H., & Titiunik, R. (2017). rdrobust: Software for regression discontinuity designs. *Stata Journal*, Forthcoming.

Calonico, S., Cattaneo, M. D., & Titiunik, R. (2014). Robust nonparametric confidence intervals for regression-discontinuity designs. *Econometrica, 82*(6), 2295–2326.

Calonico, S., Cattaneo, M. D., & Titiunik, R. (2015). rdrobust: An R package for robust nonparametric inference in regression-discontinuity designs. *R Journal, 7*(1), 38–51.

Card, D., Dobkin, C., & Maestas, N. (2009). Does medicare save lives? *Quarterly Journal of Economics, 124*(2), 597–636.

Card, D., Lee, D. S., & Pei, Z. (2009). Quasi-experimental identification and estimation in the regression kink design. *Working Paper*, Princeton University.

Card, D., Lee, D. S., Pei, Z., & Weber, A. (2015). Inference on causal effects in a generalized regression kink design. *Econometrica, 83*(6), 2453–2483.

Card, D., Lee, D. S., Pei, Z., & Weber, A. (2017). Regression kink design: Theory and practice. In M. D. Cattaneo, & J. C. Escanciano (Eds.), *Regression discontinuity designs* (Vol. 38). Advances in Econometrics. Bingley: Emerald Publishing Limited.

Caughey, D., & Sekhon, J. S. (2011). Elections and the regression discontinuity design: Lessons from close US house races, 1942–2008. *Political Analysis, 19*(4), 385–408.

Cellini, S. R., Ferreira, F., & Rothstein, J. (2010). The value of school facility investments: Evidence from a dynamic regression discontinuity design. *Quarterly Journal of Economics, 125*(1), 215–261.

Davidson, R., & Flachaire, E. (2008). The wild bootstrap, tamed at last. *Journal of Econometrics, 146*(1), 162–169.

Erikson, R. S., & Titiunik, R. (2015). Using regression discontinuity to uncover the personal incumbency advantage. *Quarterly Journal of Political Science, 10*, 101–119.

Fan, J., & Gijbels, I. (1992). Variable bandwidth and local linear regression smoothers. *Annals of Statistics, 20*(4), 1669–2195.

Frandsen, B. (2017). Party bias in union representation elections: Testing for manipulation in the regression discontinuity design when the running variable is discrete. In M. D. Cattaneo, & J. C. Escanciano (Eds.), *Regression discontinuity designs* (Vol. 38). Advances in Econometrics. Bingley: Emerald Publishing Limited.

Fujiwara, T. (2011). A regression discontinuity test of strategic voting and Duverger's Law. *Quarterly Journal of Political Science, 6*(3–4), 197–233.

Fujiwara, T. (2015). Voting technology, political responsiveness, and infant health: Evidence from Brazil. *Econometrica, 83*(2), 423–464.

Gaujoux, R. (2014). *doRNG: Generic reproducible parallel backend for foreach loops*. R package version 1.6.

Hahn, J., Todd, P., & Van der Klaauw, W. (2001). Identification and estimation of treatment effects with a regression-discontinuity design. *Econometrica, 69*(1), 201–209.

Hall, P. (1993). On edgeworth expansion and bootstrap confidence bands in nonparametric curve estimation. *Journal of the Royal Statistical Society*, 55(1), 291–304.

Hall, P., & Martin, M. A. (1988). On bootstrap resampling and iteration. *Biometrika, 75*(4), 661–671.

Imbens, G. W., & Kalyanaraman, K. (2012). Optimal bandwidth choice for the regression discontinuity estimator. *Review of Economic Studies, 79*(3), 933–959.

Imbens, G. W., & Lemieux, T. (2008). Regression discontinuity designs: A guide to practice. *Journal of Econometrics, 142*(2), 615–635.

Jacob, B. A., & Lefgren, L. (2004). Remedial education and student achievement: A regression-discontinuity analysis. *Review of Economics and Statistics, 86*(1), 226–244.

Keele, L., Lorch, S., Passarella, M., Small, D., & Titiunik, R. (2017). An overview of geographically discontinuous treatment assignments with an application to children's health insurance. In M. D. Cattaneo, & J. C. Escanciano (Eds.), *Regression discontinuity designs* (Vol. 38). Advances in Econometrics. Bingley: Emerald Publishing Limited.

Keele, L. J., & Titiunik, R. (2015). Geographic boundaries as regression discontinuities. *Political Analysis, 23*(1), 127–155.

Lee, D. S. (2008). Randomized experiments from non-random selection in U.S. House Elections. *Journal of Econometrics, 142*(2), 675–697.

Lee, D. S., & Card, D. (2008). Regression discontinuity inference with specification error. *Journal of Econometrics, 142*(2), 655–674.

Lee, D. S., & Lemieux, T. (2010). Regression discontinuity designs in economics. *Journal of Economic Literature, 48*(2), 281–355.

Ludwig, J., & Miller, D. L. (2007). Does head start improve children's life chances? Evidence from a regression discontinuity design. *Quarterly Journal of Economics, 122*(1), 159–208.

MacKinnon, J. G. (2013). Thirty years of heteroskedasticity-robust inference. In Xiaohong Chen & Norman R. Swanson (Eds.), *Recent advances and future directions in causality, prediction, and specification analysis* (pp. 437–461). New York: Springer.

MacKinnon, J. G., & White, H. (1985). Some heteroskedasticity-consistent covariance matrix estimators with improved finite sample properties. *Journal of Econometrics, 29*(3), 305–325.

Mammen, E. (1993). Bootstrap and wild bootstrap for high dimensional linear models. *The Annals of Statistics, 23*(1), 255–285.

Neumann, M. H. (1997). Pointwise confidence intervals in nonparametric regression with heteroskedastic error structure. *Statistics: A Journal of Theoretical and Applied Statistics, 29*(1), 1–36.

Porter, J. (2003). *Estimation in the regression discontinuity model.* Unpublished Manuscript, Department of Economics, University of Wisconsin, Madison.

R Core Team. (2015). *R: A language and environment for statistical computing.* R Foundation for Statistical Computing, Vienna, Austria.

Revolution Analytics, & Weston, S. (2015a). *doParallel: Foreach Parallel Adaptor for the 'parallel' Package.* R package version 1.0.10.

Revolution Analytics, & Weston, S. (2015b). *Foreach: Provides Foreach Looping Construct for R.* R package version 1.4.3.

Schmieder, J. F., Von Wachter, T., & Bender, S. (2012). The effects of extended unemployment insurance over the business cycle: Evidence from regression discontinuity estimates over 20 years. *Quarterly Journal of Economics, 127*(2), 701–752.

Thistlethwaite, D. L., & Campbell, D. T. (1960). Regression-discontinuity analysis: An alternative to the ex post facto experiment. *Journal of Educational Psychology*, *51*(6), 309.

Urquiola, M., & Verhoogen, E. (2009). Class-size caps, sorting, and the regression-discontinuity design. *The American Economic Review*, *99*(1), 179–215.

Van der Klaauw, W. (2002). Estimating the effect of financial aid offers on college enrollment: A regression-discontinuity approach. *International Economic Review*, *43*(4), 1249–1287.

Van der Vaart, A. W. (2000). *Asymptotic statistics. Cambridge Series in Statistical and Probabilistic Mathematics* (Vol. 3). Cambridge: Cambridge University Press.

APPENDIX A: MATHEMATICAL APPENDIX

Let e_p be the selection vector with 1 in element $p+1$ and 0 everywhere else and assume, with some abuse of notation, that the dimension of e_p adapts to make matrix and vector operations conformable. Much of the theory in this appendix applies to both sides of the cutoff symmetrically, so we will use a "•" as a placeholder for either + or − in equations; that is, the equation

$$\widehat{\mu}_{\bullet}(h) = \arg\min_{\mu} \min_{\beta} \sum_{i=1}^{n} (Y_i - \mu - \beta X_i)^2 K_{\bullet,h}(X_i)$$

would define both $\widehat{\mu}_{+}(h)$ and $\widehat{\mu}_{-}(h)$ as

$$\widehat{\mu}_{+}(h) = \arg\min_{\mu} \min_{\beta} \sum_{i=1}^{n} (Y_i - \mu - \beta X_i)^2 K_{+,h}(X_i),$$

$$\widehat{\mu}_{-}(h) = \arg\min_{\mu} \min_{\beta} \sum_{i=1}^{n} (Y_i - \mu - \beta X_i)^2 K_{-,h}(X_i).$$

Define the following additional notation:[21] $r_p(x) = (1, x, \ldots, x^p)'$, $\mathbb{1}_{+}(x) = \mathbb{1}\{x \geq 0\}$, $\mathbb{1}_{-}(x) = \mathbb{1}\{x \leq 0\}$,

$$\Gamma_{\bullet,p}(h) = \frac{1}{n}\sum_{i=1}^{n} r_p(X_i/h) r_p(X_i/h)' K_{\bullet,h}(X_i),$$

$$\mathfrak{B}_{\bullet}(h) = e_0'(\Gamma_{\bullet,1}(h))^{-1} \frac{1}{n}\sum_{i=1}^{n} (X_i/h)^2 r_1(X_i/h) K_{\bullet,h}(X_i),$$

$$\Psi_{\bullet,p,q}(h,b) = \frac{1}{n}\sum_{i=1}^{n} r_p(X_i/h) r_q(X_i/b)' K_{\bullet,h}(X_i) K_{\bullet,b}(X_i) \mathbb{V}(Y_i|X_i)$$

as well as their population counterparts,

$$\begin{aligned}
\tilde{\Gamma}_{\bullet,p}(h) &= \int_{-\infty}^{\infty} r_p(u) r_p(u)' K(u) f(hu) \mathbb{1}_{\bullet}(u)\,\mathrm{du} \\
\Gamma_{\bullet,p} &= \int_{-\infty}^{\infty} r_p(u) r_p(u)' K(u) \mathbb{1}_{\bullet}(u)\,\mathrm{du} \\
\tilde{\mathfrak{B}}_{\bullet}(h) &= e_0' \widehat{\Gamma}_{\bullet,1}(h)^{-1} \int_{-\infty}^{\infty} r_1(u) u^2 K(u) f(hu) \mathbb{1}_{\bullet}(u)\,\mathrm{du} \\
\mathfrak{B}_{\bullet} &= e_0' \Gamma_{\bullet,1}^{-1} \int_{-\infty}^{\infty} r_1(u) u^2 K(u) \mathbb{1}_{\bullet}(u)\,\mathrm{du}
\end{aligned}$$

$$\tilde{\Psi}_{\bullet,p,q}(h,b) = \int_{-\infty}^{\infty} r_p\left(\frac{m}{h}u\right) r_q\left(\frac{m}{b}u\right)' K\left(\frac{m}{b}u\right) K\left(\frac{m}{h}u\right) \mathbb{V}(Y_i|X_i{=}um) f(um) \mathbb{1}_{\bullet}(u)\,\mathrm{du}$$

$$\Psi_{\bullet,p} = \int_{-\infty}^{\infty} r_p(u) r_p(u)' K(u)^2 \mathbb{1}_{\bullet}(u)\,\mathrm{du}.$$

Remember that f is the pdf of the running variable. Lemmas SA1 and SA2 of CCT imply that all of these terms have well-defined limits under Assumptions 1 and 2 (for some, after normalization), namely that the following relationships hold:

$$\Gamma_{\bullet,p}(h) = \tilde{\Gamma}_{\bullet,p}(h) + o_p(1) \qquad \tilde{\Gamma}_{\bullet,p}(h) = f(0)\,\Gamma_{\bullet,p} + o(1), \tag{A.1}$$

$$\mathfrak{B}_{\bullet}(h) = \tilde{\mathfrak{B}}_{\bullet}(h) + o_p(1) \qquad \tilde{\mathfrak{B}}_{\bullet}(h) = \mathfrak{B}_{\bullet} + o_p(1), \tag{A.2}$$

$$\frac{hb}{m}\Psi_{\bullet,p,q}(h,\,b) = \tilde{\Psi}_{\bullet,p,q}(h,\,b) + o_p(1) \quad \tilde{\Psi}_{\bullet,p,p}(h,\,h) = f(0)\,\sigma_{\bullet}^2(0)\,\Psi_{\bullet,p} + o(1). \tag{A.3}$$

Implicitly, the right column of equations holds as $h \to 0$.

Also let $\widehat{\beta}_{\bullet,p}(h)$ be the coefficient estimators from the weighted regression of Y_i on $r_p(X_i)$. As in CCT, this implies that

$$\widehat{\beta}_{\bullet,p}(h) = H_p(h) \Gamma_{\bullet,p}(h)^{-1} \frac{1}{n} \sum_{i=1}^{n} r_p(X_i/h) Y_i K_{\bullet,h}(X_i)$$

with $H_p(h) = \text{diag}\left(1, h^{-1}, ..., h^{-p}\right)$. These coefficients are related to the quantities of interest by

$$\widehat{\mu}^{(\nu)}_{\bullet,\,p}(h) = \nu! e_\nu' \widehat{\beta}_{\bullet,\,p}(h).$$

for $\nu = 0, ..., p$ and similar arguments to those used above imply that

$$H_p(h)^{-1}\left(\widehat{\beta}_{\bullet,\,p}(h) - \beta_{\bullet,\,p}(h)\right) = O_p\left(1/\sqrt{nh}\right).$$

Finally, define the variance components

$$\begin{aligned} V(h) = \frac{1}{n} e_0' \Gamma_{-,\,1}(h)^{-1} \Psi_{-,\,1,\,1}(h,\,h) \Gamma_{-,\,1}(h)^{-1} \\ + \Gamma_{+,\,1}(h)^{-1} \Psi_{+,\,1,\,1}(h,\,h) \Gamma_{+,\,1}(h)^{-1} e_0 \end{aligned} \tag{A.4}$$

and

$$\begin{aligned} C(h,b) = \frac{h^4}{nb^4} e_2 [\Gamma_{+,2}(b)^{-1} \Psi_{+,2,2}(b,b) \Gamma_{+,2}(b)^{-1} \mathfrak{B}_+(h)^2 \\ + \Gamma_{-,2}(b)^{-1} \Psi_{-,2,2}(b,b) \Gamma_{-,2}(b)^{-1} \mathfrak{B}_-(h)^2] e_2 \\ - \frac{2h^2}{nb^2} e_0 [\Gamma_{+,1}(h)^{-1} \Psi_{+,1,2}(h,b) \Gamma_{+,2}(b)^{-1} \mathfrak{B}_+(h) \\ + \Gamma_{-,1}(h)^{-1} \Psi_{-,1,2}(h,b) \Gamma_{-,2}(b)^{-1} \mathfrak{B}_-(h)^2] e_2. \end{aligned} \tag{A.5}$$

See CCT for more on the motivation and derivation of these formulas.

A.1. Proof of Theorem 1

We have

$$\widehat{\tau}(h) - \Delta^*(h,\,b) - \tau = (\widehat{\tau}(h) - \mathbb{E}\,\widehat{\tau}(h)) + (\mathbb{E}\,\widehat{\tau}(h) - \tau) - \left(\mathbb{E}^*\widehat{\tau}^*(h) - \tau^*\right).$$

By construction,

$$Y_i^* = \begin{cases} r_2\left(X_i/b\right)' H_2(b)^{-1} \beta^*_{+,\,2} + \varepsilon_i^*, & X_i \geq 0 \\ r_2\left(X_i/b\right)' H_2(b)^{-1} \beta^*_{-,\,2} + \varepsilon_i^*, & X_i \leq 0, \end{cases}$$

which, along with the definitions at the beginning of the appendix, ensures that

$$\mathbb{E}^* \widehat{\mu}^*_{\bullet,1}(h) - \mu^*_\bullet = h^2 \mu^{*(2)}_\bullet \mathfrak{B}_\bullet(h)/2$$

almost surely and, consequently,

$$\mathbb{E}^* \widehat{\tau}^*(h) - \tau^* = h^2 \mu^{*(2)}_+ \mathfrak{B}_+(h)/2 - h^2 \mu^{*(2)}_- \mathfrak{B}_-(h)/2.$$

Equation (4) holds because $\mu^{*(2)}_\bullet = \widehat{\mu}^{(2)}_{\bullet,2}(b)$ almost surely.

CCT's Lemma A1 implies that

$$\mathbb{E}\widehat{\tau}(h) - \tau = h^2 \mu^{(2)}_+ \mathfrak{B}_+(h)/2 - h^2 \mu^{(2)}_- \mathfrak{B}_-(h)/2 + O_p(h^3)$$

as well, giving

$$\begin{aligned} &\widehat{\tau}(h) - \mathbb{E}\widehat{\tau}(h) + (\mathbb{E}\widehat{\tau}(h) - \tau) - \left(\mathbb{E}^*\widehat{\tau}^*(h) - \tau^*\right) \\ &\quad = \widehat{\tau}(h) - \mathbb{E}\widehat{\tau}(h) + h^2 \left(\left(\mu^{*(2)}_- - \mu^{(2)}_-\right) \mathfrak{B}_-(h)/2 - \left(\mu^{*(2)}_+ - \mu^{(2)}_+\right) \mathfrak{B}_+(h)/2 \right) + O_p\left(h^3\right) \\ &\quad = \widehat{\tau}(h) - \mathbb{E}\widehat{\tau}(h) + h^2 \left(\widehat{\mu}^{(2)}_{-,2}(b) - \mu^{(2)}_-\right) \mathfrak{B}_-(h)/2 \\ &\quad\quad - h^2 \left(\widehat{\mu}^{(2)}_{+,2}(b) - \mu^{(2)}_+\right) \mathfrak{B}_+(h)/2 + O_p\left(h^3\right) \end{aligned} \tag{A.6}$$

Asymptotic normality then follows from normality of $\widehat{\tau}(h) - \mathbb{E}\widehat{\tau}(h)$ and $\widehat{\mu}^{(2)}_{\bullet,2}(b) - \mu^{(2)}_\bullet$ using similar arguments to CCT's Lemma SA4.D. ■

A.2. Proof of Theorem 2

Repeat the steps from Theorem 1's proof through Eq. (A.6) for the iterated bootstrap to get

$$
\begin{aligned}
\widehat{\tau}^*(h) - \Delta^{**}(h,b) - \tau^* &= \left(\widehat{\tau}^*(h) - \mathbb{E}^*\widehat{\tau}^*(h)\right) + \left(\mathbb{E}^*\widehat{\tau}^*(h) - \tau^*\right) \\
&\quad - \left(\mathbb{E}^{**}\widehat{\tau}^{**}(h) - \tau^{**}\right) \\
&= \widehat{\tau}^*(h) - \mathbb{E}^*\widehat{\tau}^*(h) + h^2\left(\widehat{\mu}^{*(2)}_{-,2}(b) - \mu^{*(2)}_{-}\right)\mathfrak{B}_-(h)/2 \\
&\quad - h^2\left(\widehat{\mu}^{*(2)}_{+,2}(b) - \mu^{*(2)}_{+}\right)\mathfrak{B}_+(h)/2 \\
&= e_0'\Gamma_{+,1}(h)^{-1}\left(\frac{1}{n}\sum_{i=1}^{n} r_1\left(X_i/h\right)K_{+,h}(X_i)\varepsilon_i^*\right) \\
&\quad - e_0'\Gamma_{-,1}(h)^{-1}\left(\frac{1}{n}\sum_{i=1}^{n} r_1\left(X_i/h\right)K_{-,h}(X_i)\varepsilon_i^*\right) \\
&\quad + \left(h/b\right)^2 \cdot e_2'\Gamma_{-,2}(b)^{-1}\left(\frac{1}{n}\sum_{i=1}^{n} r_2\left(X_i/b\right)K_{-,b}(X_i)\varepsilon_i^*\right)\mathfrak{B}_-(h) \\
&\quad - \left(h/b\right)^2 \cdot e_2'\Gamma_{+,2}(b)^{-1}\left(\frac{1}{n}\sum_{i=1}^{n} r_2\left(X_i/b\right)K_{+,b}(X_i)\varepsilon_i^*\right)\mathfrak{B}_+(h)
\end{aligned}
$$

with the last equality holding from the relationship between the coefficient estimates discussed at the end of the introduction to the appendix. We will prove convergence of the variance first, then prove asymptotic normality.

To prove convergence in probability of the bootstrap variance, note that the wild bootstrap ensures

$$
\mathbb{V}^*\left(\sum_{i=1}^{n} R(X_i)\varepsilon_i^*\right) = \sum_{i=1}^{n} R(X_i)^2\widehat{\varepsilon}_i^2.
$$

for any measurable function $R(\cdot)$. Given this expression for the iterated bootstrap and the formula for the variance conditional on X presented at the end of the introduction to the appendix, it suffices to show that

$$
\frac{1}{n}\sum_{i=1}^{n} r_1\left(X_i/h\right)r_1\left(X_i/h\right)' K_{\bullet,\,h}(X_i)^2\widehat{\varepsilon}_i^2 = \frac{1}{h}\widehat{\Psi}_{\bullet,\,1,\,1}(h,\,h) + o_p\left(\frac{1}{h}\right), \tag{A.7}
$$

$$\frac{1}{n}\sum_{i=1}^{n} r_1(X_i/h) r_2(X_i/b)' K_{\bullet,h}(X_i) K_{\bullet,b}(X_i) \widehat{\varepsilon}_i^2 = \frac{m}{hb} \widehat{\Psi}_{\bullet,1,2}(h,b) + o_p\left(\frac{m}{hb}\right), \tag{A.8}$$

$$\frac{1}{n}\sum_{i=1}^{n} r_2(X_i/b) r_2(X_i/b)' K_{\bullet,b}(X_i)^2 \widehat{\varepsilon}_i^2 = \frac{1}{b} \widehat{\Psi}_{\bullet,2,2}(b,b) + o_p\left(\frac{1}{b}\right). \tag{A.9}$$

But these convergence results follow from CCT's results on $\widehat{\mu}^{(\nu)}_{\bullet,p}$, Lemmas SA3 and SA4 in particular, and standard arguments on the convergence of residuals to the population error term.

Normality also follows standard arguments for the wild bootstrap. Conditional on the regressors and residuals, $\{R(X_i)\widehat{\varepsilon}_i e_i^*\}$ is a sequence of independent and mean zero random variables for any measurable function R, each with qth absolute moment $|R(X_i)\widehat{\varepsilon}_i|^q \mathbb{E}|e_i^*|^q$. The result then holds as a consequence of the Lindeberg-Feller CLT if we establish that Liapunov's condition holds in probability, specifically that

$$\left(\frac{1}{n}\right)^{1+\delta/2} \sum_i |(R_+(X_i) - R_-(X_i))\widehat{\varepsilon}_i|^{2+\delta} \to^p 0 \tag{A.10}$$

with

$$R_\bullet(X_i) = h^{1/2} \cdot \gamma_1' r_1(X_i/h) K_{\bullet,h}(X_i) + b^{1/2} \cdot \gamma_2' r_2(X_i/b) K_{\bullet,b}(X_i)$$

for arbitrary nonzero vectors γ_1 and γ_2. (See, e.g., van der Vaart, 2000, Proposition 2.27 and Theorem 23.4.) But similar arguments to those used for the variance establish that

$$\frac{1}{n}\sum_i |(R_+(X_i) - R_-(X_i))\widehat{\varepsilon}_i|^{2+\delta} = O_p\left(h^{-\delta/2}\right) + O_p\left(b^{-\delta/2}\right). \tag{A.11}$$

Since $nb \to \infty$ and $nh \to \infty$ by assumption, Eq. (A.11) implies that Eq. (A.10) holds and completes the proof. ■

THE DEVIL IS IN THE TAILS: REGRESSION DISCONTINUITY DESIGN WITH MEASUREMENT ERROR IN THE ASSIGNMENT VARIABLE

Zhuan Pei[a,b] and Yi Shen[c]

[a]*Department of Policy Analysis and Management, Cornell University, Ithaca, NY, USA*
[b]*IZA, Bonn, Germany*
[c]*Department of Statistics and Actuarial Science, University of Waterloo, Waterloo, ON, Canada*

ABSTRACT

Identification in a regression discontinuity (RD) design hinges on the discontinuity in the probability of treatment when a covariate (assignment variable) exceeds a known threshold. If the assignment variable is measured with error, however, the discontinuity in the relationship between the probability of treatment and the observed mismeasured assignment variable may disappear. Therefore, the presence of measurement error in the assignment variable poses a challenge to treatment effect

Regression Discontinuity Designs: Theory and Applications
Advances in Econometrics, Volume 38, 455–502

ISSN: 0731-9053/doi:10.1108/S0731-905320170000038019

identification. This chapter provides sufficient conditions to identify the RD treatment effect using the mismeasured assignment variable, the treatment status and the outcome variable. We prove identification separately for discrete and continuous assignment variables and study the properties of various estimation procedures. We illustrate the proposed methods in an empirical application, where we estimate Medicaid takeup and its crowdout effect on private health insurance coverage.

JEL classifications: C10; C18

Keywords: Regression discontinuity design; measurement error

1. INTRODUCTION

Over the past two decades, studies in economics have relied on the regression discontinuity (RD) design to evaluate the effects of a wide range of policy programs. In an RD design, treatment is assigned based on whether an observed covariate (called the "assignment" or "running" variable) exceeds a known threshold. Provided that agents just above and just below the threshold share the same baseline characteristics, any difference in the outcomes between these two groups can be attributed to the causal effect of the treatment.[1]

A classical RD design depends crucially on the econometrician's ability to observe an accurate measure of the assignment variable. In many cases, however, only a noisy version of the assignment variable is available. This scenario is likely to occur when survey data are used, where the value of the assignment variable comes from self-reporting as opposed to an administrative source.

A typical example is an application that uses income as an assignment variable to study the effect of means-tested transfer programs where eligibility depends on whether income falls below a certain threshold. However, most administrative data cannot be used for an RD because they only include the treatment population, namely those who enroll in the program, and contain little information on the various outcomes for applicants who are denied benefits. Therefore, practitioners may be forced to rely on survey data in order to apply an RD design. For instance, Schanzenbach (2009) uses the Early Childhood Longitudinal Study to study the effect of school lunch on obesity and compares obesity rates for children below and above the reduced-price lunch cutoff for the National School Lunch Program.[2]

de la Mata (2012) estimates the effect of Medicaid/CHIP coverage on children's health care utilization and health outcomes with the Panel Study of Income Dynamics (PSID) and its Child Development Study (CDS) supplement. Koch (2013) uses the Medical Expenditure Panel Survey (MEPS) to study health insurance crowdout by focusing on income cutoffs in the Children's Health Insurance Program (CHIP).

The above studies all use income data gathered from surveys as their assignment variable in the RD analyses, but measurement error in surveyed income has been widely documented (see Bound, Brown, & Mathiowetz, 2001, for a review), and the presence of this measurement error threatens the identification in an RD design. Even if there is perfect compliance with the discontinuous treatment assignment rule (i.e., in a sharp RD design), there may *not* be a discontinuity in the probability of treatment conditional on the *observed* noisy assignment variable. Instead of a step function, the first-stage relationship – the probability of treatment conditional on the noisy assignment variable – is smoothly S-shaped. This lack of discontinuity complicates the identification and estimation of the program effect.

In this chapter, we study the identification and estimation of the RD treatment effect in the presence of classical measurement error in the assignment variable. We consider separately the cases where the assignment variable and measurement error are discrete and continuous. In the discrete case (e.g., binned income), we show that when the assignment variable is bounded, not only can we identify the first-stage and outcome conditional expectation functions without specifying the measurement error distribution, but we can also identify the true assignment variable distribution by exploiting the tail behavior of its observed counterpart. This is advantageous in the RD context, since a key appeal is the design's testability, which entails the smoothness of the true assignment variable distribution. In addition, recovery of the true assignment variable distribution may allow for the discovery of bunching and the estimation of meaningful economic quantities, even when the RD design is invalid.[3] The identification result leads to a simple minimum-distance estimation procedure for the true assignment variable distribution, and the estimation of the first-stage and outcome relationships follows from an application of Bayes' Theorem. Standard reasoning implies that the resulting estimators are efficient, $\sqrt{N}$-consistent, and asymptotically normal.

We also explore the case where the assignment variable is continuous and propose three alternative approaches to semiparametrically or nonparametrically identify the RD treatment effect. The first approach assumes normal measurement error while remaining agnostic about the true assignment variable distribution; identification follows from the result of Schwarz and

Bellegem (2010). The second approach is a novel identification-at-infinity strategy that exploits the tail behavior of the observed assignment variable distribution. The third approach adapts the nonparametric simulation-extrapolation (SIMEX) method of Carroll, Maca, and Ruppert (1999), which assumes that the variance of the measurement error is known. The last two approaches do not identify the true assignment variable distribution, but recover the RD treatment effect parameter under perfect compliance.

We illustrate our proposed methods in an empirical application where we study the takeup of Medicaid, a government health insurance program in the United States, and the extent to which it crowds out private health insurance. Since Medicaid eligibility is only available to families with income below a strict cutoff, we apply an RD design exploiting this policy discontinuity. We show that the noisy income measures in the Survey of Income and Program Participation (SIPP) indeed lead to a smooth first-stage relationship between Medicaid coverage and income, and then apply two modeling approaches to recover the true income distribution and the first-stage and outcome relationships. The first approach discretizes income and uses the proposed minimum distance estimation strategy; the second approach treats income as continuous and adopts a parametric maximum likelihood estimation (MLE) framework. The two methods yield similar findings: Medicaid takeup rate for the barely eligible is between 10% and 25%, and there is little evidence of sorting around the threshold and private insurance crowdout. However, the less parametric discrete approach yields much noisier estimates, limiting its empirical appeal.

Several related studies have attempted to tackle the identification challenge posed by measurement error in RD designs. Hullegie and Klein (2010) adopt a *parametric* Berkson measurement error specification, in which the true assignment variable is the sum of the observed assignment variable and an independent normally distributed measurement error. This specification is attractive in that it can be easily implemented in practice, but it implies that the distribution of the true assignment variable is smooth and precludes the testing of density discontinuity. In this chapter, we adopt the more conventional classical measurement error model (Bound et al., 2001), which allows for nonsmoothness in the assignment variable distribution. Dong (2015) considers rounding (nonclassical) error in the assignment variable typically encountered in age-based RD designs. In three recent studies, Yu (2012) and Yanagi (2014) consider the identification problem assuming "small" measurement error variance; Davezies and Le Barbanchon (2014) assume the availability of an auxiliary dataset so that the true assignment variable distribution can be observed (for the treated population). In contrast to these studies, the measurement error distribution in our setup is not restricted to have small second moments, nor do we assume the observability of the true assignment variable distribution from an auxiliary dataset.[4]

The remainder of the chapter is organized as follows. Section 2 introduces the statistical model. Sections 3 and 4 study the identification and estimation in the discrete and continuous cases, respectively. Section 5 provides an empirical illustration by employing the proposed methods to estimate Medicaid takeup and crowdout. Section 6 concludes.

2. BASELINE STATISTICAL MODEL

In a conventional sharp RD design, the econometrician observes the assignment variable X^*, eligibility/treatment $D^* = 1_{[X^* < 0]}$, and outcome Y:

$$Y = y(D^*, X^*, \varepsilon). \tag{1}$$

The function y is continuous in its second argument, ε is the unobserved error term, and the eligibility threshold is normalized to zero.[5] A standard result (Hahn, Todd, & der Klaauw, 2001) is that the treatment effect $\delta_{\text{sharp}} = E[y(1, 0, \varepsilon) - y(0, 0, \varepsilon)|X^* = 0]$ is identified by

$$\delta_{\text{sharp}} = \lim_{x^* \uparrow 0} E[Y|X^* = x^*] - \lim_{x^* \downarrow 0} E[Y|X^* = x^*]$$

when the conditional expectation $E[y(D^*, X^*, \varepsilon)|X^* = x^*]$ is continuous at $x^* = 0$ for $D^* = 0, 1$.

An important restriction of the sharp design is the assumption of the observability of program *eligibility* D^* (or equivalently, perfect compliance with the treatment assignment rule). In most applied contexts (such as those in the studies cited above), this assumption is rarely satisfied. In all social programs, for example, the takeup rate of entitlement programs among eligible individuals and families is not 100%.[6] The imperfect compliance with the treatment assignment rule gives rise to the so-called fuzzy RD design, where the outcome

$$Y = y(D, X^*, \varepsilon) \tag{2}$$

depends on the actual program participation status D. It is a well-known result that under independence (Hahn et al., 2001) or smoothness assumptions (DiNardo & Lee, 2011), the ratio,

$$\delta_{\text{fuzzy}} = \frac{\lim_{x^* \uparrow 0} E[Y|X^* = x^*] - \lim_{x^* \downarrow 0} E[Y|X^* = x^*]}{\lim_{x^* \uparrow 0} E[D|X^* = x^*] - \lim_{x^* \downarrow 0} E[D|X^* = x^*]},$$

is the average treatment effect of D on Y for the "complier" population that takes up benefit when eligible.

In this chapter, we study a variant of these RD models, in which we only observe a noisy version of X^*. Let X denote this observed assignment variable, and $u \equiv X - X^*$ the measurement error. As mentioned in Section 1, a key assumption that distinguishes this study from Hullegie and Klein (2010) is that the measurement error is independent of the true assignment variable as opposed to the observed assignment variable. Formally,

Assumption 1 (Independence). $X^* \perp\!\!\!\perp u$.

In the sharp design, the econometrician observes the joint distribution of (D^*, X, Y) with

$$\begin{aligned} D^* &= 1_{[X^* < 0]}, \\ X &= X^* + u, \end{aligned} \tag{3}$$

and Y is defined by Eq. (1). In the fuzzy design, D is observed instead of D^* and Y is defined by Eq. (2).

In both the sharp and fuzzy designs, the presence of u poses an identification challenge. For simplicity, we first focus on the sharp design (3). As seen in Fig. 1, the so-called first-stage relationship with the noisy X, $E[D^*|X]$, is smooth at $X = 0$, and we can no longer rely on the discontinuity in that relationship to identify a treatment effect.[7] In a way, this is an extreme form of the attenuation bias, which is typically associated with measurement error. Because of this lack of first-stage discontinuity, we cannot treat the problem as a fuzzy design either. In fact, the same force that smoothes out the first-stage relationship also smoothes out the outcome function $E[Y|X]$, and the fuzzy RD estimand simply becomes undefined. That said, it is possible to naively perform local regressions on both sides of the threshold and obtain nonzero discontinuity estimates for both $E[D^*|X]$ and $E[Y|X]$ through misspecification. Whether the resulting ratio

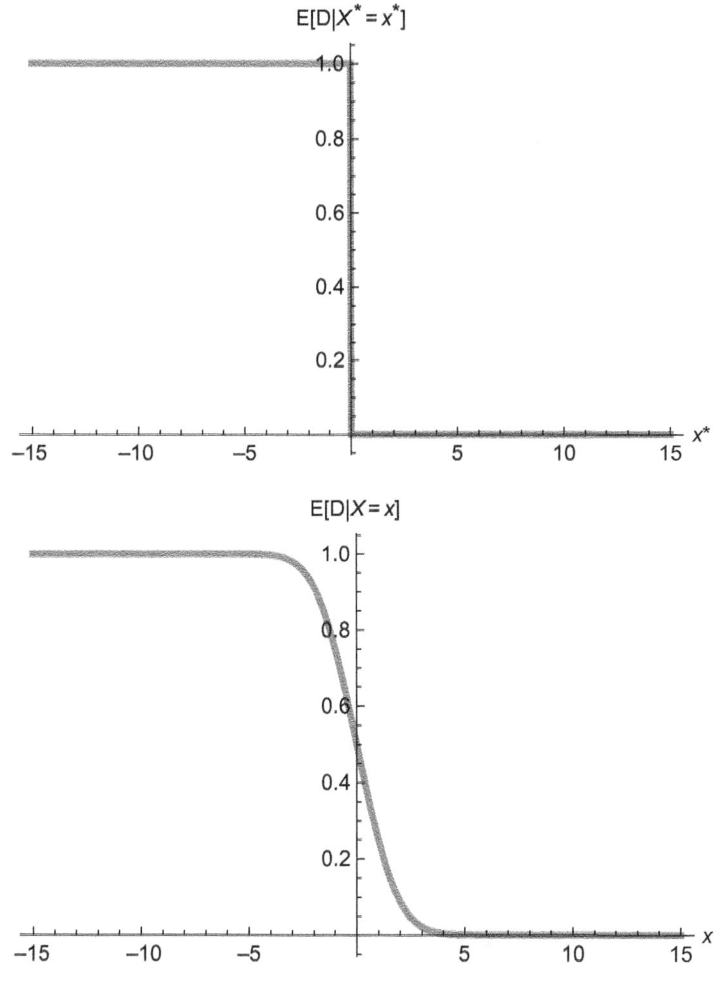

Fig. 1. Theoretical Effect of Smooth Measurement Error in the Assignment Variable. *Notes*: The upper panel plots the true first-stage relationship $E[D|X^* = x^*]$ in a sharp RD. The lower panel plots the observed first-stage relationship $E[D|X = x]$. The lower panel is generated by assuming that X^* and u are both normally distributed.

attenuates or exaggerates the true RD treatment effect depends on the particular specification adopted.

The issue of attenuation, or the lack thereof, can also be seen by following an alternative definition of the sharp RD estimand: $\tilde{\delta}_{\text{sharp}} = \lim_{x\uparrow 0}[Y|X = x, D^* = 1] - \lim_{x\downarrow 0}[Y|X = x, D^* = 0]$. When X is measured without error, $\tilde{\delta}_{\text{sharp}}$ coincides with the conventional sharp RD estimand. However, unlike the conventional sharp RD estimand, which collapses to zero even when X is measured with very small error, $\tilde{\delta}_{\text{sharp}}$ gradually drifts apart from δ_{sharp} as the measurement error variance increases. In fact, this is an attractive property of $\tilde{\delta}_{\text{sharp}}$, which we exploit in Section 4. For now, we argue that $\tilde{\delta}_{\text{sharp}}$ is not necessarily attenuated, and the direction of the bias in $\tilde{\delta}_{\text{sharp}}$ depends on the particular functional form in the relationship between X^* and Y. Suppose, for example, Y is linear in X^* for both the treatment and control groups: $E[Y|X^*, D^* = 1] = \psi_0 + \psi_1 X^*$ and $E[Y|X^*, D^* = 0] = \psi_2 + \psi_3 X^*$. If we project Y on X within each group, standard OLS results indicate that the slope estimators $\widehat{\psi}_1$ and $\widehat{\psi}_3$ are both attenuated. This in turn implies that if $\psi_1, \psi_3 < 0$, then $\widehat{\psi}_0$ is biased upward and $\widehat{\psi}_2$ downward, and the resulting RD estimator $\widehat{\psi}_0 - \widehat{\psi}_2$ is biased upward. When $\delta_{\text{sharp}} = \psi_0 - \psi_2 > 0$, we actually arrive at an exaggerated, rather than attenuated, RD estimate! In short, the association of classical measurement error with attenuation comes from slope estimation in an error-in-variable model and does not generalize to intercept estimation, as is the case for RD.

In the rest of this chapter, we study whether we can recover the RD treatment effect in the presence of u. The first step in our quest is the identification of the distribution of X^*. As mentioned in the Introduction, not only is the identification of the X^* distribution used to recover the RD treatment effect parameter, but it can also be used to test the validity of the RD design. However, it is not possible to identify the X^* distribution from the observed X distribution absent any other information. In the presence of measurement error, economists have traditionally relied on repeated measures of the true explanatory variable (Ashenfelter & Krueger, 1994; Black, Berger, & Scott, 2000; Hausman, Newey, Ichimura, & Powell, 1991; Li & Vuong, 1998; Schennach, 2004). However, such measures may not be available in the data. What is helpful in a sharp RD context is that observed program eligibility $D^* = 1_{[X^* < 0]}$, which is a deterministic function of X^*, is informative of the value of X^*. Therefore, it becomes an interesting question as to whether and under what additional assumptions we can use the joint distribution of (X, D^*) to identify the X^* distribution and δ_{sharp}. In a fuzzy RD model, program takeup D is no longer a deterministic function of true assignment variable X^*. As a consequence, we will impose additional assumptions on the u distribution for identification.

As pointed out by a referee, another deviation from model (3) that may arise in practice results from our inability to observe the true RD threshold. For example, the econometrician may not have all the information to precisely construct a family's eligibility threshold for a means-tested program. While it is difficult to disentangle the measurement error in threshold from that in income or to identify the true threshold a particular family faces, the problem of identifying the RD treatment effect is the same as above. To see this, suppose the true eligibility assignment mechanism is $D^* = 1_{[W^* < c^*]}$, where W^* is, say, the actual family income and c^* the monetary eligibility threshold (note that we use the income normalization $X^* = W^* - c^*$ above, and the eligibility threshold is normalized to zero). Suppose the econometrician only sees proxies $W = W^* + u$ and $c = c^* + v$, but not W^* and c^* directly. In this case, we can rewrite the eligibility assignment mechanism as $D^* = 1_{[X-u+v<0]} = 1_{[X-\tilde{u}<0]}$ where $\tilde{u} \equiv u - v$. Suppose Assumption 1 holds and that $X^* \perp\!\!\!\perp v$, then $X^* \perp\!\!\!\perp \tilde{u}$, and we have a model isomorphic to Eq. (3).

3. DISCRETE ASSIGNMENT VARIABLE AND MEASUREMENT ERROR

In this section, we study the identification of the assignment variable distribution and the RD treatment effect parameter when X^* and u are discrete. A discrete assignment variable setup may appear odd given the continuity assumption in identifying the treatment effect in a sharp RD design, but it is necessary in many policy contexts where an RD design appears compelling (Lee & Card, 2008). Even if the assignment variable is continuous (e.g., income), the discretization of X^* can be thought of as a binned-up approximation (this is common practice in graphical presentations of most RD applications). We should point out, however, that if we assume independence between the underlying *continuous* assignment variable and measurement error, then their respective discretized versions are not going to be independent.[8] However, we can also start with the assumption that the discretized X^* and u are independent, in which case their continuous versions will not be. Which one of these assumptions is correct? We are inclined to believe that *neither* assumption is likely to hold empirically, but it is important to gauge whether they reasonably approximate the data. We discuss an overidentification test to that effect in Section 3.4.

It is also worth noting that when the discrete X and X^* have the same support (e.g., when they denote age/birthdate, student enrollment), the measurement error cannot be independent to the true assignment variable – the extreme form of this is the misclassification error in a binary variable, which is known to be mean-reverting. Identification of the assignment variable distribution as well as the RD conditional expectation functions with this type of nonclassical measurement error is a corollary of the problem nicely solved by Chen, Hu, and Lewbel (2009).[9] Under appropriate rank conditions and shape restrictions on the conditional expectation function, Chen et al. (2009) employ the observed conditional characteristic function of Y and its derivative for identification.

3.1. Identification of the True Assignment Variable Distribution: Perfect Compliance

In this subsection, we provide sufficient conditions for identifying the distributions of $X^*$and u from the joint distribution of X and D^* in model (3), where X^* and u are *discrete and bounded*. The identification result is shown in two steps: (1) identification of the support of X^* and u and (2) identification of the probability mass at each point in the supports of X^* and u. In addition to the independence between X^* and u, the identification result relies on the assumption of positive mass in the X^* distribution around the threshold, along with a technical rank condition discussed in detail below.

Denote the support of any random variable Z by support_Z, and denote $L_Z \equiv \min\{\text{support}_Z\}$ and $U_Z \equiv \max\{\text{support}_Z\}$ for a discrete and bounded Z. Without loss of generality, we consider the case where the supports of X^* and u belong to the set of integers. Formally, the discrete and bounded support assumption is written as

Assumption DB (Discrete and Bounded Support). $\text{support}_{X^*} \subseteq \{L_{X^*}, L_{X^*}+1, \ldots, U_{X^*}-1, U_{X^*}\}$ and $\text{support}_u \subseteq \{L_u, L_u+1, \ldots, U_u-1, U_u\}$, where $L_{X^*}, U_{X^*}, L_u, U_u \in \mathbb{Z}$.

The assumption of independence between X^* and u gives strong implications relating their respective supports to the observed support of X, conditional on D^*. Specifically,

$$\begin{aligned} \min\{\text{support}_{X|D^*=d}\} &= \min\{\text{support}_{X^*|D^*=d}\} + \min\{\text{support}_u\}, \\ \max\{\text{support}_{X|D^*=d}\} &= \max\{\text{support}_{X^*|D^*=d}\} + \max\{\text{support}_u\} \quad \text{for } d = 0, 1, \end{aligned} \tag{4}$$

which imposes four restrictions on six unknowns. In order to identify the supports of X^* and u, we impose the additional assumption

Assumption 2 (Threshold Support). $-1, 0 \in \text{support}_{X^*}$.

Assumption 2 states that there exist agents with X^* right at and below the eligibility threshold 0. This is not a strong assumption and must be satisfied in all valid RD designs because the quasi-experimental variation of an RD design comes from agents around the threshold. The addition of this weak assumption is sufficient for identifying the supports.

Lemma 1 (Support Identification in a Sharp Design). Under Assumptions DB, 1, and 2, $L_{X^*}, U_{X^*}, L_u,$ and U_u are identified.

Proof of Lemma 1 is in Section A.1.1 of the online Supplemental Material (Pei & Shen, 2016). Intuitively, individuals who are in the program ($D^* = 1$) but appear ineligible ($X \geqslant 0$) have a positive measurement error $u > 0$. Analogously, those with $D^* = 0$ but $X < 0$ have a negative measurement error $u < 0$. This is essentially the insight behind Eq. (4) and the proof of Lemma 1.

With the support of X^* identified, we next derive the identification of the probability mass of X^* at every point in its support. Denote the probability mass of X^* by p_k at each integer k, and denote that of u by m_l at each integer l. Let the conditional probability masses of the observed assignment variable X be $q_j^1 \equiv \Pr(X = j | D^* = 1)$ and $q_j^0 \equiv \Pr(X = j | D^* = 0)$, for j in the support of the X distribution. Denote the marginal probabilities of D^* by $r^1 \equiv \Pr(D^* = 1)$ and $r^0 \equiv \Pr(D^* = 0)$.

Under the independence assumption of X^* and u, the distribution of $X|D^*$ is the convolution of the distribution of $X^*|D^*$ and that of u. In particular,

$$\begin{aligned} q_j^1 &= \frac{\sum_{k<0} p_k m_{j-k}}{\sum_{k<0} p_k}, \\ q_j^0 &= \frac{\sum_{k \geqslant 0} p_k m_{j-k}}{\sum_{k \geqslant 0} p_k}. \end{aligned} \tag{5}$$

Additionally, the marginal probabilities of D give rise to two more restrictions on the parameters of interest:

$$
\begin{aligned}
r^1 &= \sum_{k<0} p_k, \\
r^0 &= \sum_{k\geqslant 0} p_k.
\end{aligned}
\tag{6}
$$

Note that $r^1, r^0 > 0$ under Assumption 2, and the q_j^1 and q_j^0's are thus well defined. Note also that $\sum_k p_k = 1$ follows from $r^1 + r^0 = 1$ and Eq. (6), and $\sum_l m_l = 1$ follows from $\sum_j (q_j^1 r^1 + q_j^0 r^0) = 1$, and they are therefore redundant constraints.

Taken together, Eqs. (5) and (6) represent $2K_u + K_{X^*}$ restrictions on $K_u + K_{X^*}$ parameters, where $K_{X^*} = |\{L_{X^*}, L_{X^*}+1, \ldots, U_{X^*}-1, U_{X^*}\}|$ and $K_u = |\{L_u, L_u + 1, \ldots, U_u - 1, U_u\}|$ denote the number of probability mass points to be identified in the X^* and u distributions. Even though there are more constraints than parameters, it is not clear that the X^* distribution is always identified because of the nonlinearity in Eq. (5). To formally investigate the identifiability of the parameters, we introduce the following notations: let $p_k^1 \equiv \frac{p_k}{r^1}$ for $k<0$ and $p_k^0 \equiv \frac{p_k}{r^0}$ for $k \geqslant 0$. Define $Q^1(t) \equiv \sum_j q_j^1 e^{tj}$, $Q^0(t) \equiv \sum_j q_j^0 e^{tj}$, $P^1(t) \equiv \sum_k p_k^1 e^{tk}$, $P^0(t) \equiv \sum_k p_k^0 e^{tk}$, and $M(t) \equiv \sum_l m_l e^{tl}$ to be the moment generating functions (MGFs) of the random variables, $X|D=1$, $X|D=0$, $X^*|D=1$, $X^*|D=0$ and u.[10] It is a well-known result that the MGF of the sum of two independent random variables is the product of the MGFs of the two variables (see, for example, Grinstead & Snell, 1997, Chapter 10). Consequently, Eqs. (5) and (6) can be compactly represented as

$$
\begin{aligned}
Q^d(t) &= P^d(t)M(t) \text{ for all } t \neq 0 \quad \text{and} \quad d = 0, 1, \\
P^1(0) &= P^0(0) = 1.
\end{aligned}
\tag{7}
$$

Because $P^1(t)$ and $P^0(t)$ are everywhere positive, Eq. (7) implies

$$
M(t) = \frac{Q^1(t)}{P^1(t)} = \frac{Q^0(t)}{P^0(t)}.
$$

It follows that

$$
P^0(t)Q^1(t) = P^1(t)Q^0(t),
$$

which eliminates the nuisance parameters associated with the measurement error distribution.[11] Matching the coefficients in $P^0(t)Q^1(t)$ to those in

$P^1(t)Q^0(t)$ along with the constraint $P^1(0) = P^0(0) = 1$ results in the following *linear* system of equations in terms of the p_k^1's and p_k^0's:

$$\underbrace{\begin{bmatrix} q_{U_u-1}^1 & 0 & \cdots & 0 & -q_{U_{X^*}+U_u}^0 & 0 & \cdots & 0 \\ q_{U_u-2}^1 & q_{U_u-1}^1 & \cdots & 0 & -q_{U_{X^*}+U_u-1}^0 & -q_{U_{X^*}+U_u}^0 & \cdots & 0 \\ \vdots & q_{U_u-2}^1 & \cdots & \vdots & \vdots & -q_{U_{X^*}+U_u-1}^0 & \cdots & \vdots \\ q_{L_u+L_{X^*}}^1 & \vdots & \cdots & 0 & -q_{L_u}^0 & \vdots & \cdots & 0 \\ 0 & q_{L_u+L_{X^*}}^1 & \cdots & q_{U_u-1}^1 & 0 & -q_{L_u}^0 & \cdots & -q_{U_{X^*}+U_u}^0 \\ \vdots & 0 & \cdots & q_{U_u-2}^1 & \vdots & 0 & \cdots & -q_{U_{X^*}+U_u-1}^0 \\ \vdots & \vdots & \cdots & \vdots & \vdots & \vdots & \cdots & \vdots \\ 0 & 0 & \cdots & q_{L_u+L_{X^*}}^1 & 0 & 0 & \cdots & -q_{L_u}^0 \\ 1 & 1 & \cdots & 1 & 0 & 0 & \cdots & 0 \\ 0 & 0 & \cdots & 0 & 1 & 1 & \cdots & 1 \end{bmatrix}}_{\mathbf{Q}:(K_{X^*}+K_u)\times K_{X^*}} \underbrace{\begin{bmatrix} p_{U_{X^*}}^0 \\ p_{U_{X^*}-1}^0 \\ \vdots \\ p_0^0 \\ p_{-1}^1 \\ p_{-2}^1 \\ \vdots \\ p_{L_{X^*}}^1 \end{bmatrix}}_{\mathbf{p}:K_{X^*}\times 1} = \underbrace{\begin{bmatrix} 0 \\ 0 \\ \vdots \\ 0 \\ 0 \\ 1 \\ 1 \end{bmatrix}}_{\mathbf{b}:(K_{X^*}+K_u)\times 1}. \tag{8}$$

Standard results in linear algebra can be invoked to provide identification of the probability masses. Denote system (8) with the compact notation $\mathbf{Qp} = \mathbf{b}$, where $\mathbf{Q}$ is the $(K_{X^*} + K_u) \times K_{X^*}$ *data* matrix, $\mathbf{p}$ the $K_{X^*} \times 1$ *parameter* vector and $\mathbf{b}$ the $(K_{X^*} + K_u) \times 1$ *constant* vector of 0's and 1's. The parameter vector $\mathbf{p}$ is identified if and only if $\mathbf{Q}$ is of full rank. Note that there are more rows than columns in $\mathbf{Q}$, that is, $K_{X^*} + K_u > K_{X^*}$, and we introduce the following assumption

Assumption 3 (Full Rank). $\mathbf{Q}$ in Eq. (8) has full-column rank.

Note that Assumption 3 is not always satisfied, and an example is provided in Section A.2 of the online Supplemental Material. At the same time, Assumption 3 is directly testable because $\text{rank}(\mathbf{Q}) = \text{rank}(\mathbf{Q}^{\mathrm{T}}\mathbf{Q})$, and $\mathbf{Q}$ is of full rank if and only if $\det(\mathbf{Q}^{\mathrm{T}}\mathbf{Q}) \neq 0$. The distribution of the determinant estimator can be obtained by the delta method because it is a polynomial in the q_j^0's and q_j^1's, the observed probability masses.

With Assumption 3, the $\mathbf{p}$ vector is identified. Because $p_k = r^1 p_k^1$ for $k < 0$ and $p_k = p_k^0 r^0$ for $k \geqslant 0$ and because r^1 and r^0 are observed, uniqueness of the p_k^1's and the p_k^0's implies the uniqueness of the p_k's. Although parameters of the measurement error distribution are eliminated in Eq. (8), they are identified after the identification of the p_k's as shown in Section A.3 of the online Supplemental Material. Formally, the identification of probability masses is summarized in the following Lemma.

> **Lemma 2** (Probability Mass Identification in a Sharp Design). Suppose $(\{p_k\}, \{m_l\})$ solves the system of equations consisting of Eqs. (5) and (6). Then $(\{p_k\}, \{m_l\})$ is the unique solution if and only if Assumption 3 holds.

Combining Lemmas 1 and 2 implies the identification of the true underlying distributions of model (3):

> **Proposition 1**. Under Assumptions DB, 1, 2, and 3, the distributions of X^* and u are identified.

Intuitively, the identification result extracts information about u from the tails of the X^* distribution in the treatment and control groups. The discrete and bounded assumption reduces the dimensionality of the identification problem and fully specifies the statistical model with a finite number of parameters, even though we are completely agnostic about the shapes of the X^* and u distributions over their respective supports. After we have pinned down the set of parameters, we identify the X^* distribution by starting from the tails and working our way in, using the independent measurement error assumption.

When we relax boundedness, we no longer have the luxury of a finite-dimensional parameter space. Consequently, the distributions of X^* and u are in general not identified from model (3), as shown through a constructive example in Section A.4 of the online Supplemental Material.

3.2. Identification of the True Assignment Variable Distribution: Imperfect Compliance

As mentioned in Section 2, the assumption of perfect compliance or equivalently the observability of eligibility ($D^* = 1_{[X^*<0]}$) is not often satisfied, as is the case in almost all means-tested social programs. Instead, only a measure of program participation D may be available. In this subsection, we consider the more realistic case of imperfect compliance, or a fuzzy design, for discrete and bounded assignment variable X^* and measurement error u. We set out to identify the X^* distribution from the observed joint distribution (X, D).

The difficulty with imperfect compliance is that rather than having benefit receipt D as a deterministic step function of X^*, $\Pr(D = 1|X^*)$ is potentially an unrestricted function in X^*, even though eligibility is still given by $D^* = 1_{[X^* < 0]}$. In the extreme, program participation D could be independent from X^* and therefore would not provide any additional information. If this was the case, uniquely deconvolving X^* and u from the observed joint distribution (X, D) would not be possible. In many programs (typically means-tested transfer programs), however, it is the case that if an agent's true assignment variable is above the eligibility threshold, she is forbidden from participating in the program, that is

Assumption 4 (One-sided Fuzzy). $\Pr(D = 1|X^* = x^*) = 0$ for $x^* \geqslant 0$.

Under Assumption 4, any agent with $D = 1$ has true assignment variable $X^* < 0$. It follows that the upper end point in the $X|D = 1$ distribution identifies U_u, provided that Assumptions 1 and 2 hold for the $D = 1$ population. Unlike the perfect compliance scenario, where $X^* \perp\!\!\!\perp u$ conditional on D^* directly follows from Assumption 1, an additional assumption is needed to ensure the independence between X^* and u conditional on $D = 1$:

Assumption 1F (Strong Independence). $u \perp\!\!\!\perp X^*, D$.

Identification also requires an extension of Assumption 2:

Assumption 2F (Threshold Support: Fuzzy). $-1 \in \text{support}_{X^*|D=1}$ and $0 \in \text{support}_{X^*}$.

Remark 1. A weaker version of Assumption 1F, that $u \perp\!\!\!\perp X^*$ conditional on D, suffices for the results below. However, it may be difficult to find situations where this weaker assumption is empirically justified but Assumption 1F is not. In fact, this weaker assumption does not even imply that $X^* \perp\!\!\!\perp u$ unconditionally as in Assumption 1. Therefore, we propose the stronger Assumption 1F.

Remark 2. Although Assumption 2F is stronger than Assumption 2, it needs to be satisfied in a valid fuzzy RD design without measurement error. That is, there must be nonzero takeup just below the eligibility cutoff, without which a first-stage discontinuity does not exist.

Even with Assumptions IF and 2F, we still need to distinguish between nontakeup and ineligibility. An individual with $D = 0$ and $X = -1$ could have true income $X^* = 1$ (with an implied measurement error $u = -2$) and is not program eligible; or she could be eligible with income $X^* = -1$ (with an implied measurement error $u = 0$) but chooses not to participate in the program. On the one hand, if every observation with $D = 0$ is treated as ineligible, then the lower end point in the support of u, L_u, is that in the $X|D = 0$ distribution. On the other hand, if every observation with $D = 0$ is treated as an eligible nontakeup, then L_u is 0. Clearly, the two treatments imply different distributions. However, if the researcher believes that the identified length of the right tail in the u distribution sheds light on the length of its left tail, it may be reasonable to assume the following.

Assumption 5 (Symmetry in Support). $L_u = -U_u$.

Symmetry in support is weaker than imposing symmetry in the measurement error distribution, as is conventional in the literature. With the additional Assumptions 4, 1F, 2F, and 5, the supports of the X^* and u are identified:

> **Lemma 1F** (Support Identification in a Fuzzy Design). Under Assumptions DB, 1F, 2F, 4, and 5, the support of u and the support of X^* conditional on $D = 0$ and $D = 1$ are identified.

The proof of Lemma 1F is provided in Section A.1.1 of the online Supplemental Material. The intuition is similar to the sharp case: thanks to Assumption 4, the right tail of $X^* \geqslant 0$ in the $D = 1$ population provides information on the length of the right tail of the u distribution. The length of the left tail of the measurement error distribution is then obtained by symmetry (Assumption 5). The identification of probability masses can proceed analogously as in Section 3.1, but the matrix $\mathbf{Q}$ needs to be modified to a matrix with larger dimensions, which we denote by $\mathbf{Q_F}$. With the corresponding rank assumption

Assumption 3F (Full Rank: Fuzzy). $\mathbf{Q_F}$ has full-column rank,

we have the following identification result in the fuzzy case:[12]

> **Proposition 1F**. Under Assumptions DB, 1F, 2F, 3F, 4, and 5, the distributions of X^* conditional on D and u are identified.

> *Remark 3*. Identification is possible in the absence of Assumptions 2F, 4, and 5, provided that we know the values of U_u and L_u. In this case, $L_{X^*|D=d}$ and $U_{X^*|D=d}$ can be recovered using this knowledge, and the probability masses are identified if the full rank condition is satisfied. In practice, this observation has little practical value since researchers rarely – if at all – know the true values of U_u and L_u.

Remark 4. Related to Remark 3, identification can be obtained with only the independence assumption (Assumption 1/1F) if we have explicit knowledge of the marginal distribution of X^*, say from a validation sample.[13] This is because, as it is easy to show, the marginal distribution of u is identified from the marginal distribution of X and X^* by an overidentified linear system of equations. It follows that the distribution of X^* conditional on D^* (D) in the sharp (fuzzy) case is identified from the observed $X|D^*$ ($X|D$) distribution and the identified u distribution.

In practice, however, a researcher is unlikely to obtain the marginal distribution of X^* in the case of a transfer program even if she has access to administrative earnings data. First, commonly used administrative earnings records are of *quarterly or annual* frequency, but program eligibility is usually based on *monthly* income. Second, the income used for determining program eligibility is typically earnings after certain deductions (child care or work related expenses, for example) plus unearned income. In that sense, the administrative earnings records are also a noisy version of the income for program eligibility determination, not to mention the fact that they may not perfectly measure true earnings either (Abowd & Stinson, 2013).

Remark 5. One might question our implicit assumption that benefit receipt D is accurately measured.[14] Errors in reporting program participation status in means-tested transfer programs have been documented in validation studies of survey data. Marquis and Moore (1990) estimate that the cash welfare underreporting rate (i.e., actual benefit recipients who report no benefits) in the 1984 SIPP panel could be as high as 50%. Underreporting of Medicaid coverage appears to be less severe – Card, Hildreth, and Shore-Sheppard estimate that the overall error rate in the 1990–1993 SIPP Medicaid status is 15% for the state of California.

Underreporting, however, does not pose a threat to the identification of the X^* distribution, provided that those with $D = 1$ indeed received benefits and were therefore eligible. It follows that the supports of $X^*|D$ and u are identified correctly and that probability masses can be recovered, as long as Assumption 1F holds. It will be problematic, however, if those who do not take part in the program mistakenly report participation. Fortunately, the rate of false-positive reporting of participation in transfer programs is very small empirically – around 0.2% in the Marquis and Moore (1990) study and 1.5% in Card et al. (2004). This suggests that the reporting error in D does not pose a big threat to the procedure above when applying an RD design using benefit discontinuities at the eligibility cutoff. Furthermore, trimming procedures can be undertaken to correct for the false-positive reporting problem, which we discuss in Section 3.5.

3.3. Identification of the Conditional Expectation Functions and the RD Treatment Effect

In this subsection, we show that the conditional expectation function $E[Y|X^*]$ in a sharp design and $E[Y|X^*]$ and $E[D|X^*]$ in a one-sided fuzzy design can be identified under conditional versions of Assumptions 1, 2, and 3. In essence, these assumptions allow the performance of the deconvolution exercise detailed in the previous subsections for each value of Y. In a sharp design, for example, once we identify $\Pr(X^*|Y)$ for each value of Y, we apply Bayes' Theorem to recover $\Pr(Y|X^*)$, which we then integrate to obtain $E[Y|X^*]$. Finally, as with any RD design with a discrete assignment variable, we can parametrically extrapolate the conditional expectation functions to recover the RD treatment effect.[15] For ease of exposition, we focus on the case with a binary Y and discuss identification with a general Y distribution in the online Supplemental Material.

In the sharp RD model (1), the treatment effect is $\delta_{\text{sharp}} = \text{extp}_{c\uparrow 0} E[Y|X^* = c] - E[Y|X^* = 0]$ for discrete X^*. The first term is the left intercept of the $E[Y|X^*]$ function parametrically extrapolated using $E[Y|X^* = x^*]$ for $x^* < 0$, and we use "extp" to denote this extrapolation operator (analogous to the limit operator in the continuous X^* case). The second term $E[Y|X^* = 0]$ is directly observed from the data. In order to identify δ_{sharp}, we need to identify the conditional expectation function $E[Y|X^*]$, for which we propose the following assumptions. These assumptions imply that Assumptions 1–3 hold conditional on Y:

Assumption 1Y (Nondifferential Measurement Error). $u \perp\!\!\!\perp X^*, Y$.[16]

Assumption 2Y (Threshold Support). $-1, 0 \in \text{support}_{X^*|Y=y}$ for each $y = 0, 1$.

Note that Assumption 1Y and Assumption 2Y allow the formulation of the conditional counterparts of Eq. (8): $\mathbf{Q}_Y \mathbf{p}_Y = \mathbf{b}_Y$ for $Y = 0, 1$, where $\mathbf{Q}_Y$ and $\mathbf{p}_Y$ consist of probability masses of q_j^1, q_j^0, p_k^1 and p_k^0 conditional on Y.

Assumption 3Y (Full Rank). The matrix $\mathbf{Q}_Y$ is of full rank for $Y = 0, 1$.

Proposition 2. Under Assumptions DB, 1Y, 2Y, and 3Y, the conditional expectation function $E[Y|X^*]$ is identified for model (1).

As stated at the beginning of this subsection, the proof of Proposition 2 (and of Proposition 2F) follows from an application of the Bayes' Theorem. The details are provided in Section A.1.3 of the online Supplemental Material. Once we pin down $E[Y|X^* = x^*]$, δ_{sharp} is subsequently identified by parametric extrapolation.

Analogous to a sharp design, we can obtain the identification of δ_{fuzzy}, which in the discrete case is defined as $\delta_{\text{fuzzy}} \equiv \frac{\text{extp}_{x^* \uparrow 0} E[Y|X^* = x^*] - E[Y|X^* = 0]}{\text{extp}_{x^* \uparrow 0} E[D|X^* = x^*] - E[D|X^* = 0]}$. The only difference is the need to recover the first-stage relationship $E[D|X^*]$ in a fuzzy design. Again, assumptions underpinning Proposition 1F are extended to hold conditional on Y:

Assumption 1FY (Strong Independence and Nondifferential Measurement Error: Fuzzy). $u \perp\!\!\!\perp X^*, D, Y$.

Assumption 2FY (Threshold Support: Fuzzy). $-1 \in \text{support}_{X^*|D=1,Y=y}$ and $0 \in \text{support}_{X^*|Y=y}$ for each $y = 0, 1$.

Assumption 3FY (Full Rank: Fuzzy). The matrix $\mathbf{Q}_{FY}$ is of full rank for $Y = 0, 1$.[17]

Proposition 2F. Under Assumptions DB, 1FY, 2FY, 3FY, 4, and 5, the conditional expectation functions $E[Y|X^*]$ and $E[D|X^*]$ are identified for model (2).

Remark 6. Propositions 2 and 2F can be easily generalized from the formulation with a binary Y – see Section A.5 of the online Supplemental Material for details.

3.4. Estimators of the Assignment Variable Distribution and the RD Treatment Effect

In this subsection, we propose estimators for the assignment variable distribution and the RD treatment effect in the discrete and bounded case. As with identification, estimation of the X^* distribution follows two steps: estimation of its support and estimation of the probability masses at each point in its support. Support estimation follows the identification results in the proofs of Lemmas 1 and 1F, which derive from Eq. (4), with the population quantities replaced by sample analogs. We abstract away from the sampling error of support and simply assume that the sample is large enough to reveal the true support of the distribution. We present the case in the sharp design setting where we omit the F subscript for notational convenience. Results in the fuzzy case follow by replacing D^* by D.

Given the specification of probability model, the likelihood function can be explicitly written out by using the p_k^1's, p_k^0's, m_l's and the marginal probabilities r^1 and r^0. Formally, the likelihood for the joint distribution (X,D^*) is

$$\begin{aligned} L(X_i, D_i^*) &= L(X_i|D_i^*)L(D_i^*) \\ &= \left\{ (\sum_k p^1_{X_i-k} m_k) r^1 \right\}^{D_i^*} \left\{ (\sum_k p^0_{X_i-k} m_k) r^0 \right\}^{1-D_i^*}. \end{aligned} \tag{9}$$

Researchers can directly maximize the logarithm of Eq. (9), and the resulting estimators are efficient provided that the parameters are in the interior of the parameter space, that is, strictly between zero and one. However, the analytical solutions to maximizing the log likelihood do not appear to exist, and numerically optimizing (9) may become computationally intensive as the number of points in the supports of X^* and u increases.

An alternative strategy relies on the identification Eq. (8), which fits nicely into a standard minimum distance framework $f(\mathbf{q}, \mathbf{p}) = \mathbf{Qp} - \mathbf{b} = 0$ (Kodde, Plam, & Pfann, 1990) from which an estimator of $\mathbf{p}$ can be obtained easily.[18] Because of the linearity in Eq. (8), the parameter vector of interest $\mathbf{p}$ can be estimated analytically once an estimator of $\mathbf{q}$ is obtained. Estimation proceeds according to the following steps.

1. Obtain the estimators $\widehat{q}_j^1 = \frac{\sum_i 1_{[X_i=j]} \cdot 1_{[D_i^*=1]}}{\sum_i 1_{[D_i^*=1]}}$, $\widehat{q}_j^0 = \frac{\sum_i 1_{[X_i=j]} \cdot 1_{[D_i^*=0]}}{\sum_i 1_{[D_i^*=0]}}$, and $\widehat{r}^1 = \frac{1}{N} \sum 1_{[D_i^*=1]}$ (N denotes the sample size), as well as $\widehat{\boldsymbol{\Omega}}$, which is a consistent estimator for the asymptotic variance covariance matrix $\boldsymbol{\Omega}$ of $\widehat{\mathbf{q}}$, (i.e., $\sqrt{N}(\widehat{\mathbf{q}} - \mathbf{q}) \Rightarrow N(0, \boldsymbol{\Omega})$). Since $X|D^* = d$ follows a multinomial distribution for each d, $\widehat{\boldsymbol{\Omega}}$ is a block-diagonal matrix $\widehat{\boldsymbol{\Omega}} = \begin{bmatrix} \widehat{\boldsymbol{\Omega}}^{11} & \mathbf{0} \\ \mathbf{0} & \widehat{\boldsymbol{\Omega}}^{00} \end{bmatrix}$, and

$$\widehat{\Omega}_{ij}^{dd} = \begin{cases} (1-\widehat{q}_i^d)\widehat{q}_i^d/(d\widehat{r}^1 + (1-d)(1-\widehat{r}^1)) & \text{if} \quad i=j \\ -\widehat{q}_i^d\widehat{q}_j^d/(d\widehat{r}^1 + (1-d)(1-\widehat{r}^1)) & \text{if} \quad i \neq j; \end{cases}$$

2. Form the estimator of the $\mathbf{Q}$ matrix under perfect compliance in Eq. (8), $\widehat{\mathbf{Q}}$, by replacing the q_j^1 and q_j^0 in $\mathbf{Q}$ with their estimators;
3. Derive a consistent estimator of $\mathbf{p}$: $\widehat{\mathbf{p}} = \arg\min_{\mathbf{p}} f(\widehat{\mathbf{q}}, \mathbf{p})' f(\widehat{\mathbf{q}}, \mathbf{p}) = (\widehat{\mathbf{Q}}'\widehat{\mathbf{Q}})^{-1}(\widehat{\mathbf{Q}}'\mathbf{b})$;
4. Compute the optimal weighting matrix $\widehat{\mathbf{W}} = (\widehat{\nabla_{\mathbf{q}} f}\widehat{\Omega}\widehat{\nabla_{\mathbf{q}} f}')^{-1}$ where $\widehat{\Omega}$ is a consistent estimator for the variance covariance matrix of the $\mathbf{q}$ derived in step 1.[19] $\widehat{\nabla_{\mathbf{q}} f}$ is a consistent estimator for $\nabla_{\mathbf{q}} f$, the Jacobian of f with respect to $\mathbf{q}$. Because $\nabla_{\mathbf{q}} f$ depends on $\mathbf{p}$, step 3 is necessary for first obtaining a consistent estimate of $\mathbf{p}$. Since f is linear in $\mathbf{q}$, $\widehat{\nabla_{\mathbf{q}} f}$ can be computed analytically;
5. Arrive at the optimal estimator of $\mathbf{p}$: $\widehat{\mathbf{p}}_{\text{opt}} = \left(\widehat{\mathbf{Q}}'\widehat{\mathbf{W}}\widehat{\mathbf{Q}}\right)^{-1}\left(\widehat{\mathbf{Q}}'\widehat{\mathbf{W}}\mathbf{b}\right)$.

Provided that the true parameter lies in the interior of the parameter space:

> **Assumption 6 (Interior).** $\mathbf{p} \in (0,1)^K$ where K is the length of $\mathbf{p}$.

The derivation of the asymptotic distribution of $\widehat{\mathbf{p}}_{\text{opt}}$ is standard. Specifically,

> **Proposition 3.** Under Assumptions DB, 1, 2, 3, and 6 for the sharp case, $\sqrt{N}(\widehat{\mathbf{p}}_{\text{opt}} - \mathbf{p}) \Rightarrow \mathcal{N}(\mathbf{0}, \Sigma)$ where $\Sigma^{-1} \equiv \mathbf{Q}'(\nabla_{\mathbf{q}} f \Omega \nabla_{\mathbf{q}} f')^{-1}\mathbf{Q}$ with Ω being the asymptotic variance matrix of $\mathbf{q}$ and $\nabla_{\mathbf{q}} f \equiv \nabla_{\mathbf{q}}(\mathbf{Qp} - \mathbf{b})$.

Analogously, for the imperfect compliance case, we have

> **Proposition 3F.** Under Assumptions DB, 1F, 2F, 3F, 4, 5, and 6 for the fuzzy case, $\sqrt{N}(\widehat{\mathbf{p}}_{\mathbf{F}\text{opt}} - \mathbf{p}_{\mathbf{F}}) \Rightarrow \mathcal{N}(\mathbf{0}, \Sigma_{\mathbf{F}})$ where $\Sigma_{\mathbf{F}}^{-1} \equiv \mathbf{Q}_{\mathbf{F}}'(\nabla_{\mathbf{q}_{\mathbf{F}}} f_{\mathbf{F}} \Omega_{\mathbf{F}} \nabla_{\mathbf{q}_{\mathbf{F}}} f_{\mathbf{F}}')^{-1}\mathbf{Q}_{\mathbf{F}}$ with $\Omega_{\mathbf{F}}$ being the asymptotic variance matrix of $\mathbf{q}_{\mathbf{F}}$ and $\nabla_{\mathbf{q}_{\mathbf{F}}} f_{\mathbf{F}} \equiv \nabla_{\mathbf{q}_{\mathbf{F}}}(\mathbf{Q}_{\mathbf{F}}\mathbf{p}_{\mathbf{F}} - \mathbf{b}_{\mathbf{F}})$.

The main conclusion of Kodde et al. (1990) shows that $\widehat{\mathbf{p}}_{\text{opt}}$ is efficient – it has the same asymptotic variance as the MLE estimator – if $\mathbf{p}$ is exactly or over-identified by $f(\mathbf{q}, \mathbf{p}) = 0$ and $\widehat{\mathbf{q}}$ is the MLE estimator. Since both conditions are satisfied in our setup, we have obtained a computationally inexpensive estimator without sacrificing efficiency. Also note that the assumptions can be jointly tested by the overidentifying restrictions as is standard for minimum distance estimators. In particular, the test statistic

$N{\cdot}(f(\widehat{\mathbf{q}},\widehat{\mathbf{p}}_{\text{opt}})'\widehat{\mathbf{W}}f(\widehat{\mathbf{q}},\widehat{\mathbf{p}}_{\text{opt}}))$ in the sharp case or $N{\cdot}(f(\widehat{\mathbf{q}}_{\mathbf{F}},\widehat{\mathbf{p}}_{\mathbf{F}\text{opt}})'\widehat{\mathbf{W}}_{\mathbf{F}}f(\widehat{\mathbf{q}}_{\mathbf{F}},\widehat{\mathbf{p}}_{\mathbf{F}\text{opt}}))$ in the fuzzy case follows a χ^2-distribution with degrees of freedom equal to K_u, the number of points in the support of the measurement error, when assumptions in Propositions 3 or 3F are satisfied.[20]

A concern arises because the optimal estimators of p_k^1 and p_k^0 may never sum to one by following the procedure above. Therefore, we modify the estimation strategy and incorporate the matrix constraint $\mathbf{Rp} = \mathbf{c}$ with $\mathbf{c} = \begin{bmatrix} 1 \\ 1 \end{bmatrix}$ and $\mathbf{R} = \begin{bmatrix} 1 & \cdots & 1 & 0 & \cdots & 0 \\ 0 & \cdots & 0 & 1 & \cdots & 1 \end{bmatrix}$. In step 3, the consistent estimator is instead $\widehat{\mathbf{p}} = (\widehat{\mathbf{Q}}'\widehat{\mathbf{Q}})^{-1}\mathbf{R}'\{\mathbf{R}(\widehat{\mathbf{Q}}'\widehat{\mathbf{Q}})^{-1}\mathbf{R}'\}^{-1}\mathbf{c}$,[21] and in step 5, $\widehat{\mathbf{p}}_{\text{opt}} = (\widehat{\mathbf{Q}}'\widehat{\mathbf{W}}\widehat{\mathbf{Q}})^{-1}\mathbf{R}'\{\mathbf{R}(\widehat{\mathbf{Q}}'\widehat{\mathbf{W}}\widehat{\mathbf{Q}})^{-1}\mathbf{R}'\}^{-1}\mathbf{c}$. The asymptotic variance matrix of $\widehat{\mathbf{p}}_{\text{opt}}$ is given by $\mathbf{T}((\mathbf{QT})'\mathbf{W}(\mathbf{QT}))^{-1}\mathbf{T}'$, where $\mathbf{T}$ is a matrix whose columns are the first $K-2$ eigenvectors of the projection matrix $\mathbf{I}-\mathbf{R}'(\mathbf{RR}')^{-1}\mathbf{R}$. The computation for $\widehat{\mathbf{W}}$ is unaltered by the imposition of the linear constraints $\mathbf{Rp} = \mathbf{c}$.

In order to construct the asymptotic distribution of the RD treatment effect estimators, we need to estimate the variance matrix of $E[\widehat{Y|X^*}]$ in the sharp RD case and $E[\widehat{D|X^*}]$ and $E[\widehat{Y|X^*}]$ in the fuzzy RD case for each point in the support of X^*. As shown in Eqs. (A.5)–(A.8) of the online Supplemental Material, the estimands are differentiable functions of $\Pr(X^* = x^*|D = d, Y = y)$, $\Pr(D = d|Y = y)$ and $\Pr(Y = 1)$ for $d, y = 0, 1$. The delta method can be directly applied, where the Jacobian of the transformations is derived analytically. The general expressions of the estimators, $\widehat{\delta}_{\text{sharp}}$ and $\widehat{\delta}_{\text{fuzzy}}$, cannot be obtained because they depend on the functional forms of $E[Y|X^*]$ and $E[D|X^*]$, which vary from application to application.

Finally, we note that the restriction to a binary Y can be relaxed. First, estimation proceeds in exactly the same way when Y takes on a finite number of values. Second, we show in Section A.5 of the online Supplemental Material that our estimators can be further generalized to the case where Y is continuously distributed, though the procedure is likely to be computationally burdensome and numerically sensitive. In the numerical and empirical illustrations below, therefore, we adhere to a binary outcome variable.

3.5. Potential Issues in Practical Implementation

There are several issues in implementing the procedure described in Section 3.4. First of all, in order to have realistic support for the true assignment variable, the maximum value of X needs to be significantly larger for the $D = 0$ group than for the $D = 1$ group, since their difference is the upper end point in the X^* distribution. Also, the left tail of the X distribution may need to be significantly longer that of the right tail for the $D = 1$ group, since their difference is the lower bound of the X^* distribution (following Assumption 5). Because symmetry is a functional form assumption, which may not hold when the assignment variable is measured in levels (e.g., in the case of income), a transformation may be needed. In practice, we use a Box-Cox transformation and experiment with various transformation parameters. The overidentification test mentioned in the previous subsection can help us decide which transformation parameters to be used.

It is also possible, as mentioned in Remark 5, that an ineligible individual with a very large observed X may actually report program participation ($D = 1$) by mistake. If this is the case, the supports will not be correctly identified, and using a Box-Cox transformation will not be sufficient to correct the problem. A trimming procedure could be adopted in practice, which drops outliers in both the left and right tails of the $X|D = 1$ and $X|D = 0$ distributions. As with the case of transformation parameters, we try out several trimming percentages and examine the sensitivity of the empirical results. Finally, a quadratic programming routine with inequality constraints can be used in practice to guarantee nonnegativity of the probability masses.

3.6. Illustration with a Simple Numerical Example

In this subsection, we illustrate the proposed estimation procedure in Section 3.4 with a simple numerical example. We focus on the more complicated fuzzy case and show that the true first-stage and outcome functions as well as the X^* distribution can indeed be recovered when the assumptions in Propositions 2F and 3F are met. In the baseline example, we generate X^* following a uniform distribution on the set of integers from -10 to 10. u follows a uniform distribution between -3 and 3 and is therefore symmetric in its support (Assumption 5). The true first-stage relationship is given by

$$E[D|X^*] = (\alpha_{D^*X^*}X^* + \alpha_{D^*})1_{[X^*<0]} = \alpha_{D^*}D^* + \alpha_{D^*X^*}D^*X^*, \tag{10}$$

which reflects the one-sided fuzzy assumption (Assumption 4), and the size of the first-stage discontinuity is α_{D^*}. The outcome response function is given by the simple constant treatment effect specification

$$E[Y|X^*, D] = \delta_0 + \delta_1 X^* + \delta_{\text{fuzzy}} D, \tag{11}$$

where the treatment effect to be identified is δ_{fuzzy}. Note that Eqs. (10) and (11) together imply that the outcome relationship is

$$E[Y|X^*] = \beta_0 + \beta_{D^*} D^* + \beta_1 X^* + \beta_{D^*X^*} D^* X^*, \tag{12}$$

where $\beta_0 = \delta_0$, $\beta_{D^*} = \alpha_{D^*} \cdot \delta_{\text{fuzzy}}$, $\beta_1 = \delta_1$ and $\beta_{D^*X^*} = \alpha_{D^*X^*} \cdot \delta_{\text{fuzzy}}$.

Figs. 2 and 3 present graphical results based on a sample of 25,000 generated observations under the parameter values $\alpha_{D^*X^*} = -0.01$, $\alpha_{D^*} = 0.8$, $\delta_0 = 0.15$, $\delta_1 = -0.01$, and $\delta_{\text{fuzzy}} = 0.6$, with the implied coefficients in Eq. (12) being $\beta_0 = 0.15$, $\beta_{D^*} = 0.48$, $\beta_1 = -0.01$ and $\beta_{D^*X^*} = -0.006$. We choose $N = 25,000$ because it is about the average sample size in the relevant studies – 45,722 in Hullegie and Klein (2010), 34,949 in Koch (2013), 11,541 in Schanzenbach (2009), and 2,954 in de la Mata (2012). The top and bottom panels in Fig. 2 plot the *observed* first-stage and outcome relationships, that is, $E[D|X]$ and $E[Y|X]$, respectively. Note that there is no visible discontinuity at the threshold, and the estimated first-stage and outcome discontinuities based on these observed relationships cannot identify the true parameter values of α_{D^*} and β_{D^*}, which are 0.8 and 0.48 respectively.

Fig. 3 plots the *estimated* first-stage and outcome relationships based on procedures developed in Section 3.4 against the actual (10) and (11) specified with the parameter values above. As is evident from the graphs, the proposed procedures can correctly recover the $E[D|X^*]$ and $E[Y|X^*]$ of the underlying RD design. $\widehat{\delta}_{\text{fuzzy}}$, the RD treatment effect parameter, is obtained by fitting another linear minimum distance procedure on the estimated $E[D|X^*]$ and $E[Y|X^*]$ (as well as their estimated variance matrices) with the parametric restrictions (10) and (12). In 1,000 repeated samples, the average point estimate for α_{D^*} is 0.75 (true parameter value is 0.8), the average standard error is 0.063, and the coverage rate of the 95% confidence interval is 97%; the average point estimate for β_{D^*} is 0.48 (true parameter value is 0.48), the average standard error is 0.075 and the coverage rate of the 95% confidence interval is 98%.

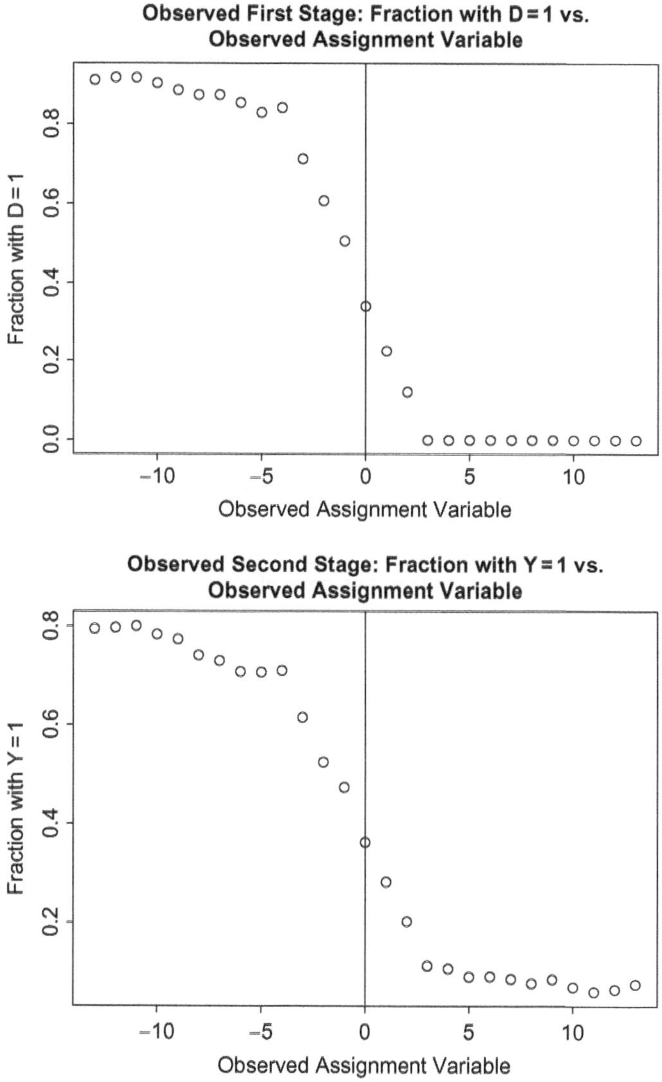

Fig. 2. Observed First-Stage and Outcome Relationships: Expectation of D and Y Conditional on the Noisy Assignment Variable X. *Notes*: Illustrative example is based on a sample of size $N = 25,000$. X^* and u are uniformly distributed on the set of integers in $[-10, 10]$ and $[-3, 3]$, respectively. The true first-stage and outcome response functions are $E[D|X^*] = (-0.01X^* + 0.8)D^*$ and $E[Y|X^*, D] = 0.15 + 0.6D - 0.01X^*$, respectively, which imply a true second-stage equation of $E[Y|X^*] = 0.15 + 0.48D^* - 0.01X^* - 0.006D^*X^*$. Plotted are $E[D|X]$ and $E[Y|X]$ respectively where $X = X^* + u$.

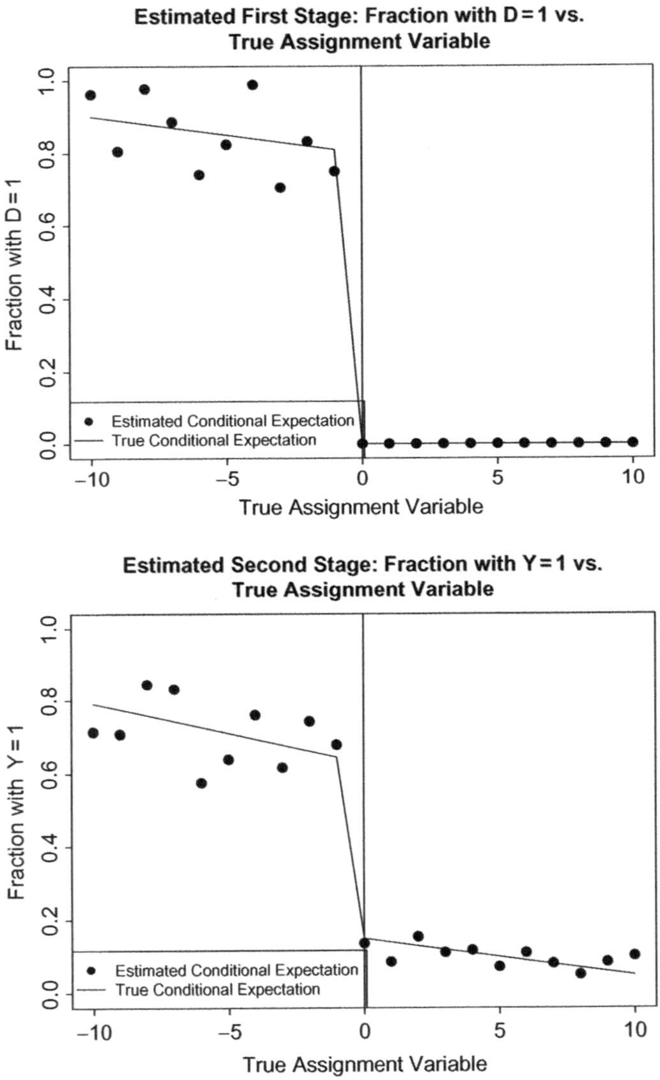

Fig. 3. Estimated First Stage and Outcome Relationships: Expectation of D and Y Conditional on the True Assignment Variable X^*. *Notes*: Same data-generating process as Fig. 2. Plotted are the estimated $E[D|X^*]$ and $E[Y|X^*]$ following procedures developed in Section 3.4 against the true conditional expectations.

To gauge the performances of the estimators in adverse settings, we test their sensitivity to (1) the violation of symmetry and (2) larger supports in X^* and u relative to the sample size. Unfortunately, the performance of the estimators quickly deteriorates when the symmetry assumption is violated, which is an important limitation of the proposed method. For example, when u is supported on the integers in $[-4, 3]$ but the lower bound of its support is erroneously assumed to be -3, the average point estimates for α_{D^*} and β_{D^*} are 0.66 and 0.39, and the coverage rates of the 95% confidence interval are 0.56 and 0.74, respectively. In comparison, the behavior of the estimators is more robust with respect to alternative specifications of support_X: as the set of points in support_X increases from 27 in the numerical example above to 37, the coverage rates are still around 95%. The coverage rates fall to around 80% and 55% when $|\text{support}_X|$ is 47 and 67, respectively.

As mentioned above, the proposed procedure can also be used to assess the degree of nonsmoothness in the X^* distribution at the eligibility threshold. In another numerical exercise, we consider two alternative specifications: (1) the specification used in Figs. 2 and 3, for which there is no discontinuity in the X^* distribution at the eligibility threshold; (2) X^* is still supported on the set of integers from -10 to 10 but with a discontinuity at the eligibility threshold: $\Pr(X^* = i) = 0.06$ for each $i<0$ and $\Pr(X^* = i) = 0.036$ for $i \geqslant 0$. Figs. 4 and 5 present the observed X and *estimated* X^* distribution for cases (1) and (2), respectively. Note that there is no obvious discontinuity in the observed X distribution at the eligibility threshold in case (2) (top panel of Fig. 5), which again illustrates the problematic nature of using the observed assignment variable X for RD analyses. In both cases, we test for the threshold discontinuity by fitting a linear minimum distance procedure on the estimated X^* distribution with the restriction

$$\Pr(X^* = x^*) = \gamma_0 + \gamma_{D^*}D^* + \gamma_1 X^* + \gamma_{D^*X^*}D^*X^*.$$

Let $\gamma_{D^*}^{(1)}$ and $\gamma_{D^*}^{(2)}$ be the coefficients of γ_{D^*} in cases (1) and (2) respectively, and based on the specifications above, $\gamma_{D^*}^{(1)} = 0$ and $\gamma_{D^*}^{(2)} = 0.024$. In 10,000 repeated samples with 25,000 observations, the average value of $\gamma_{D^*}^{(1)}$ is -0.003, and the coverage rate of the 95% confidence interval is 89%; the average value of the $\gamma_{D^*}^{(2)}$ estimates is 0.031, and the coverage rate of the 95% confidence interval is 87%. The coverage rates for the density

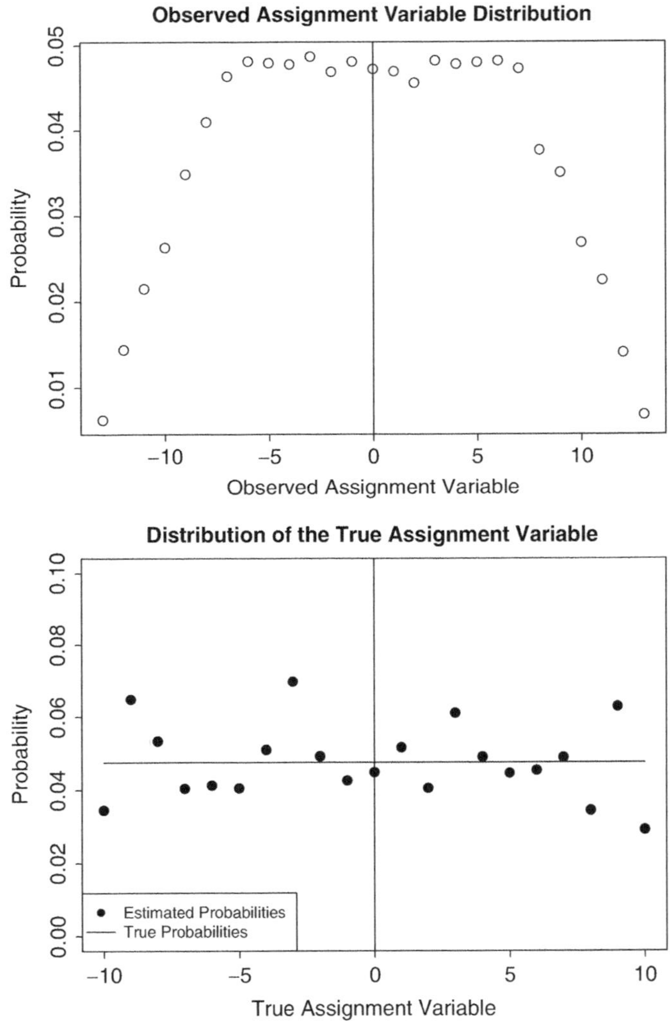

Fig. 4. Assignment Variable Distribution with Uniform X^* Distribution: Observed versus Estimated. *Notes*: Illustrative example is based on a sample of size $N = 25{,}000$. X^* and u are uniformly distributed on the set of integers in $[-10,10]$ and $[-3,3]$, respectively. Plotted are the distributions of X and X^*, with the latter against the true uniform distribution specified.

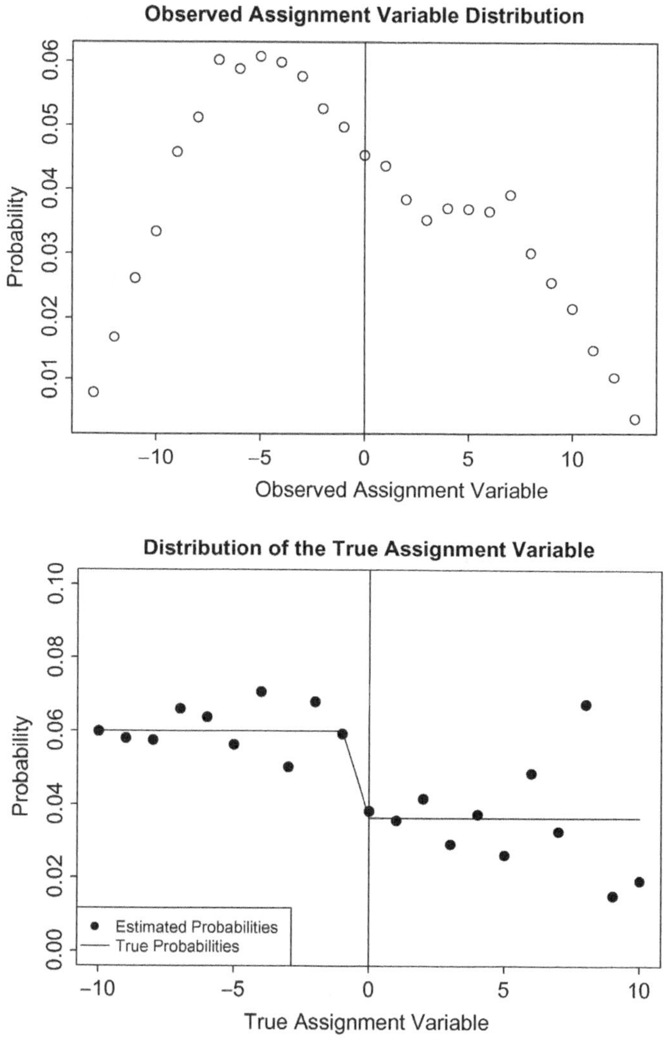

Fig. 5. Assignment Variable Distribution when True X^* Distribution Is Not Smooth: Observed versus Estimated. *Notes*: Illustrative example is based on a sample of size $N = 25,000$. X^* is supported on the set of integers in $[-10, 10]$ with $\Pr(X^* = i) = 0.06$ for $i<0$ and $\Pr(X^* = i) = 0.036$ for $i \geq 0$. Plotted are the distributions of X and X^*, with the latter against the true distribution specified above.

discontinuity confidence intervals improve to 90% or higher as the sample size exceeds 50,000. Overall, this simple numerical example verifies that the true assignment variable distribution and the RD treatment effect parameter can indeed be recovered using the proposed method, when the model assumptions are met.

4. CONTINUOUS ASSIGNMENT VARIABLE AND MEASUREMENT ERROR

In this section, we study the identification in an RD design when X^* and u are continuously distributed and discuss the sharp and fuzzy cases separately in the two subsections below.[22] Before we proceed, we introduce the analog of Assumption 2 in the continuous case, a standard assumption in the RD literature.

Assumption 2C (Positive Density at Threshold). Let the p.d.f. of X^* be f_{X^*}. There exists $a>0$ such that $f_{X^*}(x^*)>0$ for $x^* \in (-a, a)$.

Assumption 2C ensures that the conditional expectation functions are well defined in the sharp and fuzzy RD estimands as defined in Section 2.

4.1. Identification under Perfect Compliance

In this subsection, we focus on the case of perfect compliance and consider three distinct approaches.

Approach 1. In the first approach, we assume that u follows a normal distribution with mean zero and an unknown variance σ^2. Along with the classical measurement error assumption, the normality of u allows the distribution X^* to be identified. Provided that Assumption 2C holds and that the measurement error is nondifferential, the RD treatment effect is also identified.

Assumption 7 (Normality). $u \sim \phi\left(0, \sigma^2\right)$.

Proposition 4. (a) Under Assumptions 1 and 7, the distributions of X^* and u are identified. (b) Under Assumptions 1Y, 2C, and 7, δ_{sharp} is identified.

Proposition 4(a) is a corollary of Theorem 2.1 of Schwarz and Bellegem (2010), and the proof of Proposition 4(b) is similar to that of Proposition 2. Details are provided in Section A.1.4 of the online Supplemental Material.

Approach 2. In the second approach, we show that normality of the measurement error can be weakened without sacrificing the identification of the RD treatment effect under perfect compliance. Specifically, an identification-at-infinity strategy can be applied with a regularity condition governing the tail behavior of the measurement error.

Proposition 5. If Assumption 2C holds, f_{X^*} is continuous at 0, and the c.d.f. of u, F_u, satisfies

$$\lim_{x\to\infty}\frac{1-F_u(x+v)}{1-F_u(x)}=0 \quad \text{for all } v>0, \tag{13}$$

$$\lim_{x\to-\infty}\frac{F_u(x-v)}{F_u(x)}=0 \quad \text{for all } v>0, \tag{14}$$

then the RD treatment effect is identified:

$$\lim_{x\to\infty}E[Y|X=x,D^*=1]-\lim_{x\to-\infty}E[Y|X=x,D^*=0]=\delta_{\text{sharp}}. \tag{15}$$

We prove Proposition 5 in Section A.1.5 of the online Supplemental Material. Intuitively, when we see an observation with $X\gg 0$ and $D^*=1$, it is because X^* is close to zero or because u is very large. Condition (13) states that the right tail of the measurement error distribution needs to be "light" enough and makes u unlikely to be very large. Therefore, as X becomes large in the treatment group, we end up with observations for which X^* is just below zero. While the Laplace distribution violates (13) and (14), almost all of the common distributions whose tails are comparable to or lighter than the normal distribution satisfy these tail conditions.[23]

A visual illustration of the identification result in Proposition 5 is presented in Fig. 6. We use the same conditional expectation function for $E[Y|X^*]$ as in Section 3.6. X^*, u and the error term in the outcome equation are all normally distributed. We split the simulated sample of 25,000 observations into two groups by the value of D^*. In the upper and lower panels, we show the raw scatter plot for the $D^*=1$ and $D^*=0$ groups,

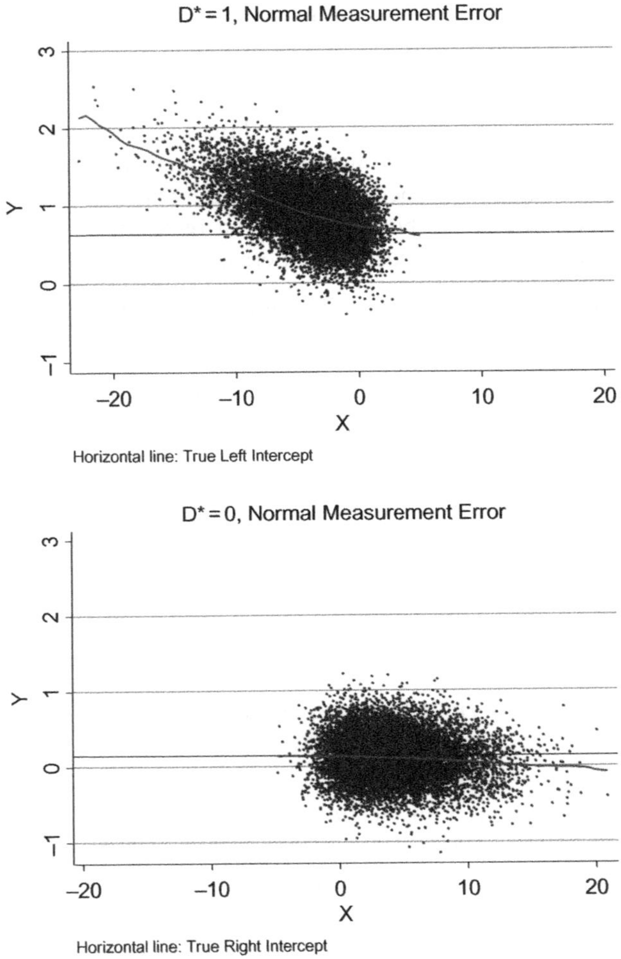

Fig. 6. Identification-at-Infinity Illustration. *Notes*: Scatter plot with 25,000 simulated observations (dots) and local constant smoothing (curve). Local smoother is implemented with the `lpoly` command in Stata with default options. The data generating process is $Y = 0.15 + 0.48D^* - 0.01X^* - 0.006D^*X^* + \varepsilon$, where $X^* \sim N(0, 25)$, $\varepsilon \sim N(0, 0.09)$ and $u \sim N(0, 2)$.

respectively. In each panel, we impose fit obtained from a simple local regression. As X becomes large in the $D^* = 1$ group and as X becomes small in the $D^* = 0$ group, we can see that the averages of the Y's (represented by the curved lines) approach the true intercepts of the $E[Y|X^*]$ function (represented by the horizontal lines).

We now discuss the relationship between the identification-at-infinity strategy proposed here and recent papers by Yu (2012) and Yanagi (2014) that study the same question. Yu (2012) assumes that the measurement error variance shrinks to zero as the sample size increases and proposes using a trimmed sample to recover the RD treatment effect, in which observations with $X \geqslant 0$ in the $D^* = 1$ group and observations with $X < 0$ in the $D^* = 0$ group are dropped. Specifically, the RD treatment effect estimand is the difference between the left intercept of $E[Y|X = x, D^* = 1]$ and the right intercept of $E[Y|X = x, D^* = 0]$. As mentioned in Section 2, dividing the population into treatment and control and separately estimating the intercept in each is attractive, because it results in a much smaller bias than naively fitting $E[Y|X]$ with local regressions. Our approach builds on the idea of Yu (2012) in this regard. The difference is that we put no restriction on the measurement error variance, and it may result in a smaller bias as a result. To see this, we illustrate the strategy of Yu (2012) in Fig. 7, and the bias appears to be somewhat larger as compared to that in Fig. 6.

Assuming bounded measurement error u, Yanagi (2014) creatively applies a small error variance approximation (Chesher, 1991) to $E[Y|X^* = x, D^* = d]$ in a sharp design:

$$\begin{aligned} E[Y|X^* = x, D^* = d] \quad = \quad & E[Y|X = x, D^* = d] \\ & -\sigma^2\left(\log^{(1)} f_{X|D}(x|d)\right)E^{(1)}[Y|X = x, D^* = d] \\ & -\frac{\sigma^2}{2}E^{(2)}[Y|X = x, D^* = d] + o\left(\sigma^2\right), \end{aligned} \tag{16}$$

where σ^2 again denotes the variance of u. For the approximation to work, however, the derivatives of the conditional expectation and density functions need to be estimated. Moreover, σ needs to be small and known.

To illustrate how our identification results relate to Yanagi (2014), first note that the proof of Proposition 5 can be carried over to the case with bounded measurement error u. In fact, when support_u is bounded, the tail conditions (13) and (14) are no longer needed. Furthermore, we can identify $\text{support}_u \equiv [\underline{u}, \overline{u}]$ and identify the RD treatment effect with

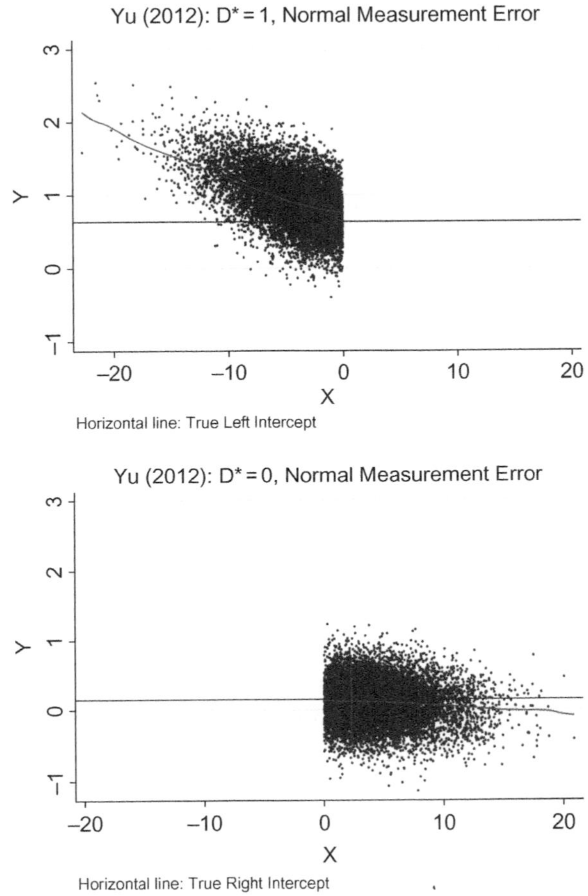

Fig. 7. Illustration of Identification in Yu (2012). *Notes*: Same data and local smoothing method as Fig. 6. Observations with $X \geqslant 0$ and $D = 1$ and with $X < 0$ and $D = 0$ are dropped per Yu (2012).

$$\lim_{x \to \bar{u}} E[Y|X = x, D^* = 1] - \lim_{x \to \underline{u}} E[Y|X = x, D^* = 0] = \delta_{\text{sharp}}. \tag{17}$$

This "identification-at-boundary" strategy of Eq. (17) has the advantage of avoiding the derivatives in Eq. (16), which may be hard to estimate in practice. Furthermore, it does not place additional restrictions on the measurement error distribution, nor does it require σ to be known.

Approach 3. When σ *is* known (e.g., from an external source), we can also apply the SIMEX strategy of Cook and Stefanski (1994) and Carroll et al. (1999) to recover the RD treatment effect. It is the final approach we propose in this subsection.[24] The idea behind the method is most simply illustrated in the linear case where the conditional expectation function $E[Y|X^* = x^*, D^* = 1]$ is equal to $\psi_0 + \psi_1 X^*$. When we regress Y on X within the $D^* = 1$ population, the least squares slope estimator converges to the expression with the well-known attenuation factor, $\psi_1 \cdot \frac{\text{var}(X^*|D^*=1)}{\text{var}(X^*|D^*=1)+\sigma^2}$. The attenuation problem worsens as σ^2 increases, and the SIMEX approach of Cook and Stefanski (1994) makes use of this observation: it traces out the estimate as a function of the degree of measurement error contamination via simulation, and extrapolates this function to recover the true parameter ψ_1 as well as the target conditional expectation function. Formally, for any $\lambda > 0$, we can add additional noise $\tilde{u}_\lambda$ with variance $\sigma^2\lambda$ to X. The new measurement error is $u + \tilde{u}_\lambda$ with variance $\sigma^2(1 + \lambda)$, and the corresponding population slope parameter is $g(\lambda) \equiv \psi_1 \cdot \frac{\text{var}(X^*|D^*=1)}{\text{var}(X^*|D^*=1)+\sigma^2(1+\lambda)}$. By choosing different λ's, we obtain the value of the g function at various points, which we can then use to extrapolate and recover $g(-1) = \psi_1$.

The linearity of $E[Y|X^* = x^*, D^* = 1]$ in the example above can be relaxed. As long as this target function is smooth, we can apply the nonparametric procedure proposed by Carroll et al. (1999). Specifically, for each λ we choose, we can estimate the value of the conditional expectation function of $\mu_d(\lambda) = E[Y|X + \tilde{u}_\lambda = 0, D^* = d]$ via a local linear regression, and use polynomials to extrapolate $\mu_d(\lambda)$ back to $\lambda = -1$ and recover the left ($d = 1$) and right ($d = 0$) intercepts in the RD estimand. The difference between μ_1 and μ_0 at $\lambda = -1$ identifies δ_{sharp}. In the next subsection, we consider whether and how the three approaches can be extended to a design with imperfect compliance.

4.2. Identification under Imperfect Compliance

Only one of the three approaches proposed in Section 4.1 carries over to an RD design with imperfect compliance. With the one-sided fuzzy assumption, Approach 1 still identifies the X^* distribution and the RD treatment effect δ_{fuzzy}. The idea is similar in spirit to that of Proposition 1F, which relies on Assumptions DB, 1F, 2F, 3F, 4, and 5. Analogous to Proposition 2F, we can identify the X^* distribution for each value of D and

Y under strong independence and nondifferential measurement error, which allows the identification of $E[D|X^*]$ and $E[Y|X^*]$ by Bayes' Theorem and hence the RD treatment effect.

> **Proposition 4F**. (a) Under Assumptions 1F, 4, and 7, the distributions of X^* and u are identified. (b) Under Assumptions 1FY, 2C, 4, and 7, δ_{fuzzy} is identified.

> *Remark 7*. Assumption 4 (one-sided fuzzy) is no longer needed if the researcher knows what σ is, which parallels the redundancy of Assumption 4 in Proposition 2F when the support of the measurement error is known.

> *Remark 8*. The normal measurement error assumption plays a key role in Proposition 4F: it encapsulates symmetry (a stronger version of Assumption 5) and the tail restriction normality imposes renders the continuous analog of Assumption 2F unnecessary in the identification of the X^* distribution.

Unfortunately, Approaches 2 and 3 do not carry over to the fuzzy case, even with the one-sided fuzzy assumption. For the identification-at-infinity/boundary strategy (Approach 2), as X approaches $-\infty$, the control group $(D = 0)$ consists of both compliers with X^* close to the threshold and never takers with a large negative X^*. Absent strong assumptions, it is not possible to disentangle the two groups and identify $\lim_{x^* \downarrow 0} E[Y|X^* = x^*]$. The SIMEX strategy (Approach 3) will not recover the discontinuity in the first-stage relationship $E[D|X^*]$. Without observing D^*, it is not possible to extrapolate the left and right intercepts in the first-stage relationship from separate regressions. Applying SIMEX to only the observed first-stage relationship $E[D|X]$ also will not work because the target function $E[D|X^*]$ is potentially discontinuous at zero, a violation of the underlying smoothness assumption of SIMEX. Despite being more nonparametric, the inability of Approaches 2 and 3 to identify the RD treatment effect in a fuzzy design greatly limits their practical relevance.

4.3. Estimation

In this subsection, we discuss estimation for the three approaches introduced in Section 4.1. For Approach 1, Schwarz and Bellegem (2010) propose a conceptual minimum distance framework. Applied to our context, it will select the estimators for the distributions of $X^*|D^* = 1$ and u to fit the observed characteristic function of $X|D^* = 1$. The fact that the distribution of X^* given $D^* = 1$ has no support on the entire interval of $(0, \infty)$ satisfies the regularity condition of Schwarz and Bellegem (2010) , which proves the consistency of the conceptual estimators. As for practical implementation, Schwarz and Bellegem (2010) suggest discretizing the $X^*|D^* = 1$ distribution with the number of support points increasing as $n \to \infty$, but defer the details to future research. Assuming that the Lapalace transform of the target density decays sufficiently quickly, Matias (2002) proposes an explicit estimator for σ and shows that the rate of convergence is slower than $\log n$. Matias (2002) also proves that the minimax mean squared error of the pointwise density estimator cannot decrease faster than $(\log n)^{-1}$ uniformly over a set of regular densities, which makes is difficult to apply empirically.

For Approach 2, one can adapt the estimator proposed by Andrews and Schafgans (1998) as a starting point for estimation and inference. Let $s(x)$ be a weight function that takes the value 0 for $x \leq 0$, the value 1 for $x \geq b$ for some positive constant b, and is smoothly increasing between 0 and b with bounded derivatives. γ_n is the bandwidth parameter, which goes to infinity with the sample size n. The Andrews-Schafgans estimator is defined as

$$\widehat{\kappa}_1 = \frac{\sum_{i=1}^{N} Y_i D_i s(X_i - \gamma_N)}{\sum_{i=1}^{N} D_i s(X_i - \gamma_N)},$$

where $\kappa_1 \equiv \lim_{x \to \infty} E[Y|X = x, D^* = 1]$ is the quantity to be estimated. The consistency result $\widehat{\kappa}_1 \to^p \kappa_1$ in Andrews and Schafgans (1998) is established under similar assumptions as in this chapter, along with additional regularity conditions requiring (1) finite moments of the random variables and (2) an upper bound on the rate at which γ_N tends to infinity, which depends on the upper tail behavior of the X distribution. Furthermore, with one extra condition setting a lower bound for γ_N as $N \to \infty$, Andrews and Schafgans (1998) also show that the distribution of the estimator $\widehat{\kappa}_1$ is asymptotically normal and centered around κ_1. The rate of convergence is $\sqrt{N}$ multiplied by a factor that depends on the tail of the X distribution and the exact choice of γ_N.

For Approach 3, Carroll et al. (1999) show that the error of their local linear nonparametric SIMEX estimator with bandwidth h is of order $O_p\{h^2 + (Nh)^{-\frac{1}{2}}\}$, provided that the polynomial extrapolant is exact. However, the performance of the estimator is sensitive in practice. Berry, Carroll, and Ruppert (2002) note that alternative choices of smoothing parameters may give rise to great instability in estimates, and Staudenmayer and Ruppert (2004) find that the SIMEX estimates are not robust to alternative values within the 95% confidence interval for σ, which is estimated in an external validation study. Given the challenges in estimation described in this section, we will adopt a parametric approach in the subsequent empirical illustration.

5. EMPIRICAL APPLICATION: MEDICAID TAKEUP AND CROWDOUT

As one of the largest entitlement programs in the United States, Medicaid has received considerable attention in policy discussions. An important debate is on the extent to which Medicaid eligibility crowds out private insurance – that is, what fraction of eligible individuals drop private insurance and enroll in Medicaid. In a seminal paper, Cutler and Gruber (1996) use simulated instruments and estimate that Medicaid reduced private insurance coverage for children by 30 to 40 percent when Medicaid eligibility greatly expanded between 1988 and 1993. This high crowdout rate, however, is not the consensus of the literature. Several subsequent studies (Card & Shore-Sheppard, 2004; Thorpe & Florence, 1998; Yazici & Kaestner, 2000) find smaller effects for the same time period using various research designs. As noted by Shore-Sheppard (2008), the question of crowdout has produced less consensus than disagreement.

In this section, we illustrate our proposed methods by using them to estimate Medicaid takeup and private insurance crowdout. Specifically, we apply an RD design by exploiting the Medicaid income eligibility cutoff rule. We use microdata constructed by Card and Shore-Sheppard (2004) from the full panel research files of the 1990–1993 SIPP and make further restrictions to arrive at our analysis sample, as detailed in Section B.1 of the online Supplemental Material.[25]

We apply two modeling approaches to estimate Medicaid takeup and private insurance crowdout around the income eligibility threshold. First, we use the proposed method in Section 3 based on a discretized income measure. Because of the difficulty in implementing the semi/nonparametric estimators in Section 4, in our second modeling strategy we treat income as continuous and adopt a parametric MLE framework. In particular, we specify (1) X^* as a function of a normal random variable that allows for bunching at the income eligibility threshold, (2) u to be mean zero normal, and (3) the one-sided fuzzy first stage and outcome relationships to be logistic in polynomials of X^* and D. Details of this parametric formulation are provided in Section B.2 of the online Supplemental Material.

As mentioned in Section 3.5, we need to transform the income variable in order for Assumptions 1FY (strong independence) and 5 (symmetry in the support of u) to plausibly hold. Through experimentation, we find that a Box-Cox transformation with parameter ρ between 0.3 and 0.35 appears to be consistent with these assumptions as indicated by the overidentification tests. Therefore, we present the main results using $\rho = 0.33$, but also present estimates under other values of ρ as robustness checks. To help understand the meaning of the transformed assignment variable, Table A1 of the online Supplemental Material provides a mapping between the transformed income measures that are normalized against eligibility cutoffs and the actual family income amount. For example, a family with a child facing the 100% federal poverty line (FPL) has an actual monthly income of \$1,117 (1991 dollars) if the transformed and normalized income measure is 0; the family's actual income is \$1,231 if the transformed and normalized income measure takes on the value 1. In the remainder of the section, we refer to this transformed and normalized income variable simply as income.

Fig. 8(a–c) plots the density of the *observed* income, the first-stage relationship between Medicaid coverage and income, and the outcome relationship between private insurance coverage and income. The distribution of income in Fig. 8(a) is approximately symmetric and shows no visible sign of bunching around the cutoff. In Fig. 8(b), Medicaid coverage decreases as income rises, but there is no discontinuous drop-off at the cutoff, which is to be expected given the presence of measurement error. Correspondingly, Fig. 8(c) shows that while private insurance coverage increases with income, the relationship is again smooth through the cutoff.

Fig. 8(d–f) plots the *estimated true* income density, first-stage, and outcome relationship. Estimates from the discrete and continuous models are juxtaposed together for ease of comparison. In Fig. 8(d), while the discrete approach yields a much noisier estimate of the density than the parametric continuous approach, the two models roughly agree for much of the distribution, especially near the cutoff. Turning to Fig. 8(e), we again find that the first-stage estimated using the discrete model displays a similar trend as

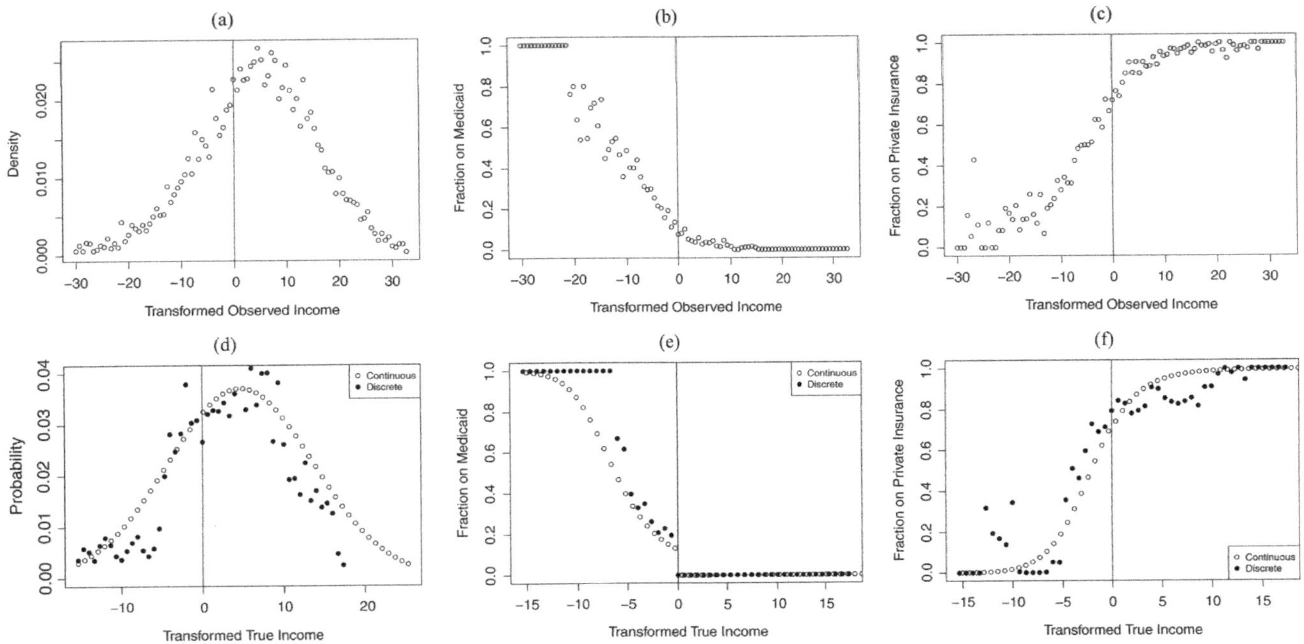

Fig. 8. Observed and Estimated Income Distributions and First-Stage and Outcome Relationships. (a) Distribution of Observed Income. (b) Fraction on Medicaid versus Observed Income. (c) Fraction on Private Insurance versus Observed Income. (d) Distribution of True Income. (e) Fraction on Medicaid versus True Income. (f) Fraction on Private Insurance versus True Income. *Notes*: Income is derived from a Box-Cox transformation of the actual family income with parameter 0.33, and is normalized against the transformed Medicaid Eligibility Threshold. Top and Bottom 1% of the normalized income is trimmed. Zero is the cutoff and a child is Medicaid eligible if the normalized income is less than zero. See Table A1 for a mapping between the transformed and actual income values for various Medicaid cutoffs. Fig. 8a through 8c plots the distribution of and first-stage and outcome relationships with the observed income variable. Fig. 8d through 9f plot the estimated distribution of and first-stage and outcome relationships with the true income variable, where solid and hollow points represent estimates from the discrete and continuous modeling approaches, respectively.

the continuous model near the cutoff, albeit with more noise. The graphical evidence suggests that Medicaid coverage increases about 10–20 percentage points at the income threshold.[26] Finally, the behavior of the discrete model estimates in Fig. 8(f) matches their continuous counterparts close to the threshold, and no visible discontinuity is detected.

The numerical discontinuity estimates are presented in Tables A2 and A3 of the online Supplemental Material for the discrete and continuous models, respectively. Consistent with visual evidence, in Table A2 we generally find a statistically insignificant discontinuity in the income distribution, a significant discontinuity in Medicaid coverage between 15 and 25 percentage points, and insignificant discontinuity estimates for private insurance coverage. The resulting crowdout estimates are insignificant as well, but they are very imprecise – the 95% confidence intervals contain practically all estimates from the literature. In column (5), we report the p-values from the overidentification tests: since none of the models in the table are formally rejected at the 5% level, the model appears to fit the data reasonably well.

For brevity, we only present the continuous model results using the Box-Cox transformation parameter $\lambda = 0.33$ and 1% trimming in Table A3. We find no evidence of bunching in the income distribution, a statistically significant 12.5% Medicaid takeup rate just below the eligibility cutoff, and a small and statistically insignificant crowdout effect. The parsimonious specification in the continuous model leads to much improved precision in the estimates, and even the upper bound of the 95% confidence interval excludes the crowd-out estimates from Cutler and Gruber (1996). Finally, as documented in Section B.3 of the online Supplemental Material, the continuous model provides reasonable fit to the data.

To summarize, the two models are broadly consistent in their estimates. We find that the Medicaid takeup rate just below the eligibility cutoff falls between 10% and 25% and that there is little evidence of income sorting around the threshold and the private insurance crowdout. However, due to the lack of functional form restrictions imposed by the discrete model, its estimates are quite imprecise.

6. CONCLUSION

This chapter investigates identification and estimation in an RD design where the assignment variable is measured with error. This is a challenging problem because the presence of measurement error may smooth out the first-stage discontinuity and eliminate the source of identification. We provide conditions for the identification of the true assignment variable distribution and the RD treatment effect, using only the mismeasured assignment variable and the binary treatment status. Our results contribute to the measurement error literature and shed light on the limitations of certain RD applications.

We first study the case where the assignment variable and the measurement error are discrete, and propose sufficient conditions for identification in both the sharp and fuzzy designs. A simple estimation procedure is proposed using a minimum distance framework. Following standard arguments, the resulting estimators are $\sqrt{N}$-consistent, asymptotically normal, and efficient. A numerical example verifies that when the model assumptions are met, the true assignment variable distribution and the RD treatment effect parameter can indeed be recovered using the proposed method.

We also explore the case where the assignment variable and measurement error are continuous and propose three different identification approaches. The first approach assumes normality of the measurement error, and identifies the assignment variable distribution and the RD treatment effect in sharp and one-sided fuzzy designs. The second approach adopts a novel identification-at-infinity strategy, and the third approach relies on the SIMEX method of Carroll et al. (1999), both of which identify the RD treatment effect in a sharp design. Because the first approach accommodates a fuzzy design, it potentially works for a larger range of applications, but it is handicapped by the slow convergence rate of the semiparametric estimator.

In our empirical application, we apply an RD design to estimate Medicaid takeup and private insurance crowdout using SIPP data. We exploit the discontinuity in the eligibility formula with respect to family income, but because income is measured with error, the first-stage relationship between Medicaid takeup and reported income is not discontinuous at the eligibility cutoff. We use two approaches to recover the true income distribution and the RD treatment effect: our proposed method based on discretized income and a *parametric* MLE framework that treats income as continuous. The two approaches yield similar results: Medicaid takeup rate for the barely eligible is between 10% and 25% and there is little evidence of income sorting around the threshold and of private insurance crowdout. However, the estimates from the discrete approach are quite imprecise, which by itself is unlikely to deliver convincing policy conclusions. Therefore, we believe the parametric approaches adopted by this chapter for classical measurement error and by Hullegie and Klein (2010) for Berkson measurement error to be more appealing alternatives for empirical research.

NOTES

1. See Lee and Lemieux (2010) for a survey of the RD literature.

2. In addition to the RD approach, Schanzenbach (2009) also conducts analyses using other research designs.

3. As a caveat, however, the detection of nonsmoothness or bunching in the assignment variable distribution when the variable itself is discrete will not entail a nonparametric procedure. Rather, it will depend on the parametric functional form subjectively specified by the researcher.

4. A related but distinct problem is heaping in the assignment variable (Barreca, Guldi, Lindo, & Waddell, 2011; Almond, Kowalski, & Williams, 2011; Barreca, Lindo, & Waddell, 2016). In the heaping setup, treatment assignment is based on the observed assignment variable. For the problem at hand, we do not observe the variable-determining treatment.

5. Note that in some applications $D^* = 1_{[X^* \geqslant 0]}$ is the treatment-determining mechanism. However, the motivating examples in Section 1 follow $D^* = 1_{[X^* < 0]}$, that is, program eligibility depends on whether the assignment variable (income) falls *below* a known cutoff.

6. See Currie (2006) for a survey on benefit takeup in social programs.

7. This is in contrast to the case where a first-stage discontinuity exists despite the presence of measurement error in the assignment variable for a subset of observations. Battistin, Brugiavini, Rettore, and Weber (2009) show that a fuzzy design can still be applied to identify the causal effect in that case. See Card, Lee, Pei, and Weber (2015) for an analogous result in the regression kink design.

8. Dong (2015) makes a similar point in the case of age rounding.

9. We thank a referee for suggesting this reference.

10. Because of the bounded support assumption, the defined moment generating functions always exist and are positive for all t.

11. The independence assumption implies that the measurement error distribution is invariant with respect to treatment status, and it plays a key role in the derivation above. This is certainly a strong restriction, since it is possible that the treatment status may affect measurement error, but we cannot achieve identification without this restriction.

12. The exact form of $\mathbf{Q}_\mathbf{F}$, along with the proof of Proposition 1F, is provided in Section A.1.2 of the online Supplemental Material.

13. This is the starting point of the approach taken by Davezies and Le Barbanchon (2014).

14. For formal studies on the identification and estimation of regression models/ treatment effects under misclassification of the treatment variable, see the seminal papers by Mahajan (2006) and Lewbel (2007).

15. As pointed out by Lee and Card (2008), there may be a misspecification error in the parametric extrapolation of $E[Y|X^* = 0, D = 1]$. In this chapter, we abstract away from this issue.

16. Nondifferential measurement error is a common assumption in the literature (Carroll, Ruppert, Stefanski, & Crainiceaunu, 2006). As with Assumption 1F, a weaker version of Assumption 1Y, $X^* \perp\!\!\!\perp u$ conditional on Y, also delivers the

identification results. However, we adopt Assumption 1Y for its simplicity in economic interpretation, and the same goes for Assumption 1FY.

17. Just like $\mathbf{Q}_Y$ is the conditional analog of $\mathbf{Q}$, $\mathbf{Q}_{\mathbf{F}Y}$ is the conditional analog of $\mathbf{Q}_\mathbf{F}$.

18. Analogous to $\mathbf{p}$, which contains the parameters p_k^0's and p_k^1's, $\mathbf{q}$ is the vector that contains all the parameters q_j^0's and q_j^1's .

19. It turns out that $\nabla_\mathbf{q} f \Omega \nabla_\mathbf{q} f'$ is singular, which is analogous to the issue raised in Satchachai and Schmidt (2008). Satchachai and Schmidt (2008) advise against using the generalized inverse, which is confirmed by our own numerical investigation. Instead, they propose dropping one or more restrictions, but state that the problem of which restrictions to drop has yet to be solved.

20. Identification in Section 3.2 is established under the symmetry assumption (Assumption 5), and the model is not identified in its absence. Therefore, this overidentification test may not have power against the violation of Assumption 5.

21. For a clear exposition of this result, see the "makecns" entry in Stata (2010).

22. A case not considered in this chapter is when X^* is continuous but u is discrete (or vice versa). A starting point in this setting is Dong and Lewbel (2011).

23. When u follows a Laplace distribution, for example, $f_u(v) = \frac{1}{2}e^{-|v|}$ for $v \in \mathbb{R}$, $\lim_{x \to \infty} E[Y|X = x, D^* = 1] = \int_0^\infty E[Y|X^* = -v]e^{-v}\text{dv}$, which is generally not equal to $\lim_{x^* \uparrow 0} E[Y|X^* = x^*]$.

24. We thank a referee for suggesting this approach. See Chapter 5 of Carroll et al. (2006) for a detailed overview of the SIMEX method in the measurement error context.

25. Card and Shore-Sheppard (2004) study public health insurance coverage and private insurance crowdout by combining RD and difference-in-differences designs and leverage off both the age and income eligibility rules for Medicaid.

26. The one-sided fuzzy assumption is plausible in the Medicaid context. According to CMS (2014), only 3.1% of the cases were incorrectly assigned eligibility status when families applied to public health insurance, and the trimming we conduct further alleviates the concern. We should note, however, that the ongoing Medicaid participants might not have their income eligibility recertified every month, potentially casting doubt on the one-sided fuzzy assumption, but there is little evidence of income rebounding for these participants as documented by Pei (2017).

ACKNOWLEDGMENTS

We thank two anonymous referees, Orley Ashenfelter, Eric Auerbach, Matias Cattaneo, Eleanor Choi, Damon Clark, Kirill Evdokimov, Hank Farber, Marjolaine Gauthier-Loiselle, Jian Kang, Marta Lachowska, Lars Lefgren, Pauline Leung, Jia Li, Carl Lieberman, Andrew Marder, Alex Mas, Jordan Matsudaira, Alexander Meister, Stephen Nei, Andrew Shephard, Lara Shore-Sheppard, and especially Bo Honor and David Lee for helpful comments. We have also benefited from helpful suggestions given by the participants of the Princeton Labor Seminar, Princeton Political Science Methodology Seminar and the Advances in Econometrics conference at the University of Michigan. We thank David Card and Lara Shore-Sheppard for graciously sharing their data. Zhuan Pei gratefully acknowledges financial support from the Richard A. Lester Fellowship. Finally, we thank Yue Fang, Suejin Lee, Katherine Wen and Le Wu for outstanding research assistance.

REFERENCES

Abowd, J. M., & Stinson, M. H. (2013). Estimating measurement error in annual job earnings: A comparison of survey and administrative data. *The Review of Economics and Statistics*, *95*(5), 1451–1467.

Almond, D., Doyle, J. J., Jr., Kowalski, A. E., & Williams, H. (November 2011). The Role of Hospital Heterogeneity in Measuring Marginal Returns to Medical Care: A Reply to Barreca, Guldi, Lindo, and Waddell. *Quarterly Journal of Economics*, *126*(4), 2125–2131.

Andrews, D. W. K., & Schafgans, M. M. A. (1998). Semiparametric estimation of the intercept of a sample selection model. *Review of Economic Studies*, *65*, 497–517.

Ashenfelter, O., & Krueger, A. (1994). Estimates of the economic return to schooling from a new sample of twins. *The American Economic Review*, *84*(5), 1157–1173.

Barreca, A. I., Guldi, M., Lindo, J. M., & Waddell, G. R. (2011). Saving babies? Revisiting the effect of very low birth weight classification. *Quarterly Journal of Economics*, *126*(4), 2117–2123.

Barreca, A. I., Lindo, J. M., & Waddell, G. R. (2016). Heaping-induced bias in regression-discontinuity designs. *Economic Inquiry*, *54*(1), 268–293.

Battistin, E., Brugiavini, A., Rettore, E., & Weber, G. (2009). The retirement consumption puzzle: Evidence from a regression discontinuity approach. *American Economic Review*, *99*(5), 2209–2266.

Berry, S. M., Carroll, R. J., & Ruppert, D. (2002). Bayesian smoothing and regression splines for measurement error problems. *Journal of the American Statistical Association*, *97*(457), 160–169.

Black, D. A., Berger, M. C., & Scott, F. A. (2000). Bounding parameter estimates with non-classical measurement error. *Journal of the American Statistical Association*, *95*(451), 739–748.

Bound, J., Brown, C., & Mathiowetz, N. (2001). Measurement error in survey data. In J. J. Heckman & E. Leamer (Eds.), *Handbook of econometrics* (Vol. 5, pp. 3705–3843). Amsterdam: Elsevier.

Card, D., Hildreth, A. K. G., & Shore-Sheppard, L. D. (2004). The Measurement of Medicaid Coverage in the SIPP: Evidence from a comparison of matched records. *Journal of Business & Economic Statistics*, *22*(4), 410–420.

Card, D., Lee, D. S., Pei, Z., & Weber, A. (2015). Inference on causal effects in a generalized regression kink design. *Econometrica*, *83*(6), 2453–2483.

Card, D., & Shore-Sheppard, L. D. (2004). Using discontinuous eligibility rules to identify the effects of the federal Medicaid expansions on low-income children. *The Review of Economics and Statistics*, *86*(3), 752–766.

Carroll, R. J., Maca, J. D., & Ruppert, D. (1999). Nonparametric regression in the presence of measurement error. *Biometrika*, *86*(3), 541–554.

Carroll, R. J., Ruppert, D., Stefanski, L. A., & Crainiceaunu, C. M. (2006). *Measurement error in nonlinear models: A modern perspective* (2nd ed.). Boca Raton, FL: Chapman & Hall/CRC.

Chen, X., Hu, Y., & Lewbel, A. (2009). Nonparametric identification and estimation of non-classical errors-in-variables models without additional information. *Statistica Sinica*, *19*, 949–968.

Chesher, A. (1991). The effect of measurement error. *Biometrika*, *78*(3), 451–462.

CMS. (2014). *Medicaid and CHIP 2014 Improper Payments Report*. Technical Report, Center for Medicare and Medicaid Services.

Cook, J. R., & Stefanski, L. A. (1994). Simulation-extrapolation estimation in parametric measurement error models. *Journal of the American Statistical Association*, *89*(428), 1314–1328.

Currie, J. (2006). The take-up of social benefits. In A. Auerbach, D. Card, & J. Quigley (Eds.), *Public policy and the income distribution* (pp. 80–148). New York, NY: Russell Sage.

Cutler, D. M., & Gruber, J. (1996). Does public insurance crowd out private insurance. *Quarterly Journal of Economics*, *111*(2), 391–430.

Davezies, L., & Le Barbanchon, T. (2014, November). *Regression discontinuity design with continuous measurement error in the running variable*. CEPR Discussion Paper No. DP11775.

de la Mata, D. (2012). The effect of Medicaid eligibility on coverage, utilization, and children's health. *Health Economics*, *21*(9), 1061–1079.

DiNardo, J., & Lee, D. S. (2011). Program evaluation and research designs. In O. Ashenfelter & D. Card (Eds.), *Handbook of labor economics* (Vol. 4, Part A, pp. 463–536). Amsterdam: Elsevier.

Dong, Y. (2015). Regression discontinuity applications with rounding errors in the running variable. *Journal of Applied Econometrics*, *30*, 422–446.

Dong, Y., & Lewbel, A. (2011). Nonparametric identification of a binary random factor in cross section data. *Journal of Econometrics*, *163*(2), 163–171.

Grinstead, C. M., & Snell, J. L. (1997, July). *Introduction to probability* (2nd revised ed.). Providence, RI: American Mathematical Society.

Hahn, J., Todd, P., & der Klaauw, W. V. (2001). Identification and estimation of treatment effects with a regression-discontinuity design. *Econometrica*, *69*(1), 201–209.

Hausman, J. A., Newey, W. K., Ichimura, H., & Powell, J. L. (1991). Identification and estimation of polynomial errors-in-variables models. *Journal of Econometrics*, *50*(3), 273–295.

Hullegie, P., & Klein, T. J. (2010). The effect of private health insurance on medical care utilization and self-assessed health in Germany. *Health Economics*, *19*(9), 1048–1062.

Kleven, H. J., & Waseem, M. (2013). Using notches to uncover optimization frictions and structural elasticities: Theory and evidence from Pakistan. *Quarterly Journal of Economics*, *128*(2), 669–723.

Koch, T. G. (2013). Using RD design to understand heterogeneity in health insurance crowd-out. *Journal of Health Economics*, *32*(3), 599–611.

Kodde, D. A., Plam, F. C., & Pfann, G. A. (1990). Asymptotic least-squares estimation efficiency considerations and applications. *Journal of Applied Econometrics*, *5*(3), 229–243.

Lee, D. S., & Card, D. (2008). Regression discontinuity inference with specification error. *Journal of Econometrics*, *142*(2), 655–674. The regression discontinuity design: Theory and applications.

Lee, D. S., & Lemieux, T. (2010). Regression discontinuity designs in economics. *Journal of Economic Literature*, *48*, 281–355.

Lewbel, A. (2007). Estimation of average treatment effects with misclassification. *Econometrica*, *75*(2), 537–551.

Li, T., & Vuong, Q. (1998). Nonparametric estimation of the measurement error model using multiple indicators. *Journal of Multivariate Analysis*, *65*(2), 139–165.

Mahajan, A. (2006). Identification and estimation of regression models with misclassification. *Econometrica*, *74*(3), 631–665.

Marquis, K. H., & Moore, J. C. (1990). *Measurement errors in SIPP program reports.* Technical Report, U.S. Census Bureau.

Matias, C. (2002). Semiparametric deconvolution with unknown noise variance. *ESAIM: Probability and Statistics*, *6*, 271–292.

Pei, Z. (2017). Eligibility recertification and dynamic opt-in incentives in income-tested social programs: Evidence from Medicaid/CHIP. *American Economic Journal: Economic Policy*, *9*(1), 241–276.

Pei, Z., & Shen, Y. (2016). *Supplemental material to "The Devil is in the Tails: Regression Discontinuity Design with Measurement Error in the Assignment Variable"*. Retrieved from https://sites.google.com/site/peizhuan/files/rdme-supplemental-material.pdf.

Ruckdeschel, P., & Khol, M. (2014). General purpose convolution algorithm in S4 classes by means of FFT. *Journal of Statistical Software*, *59*(4), 1–25.

Saez, E. (2010). Do taxpayers bunch at kink points. *American Economic Journal: Economic Policy*, *2*(3), 180–212.

Satchachai, P., & Schmidt, P. (2008). GMM with more moment conditions than observations. *Economics Letters*, *99*(2), 252–255.

Schanzenbach, D. W. (2009). Do school lunches contribute to childhood obesity. *Journal of Human Resources*, *44*(3), 684–709.

Schennach, S. M. (2004). Estimation of nonlinear models with measurement error. *Econometrica*, *72*(1), 33–75.

Schwarz, M., & Bellegem, S. V. (2010). Consistent density deconvolution under partially known error distribution. *Statistics and Probability Letters*, *80*, 236–241.

Shore-Sheppard, L. D. (2008). Stemming the tide? The effect of expanding Medicaid eligibility on health insurance coverage. *The B.E. Journal of Economic Analysis & Policy: Advances*, *8*(2), Article 6.

Stata. (2010). *Base reference manual: Stata 11 documentation.* College Station, TX: Stata Press.

Staudenmayer, J., & Ruppert, D. (2004). Local polynomial regression and simulation-extrapolation. *Journal of the Royal Statistical Society: Series B (Statistical Methodology)*, *66*(1), 17–30.

Thorpe, K. E., & Florence, C. S. (1998). Health insurance among children: The role of expanded Medicaid coverage. *Inquiry*, *35*(4), 369–379.

Yanagi, T. (2014, December). *The effect of measurement error in the sharp regression discontinuity design*. KIER Discussion Paper No. 910.

Yazici, E. Y., & Kaestner, R. (2000). Medicaid expansions and the crowding out of private health insurance among children. *Inquiry*, *37*(1), 23–32.

Yu, P. (2012, April). *Identification of treatment effects in regression discontinuity designs with measurement error*. Unpublished manuscript.

INDEX

Printed and bound by CPI Group (UK) Ltd, Croydon, CR0 4YY
13/09/2026
14961031-0004